Asia-Pacific Symposium on Mangrove Ecosystems

Developments in Hydrobiology 106

Series editor

H. J. Dumont

Asia-Pacific Symposium on Mangrove Ecosystems

Proceedings of the International Conference held at
The Hong Kong University of Science & Technology,
September 1–3, 1993

Edited by

Yuk-Shan Wong and Nora F.Y. Tam

Reprinted from Hydrobiologia, vol. 295 (1995)

Kluwer Academic Publishers
Dordrecht / Boston / London

Library of Congress Cataloging-in-Publication Data

Asia-Pacific Symposium on Mangrove Ecosystems (1993 : Hong Kong University of Science & Technology)
Asia-Pacific Symposium on Mangrove Ecosystems : proceedings of the international conference held at the Hong Kong University of Science & Technology, September 1-3, 1993 / editors, Yuk-Shan Wong, Nora F.Y. Tam.
p. cm. -- (Developments in hydrobiology ; 106)
"Reprinted from Hydrobiologia, vol. 295 (1995)."
ISBN 0-7923-3291-1 (HB : acid free paper)
1. Mangrove swamp ecology--Congresses. 2. Mangrove swamp conservation--Congresses. 3. Mangrove swamps--Management--Congresses. I. Wong, Yuk-Shan. II. Tam, Nora F. Y. III. Title. IV. Series.
QH541.5.M27A88 1993
574.5'2642--dc20 94-43527

ISBN 0-7923-3291-1

Published by Kluwer Academic Publishers,
P.O. Box 17, 3300 AA Dordrecht, The Netherlands.

Kluwer Academic Publishers incorporates
the publishing programmes of
D. Reidel, Martinus Nijhoff, Dr W. Junk and MTP Press.

Sold and distributed in the U.S.A. and Canada
by Kluwer Academic Publishers,
101 Philip Drive, Norwell, MA 02061, U.S.A.

In all other countries, sold and distributed
by Kluwer Academic Publishers,
P.O. Box 322, 3300 AH Dordrecht, The Netherlands.

Printed on acid-free paper

Contents

Preface

Mangrove ecosystems are typical formations found in coastal deposits of mud and silt throughout the tropics and some distance into the subtropical latitudes. The total worldwide mangrove area, which is estimated at about 170,000 km^2 with some sixty species of trees and shrubs exclusive to the habitat, dominates approximately 75% of the world's coastline between latitudes 25°N and 25°S. Intertidal ecosystems such as this support genetically diverse communities of terrestrial and aquatic organisms that are of direct or indirect socioeconomic value. It is generally accepted that mangrove forests play important roles such as coastal stabilization and protection against winds and storms; producers of nutrients, forest resources and animal species of economic importance. Recently, the issues on the conservation, proper utilization and management of mangrove forests have been widely discussed in the scientific arena.

Unfortunately, over-exploitation and destruction of mangroves have become a worldwide phenomenon which seriously threatens the sustainability of these ecosystems. The deliberations of this symposium are therefore timely. Its purpose is to promote scientific exchange and regional collaborations in various aspects of research work on studies of mangroves. This special volume includes a large number of the papers presented during the symposium on three main areas, namely recent advances in mangrove ecology; application and utilization of mangrove resources, and conservation and management of the ecosystems. In addition to exchanging information on the principles and techniques, the participants had also come up with a conference recommendation which urged countries of the participants to pay attention to the protection and management of mangroves in their own countries.

We would like to take this opportunity to thank the Guest Editorial Board for their hard work in reviewing all the manuscripts and all the authors for their contribution to this special issue. Finally, we would like to thank the sponsors, namely the Hong Kong Croucher Foundation, the Caltex Green Fund, the Hong Kong Electric Co. Ltd., the International Society for Mangrove Ecosystems (Headquarters in Japan) and the World Wide Fund for Nature of Hong Kong, for their generous support of this event.

THE EDITORS

Hydrobiologia **295**: 1–4, 1995.
Y. S. Wong & N. F. Y. Tam (eds), Asia-Pacific Symposium on Mangrove Ecosystems.

Thermo-osmotic gas supply not detected in *Avicennia marina* seedlings

N. J. Skelton & W. G. Allaway
School of Biological Sciences A12, The University of Sydney, NSW 2006, Australia

Key words: *Avicennia marina*, mangrove, thermo-osmotic, ventilation, ethane, oxygen

Abstract

Ethane was used as a tracer gas to assess the likelihood of thermo-osmotically induced mass-flow in the aerenchyma of *Avicennia marina* seedlings without pneumatophores. Ethane movement was measured in darkness and with illumination at approximately 600 μ mol m^{-2} s^{-1} provided to the leaves and stem, with the expectation that leaf warming under illumination would provide a means for thermo-osmotic flow. In some seedlings the flow increased with illumination, and in others it either decreased or remained unchanged. Overall, there was no statistically significant difference in the conductance to ethane between darkened and illuminated plants, and the rates of ethane movement were consistent with an average diffusive conductance to oxygen down the stem of 0.22×10^{-9} m^3 s^{-1}. It was concluded that there was no evidence for thermo-osmotically induced flow in this case.

Introduction

Seedlings of *Avicennia marina* readily become established in anaerobic mud, and they contain large amounts of aerenchyma (Ashford & Allaway, in press). In the first year of growth they do not produce pneumatophores and the only access route for oxygen transport to the root gas space is through the stem. Seedlings of *A. marina* have been shown to be able to survive but not grow in very low oxygen concentrations (Brown *et al.*, 1969). The supply of oxygen to the roots of older mangrove plants with pneumatophores has been investigated by several people (e.g. Curran, 1985; Curran *et al.*, 1986, Andersen & Kristensen, 1988; Scholander *et al.*, 1955; Skelton, 1988). However, the mechanism of supply of oxygen to the roots of young seedlings which lack pneumatophores has generally been overlooked. In particular the question of whether the supply of oxygen to the roots of mangrove seedlings is by diffusion or if there is mass flow has not been investigated. Such a mass-flow mechanism would assist mangroves to establish in the anaerobic mud and grow until they produced pneumatophores.

Pressure reductions have been detected in root systems of large mangrove plants, and it has been suggested that this could assist the ventilation process (Scholander *et al.*, 1955). In the yellow water-lily a 'thermo-osmotic' ventilation mechanism produced mass flow of gas, pressure differences being produced when light shines on the plant: young leaves were the place of pressurisation and a substantial mass flow was generated (Dacey, 1981). Pressure-induced flow has been implicated in ventilation of other herbaceous emergent and aquatic plants (e.g. Armstrong *et al.*, 1988; Grosse & Mevi-Schutz, 1987). Grosse & Schröder (1984) used ethane as a tracer gas to detect thermo-osmotic flow in leafless alder saplings, where the stem was the site of pressurisation. In this paper we describe experiments with whole *A. marina* seedlings, using ethane to determine if the supply of oxygen to the roots is assisted by a thermo-osmotic ventilation system.

Methods

Avicennia marina propagules from Botany Bay, Sydney, Australia were grown on sand culture and were flooded to soil level with 50% sea water. The seedlings were between 5 and 6 months old at the time of experimentation and were between 110 and 150 mm high with 2 large leaves and 2 small leaves. All the plants had 2 small cable roots and many fine roots. The soil was washed from the roots of plants before exper-

Table 1. Rates of ethane movement and concentrations of ethane in the bag surrounding the shoots, in 9 seedlings of *A. marina*. Each run was done on a different day; 'control' refers to a second plant on that particular day, kept in the dark throughout the experimental period. The correlation coefficients 'r^2' for the regression lines shown in Fig. 1 are also included. The rates of ethane movement are in nmol s^{-1}, and the concentrations in mol m^{-3}.

Run	Treatment	r^2	Rate nmol s^{-1}	Source Concentration mol m^{-3}
1	Light	0.86	0.15	4.21
1	Dark	0.99	0.33	4.21
1	Control	0.98	0.23	3.99
2	Dark	0.98	0.82	3.20
2	Light	0.99	0.35	3.20
3	Dark	0.96	0.25	0.86
3	Light	0.97	0.35	0.86
3	Dark 2	0.99	0.22	0.86
4	Dark	0.97	0.90	4.52
4	Light	0.73	0.33	4.52
5	Dark	1.00	0.17	0.75
5	Light	0.99	0.20	0.75
5	Control	0.99	0.18	0.71
6	Dark	0.97	0.25	0.73
6	Light	1.00	0.22	0.73
6	Dark 2	0.93	0.22	0.73
6	Control	0.98	0.30	0.73

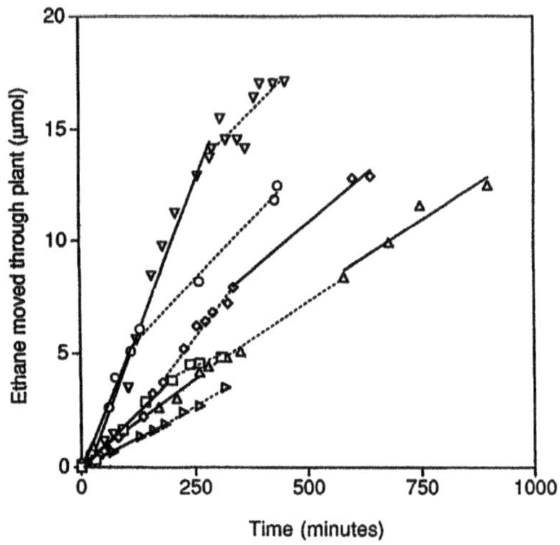

Fig. 1 A. Ethane movement through seedlings of *A. marina* from a high concentration provided at the shoot; the results show the total quantity of ethane in the polyester bag surrounding the root system at each time during the experiment. The difference in rates of flow between plants can be partially accounted for because the ethane concentration in the source (shoot-bag) was different in each case. Data collected on the same day are shown by the same symbol on the graphs. Linear regression lines are fitted to the data of each treatment (1A) and to all the data in the controls (1B). 1A – plants were kept un-illuminated for the first few hours (solid lines), then illuminated at about 600 μ mol m^{-2} s^{-1} for the next few hours (broken lines), and in two cases un-illuminated after that (solid lines). 1B – control plants, no illumination throughout.

Table 2. Conductances to ethane of *A. marina* seedlings, calculated for dark and light treatments and for the control seedlings. Where two 'dark' values are shown for any plant, that plant had been exposed to darkness, then light and then darkness a second time during the experiment. The dimensions for conductances are 10^{-9} m^3 s^{-1}.

Run	Dark	Light	Control
1	0.08	0.04	0.06
2	0.26	0.11	
3	0.29	0.41	
3	0.25		
4	0.20	0.07	
5	0.22	0.27	0.26
6	0.34	0.30	0.41
6	0.30		
mean	0.24	0.20	0.24

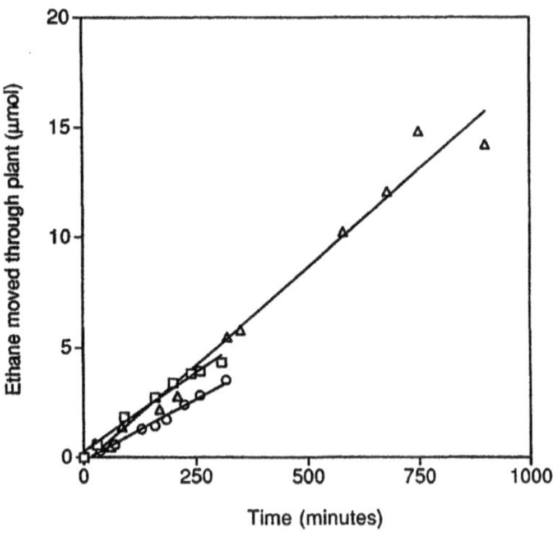

Fig. 1 B.

imentation and a thin-walled polyester bag (Glad® oven bag) placed over the root system and sealed to the hypocotyl with strips of thin rubber. Another bag was sealed over the shoot, leaving a 2 cm section of bare stem between the two bags. Five millilitres of water was injected into each bag to prevent the plants from dehydrating. The polyester bags were deliberately kept 'limp' (*i.e.* not fully inflated) to eliminate the possibility of build-up of even small pressure differentials, such as could potentially arise with rigid enclosures. The bags-plant apparatus was tested for pressure differences between the root and shoot bags with open-ended manometers using Krebs fluid, capable of detecting 5 Pa. No pressure differences were found, even when the bags were differentially illuminated. The plants were held securely by clamps.

The laboratory was maintained at 20 °C. The apparatus was allowed to equilibrate for approximately 30 min then 100 ml of ethane was injected into the bag over the shoot to make a concentration of approximately 1 in 10. Plants were treated in one of two ways: in 'treatment' plants the shoot was alternately kept un-illuminated and lighted for periods of 2 to 5.5 h; and 'control' plants were kept un-illuminated by shielding from bright light using aluminium foil. Control and treatment plants were kept under the same water bath. The light source used was a 1000 W high-pressure mercury fluorescent lamp (Philips HPLR) at approximately 600 μmol m^{-2} s^{-1} (400–700 nm), and a 65 mm deep water bath was used as a heat trap.

Gas samples of 200 μl were taken from the bags with a microsyringe through a rubber Subaseal®. Samples were taken from the root bag at regular intervals during the experiments, which lasted between 5 and 15 h depending on the rate of ethane movement detected. Samples were also taken from the shoot bag. The ethane concentration of the samples was determined immediately using a purpose-built portable gas chromatograph (Figaro type TGS-800 sensor) and ethane standards. The conductance of the plants to ethane was calculated for each seedling using a modification of Fick's law $G = J/(c_o\text{-}c_t)$, where G represents the conductance to ethane (m^3 s^{-1}), J the rate of flow of gas (mol s^{-1}), and $(c_o\text{-}c_t)$ the difference in ethane concentration between the shoot bag and the root bag (mol m^{-3}) (cf. Curran, 1985). In these experiments the shoot-bag concentration was about a thousand times the average concentration in the root bag, allowing the shoot-bag concentration to be taken as $(c_o\text{-}c_t)$ without appreciable error.

The suitability of various types of syringes and plastic bags for storage of standard gas mixtures was also investigated. Plastic syringes were found to leak ethane very quickly and were useless for standards for more than a few minutes but large plastic syringes were useful for dilution and mixing of gases. Glass syringes were useful to store gases for a few hours after which time there was significant loss. Polyester bags were found to be ethane-tight for at least 24 h but they were very fragile and had to be used with great care for standard gas concentrations and experiments. Polyethylene bags were much more durable but were not ethane-proof.

Results

The amount of ethane that had travelled through the plant was plotted against time and a regression line was fitted to the data for every illuminated and unilluminated period. The data points were linear (Fig. 1A – treatments, 1B – controls) with r^2 values all greater than 0.73 (Table 1). In some cases the line was steeper when the plants were illuminated, and in others either it was less steep or with very little change in slope. The rates of ethane flow through each plant are shown in Table 1. However, the rates can not be compared between plants because there were variations in the concentration difference of ethane across each plant (Table 1): to allow comparison the conductances for each treatment are shown in Table 2. The conductances for each of the treatments were compared using a *t* test, and no significant differences were detected at the 0.05 level of significance. The overall mean conductance to ethane was 0.23×10^{-9} m^3 s^{-1}.

Discussion

The conductance to oxygen can readily be calculated from that for ethane using a correction for the (slight) difference in diffusion coefficient of the two gases. The calculated conductance to oxygen averaged 0.22×10^{-9} m^3 s^{-1} over all experiments, about an order of magnitude less than those for the stem of much older, larger plants and two orders less than for *A. marina* pneumatophores, found by Curran *et al.* (1986). In these experiments we used whole seedlings, without cutting off the base as Curran *et al.* did, and so the measurements here must also include a restriction to diffusion through the root surfaces. Many of the fine roots (which are aerenchymatous) are likely to have been broken in taking the plants out of the soil, and we imagine that the ethane tracer gas moved out of the plant into the root-bag through these.

There was no significant effect of the light treatment on rate of ethane flow through the plants. This is in strong contrast to the results of Grosse & Schröder (1984) with *Alnus glutinosa*, where they found that illumination caused a great increase in the rate of tracer-gas flow. They measured temperatures and pressures within the stem and concluded that a thermo-osmotic pressure-flow mechanism brought about by illumination could account for the increased rate. In these *Avicennia marina* seedlings there was no evidence for a change in flow rate that could be attributed to such pressure-induced flow, even though about three times the illumination was used. We feel that it is likely that the seedlings are supplied with oxygen by diffusion.

These small seedlings without pneumatophores pose the problem for the concept of ventilation by mass flow that they only have the shoot as a single aerial organ. Mass-flow ventilation would be easier to imagine in more mature plants with pneumatophores, where thermo-osmotic flow inwards through leaves via stomata and outwards through pneumatophores via the much larger gas-spaces in the lenticels can be envisaged, and further investigation of the possibilities is required.

Acknowledgments

We thank Jörg Suckau for building the gas-chromatograph, Mark Curran for advice, and the Australian Research Council for a research grant.

References

Andersen, F. Ø. & E. Kristensen, 1988. Oxygen microgradients in the rhizosphere of the mangrove *Avicennia marina*. Mar. Ecol. Prog. Ser. 44: 201–204.

Armstrong, J., W. Armstrong & P. M. Beckett, 1988. *Phragmites australis*: a critical appraisal of the ventilating pressure concept and an analysis of resistance to pressurized gas flow and gaseous diffusion in horizontal rhizomes. New Phytol. 110: 383–389.

Ashford, A. E. & W. G. Allaway, 1995. There is a continuum of gas space in young plants of *Avicennia marina*. Hydrobiologia 295 (Dev. Hydrobiol. 106): 5–11.

Brown, J. M. A., H. A. Outred & C. F. Hill, 1969. Respiratory metabolism in mangrove seedlings. Plant Physiol. 44: 287–294.

Curran, M., 1985. Gas movements in the roots of *Avicennia marina* (Forsk.) Vierh. Aust. J. Plant Physiol. 9: 519–528.

Curran, M., M. Cole & W. G. Allaway, 1986. Root aeration and respiration in young mangrove plants (*Avicennia marina* (Forsk.) Vierh.). J. exp. Bot. 37: 1225–1233.

Dacey, J. W., 1981. Pressurized ventilation in the yellow waterlily. Ecology 62: 1137–1147.

Grosse, W. & H. Mevi-Schutz, 1987. A beneficial gas transport system in *Nymphoides peltata*. Am. J. Bot. 74: 947–952.

Grosse, W. & P. Schröder, 1984. Oxygen supply of roots by gas transport in alder-trees. Z. Naturforsch. 39c: 1186–1188.

Scholander, P. F., L. Van Dam & S. I. Scholander, 1955. Gas exchange in the roots of mangroves. Am. J. Bot. 42: 92–98.

Skelton, N. J., 1988. Gas movement and changes in the internal atmosphere of the roots of the grey mangrove *Avicennia marina* (Forsk.) Vierh. BSc thesis, The University of Sydney.

Hydrobiologia **295**: 5–11, 1995.
Y. S. Wong & N. F. Y. Tam (eds), Asia-Pacific Symposium on Mangrove Ecosystems.

There is a continuum of gas space in young plants of *Avicennia marina*

A. E. Ashford[1] & W. G. Allaway[2]
[1] *School of Biological Science, University of New South Wales, Kensington, NSW 2033, Australia;*
[2]*School of Biological Sciences, The University of Sydney, NSW 2006, Australia*

Key words: aerenchyma, *Avicennia marina*, flooding, lenticel, lignin, seedling

Abstract

Gas-spaces form a continuum throughout 10-month-old *Avicennia marina* seedlings. This has direct connection with the atmosphere via the stomata and spongy mesophyll of the leaves, and via the lenticels that occur on all the internodes and the hypocotyl. There is therefore provision for aeration of the root system prior to the development of pneumatophores. The amount of gas space is greatest in the cortex, and in elongated organs (petiole, root, stem internodes and hypocotyl) it occurs as wide elongated channels that are all interconnected. Continuity is maintained across the nodes by a mass of spongy tissue which connects up with the elongated channels above and below. In all organs the amount of gas space in the pith is much less, but there it also occurs as interconnected elongated channels. The volume of gas space is greatest in the major roots and specialised lignified cells which become collapsed may play a role in supporting the very large gas spaces.

Introduction

Marine and intertidal sediments such as mangrove swamps are generally waterlogged and anaerobic. On the other hand to function normally mangrove roots, which have high rates of respiration, require large amounts of oxygen. Roots of the grey mangrove, *Avicennia marina* (Forsk.) Vierh., in common with many other species, develop large amounts of aerenchyma (e.g. Chapman, 1976; Saenger, 1982; Tomlinson, 1986). This species produces characteristic pneumatophores that are uncovered at low tide and in contact with the atmosphere. These have lenticels and it is generally considered that they play a central role in aerating the root system. However, pneumatophores do not develop until the second year in *A. marina* seedlings and this raises the problem of how the root system is aerated during establishment and growth of the mangrove seedlings before the pneumatophores have developed.

Because they are so small, mangrove seedlings are completely submerged for long periods depending on how far down the intertidal zone they have become established. The stem and leaves may therefore be deprived of atmospheric oxygen for a long time each day. For the root system the situation is worse. Because there are no pneumatophores at this stage of development, there is no direct contact between the root system and the atmosphere at any time and the oxygen demand of the roots must be satisfied by transfer of oxygen exchanged by the above-ground parts of the system, via the hypocotyl.

We have examined the morphology of *A. marina* seedlings and the distribution of gas space within them, to investigate the continuity of aerenchyma in the seedling.

Materials and methods

Seedlings were collected from Botany Bay (34°10′S, 151°10′E) when about 10 months old. Six replicates were examined. The anatomy and distribution of gas space were first investigated by examination of sections cut by hand with a double-edged Gillette Blue^{TM} razor blade and stained with toluidine blue at pH 4.4 in sodium acetate buffer (O'Brien & McCully, 1981). This produced a general picture of the anatomy and distribution of aerenchyma throughout the seedling. Specific regions were then excised, fixed and embedded for a more detailed examination. Selected material was fixed in 3% glutaraldehyde in 0.025 M potassium phosphate buffer overnight, dehydrated in a graded ethanol series with steps of 10%, 20%, 30%, 50%, 70%, 85%, 95%, and two changes of 100% ethanol, over a period of 2 days, and embedded in (LR White) resin. Sections were cut at 1 to 1.5 μm and stained with toluidine blue as above. Regions of leaf, petiole, node, internode, hypocotyl and root were sampled in this way for detailed examination. Particular attention was given to the nodes which were in some cases serially sectioned. Photomicrographs were taken on a Leitz Orthoplan microscope on Kodak Technical Pan Film rated at 50 ASA.

Results

At 10 months the cotyledons had fallen off and there were three pairs of expanded leaves. Plants differed slightly in the length of the top (third) internode, counting the epicotyl as internode 1. The root system was small and comprised several adventitious roots arising from the base of the hypocotyl. These had given rise to two further orders of very fine lateral roots. There were no pneumatophores.

Roots

Sections through the adventitious roots (Fig. 1A–D) showed that they had a distribution of gas spaces that is typical for *A. marina* roots. The greatest volume of gas space was in the cortex. This was in the form of elongated, more or less cylindrical channels that all ran parallel to the long-axis of the organ (Fig. 1A, B). The channels were commonly separated by a single layer of small cells arranged in a regular row (Fig. 1B). Some of these cells were thin walled, non-lignified and roughly isodiametric in transverse section. Others in the row showed irregular thickened ridges on their walls (Fig. 1B, D); these stained blue with toluidine blue and appeared to be lignified. The number of these lignified cells varied in different roots and in different regions of the same root: in the roots in Fig. 1 about 50% of the cells in the row exhibited this thickening. Many of the cells had irregular shapes. Some of them appeared to have lost their contents and collapsed inwards so that their walls appeared totally flattened; cells at various stages in this process were apparent (Fig. 1B, D). In the final stage of collapse the cells formed a thin layer that separated adjacent longitudinal channels (Fig. 1B). Collapsed cells that had lignified ridges formed a two- to three-pronged scaffolding when seen in sections. There were frequent gaps between both lignified and unlignified cells in the middle-lamella region so that adjacent longitudinal channels were in continuity (Fig. 1A, C). The amount of gas space diminished in the outermost few rows of cells at the root surface (Fig. 1C) and in some roots cell division just beneath the epidermis indicated that a cork cambium had formed. Just inside the cortex, delimited by the endodermis, was the vascular cylinder with no obvious aerenchyma (Fig. 1C). A vascular cambium had developed. Internal to the vascular cylinder there was a large pith, many cells of which had somewhat thickened lignified walls. The pith also contained gas space but there was much less than in the cortex, and it was confined to the triangular spaces in the cell corners as seen in transverse section (Fig. 1C). These spaces formed narrow elongated channels, much narrower than those in the cortex (Fig. 1A). There was an increase in the size of the triangular gas spaces towards the centre of the pith. The fine roots also contained aerenchyma in the cortex.

Hypocotyl and stem

Gas space in the hypocotyl was again greatest in the cortex (Fig. 1E, F) and it similarly consisted of longitudinal channels. However there appeared to be much less gas space in the hypocotyl cortex and the channels appeared narrower and less interconnected than in the root (Fig. 1E, cf. 1A, B). Transverse sections showed that the cortical aerenchyma channels were more regular and more widely separated from one another than in the root (Fig. 2A). In contrast to the root the cells had generally not collapsed, and there were few lignified cells. Like the gas-channels in the root they were connected laterally by gaps between cells (Fig. 2A). The cortical aerenchyma was situated beneath a thick layer of collenchyma of about 7 rows of small cells with unevenly thickened walls and few to no gas spaces, and an epidermis with a thick cuticle. A cork cambium was usually developed. The central pith contained gas spaces in the cell corners that were triangular in cross section (Fig. 2B) and made up narrow elongated cylindrical channels like those of the root pith (Fig. 1F, cf. 1A). Many cells of the pith parenchyma had thickened, lignified walls with frequent pits (Fig. 2B). The stem internodes had a distribution of aerenchyma similar to that of the hypocotyl (cf. Fig. 2A, C).

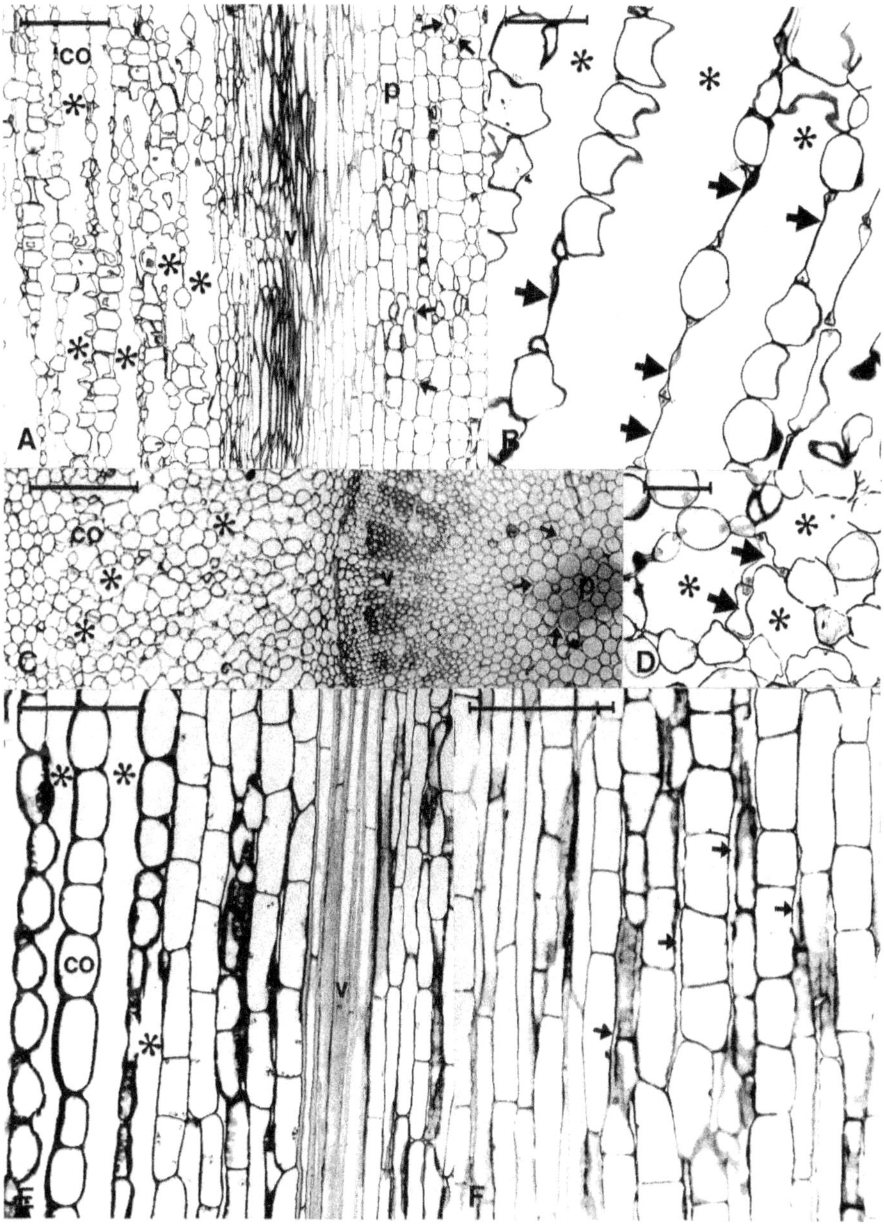

Fig 1 Thin sections of resin embedded portions of 10 month old *Avicennia marina* seedlings from Botany Bay A – Longitudinal section of first order root showing wide elongated gas channels (⋆) in the cortex (co) few or no gas spaces in the vascular tissue (v) and narrow elongated channels (arrows) in the pith (p) Scale 200 μm B – Longitudinal section of first order root cortex showing rows of cells lining elongated gas channels (⋆) Lignified collapsed cells (arrows) occur at intervals Scale 50 μm C – Transverse section of first order root showing a distribution of gas space typical for *A marina* roots Large channels (⋆) are present in the cortex and smaller triangular spaces (arrows) in the pith (p) Scale 200 μm D – Transverse section of first order root cortical gas channels (⋆) showing the appearance of collapsing lignified cells (arrows) Scale 50 μm E – Longitudinal section of hypocotyl showing wide elongate channels (⋆) in the cortex (co) Fibres in the vascular tissue are shown at v Scale 100 μm F – Longitudinal section of hypocotyl pith showing narrow elongated channels (arrows) Scale 50 μm

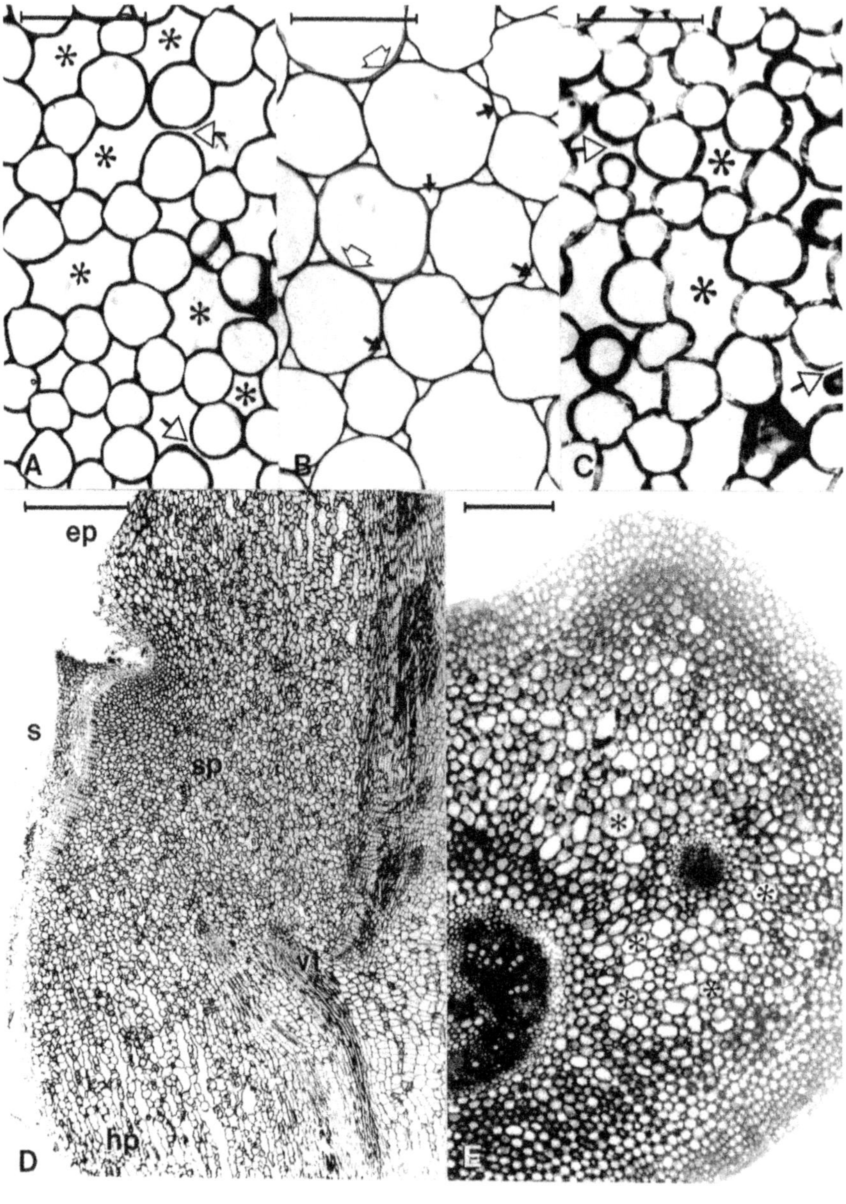

Fig 2 A, B and D thin resin sections, C and E hand-cut sections, of parts of 10-month-old *Avicennia marina* seedlings A – Transverse section of hypocotyl cortex showing arrangement of large gas channels (⋆) each surrounded by a single layer of cells, none of which is collapsed Note lateral connexions between adjacent channels (open arrows) Scale 50 μm B – Transverse section of hypocotyl pith showing small triangular gas spaces (arrows) Some of the cells have lignified walls (broad arrows) Scale 50 μm C – Hand-cut transverse section of stem cortex showing similar large gas channels (⋆) to those of the hypocotyl Some lateral connexions are also visible (open arrows) Scale 50 μm D – Longitudinal section through the cotyledonary node, showing the junction between the longitudinal channels in the cortex of the hypocotyl (hp) and the first internode or epicotyl (ep) via a mass of spongy tissue (sp) at the node Note the leaf trace (vt) towards the cotyledon, which has fallen off leaving a scar (s) Scale 200 μm E – Hand-cut transverse section of petiole showing gas spaces (⋆) in the lateral extension of the petiole, resembling the spaces in the stem cortex Scale 200 μm

At the nodes, there was a transition in the outer region between cortical tissue with longitudinally oriented cylindrical channels and a loose spongy tissue with no apparent direction of orientation (Fig. 2D). The gas spaces in this spongy aerenchyma formed a continuum of gas space between the cylindrical channels of one internode and the next. This is illustrated here at the epicotyl/hypocotyl junction of the cotyledonary node (Fig. 2D), but all nodes sampled showed a similar arrangement.

Petioles and leaves

The petiole was covered by a thick cuticle, and was roughly elliptical in cross section. It extended into lateral wings and had a partial ellipse of vascular tissue in the centre, with one or two small bundles in each of the wings (Fig. 2E). Large, longitudinally oriented gas channels occurred throughout the entire length of the petiole (Figs 2E, 3A). They were located in the cortex beneath an outer, compact region of collenchyma of about 4 layers of thick-walled small cells (Fig. 2E). The aerenchyma channels were particularly large and well developed in the lateral wings of the petiole (Fig. 2E). Each channel was surrounded by a single layer of small cells and the channels resembled those in the stem and hypocotyl (Fig. 2). There was no evidence of the cells with lignified ridges that characterised the root aerenchyma. These longitudinal channels in the petiole were continuous with the cortical channels in the stem via the spongy aerenchyma at each node. The pith inside the vascular partial-ellipse contained triangular gas spaces that were very similar to those in the pith throughout the plant (Fig. 3B).

The longitudinal channels in the petiole were continuous with the gas spaces in the spongy mesophyll, and Fig. 3C illustrates a transition region showing the change in orientation of the channels. A transverse section of the *A. marina* seedling leaf is shown in Fig. 3D. Leaves were dorsiventral with a palisade mesophyll of about 3 layers of elongate cells and a well developed spongy mesophyll, connected with the stomata on the abaxial side. The greatest volume of gas space in the leaf was within the spongy mesophyll. The palisade mesophyll was overlain on its adaxial side by a multilayered water storage parenchyma (hypodermis) of about 5 rows of larger thin-walled cells (cf. Baker, 1915; Fahn & Shimony, 1977) occupying almost half the thickness of the leaf. The cells here were tightly packed and had little aerenchyma, the intercellular spaces being filled with a material that stained pink with toluidine blue. The stomata on the abaxial surface were covered with a dense tomentum of peltate hairs. The entire leaf was covered with a cuticle, thickest on the adaxial epidermis and thinnest over the peltate abaxial hairs.

Lenticels

All of the internodes and the hypocotyl possessed lenticels to a greater or lesser extent. These varied in size and ranged in morphology from classical crater-like structures with a mass of fluffy complementary tissue in the centre to unopened pustules. The number varied from plant to plant and on different internodes. In addition there were raised broader areas with a broken surface and areas of damage, where aerenchyma was visible from the surface. In the hypocotyl the surface had a blistered or damaged appearance over extensive regions.

Discussion

In all of the organs of young *Avicennia marina* seedlings the largest amount of gas space occurs in the outer parts. In the elongated organs, namely petioles, internodes, hypocotyl and roots, it takes the form largely of highly elongate channels in the cortex. These are extensively connected to one another laterally, with the result that the gas space forms a continuum. The continuum is maintained across the nodes by a mass of spongy tissue.

This gas space continuum will have access to the atmosphere through the stomata when these are open at low tide, and this will occur when the tide is low in the daytime. The lenticels on the stem and hypocotyl provide continuity with the atmosphere at low tide at any time of day or night. In addition the leaves may be a source of photosynthetically-produced oxygen for the plant, perhaps even when it is underwater (cf. Laan & Blom, 1990).

Diffusion of oxygen through the longitudinal gas channels is remarkable: it has been measured in pneumatophores and cable roots of adult *A. marina* mangroves and found to be little different from movement through laboratory tubing (Curran, 1985). The structure is quite similar here in the seedling stem, though there is less gas space, and similar high rates of diffusion may be expected. We do not see the nodes as generating major resistance (cf. Chapman, 1976) since the gas spaces within them are extensive and greatly

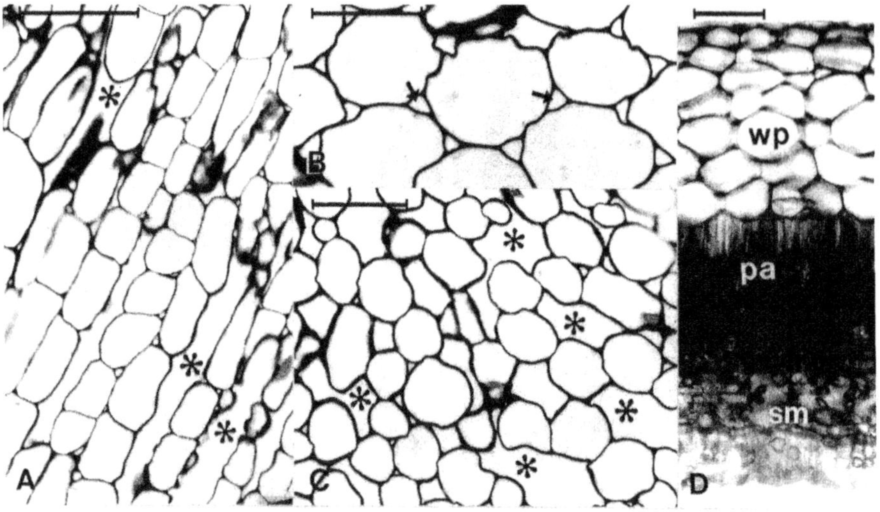

Fig. 3. A, B and C: thin resin sections, D: hand-cut section, of parts of 10-month-old *Avicennia marina* seedlings. A – Longitudinal section of petiole showing wide gas channels running longitudinally (*) Some small triangular spaces representing transverse connexions can also be seen. Scale 100 μm. B – Transverse section of petiole tissue within the vascular partial-ellipse, showing narrow triangular gas spaces (arrows) similar to those found throughout the pith. Scale 50 μm C – Transitional zone between the petiole (on the right) and lamina (on the left) showing petiolar gas channels (*) connecting with spongy mesophyll of the lamina. Scale 50 μm. D – Hand-cut transverse section of leaf, showing water parenchyma (or hypodermis, wp) on the adaxial side, three layers of palisade mesophyll (pa) and spongy mesophyll (sm). There is a large amount of gas space in the palisade and spongy mesophyll. Scale 100 μm.

interconnected: there may be some increase in diffusive resistance due to the greater tortuosity of the channels in the nodes. To visualise the situation at the nodes adequately it is necessary to use thin plastic-embedded sections, and both the extent and interconnexion of the spaces will have been underestimated in thicker hand-cut or paraffin-sections. In view of the continuity of gas space throughout the seedling, there is likely be no difficulty during low tide in aerating the root system at stages prior to the development of pneumatophores. Presumably a critical path length is reached as the seedling grows in height when this system no longer can adequately supply the roots and after this it becomes increasingly important for pneumatophores to develop. In the Botany Bay mangroves this happens during the second year.

As is often the case when aerenchyma is present, many cells had thickened, lignified walls. In *A. marina* seedlings the presence of such walls in the pith, and the band of thick-walled collenchyma around the periphery would add greatly to the structural strength of the stem. The specialised lignified strengthening cells in the root aerenchyma were in various stages of collapse suggesting that this may be a developmental sequence. This would provide a support system that maximises the amount of gas space because many of the cells have a high degree of strength but a minimum thickness. At the same time these specific cells are dead and have no oxygen demand.

The aerenchyma of *A. marina* has two distinct but complementary functions. On the one hand it is a gas-conduction tissue, allowing rapid movement of gases throughout an internal atmosphere in the plant, whether by diffusion or flow. On the other hand it provides a store of oxygen for the plant to use when cut off from the atmosphere. It can also be seen as a means of reducing oxygen demand in certain regions of the plant by reducing the number of cells per unit volume of organ. The latter may be a benefit of having so much more gas space in the root cortex than stem cortex. Nevertheless, the fact that the gas spaces in all the elongated structures in the plant, petiole, internodes, hypocotyl and roots, take the form of elongated channels is strongly suggestive that the spaces have a primary function in gas transport.

Acknowledgments

We are very grateful to Sue Bullock for preparing the Plates. We thank the University of Western Australia for facilities while the authors were on study leave, Robert Scott and Elizabeth Begg for assistance, and the Australian Research Council for a research grant. This work began as a third-year undergraduate project, and we thank the following students for material and photomicrographs: T. Baudoin, C. Harris, L. Ng, S. Schibeci and S. Wilson.

References

Baker, R. T., 1915. The Australian 'grey mangrove'. Proc. R. Soc. N.S.W. 49: 257–281.

Chapman, V. J., 1976. Mangrove Vegetation. J. Cramer, Vaduz.

Curran, M., 1985. Gas movements in roots of *Avicennia marina* (Forsk.) Vierh. Aust. J. Plant Physiol. 12: 97–108.

Fahn, A. & C., Shimony, 1977. Development of glandular and non-glandular leaf hairs of *Avicennia marina* (Forsskal) Vierh. Bot. J. Linn. Soc. 74: 37–46.

Laan, P. & C. W. P. M. Blom, 1990. Growth and survival responses of *Rumex* species to flooded and submerged conditions: the importance of shoot elongation, underwater photosynthesis and reserve carbohydrates. J. exp. Bot. 41: 775–783.

O'Brien, T. P. & M. E. McCully, 1981. The Study of Plant Structure Principles and Selected Methods. Termarcarphi Pty. Ltd., Melbourne.

Saenger, P., 1982. Morphological, anatomical and reproductive adaptations of Australian mangroves. In B. Clough (ed.), Mangrove Ecosystems in Australia: Structure, Function and Management. AIMS, Townsville: 153–191.

Tomlinson, P. B., 1986. The Botany of Mangroves. Cambridge University Press.

Hydrobiologia **295**: 13–22, 1995.
Y. S. Wong & N. F. Y. Tam (eds), Asia-Pacific Symposium on Mangrove Ecosystems.

Diurnal gas exchange characteristics and water use efficiency of three salt-secreting mangroves at low and high salinities

G. Naidoo[1] & D. J. von Willert[2]
[1]*Department of Botany, University of Durban-Westville, Private Bag X54001, Durban 4000, South Africa*
[2]*Institut für Angewandte Botanik, Westfalische Wilhelms Universität, Hindenburgplatz 55, 48143 Münster, Germany*

Key words: gas exchange, mangrove, photosynthesis, salinity

Abstract

Continuous measurements of gas exchange characteristics were made on two to nine year old hydroponically grown *Avicennia germinans* (L.) Stearn, *Aegialitis annulata* R. Br. and *Aegiceras corniculatum* (L.) Blanco maintained at 50 or 500 mol m^{-3} NaCl. In *Avicennia germinans* and *Aegialitis annulata*, CO_2 assimilation rates were initially higher at 500 mol m^{-3} NaCl and decreased gradually towards the end of the photoperiod when rates were similar to those at the lower salinity. In *Aegiceras corniculatum*, assimilation rates were higher at 50 mol m^{-3} NaCl and about 55% lower at the higher salinity. In all three species, leaf conductance and transpiration exhibited trends similar to those for CO_2 assimilation. Intercellular CO_2 concentrations were similar at both salinities in *Avicennia germinans* and *Aegialitis annulata*, but considerably higher at the lower salinity in *Aegiceras corniculatum*. Water use efficiencies (WUE), although similar between salinity treatments in *Avicennia germinans* and *Aegialitis annulata*, were greater at the higher salinity in *Aegiceras corniculatum*. Data obtained from CO_2 response curves indicated that assimilation at high salinity in *Aegiceras corniculatum* was limited by conductance, and to a lesser extent, by photosynthetic capacity. In *Avicennia germinans* and *Aegialitis annulata*, assimilation was greater at the higher salinity as indicated by increase in both the initial slope and the upper plateau of the CO_2 response data. Greater assimilation at high salinity in *Avicennia germinans* and *Aegialitis annulata* may be attributed to lower carbon losses via photorespiration and to efficient salt excretion and sequestration.

Introduction

Mangroves thrive in a stressful, intertidal environment characterised by fluctuating salinities and waterlogging. Despite these harsh conditions, survival and dominance of mangroves are not compromised (Tomlinson, 1986; Cheezeman *et al.*, 1991). Several studies indicated that mangrove species have different salinity optima for growth (Snedaker, 1982; Ball, 1988a, b; Burchett *et al.*, 1984) which may influence distributional patterns (Naidoo, 1985; Clough & Sim, 1989). Substrate salinity affects several physiological processes including leaf conductance, transpiration, CO_2 assimilation and water use efficiency (Ball & Farquhar, 1984a; Andrews & Muller, 1985; Clough & Sim, 1989; Cheezeman *et al.*, 1991).

Although gas exchange characteristics of mangroves have been investigated extensively, the effects of salinity are still inconclusive (Clough & Sim, 1989; Pezeshki *et al.*, 1990; Cheezeman *et al.*, 1991). Most studies, under growth chamber conditions, were on very young seedlings subjected to salinity treatments ranging from a few days (Ball & Farquhar, 1984a) to 3–6 months (Ball & Farquhar, 1984b; Pezeshki *et al.*, 1990; Lin & Sternberg, 1993). Results from these studies are contradictory and inconsistent. For example, CO_2 assimilation in *Avicennia marina* decreased at salinities above 250 mM when the salinity of the culture solution was varied every 2 d by 50 mM from 50 to 500 mM (Ball & Farquhar, 1984a). In another study on 6 wk old *Avicennia marina* and *Aegiceras corniculatum* seedlings, CO_2 assimilation rate decreased with increase in salinity from 50 to 500 mM NaCl over

3 months (Ball & Farquhar, 1984b). Pezeshki *et al.* (1990), however, found no significant differences in net CO_2 assimilation between plants subjected to 0 and 342 mol m^{-3} NaCl for 180 d in *Avicennia germinans*, *Laguncularia racemosa* and *Rhizophora mangle*.

Field measurements of mangrove photosynthesis are subject to great variability (Moore *et al.*, 1972; Attiwill & Clough, 1980; Andrews & Muller, 1985; Clough & Sim, 1989; Cheezeman *et al.*, 1991), are strongly inhibited by high leaf temperature and leaf to air vapour pressure deficit (Andrews & Muller, 1985) and represent a complex and highly dynamic system (Cheezeman *et al.*, 1991) that is difficult to interpret.

Thus, the effects of salinity on mangrove photosynthesis are unclear and warrant further investigation on older plants subjected to long term salinities. With the exception of Andrews & Muller (1985), all photosynthetic investigations involved spot measurements, over a few minutes, using gas exchange systems with no temperature or humidity control. This study was undertaken to evaluate and compare photosynthetic gas exchange in three salt secreting mangroves grown hydroponically over 2 to 9 years at low or high salinity. Continuous gas exchange measurements were made simultaneously on plants subjected to low or high salinity, using environmentally controlled microcuvette systems.

Materials and methods

Propagules of *Avicennia germinans* (L.) Stearn were obtained from South America and those of *Aegialitis annulata* R. Br. and *Aegiceras corniculatum* (L.) Blanco from Australia. These propagules were cultivated in Germany in an environmentally controlled glasshouse. Plants were grown in plastic pots (11 cm diameter × 8.5 cm depth) containing 1.0–1.5 cm diameter volcanic gravel. These pots were placed in larger pots (11.5 cm diameter × 9.5 cm depth) containing 50 or 500 mol m^{-3} NaCl as synthetic seawater, to which was added NH_4Cl (1 mol m^{-3}), NH_4NO_3 (1 mol m^{-3}) KH_2PO_4 (0.1 mol m^{-3}), Fe-EDTA (0.1 mol m^{-3}) and micronutrients (0.1 ml l^{-1}). Distilled water was added to compensate for evapotranspirational losses and solutions were renewed biweekly. Solution levels, indicated by floats, were maintained to one-third the pot. All plants were grown in a glasshouse set at 30 °C (day) and 18 °C (night) and supplemented with additional lighting. The age of plants used in the study varied from 2–3 years (*Avicennia germinans*) to 7–9 years (*Aegialitis annulata* and *Aegiceras corniculatum*). With increase in size, plants were transferred to larger pots.

Gas exchange measurements were made on attached leaves with an open microcuvette system (Walz, Effeltrich, Germany). Ambient air supply was separated into four paths of equal volume. The first path was directed to the reference channel of a differential infra-red gas analyzer (Binos 1, Heraeus, Hanau, Germany) for the measurement of CO_2 and water vapour. The other three paths were also permanently streamed and operated by a magnetic switch which allowed each path to open successively to the second channel of the IRGA. One of these, when directed to the IRGA, served as the reference path and was used for reading the zero position of both the CO_2 and water vapour signals. The remaining two paths contained a temperature and humidity controlled microcuvette, containing an intact leaf and two dew point mirrors (Walz), which measured the dew point temperature of the air stream entering and leaving the assimilating chamber. This configuration permitted monitoring of dew point temperature of the ambient air, as well as calculation of transpiration, when high daytime transpiration rates exceeded the range of the Binos water channel. Air flow in the two paths was controlled and measured by a mass flow meter (Tylan, Carson, Ca., USA).

The microcuvette comprised an environmental control unit of nickel construction, mounted inside a white polyethylene shield and a plexiglass leaf chamber thoroughly ventilated by a fan to give an air speed across the leaf of 3.5 m s^{-1}. The temperature of the chamber could be set to follow ambient, as measured by a ventilated PT100 resistance sensor, or set to a constant. Cooling and heating of the chamber was via Peltier elements.

Plastic putty, Terostat (Teroson, Gmbh., Heidelberg, Germany) was used to seal the leaf in the chamber. Leaf temperature was measured by a NiCr-Ni thermocouple, which was in contact with the undersurface of the leaf. The temperature of the chamber air was monitored by a radiation-shielded NTC thermistor and photosynthetic photon flux density (PPFD) by a quantum sensor (Licor Li 190S).

The gas exchange system was controlled by a data logging unit which permitted manipulation of the magnetic switches by the user. When necessary, data were transferred from the data logging unit to a microcomputer, and further converted by a special conversion programme. Photosynthesis, transpiration and internal

CO_2 concentration were calculated according to the equations of von Caemmerer & Farquhar (1981).

All gas exchange measurements were made on entire, fully expanded leaves of plants maintained at 50 or 500 mol m^{-3} NaCl for 2–9 years. Measurements were made in a temperature-controlled laboratory set at 25 °C (day) and 20 °C (night). For each species, gas exchange measurements at 50 and 500 mol m^{-3} NaCl were made simultaneously, using two plexiglass microcuvettes in parallel. After installation of leaves into the assimilation chamber, plants were allowed to stabilise to the new environmental conditions for at least 48 h before data logging commenced.

For the determination of CO_2 response curves, an additional gas mixing device permitted lowering or elevating the CO_2 concentration of the air stream. The outlet of the reference gas path was connected to an absolute IRGA to give the ambient CO_2 concentration, C_a (Farquhar & Sharkey, 1982). Carbon dioxide assimilation rate was measured as a function of the intercellular CO_2 concentration, c_i, by varying the ambient CO_2 concentration from 340 up to 600 $\mu l\ l^{-1}$, and back to 20 $\mu l\ l^{-1}$, allowing sufficient time for gas exchange characteristics to attain steady state. Data for each set of conditions were obtained from a single leaf. There were five replications per treatment.

Photorespiration was measured by determining differences in CO_2 assimilation rates at 2% and 21% O_2 while maintaining CO_2 concentrations unchanged at 340 $\mu l\ l^{-1}$. Carbon dioxide uptake at 21% O_2 was expressed as a percentage of that at 2% O_2 to indicate the magnitude of photorespiration.

Results

Diurnal trends in gas exchange

Gas exchange characteristics for each species were measured simultaneously at low and high salinities for periods ranging from 5–10 days. Although there was some variation in gas exchange from day to day, the general trends were the same over all days. The data presented were selected for being the most representative over all days.

In *Avicennia germinans* and *Aegialitis annulata*, CO_2 assimilation rates (Figs 1 & 2) were typically higher at 500 than at 50 mol m^{-3} NaCl at the onset of the photoperiod, and decreased gradually at 500 mol m^{-3} NaCl towards the end of the day, when rates were similar to those at the lower salinity. At the higher salinity, maximal rates of assimilation were in the early morning, being about 52% higher in *Avicennia germinans* and about 30% higher in *Aegialitis annulata* than at the lower salinity. At the end of the photoperiod, assimilation rates at the higher salinity were 44% lower in *Avicennia germinans* and 51% lower in *Aegialitis annulata* than their early morning maxima (Figs 1 & 2). In *Avicennia germinans*, CO_2 assimilation at 50 mol m^{-3} NaCl, although 52% lower than at 500 mol m^{-3} NaCl in the early morning, increased gradually, so that rates at the end of the photoperiod were similar. Dampened oscillations in assimilation with small amplitudes were apparent in *Avicennia germinans* (Fig. 1), especially at low salinity, during the latter part of the day.

In *Aegialitis annulata*, CO_2 assimilation at 50 mol m^{-3} NaCl remained at a fairly steady state level throughout, being about 30% lower than that at higher salinity at the beginning, but about 5% higher at the end of the photoperiod (Fig. 2). In contrast to *Avicennia germinans* and *Aegialitis annulata*, CO_2 assimilation in *Aegiceras corniculatum* was higher at 50 mol m^{-3} NaCl and reduced by about 55% at the higher salinity (Fig. 3). Unlike the other two species, assimilation was fairly steady at both salinities, but exhibited a gradual decline towards the end of the photoperiod.

Although there was strong correspondence between CO_2 assimilation, leaf conductance and transpiration, there were marked differences in internal CO_2 concentration (c_i) among the three species (Figs 1, 2 & 3). In *Avicennia germinans* and *Aegialitis annulata*, c_i was about 220 $\mu l\ l^{-1}$ initially at both salinities, and decreased gradually for the rest of the day, differences in c_i between salinities being about 20–25 $\mu l\ l^{-1}$ (Figs 1 & 2). In *Aegiceras corniculatum*, c_i was about 260 $\mu l\ l^{-1}$ at 50 mol m^{-3} NaCl, while at the higher salinity, c_i decreased from about 220 $\mu l\ l^{-1}$ in the early morning to about 150 $\mu l\ l^{-1}$ in the late afternoon (Fig. 3), being consistently lower by about 40–80 $\mu l\ l^{-1}$ at the higher salinity.

The relationship between CO_2 exchange and leaf conductance, from data obtained over several days (Fig. 4), shows that variation in assimilation between salinity treatments was matched by corresponding differences in conductance. At low conductance there was little difference in assimilation between salinity treatments in *Avicennia germinans* and *Aegialitis annulata*, while at higher conductance, assimilation was greater at high salinity. In *Aegialitis annulata*, for example, assimilation was consistently greater at the higher salinity at conductance values between 25 to

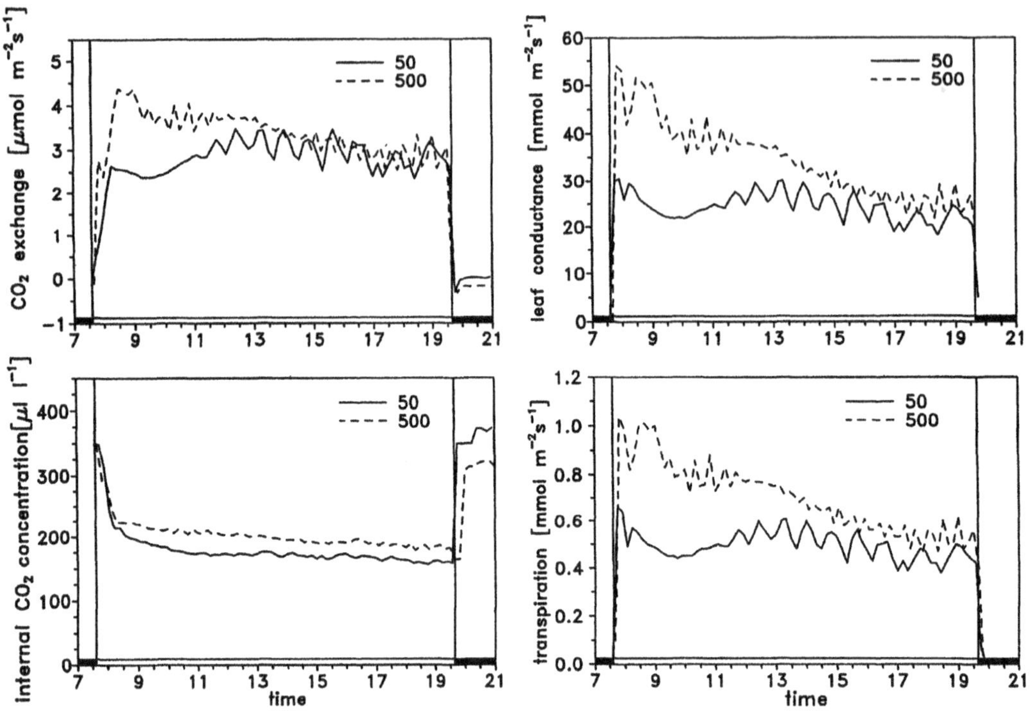

Fig. 1. Diurnal gas exchange characteristics of *Avicennia germinans* at 50 and 500 mol m^{-3} NaCl. Plants were subjected to 25 °C (day), 20 °C (night) VPD of 15 mPa Pa^{-1} (day) and 5 mPa Pa^{-1} (night) and PPFD of 400 μmol m^{-2} s^{-1}. External CO_2 concentration was 360 μl l^{-1}.

40 mmol m^{-2} s^{-1} (Fig. 4). In *Aegiceras corniculatum*, low conductance (<50 mmol m^{-2} s^{-1}) contributed to low assimilation rates at high salinity. Conductance was considerably higher (50–110 mmol m^{-2} s^{-1}) at low salinity, and contributed to higher rates of assimilation (Fig. 4).

CO_2 response characteristics

The assimilation rate, plotted as a function of c_i, indicated distinct differences among the species in response to low or high salinity (Fig. 5). In *Avicennia germinans* and *Aegialitis annulata*, the initial slope, as well as the upper plateau of the CO_2 response data were greater at 500 than at 50 mol m^{-3} NaCl, with differences between salinities being more distinct in the latter species (Fig. 5). In *Aegiceras corniculatum*, there was little difference between salinities in the initial slope of the CO_2 response curves at c_i < 140 μl l^{-1}, but at higher c_i the difference was more distinct, with assimilation being greater at the lower salinity.

Carbon dioxide response curves for all species, plotted at 50 mol m^{-3} NaCl (Fig. 6A), demonstrated that assimilation was low in *Avicennia germinans* and greater in *Aegialitis annulata* and *Aegiceras corniculatum*. At 500 mol m^{-3} NaCl, there were distinct differences among all three species, with *Aegialitis annulata* having the greatest initial slope and upper plateau and *Avicennia germinans* the least (Fig. 6B).

Water Use Efficiency (WUE)

Trends in WUE were similar for all species, being low in the early morning and increasing progressively during the course of the day (Fig. 7). In *Avicennia germinans* and *Aegialitis annulata*, there was little difference in WUE between salinity treatments. In *Aegiceras corniculatum*, WUE was greater at the higher salinity, and reduced by about 35–45% at the lower salinity.

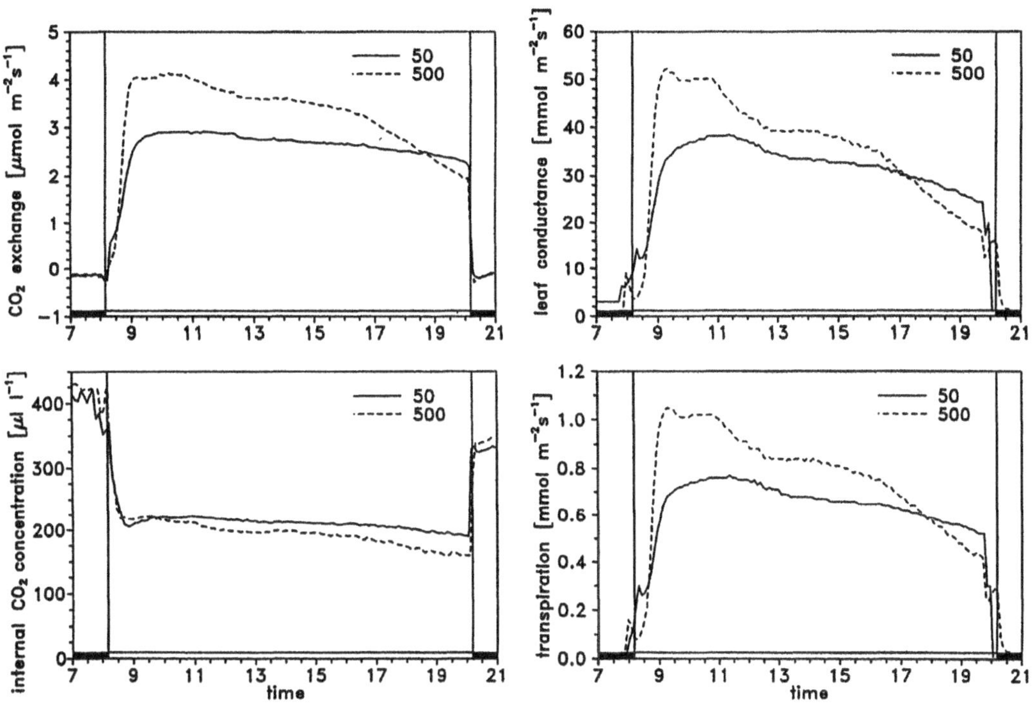

Fig. 2. Diurnal gas exchange characteristics of *Aegialitis annulata* at 50 and 500 mol m^{-3} NaCl. Other details as for Fig. 1.

Photorespiration

At low salinity, photorespiratory losses varied from 29–38% and were highest in *Avicennia germinans* (Fig. 8). High salinity reduced photorespiration by about 3% in *Avicennia germinans* and by about 7–10% in the other two species.

Discussion

Although all three species possess glands that actively secrete salt from the leaves, gas exchange characteristics clearly indicated that *Avicennia germinans* and *Aegialitis annulata* were more salt tolerant than *Aegiceras corniculatum*, and that CO_2 assimilation in these two species was not compromised by high salinity (Figs 1 to 6). High salinity induced similar gas exchange strategies in *Avicennia germinans* and *Aegialitis annulata*. In both these species, maximal conductance, transpiration and CO_2 exchange at the onset of the photoperiod at 500 mol m^{-3} NaCl was probably in response to a favourable plant water status in the early morning. Maximal stomatal opening in the early morning contributed not only to high photosynthesis, but also to excessive water loss (Figs 1 & 2), so that WUE was at its minimum (Fig. 7). High evaporative losses in the morning probably caused a water deficit which increased during the course of the day and resulted in a gradual decline in conductance, photosynthesis and transpiration (Figs 1 & 2). Decreased water loss resulted in increased WUE which was achieved at the expense of the assimilation rate. Consequently, towards the end of the day, there were little differences in gas exchange characteristics between low and high salinities in these two species (Figs 1 & 2).

The diurnal pattern of gas exchange in *Avicennia germinans* and *Aegialitis annulata* at high salinity are similar to the transpiration curves reported previously in field studies for *Avicennia marina* (Lewis & Naidoo, 1970; Leshem & Levison, 1972; Waisel *et al.*, 1986) and to trends in stomatal conductance in *Avicennia marina* and *Ceriops tagal* (Gordon, 1993).

At low salinity, gas exchange responses in *Aegialitis annulata* (Fig. 2) were at steady state probably because of greater stomatal regulation, while in *Avi-*

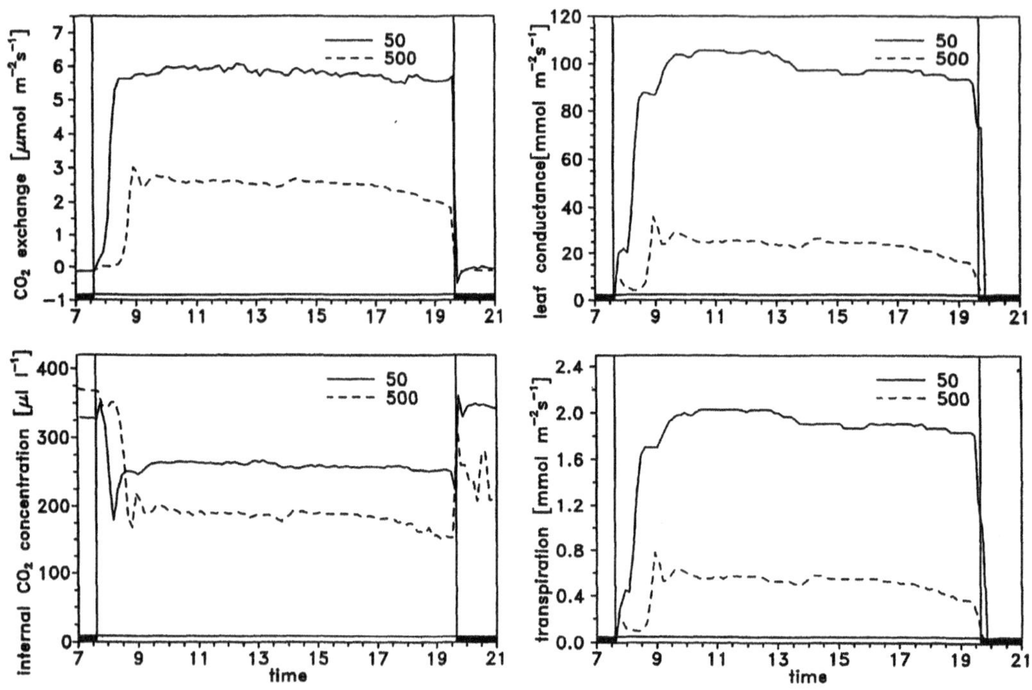

Fig. 3. Diurnal gas exchange characteristics of *Aegiceras corniculatum* at 50 and 500 mol m^{-3} NaCl. Other details as for Fig. 1.

cennia germinans there were some fluctuations during stomatal oscillation (Fig. 1).

Gas exchange responses in *Aegiceras corniculatum* reported here are in agreement with those of others (Ball & Farquhar, 1984b; Clough & Sim, 1989), confirming the sensitivity of this species to high salinity, CO_2 exchange being about 55% lower at 500 than at 50 mol m^{-3} NaCl (Fig. 3). Reduced photosynthesis at high salinity was due primarily to low conductance (Figs 3 & 4) which contributed to c_i values that were about 40–80 $\mu l\ l^{-1}$ lower than at low salinity. Low conductance at high salinity reduced transpiration (Fig. 3) and contributed to high WUE (Fig. 7). Thus, in this species, conductance limited the assimilation rate at high salinity. Reduced assimilation at salinities above 50 mol m^{-3} NaCl in *Aegiceras corniculatum* was attributed to co-limitation by stomatal conductance and reduced photosynthetic capacity by other workers (Ball & Farquhar, 1984b). In this study, however, there was no evidence for reduced photosynthetic capacity in *Aegiceras corniculatum* (Fig. 5).

Results obtained for *Avicennia marina* (Ball & Farquhar, 1984a, b), another salt secreting species, are contrary to those presented here for *Avicennia germinans* and *Aegialitis annulata*. The results of Ball & Farquhar (1984a, b) probably represent short term responses of young seedlings to high salinities. Our results are supported by those of Pezeshki *et al.* (1990), who reported no differences in conductance or net carbon assimilation in *Avicennia germinans* between non saline and 342 mol m^{-3} salinity treatments. In addition, field studies in Australia indicated that *Avicennia marina* consistently exhibited the highest CO_2 assimilation rates and conductance of 19 mangrove species, even at salinities as high as 45 g l^{-1} (Clough & Sim, 1989).

The close coordination between conductance and assimilation (Fig. 4), characterized by a steep initial slope followed by convexity at high conductance, resulted in high WUE, especially at high salinity (Fig. 7). The high WUE may be an important adaptation for salt tolerance (Andrews & Muller, 1985; Ball, 1988a; Clough & Sim, 1989) by not only minimising water loss but also reducing salt loading in shoots and energy loss in the elimination of accumu-

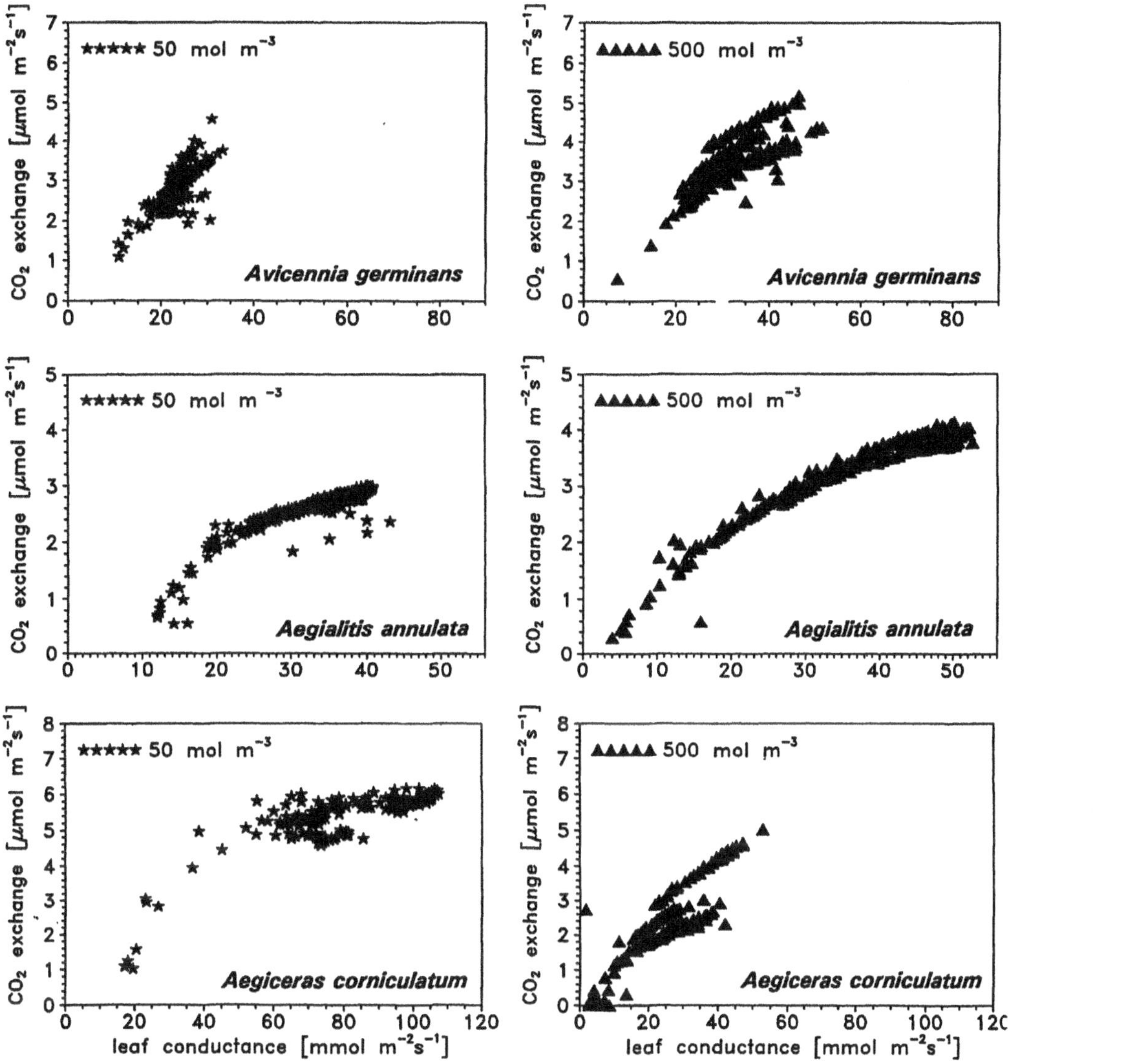

Fig. 4. Relationship between CO_2 exchange and leaf conductance in *Avicennia germinans*, *Aegialitis annulata* and *Aegiceras corniculatum* at 50 and 500 mol m^{-3} NaCl. Other details as for Fig. 1.

lated salt. High WUE, however, was at the expense of the assimilation rate.

Although assimilation was higher at 500 than at 50 mol m^{-3} NaCl in *Avicennia germinans* and *Aegialitis annulata*, plants grown at the higher salinity were shorter, but healthier, with smaller, thicker leaves. Consequently, the carbon investment per unit leaf area is higher and growth slower than for the thinner leaf at the same CO_2 uptake rate. Reduced growth at high salinity in these two salt tolerant species was attributed to the higher carbon cost of water uptake, salt secretion and osmoregulation (Ball, 1988a, b).

Changes in the shape of the CO_2 response curves for each species were used to determine the effects of salin-

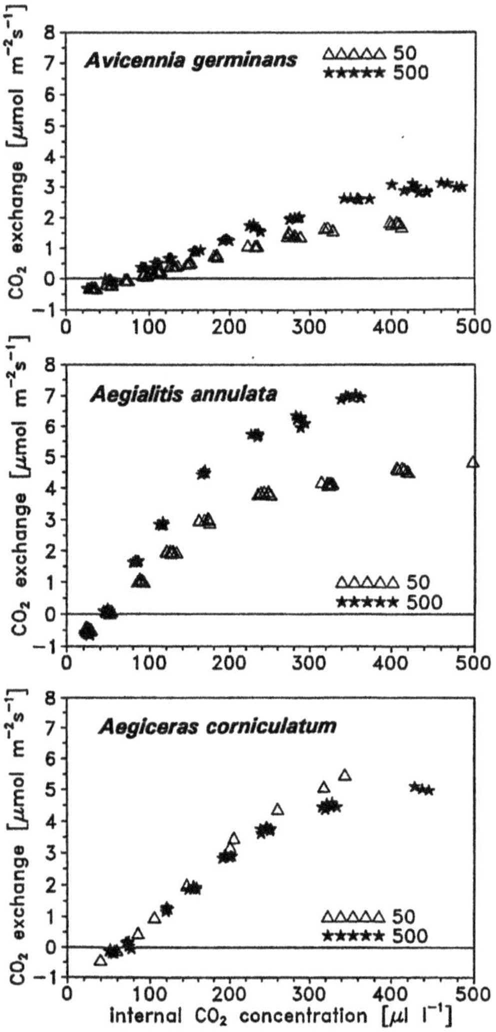

Fig. 5. Relationship between CO_2 exchange and intercellular CO_2 concentration in *Avicennia germinans*, *Aegialitis annulata* and *Aegiceras corniculatum* at 50 and 500 mol m^{-3} NaCl. Other details as for Fig. 1.

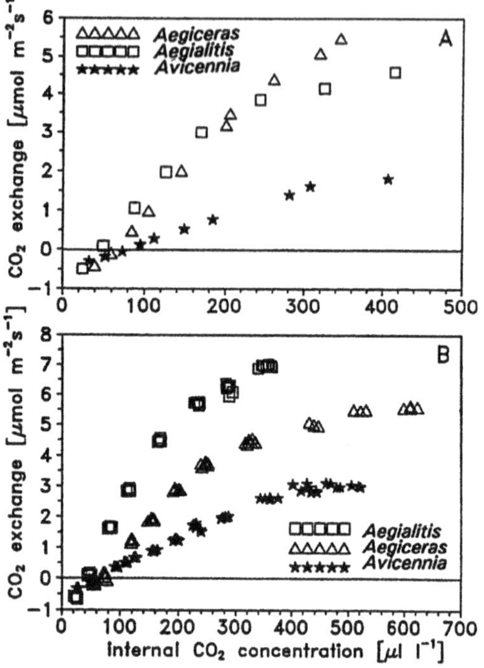

Fig. 6. Relationship between CO_2 exchange and intercellular CO_2 concentration in *Avicennia germinans*, *Aegialitis annulata* and *Aegiceras corniculatum* at 50 (A) and 500 mol m^{-3} NaCl (B). Other details as for Fig. 1.

ity on photosynthetic capacity. In *Aegiceras corniculatum*, similar linear slopes of the CO_2 response curves between salinity treatments (Fig. 5) indicated that photosynthetic CO_2 fixation was more or less unaffected by salinity. Reduction in the level of the upper plateau at the higher salinity suggested that decreased conductance limited photosynthesis in this species. Reduced photosynthesis is apparently due to the disruption of *K* uptake by high salinity leading to low leaf *K* concentrations (Robinson & Downton, 1984) which adversely influences several photosynthetic processes (Huber, 1985), including reduced photosynthetic activity (Ball *et al.*, 1988).

In *Avicennia germinans* and *Aegialitis annulata*, higher plateau levels and steeper initial slopes of the CO_2 response curves at 500 than at 50 mol m^{-3} NaCl, clearly indicated that photosynthetic capacity was greater at the higher salinity. However, there is no evidence that the chloroplast is adapted to high salinity (Ball, 1986) because of similar ionic composition of chloroplasts isolated from species differing in salt tolerance (Nobel, 1969; Harvey & Flowers, 1978; Harvey *et al.*, 1981; Robinson & Downton, 1984). High CO_2 assimilation in *Avicennia germinans* and *Aegialitis annulata* at high salinity may be due to an efficient salt excretion mechanism that maintains salt concentrations in leaves at physiologically acceptable levels. Some of the highest and most efficient rates of salt excretion in mangroves have been reported for *Avicennia* and *Aegialitis* species (Scholander *et al.*, 1962; Atkinson *et al.*, 1967; Waisel *et al.*, 1986; Boon & Allaway, 1986; Gordon, 1993). Efficient excretion as well

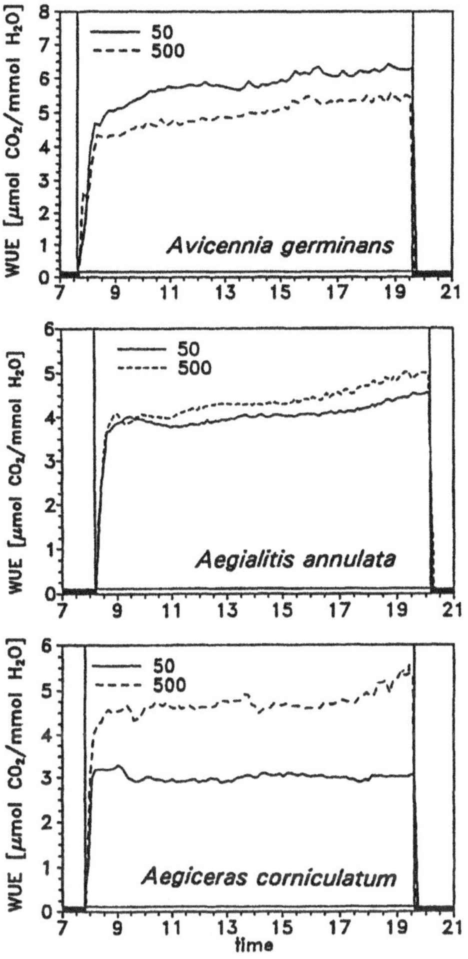

Fig 7 Diurnal water use efficiency (WUE) in *Avicennia germinans*, *Aegialitis annulata* and *Aegiceras corniculatum* at 50 and 500 mol m^{-3} NaCl. Other details as for Fig. 1.

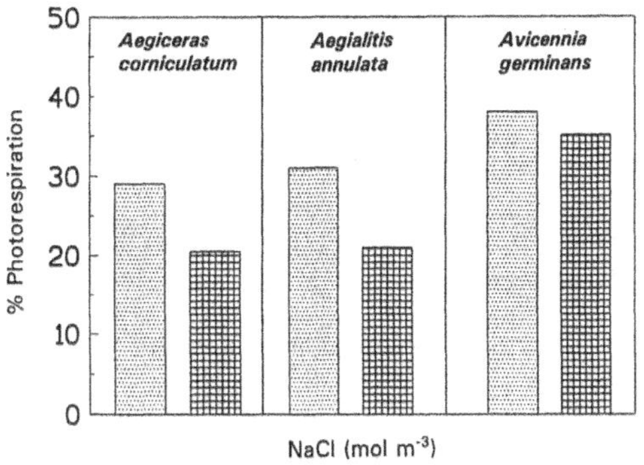

Fig 8 Photorespiration in *Avicennia germinans*, *Aegialitis annulata* and *Aegiceras corniculatum* at 50 (dots) and 500 (cross hatch) mol m^{-3} NaCl CO_2 concentration was 340 $\mu l\, l^{-1}$ and O_2 was 21% and 2% respectively There was no replication Other details as for Fig. 1.

as sequestration of salts away from sensitive metabolic sites, such as chloroplasts, may have contributed to higher rates of CO_2 assimilation at the higher salinity in these two species. In *Avicennia marina* and *Aegiceras corniculatum*, Ball & Farquhar (1984a, b) reported no effect of salinity on the CO_2 compensation point and concluded that salinity had no effect on photorespiration. Our data show conclusively that photorespiratory losses are high (29–38%) at low salinity in all three species and reduced at high salinity (Fig. 8). High CO_2 assimilation at high salinity in *Avicennia germinans* and *Aegialitis annulata* may be attributed in part to lower carbon losses via photorespiration. Although photorespiratory losses in *Aegiceras corniculatum* were also reduced at high salinity, less efficient ion sequestration and salt excretion may have contributed to reduced photosynthesis at 500 mol m^{-3} NaCl.

Although *Avicenna germinans* exhibited high salt tolerance, CO_2 exchange rates were the least of the three species (Fig. 6A & B) while photorespiratory carbon losses were the highest (Fig. 8). These results may explain the stable but slow growth of this species over a broad range of salinities. *Avicennia germinans* plants used in this study, however, were also considerably younger (2–3 years) than those of the other species (7–9 years).

Conclusion

Contrary to the general view that mangrove photosynthesis is sensitive to high salinity (Ball & Farquhar, 1984a, b), our measurements show that in *Avicennia germinans* and *Aegialitis annulata*, two highly salt tolerant species, photosynthesis is not compromised, but rather, tends to be enhanced, at 500 compared to 50 mol m^{-3} NaCl. In these two species, the diurnal course of gas exchange is altered at high salinity, being maximal in the early morning and decreasing gradually for the rest of the day with increasing WUE.

In *Aegiceras corniculatum*, assimilation at high salinity was limited by conductance and to a lesser extent, by photosynthetic capacity In *Avicennia germinans* and *Aegialitis annulata*, greater assimilation at high salinity was attributed to lower photorespiratory carbon losses and possibly to efficient salt excretion and sequestration

Acknowledgments

This work was sponsored by a research fellowship from the Alexander von Humboldt Foundation and a sabbatical grant from the Foundation for Research Development, South Africa The authors are grateful to Ute Wagner-Douglas for technical assistance and to Prof M Popp for the use of plant material

References

Andrews, T J & G J Muller, 1985 Photosynthetic gas exchange of the mangrove, *Rhizophora stylosa* Griff, in its natural environment Oecologia 65 449–455

Atkinson, M R, G P Finlay, M G Pitman, H D W Saddler & K R West, 1967 Salt regulation in the mangroves *Rhizophora mucronata* Lam and *Aegialitis annulata* R Br Aust J Biol Sci 20 589–599

Attiwill, P M & B F Clough, 1980 Carbon dioxide and water vapour exchange in the white mangrove Photosynthetica 14 40–47

Ball, M C, 1986 Photosynthesis in mangroves Wetlands (Australia) 6 12–22

Ball, M C, 1988a Ecophysiology of mangroves Trees 2 129–142

Ball, M C, 1988b Salinity tolerance in the mangroves *Aegiceras corniculatum* and *Avicennia marina* 1 Water use in relation to growth, carbon partitioning, and salt balance Aust J Plant Physiol 15 447–464

Ball M C & G D Farquhar, 1984a Photosynthetic and stomatal responses of the grey mangrove, *Avicennia marina* to transient salinity conditions Plant Physiol 74 7–11

Ball, M C & G D Farquhar, 1984b Photosynthetic and stomatal responses of two mangrove species, *Aegiceras corniculatum* and *Avicennia marina*, to long term salinity and humidity conditions Plant Physiol 74 1–6

Ball, M C, I R Cowan & G D Farquhar, 1988 Maintenance of leaf temperature and the optimization of carbon gain in relation to water loss in a tropical mangrove forest Aust J Plant Physiol 15 263–276

Boon, P I & W G Allaway, 1986 Rates and ionic specificity of salt secretion from excised leaves of the mangrove *Avicennia marina* (Forsk) Vierh Aquat Bot 26 143–153

Burchett, M D, C D Field & A Pulkownik, 1984 Salinity growth and root respiration in the grey mangrove, *Avicennia marina* Physiol Plant 60 113–118

Cheezeman, J M, B F Clough, D R Carter, C E Lovelock, Ong Jin Eong & R G Sim, 1991 The analysis of photosynthetic performance in leaves under field conditions A Case study using *Bruguiera* mangroves Photosynth Res 29 11–22

Clough, B F & R G Sim, 1989 Changes in gas exchange characteristics and water use efficiency of mangroves in response to salinity and vapour pressure deficit Oecologia 79 38–44

Farquhar, G D & T D Sharkey, 1982 Stomatal conductance and photosynthesis Annu Rev Plant Physiol 33 317–345

Gordon, D M, 1993 Diurnal water relations and the salt content of two contrasting mangroves growing in hypersaline soils in tropical-arid Australia In H Lieth & A Al Masoon (eds), Towards the rational use of high salinity tolerant plants, Vol I Kluwer Academic Publishers 193–216

Harvey, D M R & T J Flowers, 1978 Determination of the sodium, potassium and chloride ion concentrations in the chloroplasts of the halophytes *Suaeda maritima*, by non-aqueous cell fractionation Protoplasma 97 337–349

Harvey, D M R, J L Hall, T J Flowers & B Kent, 1981 Quantitative ion localization within *Suaeda maritima* leaf mesophyll cells Planta 151 555–560

Huber, S C, 1985 Role of potassium in photosynthesis and respiration In R D Munson (ed), Potassium in agriculture Am Soc Agron, Madison, Wisconsin 369–396

Leshem, Y & E Levison, 1972 Regulation mechanisms in the salt mangrove, *Avicennia marina*, growing in the Sinai littoral Oecol Plant 7 167–176

Lewis, O A M & G Naidoo, 1970 Tidal influence on the apparent transpirational rhythm of the white mangrove S Afr J Sci 66 268–270

Lin, G & LdaSL Sternberg, 1993 Effect of salinity fluctuation on photosynthetic gas exchange and plant growth of the red mangrove (*Rhizophora mangle* L) J exp Bot 258 9–16

Moore, R T, P C Miller, D Albright & L L Tieszen, 1972 Comparative gas exchange characteristics of three mangrove species in winter Photosynthetica 6 387–393

Naidoo, G, 1985 Effects of waterlogging and salinity on plant water relations and on the accumulation of solutes in three mangrove species Aquat Bot 22 133–143

Nobel, P S, 1969 Light-induced changes in the ionic contents of chloroplasts in *Pisum sativum* Biochim Biophys Acta 172 134–144

Pezeshki, S R, R D DeLaune & W H Patrick, 1990 Differential response of selected mangroves to soil flooding and salinity gas exchange and biomass partitioning Can J Forest Res 20 869–874

Robinson, S P & W J S Downton, 1984 Potassium, sodium and chloride content of isolated intact chloroplasts in relation to ionic compartmentation in leaves Arch Biochem Biophys 228 197–206

Scholander, P F, H T Hammel, E A Hemmingsen & W Garey, 1962 Salt balance in mangroves Plant Physiol 37 722–729

Snedaker, S C, 1982 Mangrove species zonation why? In D M Sen & K S Rajpurohit, (eds), Contributions to the ecology of halophytes Dr W Junk Publishers, The Hague 112–125

Tomlinson, P B, 1986 The botany of mangroves Cambridge University Press, Cambridge, 413 pp

Von Caemmerer, S & G D Farquhar, 1981 Some relationships between the biochemistry of photosynthesis and the gas exchange of leaves Planta 153 376–397

Waisel, Y, A Eshel & M Agami 1986 Salt balance of leaves of the mangrove, *Avicennia marina* Physiol Plant 67 67–72

Hydrobiologia **295**: 23–29, 1995.
Y. S. Wong & N. F. Y. Tam (eds), Asia-Pacific Symposium on Mangrove Ecosystems.

Ventilation and respiration in roots of one-year-old seedlings of grey mangrove *Avicennia marina* (Forsk.) Vierh.

M. J. Hovenden, M. Curran, M. A. Cole, P. F. E. Goulter, N. J. Skelton & W. G. Allaway*
*School of Biological Sciences A12, The University of Sydney, NSW 2006, Australia (*author for correspondence)*

Key words: aerenchyma, *Avicennia marina*, diffusion, mangrove, roots, respiration

Abstract

Avicennia marina (Forsk.) Vierh. was grown from seed for 12 months in artificially tidal tanks providing a range of duration and depth of inundation. Plant growth characteristics were measured at harvest. Root aerenchyma development was estimated by pycnometry, root respiration rates by manometry, and the oxygen supply capacity of the above-ground portions of the plant was determined using oxygen electrode chambers. The mass per plant at harvest was influenced by the extent of inundation during growth with maximal growth at intermediate-length (1.5 to 6.5 h per tide) inundation periods. Those plants that had been submerged the longest (8.5 h per tide) had the least root tissue. The oxygen conductance of the stem base plus any pneumatophores showed a maximum in plants grown under intermediate inundation. Oxygen demand and internal gas space per unit dry weight of root were independent of extent of inundation. During high tide the plants grown at inundation periods of more than about 3–5 hours per tide were likely to become anaerobic. This may constitute a physiological limit for this species at the bottom of the tidal range.

Introduction

The soils of wetlands are waterlogged and low in molecular oxygen, and thus the supply of molecular oxygen to the roots of wetland plants is restricted (Armstrong, 1978). Root aeration of wetland plants has been studied extensively (e.g. Armstrong *et al.*, 1988; Dacey, 1981; Grosse & Mevi-Schutz, 1987; Jackson & Drew, 1984; Raskin & Kende, 1983; Yamasaki, 1984). Wetland plants' roots typically contain a large proportion of aerenchyma providing an extensive internal gas space (Kozlowski, 1984) which serves to store and conduct gases throughout the root system (Chapman, 1976; Hook & McKevlin, 1988).

By the nature of their habitat, mangroves must be able to support and grow their roots through an anaerobic soil and must tolerate the effects on the roots of tidal flooding. The aeration of their roots has long been of interest (Scholander *et al.*, 1955; Troll & Dragendorff, 1931), including *Avicennia marina* in particular (Andersen & Kristensen, 1988; Brown *et al.*, 1969). In *Avicennia marina* the root system may be isolated from the atmosphere for six to seven hours in every tidal cycle (about 12.5 h). Previous work has led us to believe that the gas transport pathways available in the horizontal or cable roots of this mangrove are much larger than is necessary for the diffusive transport of oxygen, and that the large gas spaces in these roots have a storage function which is sufficient to maintain aerobic conditions within the roots while they are covered by water during high tide (Curran *et al.*, 1986). One observation we made was that although the amount of root material the plants had produced varied widely (32 g to 78 g dry mass per plant), the ratio of the respiratory demand of the root system to the volume of gas space was surprisingly constant. We calculated how long the plants could maintain their respiratory rates using the stored oxygen, and found that this too was quite constant and very close to the six-hour tidal inundation in which the plants had grown. A question which arose and which this work attempts to address asks: 'Is there a constant relationship between the amount of root and the volume of gas space in individual plants, or is the development of root gas space adaptive (*i.e.* does the volume of the gas space vary with the length of time the roots are isolated from the air by the tide)?'

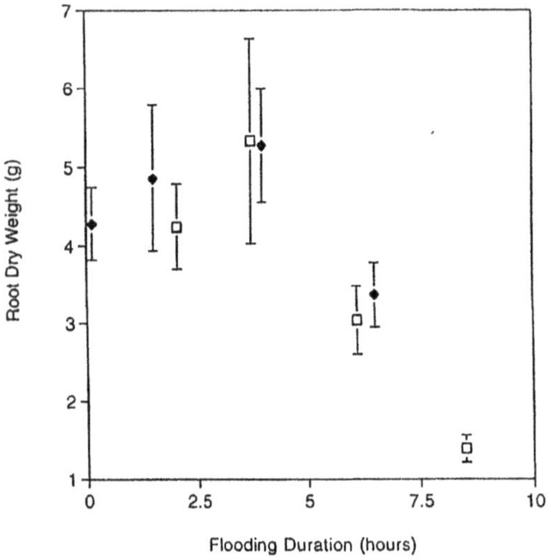

Fig. 1. Total root dry mass of one-year-old *Avicennia marina* plants versus the duration of inundation at which they were grown. Vertical lines are ± one standard error based on $n = 4$. Symbols: ♦ east bath; □ west bath (in all Figures).

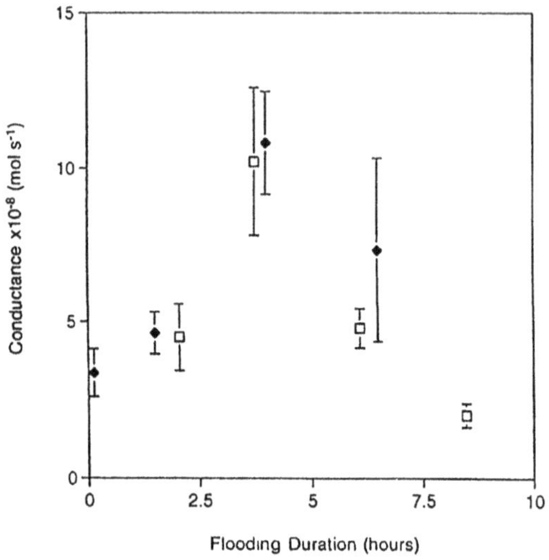

Fig. 2. The oxygen conductance of the stem base plus any pneumatophores of one-year-old *Avicennia marina* plants versus the duration of inundation at which they were grown. Vertical lines are ± one standard error based on $n = 4$. The two mean values with large error bars are the only treatments in which some plants had pneumatophores, which greatly increased their conductance.

This experiment was intended to investigate relationships between the duration of inundation at which seedlings of *A. marina* were grown and their root aeration characteristics.

Materials and methods

Plant material

Plants of *Avicennia marina* (Forsk.) Vierh. were raised from seeds collected in Botany Bay, Sydney, Australia (34°01′S 151°10′E) in spring 1986. Seeds (with pericarp attached) weighing 6.5–8.5 g were planted, one per pot, in washed river sand in plastic 20 cm pots lined with a plastic bag. Each pot was placed in a 4 l ice-cream container to limit drainage and a perforated 15 ml centrifuge tube containing Richgrow Osmocote™ slow-release fertilizer was forced vertically into the soil and renewed every six months. The ice-cream containers with their pots were immediately placed in tidal tanks, as described by Curran *et al.* (1985). Briefly, this system consisted of two outdoor tanks approximately two metres square and 900 mm deep. The tanks were side by side with a sunny northerly aspect, and where necessary to refer to individual tanks the terms 'east' and 'west' are used. Both tanks were covered by a 70% transmission shade-screen cover. Tidal change was simulated by pumping water from one tank to the other on a 12.5-hour cycle. The plants were placed on four shelves at different levels in each tank to give a range of durations of flooding, and of different depths of maximal inundation. The intention was to provide pairs of shelves on which the plants were at the same depth for different flooding-times (Table 1). Each shelf held up to 18 pots. At high tide the plants on the uppermost level in each tank had access to atmospheric oxygen via the stem. In all plants except some of those at the very deepest level some leaves and stem were exposed to the air throughout the tidal cycle, by the end of the growth period. The salinity was adjusted weekly.

After 12 months four plants, selected at random from each level in each tank for harvesting, were measured for stem height, total fresh mass, shoot fresh mass, root fresh mass and leaf area. The root tissue was segregated into three size classes, <1.5 mm, 1.5–5 mm and >5 mm diameter, and each of these was subsampled for determining the fresh/dry mass ratio, internal gas volume and respiration rate. All of the above-ground tissue was dry-weighed.

Table 1. Plant growth characters for each of the duration-depth treatments. Numbers shown are mean values with standard errors based on n=4 in parentheses. The two depth treatments of 0.00 m had the plants flooded to exactly soil level.

Bath	Duration of inundiation (h)	Inundiation depth (m)	Total plant dry wt. (g)	Shoot dry wt. (g)	Stem height (m)
W	8.50	0.50	8.72(0.22)	5.00(1.79)	0.606 (0.389)
W	6.10	0.31	12.67(1.73)	9.64(1.34)	0.474 (0.089)
W	3.75	0.12	20.87(4.50)	15.53(3.21)	0.640 (0.332)
W	2.05	0.00	13.66(1.47)	10.23(0.47)	0.749 (0.255)
E	6.50	0.50	15.66(3.00)	12.09 (2.59)	0.706 (0.686)
E	4.00	0.31	20.13(1.58)	14.86 (0.91)	0.589 (1.30)
E	1.50	0.12	19.74(3.51)	14.88(2.61)	0.690 (0.951)
E	0.13	0.00	13.10(1.79)	8.82(1.42)	0.458 (0.159)

Root respiratory rate

Measurements of root oxygen uptake were made by constant volume manometry as in Curran *et al.* (1986). Samples from each root size class from each plant were placed into three flasks and each flask immersed in a separate 20 °C water bath. Flasks had a central well which contained 200 μl of 20% KOH, a side-arm which contained a filter paper wick and 200 μl of distilled water for humidification, and about 1 g of wet, finely divided root tissue without any bathing solution. Roots maintained stable respiratory rates under these conditions for several hours. Each measurement of respiration took 50–60 minutes. The respiration rate was expressed per gram of root dry-mass.

Root gas-volume

A weighed subsample from each size class was taken from each root system and the gas volume estimated (Jensen *et al.*, 1969). The total root gas space for each root system was then calculated on a fresh weight basis. This value, when multiplied by the concentration of oxygen in the air, is an estimate of the maximum possible quantity of oxygen stored in the root system.

Root ventilation

Internal oxygen diffusion from the atmosphere to the root system was measured with Rank polarographic oxygen electrode chambers (Curran, 1985). The conductance to oxygen of the stem of the plant, and any pneumatophores that were present, were measured.

Root aerobic respiratory capacity

In an attempt to assign a value to the capacity of a root system to maintain aerobic respiration whilst isolated from the air, a variable which we have called the root aerobic respiratory capacity has been derived. For each plant, this value was calculated by dividing the total maximum oxygen storage (mol O_2) by the total respiratory demand (mol O_2 s^{-1}). The unit is time and the value represents the longest period for which the root system could maintain aerobic respiration without oxygen resupply, assuming that the respiration rate as measured in air is maintained undiminished and that the gas space initially contained air. Although neither these assumptions nor the estimation of the oxygen storage capacity are likely to hold with a high degree of accuracy, the result appears to supply a useful description of the abilities of the root system.

Statistical analysis

Data were analysed by analysis of variance, except in one set where transformation failed to stabilize the data. Cochran's test was used and non-homogeneous data were transformed following Underwood (1981). Significant differences between treatments were identified using the Student-Newman-Keuls multiple comparison procedure. $P < 0.05$ was taken to indicate statistical significance.

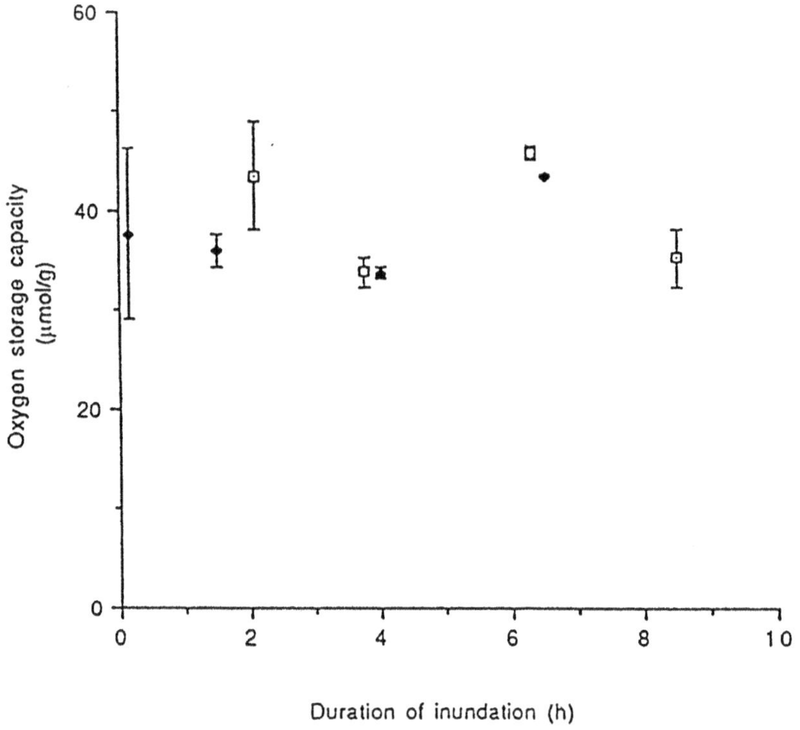

Fig. 3. The maximum oxygen storage capacity per unit dry mass of roots for one-year-old *Avicennia marina* plants versus the duration of inundation at which they were grown. Vertical lines are ± one standard error based on $n = 4$. The maximum oxygen storage is the total volume of internal gas space in the root system multiplied by the concentration of oxygen in normal air.

Results

There was a statistically significant effect of flooding on plant mass, with a tendency for plants to be heavier when they had been grown in intermediate-length tidal periods, and lighter plants at both extremes (Table 1). Plants grown with the longest inundation period had statistically significantly less root dry matter and shoot dry matter than the other plants (Fig. 1). Stem height showed no trend with inundation (Table 1).

There was also a statistically significant effect of duration of flooding on oxygen conductance through the stem and pneumatophores. The multiple-range test results were complex: there was a tendency for higher conductance at inundation periods of intermediate length, and lower conductance at both longer and shorter inundation periods (Fig. 2). Only three plants had pneumatophores, so the conductance results for most plants were due to conductance through the stem base alone. The marked effect of pneumatophore development can be seen in large error-bars of the two intermediate-range treatments.

The gas space volume and hence the total maximum amount of stored oxygen per plant showed no statistically significant effect of extent of inundation. When expressed per unit dry mass, the oxygen storage capacity of the roots was also independent of the extent of inundation at which the plant was grown (Fig. 3).

Root oxygen demand per unit dry mass did not differ significantly with extent of inundation (Fig. 4). The maximum time for which the roots could maintain aerobic respiration when isolated from the atmosphere ('root aerobic respiratory capacity') did not appear to change with the extent of inundation at which the plants were grown (Fig. 5). In this case no formal analysis was done due to the heterogeneity of the variances which failed to stabilize with any transformation. The root aerobic respiratory capacity, as defined above, was less than about three and a half hours for all treatments.

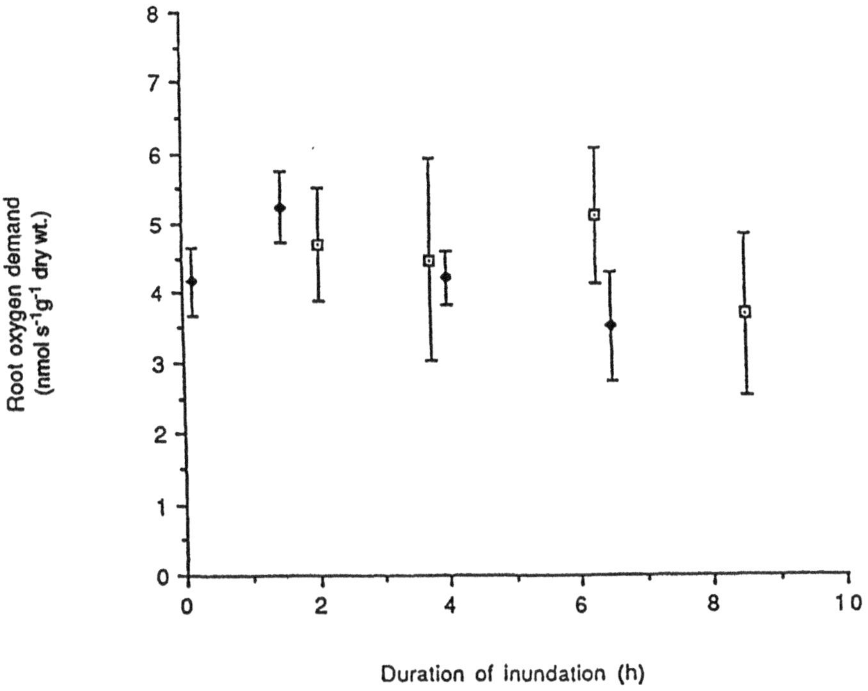

Fig. 4. Root oxygen demand per unit dry mass versus the duration of inundation at which the plants were grown. Vertical lines are ± one standard error based on $n = 4$.

Discussion

These observations were intended to determine whether the extent of inundation at which mangrove seedlings are grown influences their growth and the aeration characteristics of their roots. The design of the tanks allows only limited experimental separation of depth and duration of inundation. This is unfortunate but quite realistic in that those plants in the field which are inundated for longer periods are also usually more deeply covered than those inundated for shorter periods.

The intention in setting the time-switches for the pumps of the artificial tidal system had been to provide pairs of shelves at different depths at each of the intermediate durations, to enable depth and duration effects to be separated. In practice there was a pair at 6.1 to 6.5 hours tidal inundation (E1 and W2) and another at approximately 4 hours (E2 and W3), but the shallower E3 and W4 were not well paired for tidal duration. There were no statistically significant effects of depth in any of the three pairs. However, the depths of the members of the pairs were not greatly different, and the experiment was not specifically intended to examine depth as a factor.

The results indicate, however, that duration of inundation affected some aspects of the growth and development of *A. marina* seedlings. Plants grown at particularly long inundation periods had markedly less root and shoot tissue than those grown at shorter inundation periods.

If we are to postulate that a plant maintains aerobic roots throughout a normal tidal cycle, then the amount of oxygen stored in the aerenchyma must equal or exceed the amount respired during inundation. If the plants responded adaptively to the extent of flooding it could be expected that those growing with longer inundations would have more root internal gas space per unit dry weight. This does not occur. In plants of this age the amount of gas space (oxygen storage capacity) per unit dry mass of root tissue appears very tightly developmentally controlled, and independent of the extent of inundation (whether depth or duration).

Curran *et al.* (1986) suggested that in older plants with pneumatophores there was sufficient aerenchyma to supply the oxygen requirements of the root system

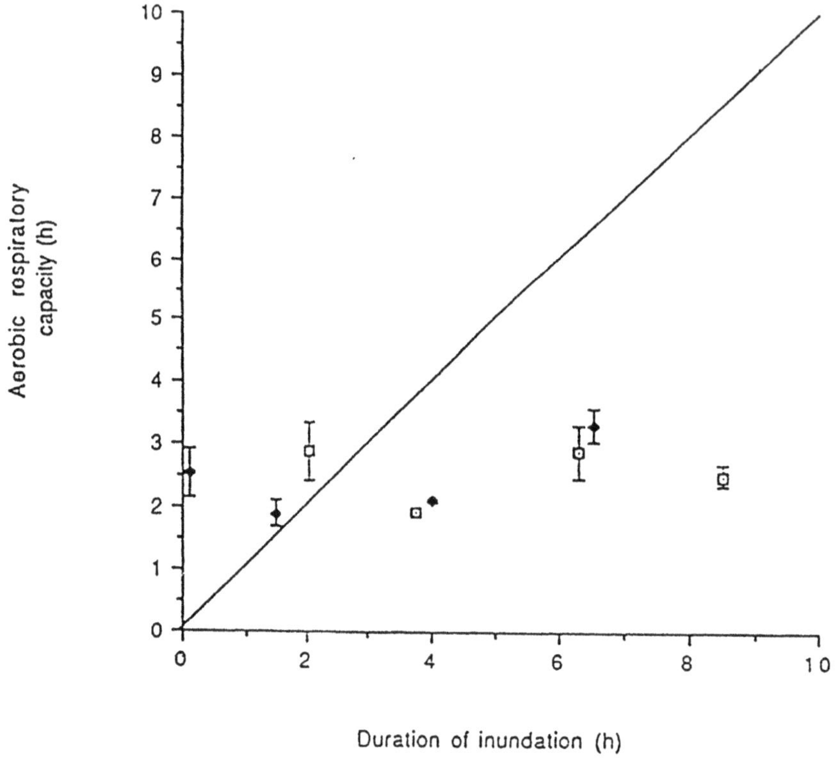

Fig. 5. The aerobic respiratory capacity for roots of one-year-old *Avicennia marina* plants versus the duration of inundation at which they were grown. Vertical lines are ± one standard error based on $n=4$. The diagonal line represents the critical value at which the maximum aerobic respiratory period equals exactly the period for which the roots are isolated from the atmosphere. Any value which falls above this line indicates that in that case the roots are always aerobic; and any value which falls below the line indicates that, if the oxygen demand remains constant, some parts of the root system must become anaerobic.

for the period during which it was isolated from the air. This is also true for the younger plants in these experiments which were grown with inundation less than about 3.5 h per tide, but plants submerged longer in each tide are unlikely to stay aerobic. This conclusion is drawn from the calculation of root aerobic respiratory capacity, which represents the longest time for which a root system can respire aerobically when isolated from the atmosphere if the oxygen consumption rate is constant and the gas space was full of air at the start. The extent of inundation had no effect on this maximum aerobic period, which averaged 1.5–3.5 h. Thus, plants grown at an inundation period of more than about 3.5 h will become anaerobic during every high tide if the root respiration rate remains constant. This effect in the seedlings may limit the distance down the shore at which adult mangroves occur, even though the larger plants have enough oxygen storage capacity to remain aerobic there (Curran *et al.*, 1986): the seedlings may not survive in the field to produce adults at excessive flooding durations.

The calculation of the maximum aerobic period relies in part on a correct estimation of oxygen-consumption rate. This was measured by Warburg respirometry at constant temperature and atmospheric conditions. This gives no information about respiratory rates at different temperatures. More importantly, it gives no information about possible effects of different oxygen and carbon dioxide concentrations in intact roots. Internal gas concentrations do change when the respiring roots are isolated from the atmosphere (Scholander *et al.*, 1955).

Other sources of oxygen that could potentially be available for the plants include the leaves and stems, which have gas spaces in continuity with those of the root system (Ashford & Allaway, in press). Plants on the top shelves were always emergent, and at least the topmost pair of leaves had reached the high-tide water surface in even the deepest-flooded plants by the end of the year's growth However for the early part of the year the latter plants would have been completely isolated from atmospheric oxygen during the high tides Supply of oxygen from the water surrounding the roots (cf Laan *et al*, 1990) seems unlikely in these plants since the soil was anaerobic and movement of soil water was highly restricted

Acknowledgments

We thank Anne Constable for assistance in collecting and planting the seeds The research was funded by the University of Sydney and the Australian Research Council

References

Andersen, F Ø & E Kristensen, 1988 Oxygen microgradients in the rhizosphere of the mangrove *Avicennia marina* Mar Ecol Progr Series 44 201–204

Armstrong, J, W Armstrong & P M Beckett, 1988 *Phragmites australis* a critical appraisal of the ventilating pressure concept and an analysis of resistance to pressurized gas flow and gaseous diffusion in horizontal rhizomes New Phytol 110 383–389

Armstrong, W, 1978 Root aeration in the wetland condition In D D Hook & R M M Crawford (eds), Plant life in anaerobic environments Ann Arbor Science, Michigan, USA 269–297

Ashford, A E & W G Allaway, 1995 There is a continuum of gas space in young plants of *Avicennia marina* Hydrobiol 295 (Dev Hydrobiol 102) 5–11

Brown, J M A, H A Outred & C F Hill, 1969 Respiratory metabolism in mangrove seedlings Pl Physiol 44 287–294

Chapman, V J, 1976 Mangrove vegetation J Cramer, Vaduz

Curran, M, 1985 Gas movements in the roots of *Avicennia marina* (Forsk) Vierh Austr J Pl Physiol 9 519–528

Curran, M, W G Allaway & M Cole, 1985 Artificial tidal system for growing mangroves Wetlands (Australia) 5 70–77

Curran, M, M Cole & W G Allaway, 1986 Root aeration and respiration in young mangrove plants (*Avicennia marina* (Forsk) Vierh) J exp Bot 37 1225–1233

Dacey, J W, 1981 Pressurized ventilation in the yellow waterlily Ecology 62 1137–1147

Grosse, W & H Mevi-Schutz, 1987 A beneficial gas transport system in *Nymphoides peltata* Am J Bot 74 947–952

Hook, D D & M R McKevlin, 1988 Use of oxygen microelec trodes to measure aeration in the roots of intact tree seedlings In D D Hook *et al* (eds), The Ecology and Management of Wetlands Volume 1 Ecology of Wetlands Croom Helm, Beckenham, Kent 467–476

Jackson, M B & M C Drew, 1984 Effects of flooding on growth and metabolism of herbaceous plants In T T Kozlowski (ed), Flooding and Plant Growth Academic Press, Orlando 47–128

Jensen, C R, R J Luxmore, S D van Grundy & L H Stolzy, 1969 Root air space measurements by a pycnometer method Agronomy Journal 61 474–475

Kozlowski, T T, 1984 Responses of woody plants to flooding In T T Kozlowski (ed), Flooding and Plant Growth Academic Press, Orlando 129–163

Laan, P, M Tosserams, C W P M Blom & B W Veen, 1990 Internal oxygen transport in *Rumex* spp and its significance for respiration under hypoxic conditions Pl Soil 122 39–46

Raskin, I & H Kende, 1983 How does deep water rice solve its aeration problem? Pl Physiol 72 447–454

Scholander, P F, L van Dam & S I Scholander, 1955 Gas exchange in the roots of mangroves Am J Bot 42 92–98

Troll, W & O Dragendorff, 1931 Uber die Luftwurzeln von *Sonneratia* Linn f und ihre biologische Bedeutung Mit einem rechnerischen Anhang von Han Fromherz Planta 13 311–473

Underwood, A J, 1981 Techniques of analysis of variance in marine biology and ecology Oceanogr Mar Biol Ann Rev 19 513–605

Yamasaki, S, 1984 Role of plant aeration in zonation of *Zizania latifolia* and *Phragmites australis* Aquat Bot 18 287–297

Hydrobiologia **295**: 31–42, 1995.
Y. S. Wong & N. F. Y. Tam (eds), Asia-Pacific Symposium on Mangrove Ecosystems.

Transport of sediment in mangrove swamps

Eric Wolanski
Australian Institute of Marine Science, PMB No. 3, Townsville, M.C., Qld. 4810, Australia

Abstract

The transport of suspended sediment in mangrove swamps is controlled by three dominant processes. First, the transport processes in the estuaries and coastal waters draining the swamp, including flocculation, tidal pumping, baroclinic circulation, trapping of the smallest particles in the turbidity maximum zone, and the effect of the mangrove tidal prism. Second, the mechanical and chemical reactions in mangrove waters destroying flocs of cohesive sediment in suspension. Third, biological processes have a dominant influence on the ultimate fate of clay particles in mangroves.

Introduction

The fate of fine sediment in mangrove swamps is of prime scientific and economic importance. It affects the navigation, shoreline erosion and stability, migration of shoals, fate of nutrients and contaminants such as heavy metals and pesticides, turbidity, primary productivity and even bird life.

The flushing of dissolved matter or non-buoyant particles in mangrove swamps is fairly well understood as it is controlled by the water circulation. In turn, the water circulation in mangrove swamps is controlled by a number of oceanographic processes driven by the hydrologic regime (freshwater runoff, evapotranspiration) and the tidal forcing (Wolanski, 1992; Wolanski *et al.*, 1992a). In contrast, the transport of buoyant material is more complex and is not simply controlled by the 'large-scale' oceanography of the mangrove swamp. Secondary flow processes prevail that not only inhibit dispersion and mixing of the floating plant litter, but aggregate the material in foam lines, i.e., negative dispersion occurs. These processes have not been studied but the aggregation is probably due to complex three-dimensional flow patterns in the tidal creeks and to buoyancy effects (Simpson & James, 1986; Wolanski & Hamner, 1988; Wolanski, 1992).

While the oceanography of mangrove swamps has received some attention, the sediment transport processes are very poorly known. Some of the key processes controlling the fate of sediment in mangroves are reviewed here. Gaps in knowledge where research is needed are identified.

Scales

There is no unique description of sediment transport in mangroves; the dominant processes vary from site to site. For example in one extreme situation some tropical estuaries have only a thin fringe of mangroves, often only 1 to 3 trees wide, lining the tidal creeks (Fig. 1a). Examples include the Norman River estuary and the South Alligator River estuary in Australia. In such systems the mangroves play a minor role in the sediment budget of the estuary. In the other extreme, the mangrove swamp covers an area larger than that of the tidal rivers and the mangroves play a dominant role in the sediment budget of the estuary. Such is the case of Coral Creek in Australia (Fig. 1b). Similar examples include the Klong Ngao estuary in Thailand, Estero Pargo in Mexico, and the Nakama-Gawa estuary in Iriomote Island, Japan.

Key processes controlling the fate of sediment in mangrove swamps include the secondary circulation in meanders, flocculation, trapping, settling, fluidisation by waves, mangrove tidal prism and biological effects. These are discussed below.

Fig 1 Photograph of (a) a small, fringing strip of mangroves between the river and the salt flats in the Norman River estuary in Australia, and (b) the vast, heavily vegetated mangrove swamp fringing Coral Creek, also in Australia

(Wolanski *et al.*, 1988), and the internal circulation is limited to the lower layer (Fig. 2b). Because of the secondary circulation, the smallest particles (clay and fine silt) aggregate on the sloping banks and only the coarsest particles (sand and gravel) remain on the bed. If the supply of fine sediment is limited, the bed can be entirely sandy and the sloping banks muddy, such as in the upper reaches of the South Alligator River estuary. In some estuaries mangrove trees colonise these muddy sloping banks and form thin strips of vegetation (Fig. 1a).

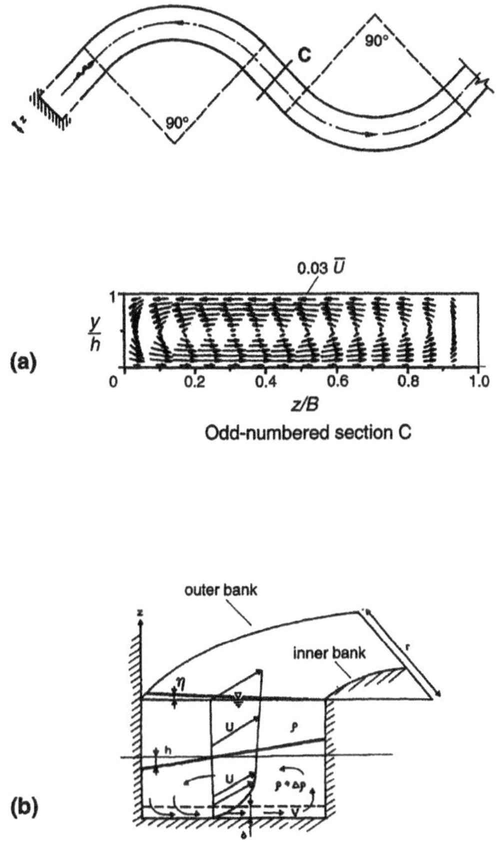

Fig. 2. (a) Sketch of the secondary circulation in a meander (adapted from Dermuren and Rodi, 1986). (b) Sketch of the secondary circulation in a meander in a sediment stratified estuary (adapted from Wolanski *et al.*, 1988).

Secondary circulation in meanders

Meanders create a secondary circulation (Dermuren & Rodi, 1986). This secondary circulation sorts the sediment by size (Fig. 2a). In the presence of a stratification generated by salinity or the suspended sediment, the density interface can be lifted several meters in the inside of a meander and can even reach the surface

Flocculation

In the freshwater region of an estuary, fine suspended particles are often not flocculated (Fig. 3a). As a result, in freshwater, for concentrations < 1 g l^{-1}, the settling velocity of the suspended sediment is usually independent of the concentration. However, at higher values of the concentration the settling velocity may increase slightly due to large particles entraining other particles in their wake (Fig. 4; Wolanski *et al.*, 1994). In the presence of much organic riverine material the inorganic particles can be attached to organic components from both living organisms such as algae, ciliates, zooplankton and bacteria, and detritus from plants and animals (Uiterwijk Winkel, 1975). The settling velocity of such aggregates can be much larger than that of individual particles.

As soon as the sediment reaches brackish water in the estuary, salt flocculation begins. Silt and clay particles are cohesive and their behaviour is controlled by flocculation. Flocculation processes are dependent on the nature of the metallic and organic coatings which affect the electrostatic stability of the suspended particles (Gibbs, 1983) Salt flocculation can destroy freshwater aggregates, if any, while forming saltwater aggregates, and this aggregation is greatly enhanced by biological activity (van Leussen, 1988).

In the saline region of estuaries dominated by clay (e.g. the Amazon and the Gironde Rivers), clay flocs form. These are typically 30 to 200 μm in size. A floc can incorporate thousands of clay particles. In early observations of flocs, the water was gathered by pumping water or from Niskin bottles through the ports. This technique however resulted in a mechanical disaggregation of the flocs (Gibbs & Konwar, 1983). It was later recognised that flocs are 'weak' and disaggregate in such sampling techniques. Special techniques have since been devised to properly observe flocs in the field and two techniques appear promising. First, *in*

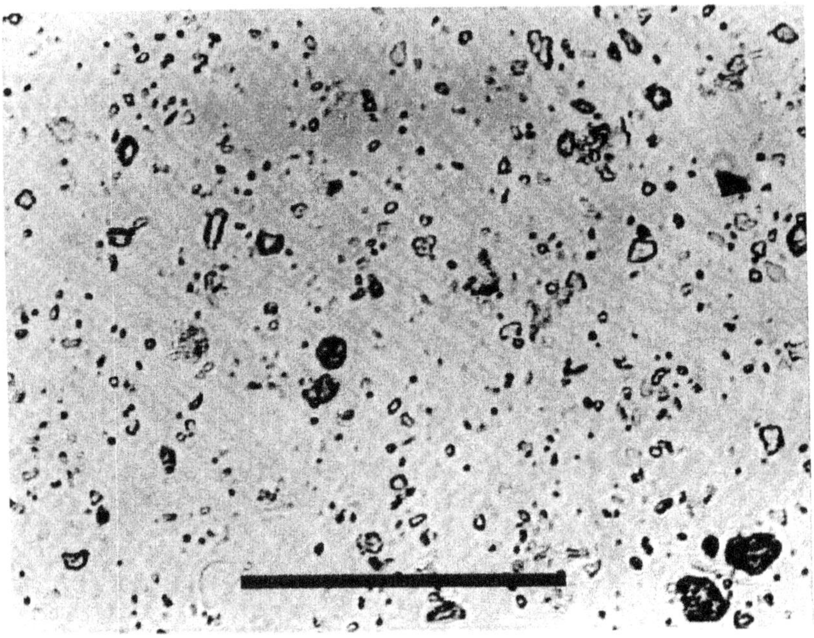

3a

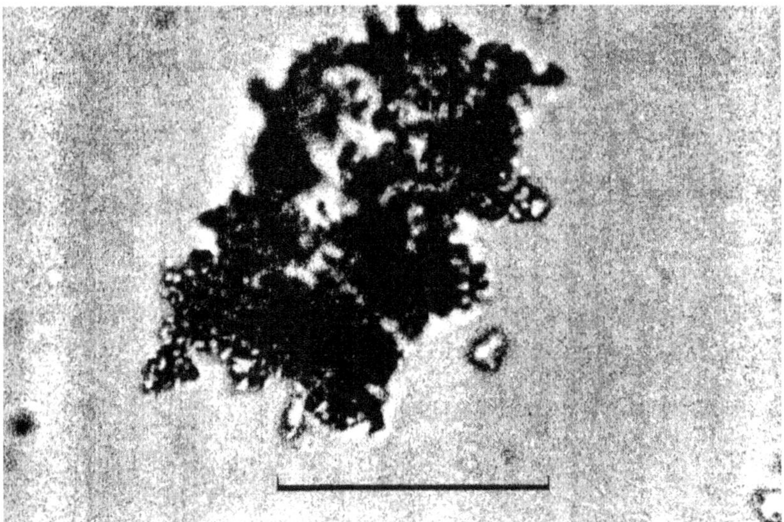

3b

situ photography of flocs in suspension (Wells, 1989; Eisma *et al.*, 1990). This technique has the advantage that the flocs are undisturbed. However this method is limited to low values of the suspended sediment concentration (SSC), typically less than 0.4 g l^{-1}. For higher concentrations, the turbidity is too high for *in situ* photography. For small values of SSC (< 0.3 g l^{-1}), this technique reveals an abundance of large clay

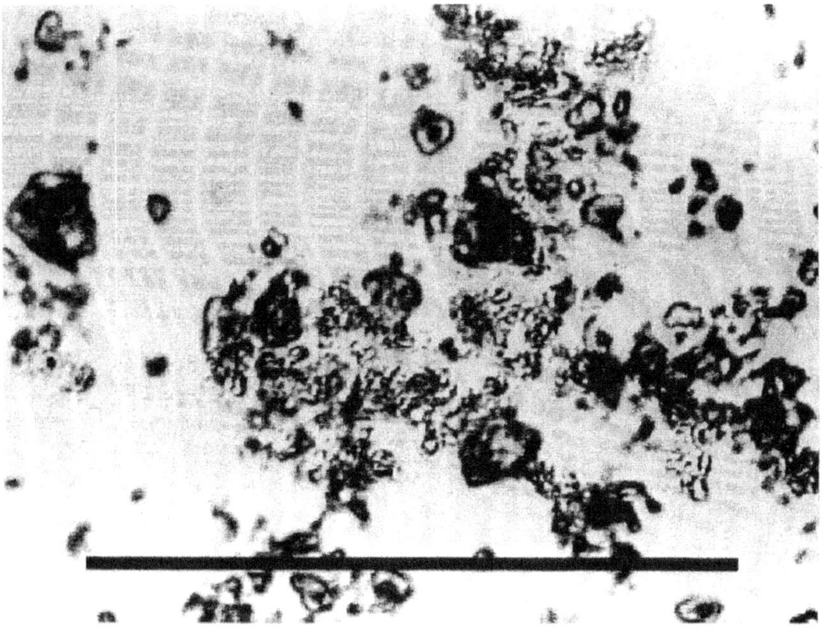

3c

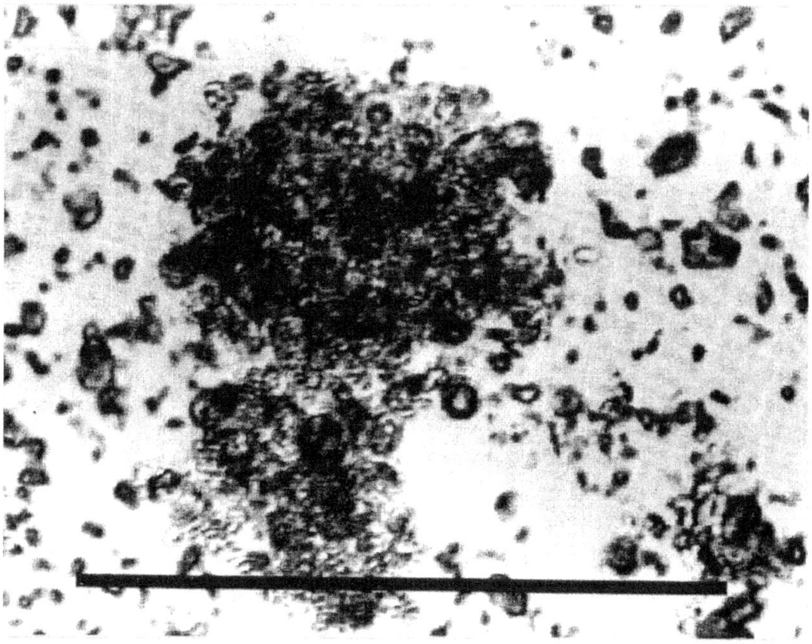

3d

3e

Fig 3 Microphotographs of estuarine sediment (a) sediment particles in suspension in the fresh water region of the Fly River estuary, in Papua New Guinea, where the sediment particles are not flocculated; (b) clay flocs in suspension in the saline region of the Amazon River plume (adapted from Gibbs & Konwar, 1986), (c) a silt-clay floc in suspension at spring tides in the saline region of the Fly River estuary; (d) a silt-clay floc in suspension at neap tides in the turbidity maximum zone of the Fly River estuary; (e) flocs of suspended matter at the mouth of Coral Creek, Australia. The unlabeled bars represent 100 μm.

aggregates ($> 100\ \mu$m) in suspension over a fluid mud bed, with typically 1700 to 5300 aggregates per litre. The aggregates account typically for 5 to 95% of the total particle mass in suspension. This aggregation process is important because large aggregate are optically inactive, i.e., have little effect on the scattering of light, as opposed to small non flocculated particles.

Mangrove-fringed tidal channels often have near-bottom SSC $> 0.4\ l^{-1}$, so other sampling techniques are necessary. An appropriate technique is to sample the water using Owen tubes or Niskin bottles. Water is not pumped or drawn through the port, but a sample is taken by immersing a special microscope slide in the sampling tube and capping the slide while still immersed (Gibbs & Konwar, 1986). This technique permits close-up observations of flocs but may break up the very large aggregates.

The morphology of flocs depends on sediment type. In clay dominant systems, the flocs appear dense (Fig. 3b) but they are 60 to 80% porous (Gibbs & Konwar, 1986). The flocs are equidimensional and do not have an elongate shape. They have smooth edges and are readily distinguished from inorganic silt and sand particles which normally have sharp edges. Some flocs are 'strong' and survive a tidal cycle without being destroyed by tidal turbulence (Gibbs *et al.*, 1989). Laboratory studies (Mehta & Partheniades, 1975) show that the 'weak' flocs are broken up by turbulence near the bottom and that the broken aggregates and particles will again participate in the aggregation process higher in the water column. The largest flocs are found near the bottom. Flocs of dimensions of 200 μm are not uncommon.

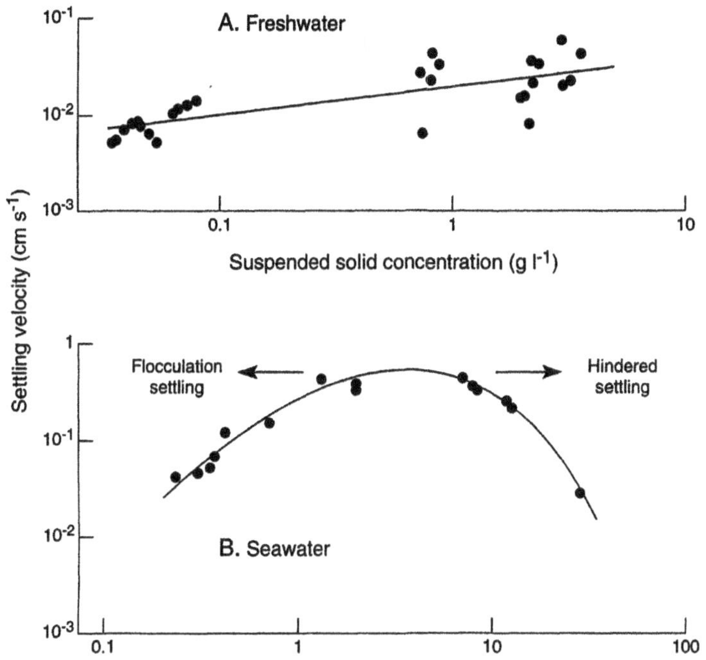

Fig. 4. Dependence of the settling velocity of suspended sediment in freshwater and in seawater, for the Fly River estuary (adapted from Wolanski & Gibbs, 1994).

In systems dominated by silt, colloidal forces are much weaker. Large silt flocs, some of them exceeding 500 μm, have been observed in the Jiaojiang estuary in China but these are destroyed when tidal velocities exceed 0.4 m s^{-1} (Li *et al.*, 1993). These large silt flocs only exist at neap tides, while clay flocs were observed at all tides.

In the presence of silt and clay in comparable volumes, both clay and silt are incorporated in the flocs. This is the case of the Fly River estuary in Papua New Guinea (Wolanski & Gibbs, 1994). The flocs are a loose matrix of clay and silt particles (Fig. 3c). These flocs are structurally 'weak' and their size varies with the tidal currents, the flocs starting to disaggregate when the tidal velocity is > 1 m s^{-1}. The floc size decreases with increasing tidal currents, and vice-versa, as the flocs periodically disaggregate and form at tidal period. The mean floc size is small, varying typically between 15 and 40 μm during spring tides. During neap the flocs are larger, frequently > 100 μm, and contain silt particles immersed in an envelope of clay particles (Fig. 3d).

In mangrove-fringed estuaries with calcareous sediment (e.g., Terminos Lagoon, Mexico) flocs are quite different, containing a core of relatively dense, opaque material surrounded by a fluffy, porous carbonate membrane. The flocs incorporate most of the suspended particles. These flocs are structurally very weak and disaggregate readily during sampling. They are very large, with occasional flocs with dimensions of a few cm. They are extremely porous, hence only slightly negatively buoyant; they also have a very large drag because of their large surface area; as a result they remain in suspension under small tidal currents of 0.2 m s^{-1} (Wolanski & Yanez-Arancibia, unpubl. data).

In mangrove-fringed coastal waters, biological productivity is generally high, and clay flocs are interwoven by fibres and threads of micro-organisms. Such flocs have been found by Wolanski (unpubl. data) at the mouth of the mangrove swamp of Coral Creek (Fig.

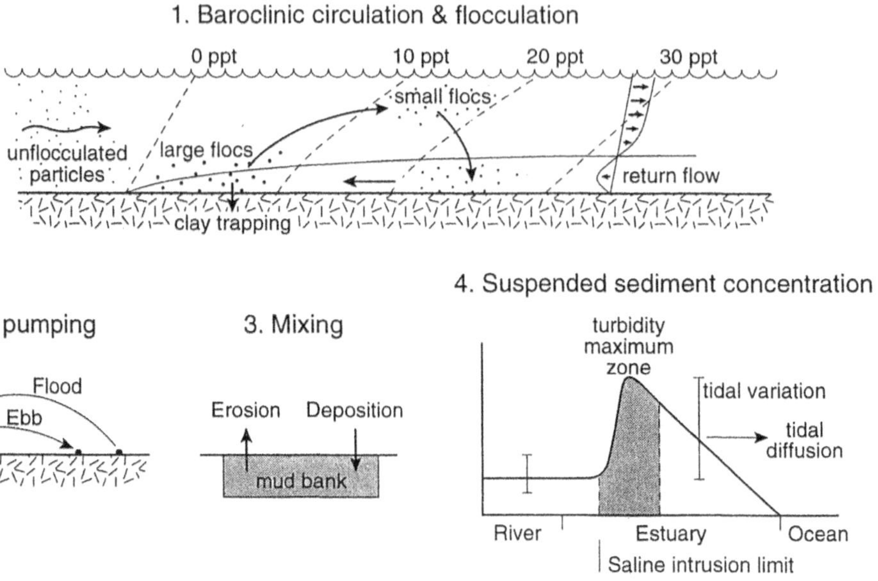

Fig. 5. Sketch of the baroclinic circulation, tidal pumping, formation of a mud bank and the turbidity maximum zone in an estuary.

3e). These flocs appear robust as the particles seem to be glued to a flexible mucus membrane that bends and twists in the turbulence and hence can sustain some turbulence and shear without floc disaggregation.

Trapping of fine sediment

Without flocculation, the clay and silt particles are likely to travel the length of an estuary as a 'wash load' with the nett river-driven currents and readily reach the coastal waters in a similar manner as a dissolved substance. Because of flocculation, the sediment pathways are different from the water pathways. Two mechanisms dominate the transport of flocculated fine sediment in the estuary, the baroclinic circulation and the tidal pumping (Fig. 5). These mechanisms lead to the formation of mud banks in estuaries.

The baroclinic circulation advects more saline water landward near the bottom of the estuary. The strength of that circulation depends on the degree of vertical mixing of salinity in the estuary (Dronkers & van Leussen, 1988; Dyer, 1986). Fine sediment particles are usually not flocculated when they enter the brackish region of the estuary. On meeting brackish water, salt flocculation is initiated at very small values of the salinity (often < 1 ppt). The largest flocs stay near the bottom. The small flocs and unflocculated particles move downstream with the nett (tide-averaged) currents and aggregate with surrounding particles. Their floc size increases as they move downstream. This brings them closer to the bottom where they are entrained upstream by the baroclinic circulation.

The tidal pumping effect results from an asymmetry between peak flood and ebb tidal current found in many shallow fan-shaped estuaries (Postma, 1961, 1967; Dyer, 1986). The flood tidal current may be of shorter duration, but with stronger peak currents than the ebb tide. Because of the strongly non-linear relationship between water velocity and sediment transport, the sediment floc is carried further upstream at flood tide than downstream at ebb tide. After a tidal cycle, the sediment floc nett displacement is to upstream.

These two effects result in forming a turbidity maximum zone. This turbidity maximum zone has been observed in many muddy estuaries world-wide (e.g. Christiansen, 1974; Krone, 1972; Wellerhaus, 1981; Dyer, 1986; Gibbs *et al.*, 1989; Postma, 1961, 1967; Riethmuller, 1988; Wolanski & Eagle, 1991; Althausen & Kjerfve, 1992; Uncles & Stevens, 1993). The turbidity maximum zone is generally most conspicuous at spring tides and least pronounced at neap tides.

The presence of suspended sediment at high concentration has important hydrodynamic effects by reducing the apparent bottom friction coefficient (Adams & Weatherly, 1981; Gust, 1976; King & Wolanski, 1994).

The turbidity maximum zone is most conspicuous near the bottom. If the estuary is strongly vertically stratified in salinity, the near surface waters, i.e., those entering the fringing mangroves, are little affected. If the salinity stratification is weak, the surface waters are also very turbid. Hence, the location of the fringing mangroves along the estuary, and the vertical stratification in salinity in the estuary, determine the SSC of the estuarine waters entering the fringing mangroves. The siltation rates in the mangroves vary with distance along the estuary.

In the Fly River estuary, the disaggregation and formation of flocs at tidal period results in the selective trapping of clay particles, as opposed to the silt particles, in the turbidity maximum zone of the estuary (Wolanski and Gibbs, 1994). This trapping results in a relative enrichment of clay, from its 20% fraction in the freshwater part of the estuary, to about 50% (by volume) near the bottom and about 100% near the surface in the turbidity maximum zone. Clay particles selectively enter the fringing mangroves and their relative fraction varies with distance along the estuary. Though the clay particles account for less than 20% of the sediment entering the estuary, they account for 50% of the suspended particles entering the fringing mangrove swamps at flood tide in the turbidity maximum zone.

Settling and compaction

The settling velocity w_s of the fine sediment varies with the floc size. For clay-dominant flocs of Chesapeake Bay, Gibbs (1985) found that w_s (cm s^{-1}) = 1.73 (floc diameter, cm)$^{0.78}$. A 10 μm floc thus settles at a velocity of about 0.01 cm s^{-1} and a 100 μm floc at about 0.03 cm s^{-1}. A 400 μm floc thus settles at the same velocity as a quartz particle of 45 μm; likewise a 40 μm floc would have the same settling velocity as a 16 μm quartz particle.

Silt-dominant flocs are extremely porous and have a settling velocity comparable to that of clay-dominant flocs 1/2 to 1/3 their sizes (Li *et al.*, 1993). Settling data on calcareous mud flocs are unavailable.

It is convenient in laboratory experiments to measure the settling velocity as a function of the SSC. The function is non-linear in seawater (Fig. 4) with a zone of flocculation settling, where w_s increases with increasing SSC, and a zone of hindered settling, where w_s decreases with increasing SSC (Mehta, 1986).

Bioturbation significantly increases the settling velocity (Stolzenbach *et al.*, 1992). Organism activity generates a fluffy, interfacial layer and causes suspended particles to collide with and stick to the interfacial sediment.

Hindered settling and the aggregation of fine sediment in the turbidity maximum zone often leads to the formation of unconsolidated fluid mud layers, with SSC > 100 g l^{-1} (e.g. Ross, 1988; Kineke, 1993). Recent studies of the Amazon shelf suggest that even at SSC values of 100 g l^{-1} the fluid mud is not stationary but still moves with the currents (Trowbridge & Kineke, 1993).

In a quiescent environment, the settled fluid mud can consolidate, i.e., de-water, to typically 8% of its volume. However, this consolidation is greatly inhibited by even small amounts of turbulence. Wolanski *et al.* (1992b) found that this may be due to the plugging of micro-channels used in the de-watering process. Thus compaction is much slower, and may be altogether stopped, in the presence of turbulence in the unconsolidated fluid mud, such as under the influence of waves.

Fluidisation by waves

Consolidated mud can be fluidised by pore pressure effects from wave loading (Maa & Mehta, 1987; Clukey *et al.*, 1985). The deeper sediment, however, is more compacted, its yield strength increasing with depth, and it can resist erosion (Parchure & Mehta, 1985; Faas, 1986).

Mud banks in coastal waters are fluidised by waves (Wells, 1983; Wells & Roberts, 1981; Jiang & Mehta, 1992). Waves are attenuated over a fluid mud bed. In Surinam, a 96% attenuation of the wave energy has been measured, without wave breaking, as the waves propagate from the ocean towards the coast over a mud flat. This wave energy loss protects the shorelines that are fronted by mud flats. These mud flats can be colonised by mangrove vegetation in a single season.

Mangrove tidal prism

The presence of a large mangrove swamp, with a surface area comparable to that of the tidal creek, significantly increases the tidal prism at spring tides. This

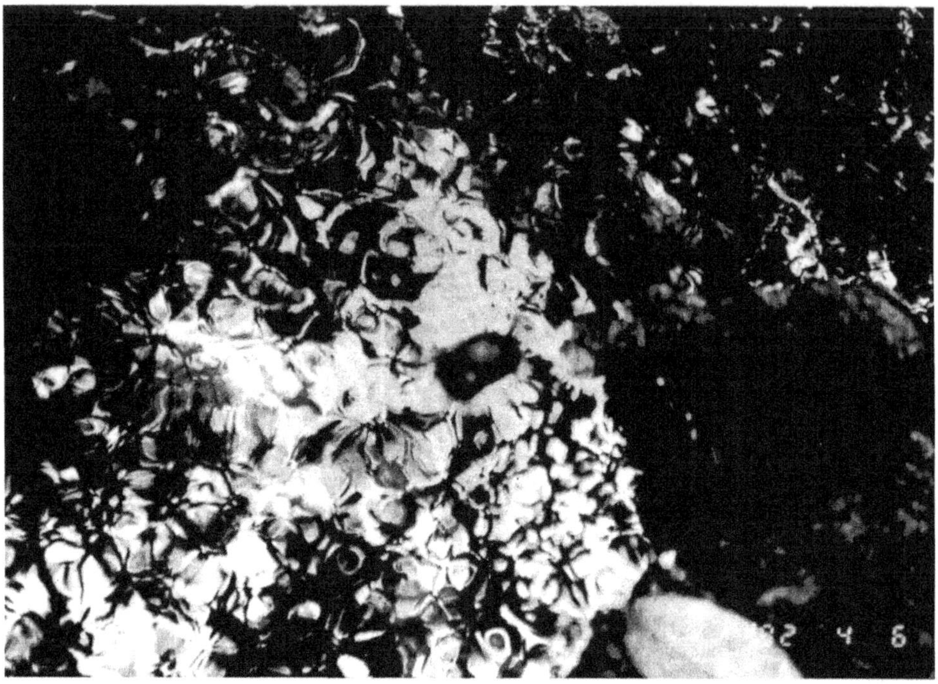

Fig. 6 Photograph of the water surface in a mangrove forest, showing the complex water circulation patterns around the vegetation and rugged substrate in the Nakama-Gawa mangrove swamp, Japan.

in turn modifies the tidal hydrodynamics from a flood-dominant to an ebb-dominant system at spring tides (Wolanski, 1992). When the bottom sediment in the estuary is mostly sand, the estuary is self-scouring, the tidal asymmetry maintaining a deep, navigable tidal channel fringed by a large area of mangrove swamp inundated only at spring tides. This explains recent measurements of a sand export of about 400 tonnes per year for a 1.5 km^2 tidal creek-mangrove swamp system (Larcombe & Ridd, 1992).

Biological effects

Microphotographs of the suspended sediment in the headwaters of Coral Creek at ebb tide show (Wolanski, unpubl. data) that the bulk of the particles are unflocculated and the very few flocs present are very small with a diameter typically < 20 μm. Since the sediment particles in the tidal creek was flocculated at flood tide (Fig. 3e) but much less at ebb tide after the water has circulated in the swamp, it can be inferred the flocs were disaggregated or settled in the swamp. The reasons for floc disaggregation are unknown and may include mechanical destruction of flocs in a turbulent flow through the vegetation (Fig. 6), and chemical reactions with the humic acids or the tannins released by roots and decomposing wood and leaves which inhibit phytoplankton in mangrove waters (Tundisi *et al.*, 1973).

At Coral Creek, the bulk of the suspended clay and silt particles are unflocculated in the headwaters. The particles settling velocities are then very small, of the order of 0.001 cm s^{-1} for clay and 0.005 cm s^{-1} for silt. A significant fraction of the silt particles and the clay flocs entering the swamp at flood tide probably settles in the mangroves. This mechanism may be efficient because some suspended material is trapped in small eddies shed by the vegetation and the rugged substrate (Fig. 6). Little is known about the efficiency of this mechanism.

Two important processes for trapping clay particles in suspension in mangroves are the particle sticking to bacterial, algal or animal mucus, and pelletisation by benthic detritivores deposit-feeding on the sediment surface. For instance, Wolanski (unpubl. data) found that a vast number of the non-flocculated particles that entered the Coral Creek mangroves at flood tide were re-exported at ebb tide to the tidal creek, sticking to the mucus at the water surface At ebb tide, this mucus was exported from the swamp to the tidal creek where it formed conspicuous foam lines Microphotograph of water in this film show thousands of small particles per cm^2, sticking to the mucus but not touching each other, implying they were not flocculated when trapped

Another possible mechanism for trapping clay particles in mangroves may be the pelletisation of the small particles in the 1 to 5 μm range by filter feeders (Haven & Morales-Alamo, 1972), such as the epiphytes growing on mangrove roots, though filter feeders are not common in mangroves

Regardless of source, mucus is common in mangrove waters, on wood rotting on the forest floor and on the mud As bacterial populations are extremely abundant and productive in mangroves (Alongi, 1988), it is likely that they may be the major source of mucus for trapping of clay particles Benthic and pelagic detritivores are known to produce mucus, but the extent of this production is unknown

Nevertheless numerous non-flocculated clay particles remain in suspension after the water has circulated in the swamp at Coral Creek These clay particles have escaped trapping in mucus or pelletisation Their fate is likely to be controlled by the oceanography since the settling velocity of clay particles is extremely small and tidal turbulence can readily keep them in suspension In dry tropical mangrove systems evapotranspiration pumps salt water upstream, the non-flocculated particles presumably travel upstream where they will ultimately settle In the presence of river runoff it may be, but no field data are available, that a significant fraction of the unflocculated clay particles will be transported downstream to coastal waters

Large aggregates, marine snow, with dimensions of mm to cm and even m, have been reported in open ocean (Honjo *et al*, 1984), and these are an assemblage of microorganisms and detritus, organic aggregates and small particles Mucus provided by zooplankton provides a nucleus for accumulation of small particles (Alldredge, 1976) These aggregates may be trapping in coastal waters the fine sediment exported from mangroves

References

Adams, C E & G Weatherly, 1981 Some effects of suspended sediment stratification on an oceanic bottom boundary layer J Geophys Res 86 4161–4172

Alldredge, A L, 1976 Discarded appendicularian houses as sources of food, surface habitats, and particulate organic matter in planktonic environments Limnol Oceanogr 21 14–23

Alongi, D M, 1988 Bacterial productivity and microbial biomass in tropical mangrove sediments Microbial Ecol 15 59–79

Althausen, J D & B Kjerfve, 1992 Distribution of suspended sediment in a partially mixed estuary, Charleston Harbor, South Carolina, U S A Est Coastal Shelf Sci 35 517–531

Christiansen, H, 1974 Uber den transport suspendierter festoffe in astuarien an beipiel der Elbemundung bei newerk Theses, Universitat Hannover, H 28

Clukey, E C, F H Kulhaway, P Liu & G Tate, 1985 The impact of wave loads and pore-pressure generation on initiation of sediment transport Geo Marine Letters 5 177–183

Dermuren, A V & W Rodi, 1986 Circulation of flow and pollutant dispersion in meandering channels J Fluid Mech 172 63–92

Dronkers, J & W van Leussen (eds), 1998 Physical processes in estuaries Springer-Verlag, Berlin, 560 pp

Dyer, K, 1986 Coastal and estuarine sediment dynamics J Wiley, New York, 342 pp

Eisma, D, T Schuhmacher, H Boekel, J van Heerwaarden, H Franken, M Laan, A Vaars, F Eugenraam & J Kalf, 1990 A camera and image analysis system for in situ observations of flocs in natural waters Netherlands J Sea Res 27 43–56

Faas, R W, 1986 Mass physical and geotechnical properties of surficial sediments and dense nearbed sediment suspensions on the Amazon continental shelf Cont Shelf Res 4 189–208

Gibbs, R J, 1983 Effect of natural coatings on the coagulation of particles Environ Sci Tech 17 237–240

Gibbs, R J, 1985 Estuarine flocs their size, settling velocity and density J Geophys Res 90 3249–3251

Gibbs, R J & L Konwar, 1983 Disruption of mineral flocs using Niskin bottles Environ Sci Tech 17, 374–375

Gibbs, R J & L Konwar, 1986 Coagulation and settling of Amazon River suspended sediment Cont Shelf Res 6 127–149

Gibbs, R J, D M Tshudy, L Konwar & J M Martin, 1989 Coagulation and transport of fine sediment in the Gironde estuary Sedimentology 36 987–999

Gust, G, 1976 Observations on turbulent-drag reduction in a dilute suspension of clay in sea water J Fluid Mech 75 29–47

Haven, D S & R Morales Alamo, 1972 Biodeposition as a factor in sedimentation of fine suspended solids in estuaries In B W Nelson (ed), Environmental framework of coastal plain estuaries Geological Society of America Memoirs 133 121–130

Honjo, S, K Doherty, Y Agrawal & V Asper, 1984 Direct optical assessment of large amorphous aggregates (marine snow) in the deep ocean Deep-Sea Res 31 67–76

Jiang, F & A J Mehta, 1992 Some observations on fluid med response to water waves In D Prandle (ed) Dynamics and exchanges in estuaries and the coastal zone American Geophysical Union, Washington D C 351–376

Kineke, G C, 1993 Fluid muds on the Amazon continental shelf Ph D dissertation, University of Washington, Seattle, 259 pp

King, E & E Wolanski, 1994 Drag reduction in turbid estuaries In C Pattiaratchi (ed) Physics of estuaries and coastal bays (in press)

Krone, R B, 1972 A field study of flocculation as a factor in estuarial shoaling processes Techn Bull 19, Committee on Tidal

Hydraulics, US Army Corps of Engineers WES, Vicksburg, 91 pp

Larcombe, P & P V Ridd, 1992 Dry season hydrodynamics and sediment transport in mangrove creeks and some implications for the interpretation of buried mangrove sequences Proc Int Conf of Physics of Estuaries & Shallow Seas Margaret River 8–10 Dec 1992, West Australia

Li, Y, E Wolanski & Q Xie, 1993 Coagulation and settling of suspended sediment in the Jiaojiang River estuary, China J Coastal Res 9 390–402

Maa, P-Y & A J Mehta, 1987 Fluid mud erosion by waves a laboratory study Cont Shelf Res 7 1269–1284

Mehta, A J , 1986 Characterization of cohesive sediment properties and transport processes in estuaries In A J Mehta (ed) Estuarine cohesive sediment dynamics Lecture Notes Coastal Estuarine Studies 14 290–325

Mehta, A J & E Partheniades, 1975 An investigation of the depositional properties of flocculated fine sediments J Hydr Res ASCE 13 361–381

Parchure, T M & A J Mehta, 1985 Erosion of soft cohesive sediment deposits J Hydr Division ASCE 111 1308–1326

Postma, H , 1961 Transport and accumulation of suspended matter in the Dutch Wadden Sea Neth J Sea Res 1 148–190

Postma, H , 1967 Sediment transport and sedimentation in the marine environment In G H Lauff (ed) Estuaries Assoc Am Sci Publ 83 Washington, D C 158–179

Riethmuller, R , H U Fanger, I Grabemann, H L Krasemann, K Ohm, J Boning, L J R Neumann, G Lang, M Marfosky & Schubert R , 1988 Hydrographic measurements in the turbidity zone of the Weser estuary In J Dronkers & W van Leussen (eds) Physical processes in estuaries Springer-Verlag, Berlin 332–344

Ross, W M , 1988 Vertical structure of estuarine fine sediment suspension Ph D thesis, Coastal & Oceanog Engin Dept , University of Gainesville, Florida

Simpson, J H & I D James, 1986 Convergent fronts in the circula tion on tidal estuaries In D A Wolfe (ed) Estuarine variability Academic Press, London

Stolzenbach, K D , K Newman & C Wong, 1992 Aggregation of fine particles at the sediment-water interface J Geophys Res 97 17889–17898

Trowbridge, J H & G C Kineke, 1993 Structure and dynamics of fluid muds over the Amazon continental shelf Mar Geol

Tundisi, J , T M Tundisi & M Kutner, 1973 Plankton studies in a mangrove environment VIII Further studies on primary pro duction, standing crop and phyto- and zooplankton and some environmental factors Intern Revue der Gesamten Hydrobiologie 58 925–940

Uiterwijk Winkel, A P B , 1975 Microbiologische apscten en het sedimenatiegedrag van rivierslib Rijkswaterstaat, Direktie Waterhuishounding en Waterbeweging, District Zuidwest, Report No 44 006 001, 60 pp

Uncles, R J & J A Stevens, 1993 Nature of the turbidity maximum in the Tamar Estuary, U K Estuarine Coastal Shelf Sci 36 413–432

van Leussen, W , 1988 Aggregation of particles, settling velocity of mud flocs A review In J Dronkers & W van Leussen (eds) Physical processes in estuaries Springer-Verlag, Berlin 347–403

Wellerhaus, S , 1981 Turbidity maximum and mud shoaling in the Weser estuary Arch Hydrobiol 92 161–198

Wells, J T , 1983 Dynamics of coastal fluid muds in low-, moderate-, and high-tide-range environments Can J Fish Aquat Sci 40 (suppl 1) 130–142

Wells, J T , 1989 In situ measurements of large aggregates over a fluid mud bed J Coastal Res 5 75–86

Wells, J T & H H Roberts, 1981 Fluid mud dynamics and shoreline stabilisation Proc Int Conf Coastal Eng , Sydney 17 1382–1401

Wolanski, E , 1992 Hydrodynamics of mangrove swamps and their coastal waters Hydrobiologia 247 141–161

Wolanski, E & W Hamner, 1988 Topographically controlled fronts in the ocean and their biological influence Science 241 177–181

Wolanski, E & M Eagle, 1991 Oceanography and sediment transport in the Fly River estuary and Gulf of Papua Proc 10th Austr Conf Coastal & Ocean Eng , Auckland 453–457

Wolanski, E & R Gibbs, 1994 Flocculation of suspended sediment in the Fly River estuary, Papua New Guinea J Coastal Res (in press)

Wolanski, E , J Chappell, P Ridd, & R Vertessy, 1988 Fluidisation of mud in estuaries J Geophys Res 93 2351–2361

Wolanski, E , Y Mazda & P Ridd, 1992a Mangrove hydrodynamics In A I Robertson & D Alongi (eds) Tropical mangrove ecosystems American Geophysical Union, Washington D C 43–62

Wolanski, E , R Gibbs, Y Mazda, A Mehta & B King, 1992b The role of turbulence in the settling of mud flocs J Coastal Res 8 35–46

Wolanski, E , B King & D Galloway, 1994 Dynamics of the turbidity maximum in the Fly River estuary, Papua New Guinea Estuarine Coastal Shelf Sci (in press)

Hydrobiologia **295**: 43–49, 1995.
Y. S. Wong & N. F. Y. Tam (eds), Asia-Pacific Symposium on Mangrove Ecosystems.

Mangroves and climate change in the Florida and Caribbean region: scenarios and hypotheses

Samuel C. Snedaker
Marine Biology and Fisheries, Rosenstiel School of Marine and Atmospheric Science, University of Miami, 4600 Rickenbacker Causeway, Miami, Florida 33149-1098, USA

Key words: climate change, mangroves, peat, precipitation, roots, sealevel rise, seawater sulfate, ultraviolet radiation

Abstract

The principal scenario concerning the potential effects of climate change on mangrove forest communities revolves around sealevel rise with emphases on coastal abandonment and inland retreat attributable to flooding and saline intrusion. However, at the decade to century scale, changes in precipitation and catchment runoff may be a more significant factor at the regional level. Specifically, for any given sealevel elevation it is hypothesized that reduced rainfall and runoff would necessarily result in higher salinity and greater seawater-sulfate exposure. This would likely be associated with decreased production and increased sediment organic matter decomposition leading to subsidence. In contrast, higher rainfall and runoff would result in reduced salinity and exposure to sulfate, and also increase the delivery of terrigenous nutrients. Consequently, mangrove production would increase and sediment elevations would be maintained. Support for this scenario derives from studies of the high production in saline mangrove impoundments which are depleted in seawater sulfate. This paper also examines other components of climate change, such as UVb, temperature, and storm frequency, and presents a suite of hypotheses and analytical protocols to encourage scientific discussion and testing.

Introduction

The conventional wisdom concerning mangroves and global climate change focuses almost exclusively on sea level rise as the most critical factor. This has been translated into two major studies. The first was a large model simulation exercise using the Sea Level Affecting Marshes Model (SLAMM) (Park *et al.*, 1989). The SLAMM simulations were based on land elevations and hypsographic curves, and computations of the land area that would be inundated with incremental increases in sea level. Other forcing factors, such as precipitation or temperature changes, or consequential effects, such as inland salinization, were not explicitly examined. Notwithstanding, the basic conclusions are highly informative based on the assumption that other potentially-interacting factors remain unchanged as sea level rises.

In the second study, Ellison & Stoddart (1991) took a more empirical approach and examined Holocene stratigraphic records and sea level curves for a number of sites worldwide, including south Florida. Based on their analyses, they concluded (p. 161) that '...mangrove ecosystems appear to be able to keep pace with rising sea-level of 8–9 cm/100 cal years, are under stress at rates between 9 and 12 cm/100 cal years, and cannot persist in their expansive mode at rates above this.' Their study (see also Ellison, 1993), however, is difficult to reconcile in south Florida since Maul & Martin (1993) report a relative sea level rise over the past 147 yrs of about 30 cm for the southern tip of Florida (Fig. 1) attributing it to sea level rise plus components attributable to post glacial rebound and subsidence. It is pertinent to note that the regional mangrove ecosystem has not 'collapsed' even though Maul and Martin report the mean rate of relative sea level rise since 1925 to be equivalent to 23 cm/100 yrs, almost twice the critical maximum threshold of Ellison & Stoddart (1991). Also, the regional mangrove ecosystem is not under stress except in a few extremely lowlying areas and semi-enclosed hypersaline habitats, and in locations that are anthropogenically influenced by development and land use changes. From a historical perspective, it was also during this period of cryptic sea level rise that the venerable field naturalist, John Henry Davis, observed the widespread accretion and expansion of mangroves in south Florida, and published several papers on the presumed land-building role of mangroves (Davis, 1938a, b, 1940, 1942, 1943).

In this paper, I will discuss several scenarios and hypotheses pertaining to how mangroves might respond to variety of components of global climate change, attempt to define the most critical components, and identify some research tasks to test the hypotheses. The work reported here is based on an ongoing project entitled, 'Effects of Global Change on Mangrove Ecosystems: Methodology Development and Assessment', supported by the U.S. Environmental Protection Agency. Since the focus is on Florida and the Caribbean, the explicit hypotheses may not have equal universal importance.

Hypothesis 1

As a consequence of global warming, mangroves will colonize poleward, thus expanding and extending their range into regions that presently have temperate climates.

This is also part of the conventional wisdom in so far as the latitudinal limits of mangroves, particularly *Rhizophora mangle*, are determined by freezing temperatures. In the 1970s, the northern distributional limit of mangroves along the east coast of the USA extended north to around the border between the states of Georgia and Florida (National Wetland Inventory 1982, and Coastal Coordinating Council 1973, cited in Lewis *et al.*, 1985). However, recent severe freezes (January 1977, January 1981, and December 1989) have caused massive mangrove mortality such that both the mangrove area and northern range limit have been significantly reduced. Thus, the hypothesis can be easily rejected, at least, for this region.

Hypothesis 2

Sealevel rise will affect mangrove species that have a horizontal root structure (e.g., *Avicennia* spp.) before it affects those that have a vertical structure (e.g., *Rhizophora* spp.).

The root system of *Rhizophora* is a conspicuous vertical structure that is in marked contrast with the horizontal orientation in *Avicennia*. In *Avicennia germinans*, the root system consists of a relatively large proximal mass of intertwined woody roots that show evidence of significant secondary thickening in older trees. Radiating outward from the proximal root mass is an array of horizontal distal roots which are referred to as cable roots. Aerating pneumatophores arise vertically from the cable root and extend above the sediment surface. Extending downward from the cable roots are elongated 'anchoring' roots. The cable roots and their related structures exhibit little or no evidence of secondary thickening or the formation of hardened woody tissue.

Of particular research interest is the maximum rooting depth of the proximal root mass relative to vertical chemistry of the substrate. Whereas the anchoring roots may extend downward for some distance, the proximal root mass is flattened along a horizontal plane that is relatively close to the sediment surface. It is hypothesized that during the development of the proximal root system, growth is confined to the sediment horizon above the deeper anaerobic sediment that is normally saturated. In other words, downward growing proximal roots are unable to penetrate and survive in the saturated anaerobic horizon, and therefore develop horizontally in the horizon at or above the mean depth of the aerobic and anaerobic interface. Conversely, the anchoring roots appear to be able to penetrate and survive below the aerobic-anaerobic sediment boundary.

With respect to sea level rise, it is hypothesized that as the mean watertable elevation rises, vertical changes in the physical/chemical sediment environment induce a stress on the proximal root mass. This could occur as a result of increased anoxia, sulfide toxicity, or some related mechanism. In any event, it may partially explain why *A. germinans* in the interiors of some mangrove islands in southwest Florida exhibit both a relatively high incidence of mortality and subtle symptoms of physiological stress in the absence of any obvious causal factor (also see Hypothesis 7, below). It might also explain why all *A. germinans* (but not *R. mangle*) are eliminated from mosquito-control impoundments that have been subjected to prolonged periods of relatively small artificial increases in the mean water elevation.

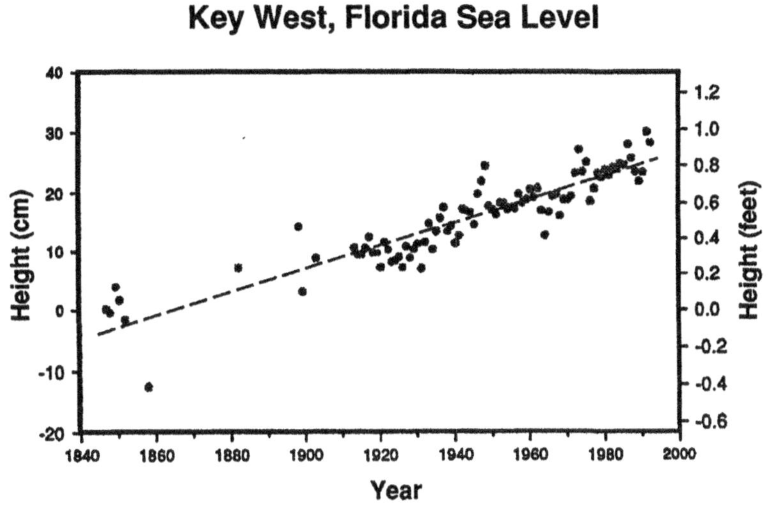

Fig. 1. Mean annual tidal heights at Key West, Florida, from 1946 through 1992. (Adapted from Maul & Martin, 1993)

Hypothesis 3

In mangrove forest areas where cyclonic storms (*i.e.*, typhoons, cyclones, hurricanes) cause widespread mass mortality, subsidence due to root decomposition will result in conversion to open water.

Based on observations in the Spring of 1993 following the widespread destruction of mangroves by Hurricane Andrew in August of 1992, the sediment surface in some mangrove areas that had previously been exposed during most low tides became permanently wetted or flooded. This was remarkable since Spring is the season of mean low sealevel in south Florida (Wanless, 1982) and there was no evidence of surface erosion. The most likely explanation for the observed subsidence is the loss of substrate mass as a consequence of root decomposition over the observational period of about seven months. This interpretation is based on the data in Table 1 which indicate that mangroves tend to have a very high root/shoot ratio with roots contributing to more than 40 percent of the total forest biomass. The data also indicate a possible difference between *Avicennia* and non-*Avicennia* mangrove species. It should also be noted that these mangrove root/shoot ratios are generally higher than ratios for other forest types that are reported in the literature.

Hypothesis 4

Mangrove species with large propagules (e.g., *Rhizophora*) will be the first to experience a reduction in recruitment. Consequently, in the absence of self-replacing recruitment, shorelines dominated by these species may be progressively abandoned.

This relatively simple hypothesis is based on the original work of Watson (1928) in Malaysia, field research and experimental testing in Panama by Rabinowitz (1978a, b, c) and in Costa Rica by Jimenez (1988). The most common shoreline dominant in south Florida is *R. mangle* which has a long slender propagule that can take root and become established in deeper water than can the other Florida species. Because it is also a shade intolerant species, regeneration is most abundant in open areas of active sedimentation and shoaling close to the fringing shoreline. Although *R. mangle* seedlings do become established in the shade of closed-canopy forests, growth and structural development do not occur except in light gaps.

In these regards, should rising sea level result in shoreline water depths too deep for *R. mangle* to become established, and canopy shade prevents in situ establishment, regeneration could fall below the level needed to balance natural mortality. Since any regeneration would likely be restricted to ephemeral light gaps in more inland areas, the current shoreline fringe forests dominated by this species might eventually disappear.

Table 1 Above- and below-ground mangrove biomass (kg/ha, dry weight), root/shoot ratios, and below-ground root biomass as a percent of total biomass

	Above ground	Below-ground	Total	R/S ratio	Roots % total
CHINA (Lin, n.d.)					
Kandella candel (20 y)	93368	69257	162625	0.742	0.426
Bruguiera sexangula (55 y)	248495	171804	420299	0.691	0.409
Rhizophora stylosa (64 y)	196212	95368	291580	0.486	0.327
PUERTO RICO (Golley *et al.*, 1962)					
Rhizophora mangle	62850	49970	112820	0.795	0.443
PANAMA (Golley *et al.*, 1974)					
Rhizophora sp.	279212	189761	468973	0.680	0.405
AUSTRALIA (Clough & Attiwill, 1982)					
Avicennia marina	90000	146000	236000	1.622	0.619
AUSTRALIA (Briggs, 1977)					
Avicennia marina	145000	147000	292000	1.014	0.503
Avicennia marina	110000	160000	272000	1.455	0.588
MEAN	153142	128645	282037	0.936	0.465
STD DEV					0.092

Hypothesis 5

Following destructive cyclonic storms, those mangrove species with significant secondary or reserve meristematic tissues (e.g., *Avicennia* spp.) will dominate over those species that lack the ability to coppice or initiate trunk sprouts (e.g., *Rhizophora* spp.).

The putative 'true' mangrove species (Tomlinson, 1986) have varying capacities for recovery and regeneration following structural damage. These include the development of new coppice growth at the root collar or from stump remains, the formation of trunk sprouts, and the refoliation of denuded branches and limbs (Hamilton & Snedaker, 1984). In a relative context, members of the Rhizophoraceae (e.g., *Rhizophora, Ceriops*) do not have a extensive capacity for recovery and regeneration since they lack significant reserve or secondary meristematic tissues. For example, *R. mangle* cannot regenerate or refoliate severed branches greater than about 2.5 cm in diameter, which is equivalent to 2–3 years of recent growth (Gill & Tomlinson, 1971). In a study of the recovery of a mixed species mangrove forest in south Florida following canopy removal, individuals of *A. germinans* and *Laguncularia racemosa* regenerated quickly whereas the *R. mangle* experienced a high incidence of mortality and extremely slow recovery (Snedaker *et al.*, 1992).

More recently, following the destruction of large areas of mangrove forests by Hurricane Andrew in south Florida, the following observations were made. *R. mangle* with severe structural damage, e.g., windthrown or uprooted, broken trunks, and/or severed limbs and branches, failed to show any evidence of recovery and eventually died. In contrast, individuals of *R. mangle* that were merely defoliated quickly developed a new complement of leaves from existing buds. The majority of individuals of *L. racemosa* that were severely damaged did not recover particularly in locations where the trees were windthrown in piles or the main trunks were broken close to the ground. However, trunk sprouts quickly developed on upright and leaning trees even if the top and all branches had been removed. Individuals of *A. germinans* that were not completely uprooted or windthrown, and which remained at least partially rooted, showed evidence of rebranching and releafing even along horizontal trunks. We therefore expect that in mixed-species areas, the surviving *Avicennia* and *Laguncularia* will dominate the developing canopy foliage, and eventually become the dominant species in those areas.

Hypothesis 6

Elevated levels of UVb radiation will affect mangroves in open areas more than mangroves in semi-closed closed canopy forests. Consequently, new recruitment will occur mainly in areas that are at least partially shaded.

The basis for this hypothesis is the frequent observation that mangroves which become established in isolated and completely-open areas fail to develop into large trees even when all habitat conditions appear otherwise to be suitable. The one unique characteristic of these isolated mangroves is that they experience dawn-to-dusk exposure to full intensity solar radiation that includes the blue and ultraviolet (UVa, b) components, as well as the red and far-red (Goodwin, 1965; Hillman, 1967; Spencer & Ksander, 1990). One hypothesis to explain the failure-to-thrive observation is that under high solar irradiance, the UVb component causes: (1) the accelerated production of the UV-absorbing pigment, anthocyanin, at the expense of the chlorophylls (Levitt, 1972; Rabino *et al.*, 1977) which results in a corresponding the decrease in the chlorophyll/carotenoid ratio, and, (2) the reduction in endogenous levels of the growth hormone, auxin (indole acetic acid) (Tevini, 1993). To the extent that the capacity for photosynthesis and growth are diminished, the developing seedling lacks the metabolic ability to fully develop as do established seedlings in marginally-shaded environment. High UV irradiance might also be a contributing factor in the dwarfing or stunting of mangroves in marginal environments.

Hypothesis 7

In the context of climate change, precipitation and catchment runoff is more important than sealevel rise since freshwater moderates salinity and limits the activity of sulfate-reducing decomposers and surface runoff is a source of terrigenous nutrients.

Although mangroves are ecologically restricted to saline intertidal environments, mangrove productivity in general increases proportionally with the availability of fresh water, principally in the form of surface runoff (Pool *et al.*, 1977). The main causal factors are the reduction of pore water salinity and the corresponding salt stress (Ahmed, 1992), and the enhanced availability of mineral nutrients entrained in the surficial runoff (Lugo *et al.*, 1976). The production and accumulation of mangrove peat also appears to be the greatest in areas that receive upland freshwater runoff (surface or subsurface) during most the year. To a limited extent, this observation is supported by the work of Lahmann (1988) who linked the low salinity of flooded mangrove impoundments along the east coast of Florida with high mangrove production, and the greatest abundance of mangrove peat.

Mangrove peat results principally from root production and natural mortality, and the subsequent accumulation/preservation of the organic remains under strongly reducing, or anoxic, conditions. Although anaerobic decomposition and remineralization occur continually (as does the aerobic decomposition of surface organic debris), the longer term rates of loss tend to balance the continual production of peat-forming roots. Thus, as long as the balance is maintained, the volume and mass of the peat body remain relatively constant. In terms of sealevel rise, as long as there is sufficient freshwater runoff to maintain an optimum low salinity (and low sulfate), nutrient balance and productivity, it is hypothesized that there would be a net accumulation of peat proportional to the rising water level, and accordingly, the mangrove zone would neither retreat, nor be overstepped or abandoned with sealevel rise (Snedaker, 1993).

If freshwater runoff, however, ceased or diminished to the point that the mangrove habitats were continually exposed to full or close-to-full strength sea water, then organic production would decline. At the same time, the increased availability of sulfate ($^{=}SO_4$, present in sea water at approximately 2.7 g kg^{-1}) to suffuse subsurface peat would necessarily lead to increased anaerobic decomposition by sulfur-reducing microorganisms, and thus, a loss of peat mass. Theoretically, the sulfate in one liter of sea water is capable of potentially fueling the anaerobic decomposition and breakdown of approximately 1.9 g dry wt of organic matter. This observation might also help to explain why there is so much freshwater peat in the world, and so little mangrove peat in marine environments.

In this regard, the apparent sealevel rise observed in the interiors of temperate (non-mangrove) coastal marshes of the eastern U.S. has been attributed to the biological decomposition of sediment organic matter (Courtney Hackney, pers. comm., University of North Carolina, Wilmington; Padgett *et al.*, 1986, Hackney, 1987). Although the causal factor (a microbial process) has not been shown to be related to the seawater-freshwater balance, and therefore the presence of seawater sulfate, the rapid rate of subsidence illustrates how quickly intertidal habitats can be degraded and dimınished by the loss of sediment organic matter In addition to the habitat loss, the induced anaerobic conversion of large areas of coastal organic substrates would contribute to the total atmospheric loading of green-house gases (Morris, 1991), notably carbon dioxide and methane Although there are no reports, either published or anecdotal, of mangroves being lost to this type of subsidence, the specific spatial pattern of incipient defoliation, decline, and mortality of mangrove areas observed in some parts of southwest Florida is similar to that described in the temperate marshes

Acknowledgments

The background research for this paper was supported by the U S Environmental Protection Agency (CR 820667-01) and the National Science Foundation (OCE-9300649) Rafael Araujo and Steve Smith assisted with the research and data syntheses, and Jane Snedaker helped assemble the final manuscript

References

Ahmed, S I, 1992 Coping with excess salt in their growth environments Osmoregulation and other survival strategies deployed by the mangroves Pak J mar Sci 1 73–86

Clough, B F & P M Attiwill, 1982 Primary productivity of mangroves, pp 213–222 In B F Clough [ed] Mangrove Ecosystems of Australia Australian National University Press, Canberra, 302 pp

Briggs, S V, 1977 Estimates of biomass in a temperate mangrove community Aust J Ecology 2 369–373

Davis, J H, 1938a The role of mangrove vegetation in land building in southern Florida American Philosopical Society Yearbook 162–164

Davis, J H, 1938b Mangroves, makers of land Nature Magazine 31 551–553

Davis, J H, 1940 The ecology and geologic role of mangroves in Florida Publication Carnegie Institution 517 303–412

Davis, J H, 1942 The ecology of the vegetation and topography of the sand keys of Florida Publication Carnegie Institution 33 113–195

Davis, J H, 1943 The natural features of Southern Florida, especially the vegetation, and the everglades Geological Bulletin Florida 25, 311 pp

Ellision, J C, 1993 Mangrove retreat with rising sea-level, Bermuda Estuar coast mar sci 37 75–87

Ellison, J C & D R Stoddart, 1991 Mangrove ecosystem collapse during predicted sealevel rise Holocene analogues and implications J Coast Res 7 151–165

Gill, A M & P B Tomlinson, 1971 Studies on the growth of red mangrove (*Rhizophora mangle* L) 3 Phenology of the shoot Biotropica 3 109–124

Golley, F B, H T Odum & R F Wilson, 1962 The structure and metabolism of a Puerto Rican red mangrove forest in May Ecology 43 9–19

Golley, F B, J T McGinnis, R G Clements, G I Child & M J Duever, 1974 Mineral Cycling in a Tropical Moist Forest Ecosystem Univ Georgia Press, Athens, 248 pp

Goodwin, T W, 1965 Chemistry and Biochemistry of Plant Pigments Academic Press, New York

Hackney, C T, 1987 Factors affecting accumulation or loss of macroorganic matter in salt marsh sediments Ecology 68 1109–1113

Hamilton, L & S C Snedaker (eds), 1984 Handbook for Mangrove Area Management, Environment and Policy Institute, East-West Center, Honolulu, Hawaii, 123 pp

Hillman, W S, 1967 The physiology of phytochrome Ann Rev Pl Physiol 18 301–324

Jimenez, J A, 1988 Dynamics and dispersion patterns of two mangrove populations on the Pacific coast of Costa Rica Ph D Dissertation Univ Miami, Coral Gables, 176 pp

Lahmann, E J, 1988 Effects of different hydrological regimes on the productivity of *Rhizophora mangle* L A case study of mosquito control impoundments at Hutchinson Island, Saint Lucie County, Florida Ph D Dissertation Univ Miami, Coral Gables, 149 pp

Levitt, J, 1972 Responses of Plants to Environmental Stresses Academic Press, New York, 607 pp

Lewis, R R, III, R G Gilmore, Jr, D W Crewz & W E Odum, 1985 Mangrove habitat and fishery resources of Florida 281–336 In William Seaman, Jr (ed) Florida Aquatic Habitat and Fishery Resources Florida Chapter, American Fisheries Society, Kissimmee, FL 543 pp

Lin, Peng n d, Element cycle and energy dynamics in three kinds of mangroves of China Manuscript, 14 pp

Lugo, A E, M Sell & S C Snedaker, 1976 Mangrove ecosystem analysis, pp 113–145 In B C Patten (ed) Systems Analysis and Simulation in Ecology, Vol IV Academic Press, New York, NY, 593 pp

Maul, G A & D M Martin, 1993 Sea level rise at Key West, 1846–1992 America's longest instrument record? Geophysical Research Letters 20 1955–1958

Morris, J T, 1991 Effects of nitrogen loading on wetland ecosystems Ann Rev Ecol and Syst 22 257–279

Padgett, D E, C T Hackney & A A de la Cruz, 1986 Growth of filamentous fungi into balsa wood panels buring in North Carolina salt marsh sediments Trans br mycol Soc 87 155–162

Park, R A, M S Trehan, P W Mausel & R C Howe, 1989 The effects of sea level rise on U S coastal wetlands and lowlands HRI Report No 164 OPPE, U S Environmental Protection Agency, Washington, D C

Pool, D J, S C Snedaker & A E Lugo, 1977 Structure of mangrove forests in Florida, Puerto Rico, Mexico and Costa Rica Biotropica 9 195–212

Rabino, I, L Mancinelli & K M Kuzmanoff, 1977 Photocontrol of anthocyanin synthesis Plant Physiol 59 569–573

Rabinowitz, D, 1978a Dispersal properties of mangrove propagules Biotropica 10 47–57

Rabinowitz, D, 1978b Mortality and initial propagule size in mangrove seedlings in Panama J Ecol 66 45–51

Rabinowitz, D, 1978c Early growth of mangrove seedlings in Panama, and an hypothesis concerning the relationship of dispersal and zonation J Biogeogr 5 113–133

Snedaker, S C, 1993 Impact on mangroves, pp 282–305 In G A Maul (ed) Climate Change in the Intra-Americas Sea

Edward Arnold, Hodder and Stoughton Publishers, Kent, UK., 389 pp.
Snedaker, S. C., M. S. Brown, E. J. Lahmann & R. J. Araujo, 1992. Recovery of a mixed-species mangrove forest in south Florida following canopy removal. J. coastal Res. 8: 919–925.
Spencer, D. F. & G. G. Ksander, 1990. Influence of temperature, light and nutrient limitation on anthocyanin content of *Potamogeton gramineus* L. Aquat. Bot 38: 357–367.
Tevini, M., 1993. Effects of enhanced UV-B radiation on terrestrial plants, pp. 125–153. In M. Tevini (ed.) UV-B Radiation and Ozono Depletion: Effects on Humans, Animals, Plants, Microorganisms, and Materials. Lewis Publishers. Boca Raton, FL., 248 pp.
Tomlinson, P. B., 1986. The Botany of Mangroves. Cambridge University Press, New York, 413 pp.
Wanless, H. R., 1982. Sea level is rising-so what? J. Sed. Petrol. 52: 1051–1054.
Watson, J. G., 1928. Mangrove forests of the Malay Peninsula. Malayan Forest Records 6: 125–149.

Hydrobiologia **295** 51–58, 1995
Y S Wong & N F Y Tam (eds) Asia Pacific Symposium on Mangrove Ecosystems

Tidal asymmetry in mangrove creeks

Yoshihiro Mazda[1], Nobuyuki Kanazawa[2] & Eric Wolanski[3]
[1]*School of Marine Science and Technology, Tokai University, 20-1, Orido 3, Shimizu, Shizuoka 424, Japan*
[2]*Shin-Nihon Meteorological & Oceanographical Consultant Co, Ltd, 3-14-5, Tamagawa, Setagaya, Tokyo 158, Japan*
[3]*Australian Institute of Marine Science, PMB No 3, Townsville MC, Qld 4810, Australia*

Key words hydrodynamics, mangrove swamp, tidal creek, tidal asymmetry, drag force, flood plain

Abstract

We model the dynamics of a tidal creek – mangrove swamp system In the creek, a tidal asymmetry prevails The ebb flow dominance at spring tides helps flush out the coarse sediment from the creek Results from the numerical model suggest that the ebb dominance is due to friction in the mangrove forest and in turn this is controlled by the density of the vegetation The tidal asymmetry of the current is negligible for a very small or a very large vegetation density, and is maximum for an intermediate vegetation density typical of that in undisturbed healthy mangroves

Introduction

Mangrove forests are classified into five types riverine forest, basin forest, fringe forest, overwash forest and dwarf forest (Lugo & Snedaker, 1974) Riverine forests exhibit the highest level of structural development (Cintron & Novelli, 1984) Such forests have a fairly flat substrate, exposed at low tide and flooded at high tide, and creeks with steep banks and a relatively flat bottom that do not dry up even at low tide The swamp area is usually larger than the creek area (Table 1) In the swamp, the roughness, that enhances resistance, is very large due to the vegetation and bioturbation (crab holes)

Physical processes are important for mangrove ecosystems (Wolanski *et al*, 1990) One key process is the asymmetry of the tidal currents in the creek The dynamics behind this asymmetry have been alluded by Wolanski *et al* (1980) In the present paper, this process is analysed using a numerical model in order to focus of the relative importance of the interaction between the geometries of the tidal creek and the swamp with densely vegetated mangroves

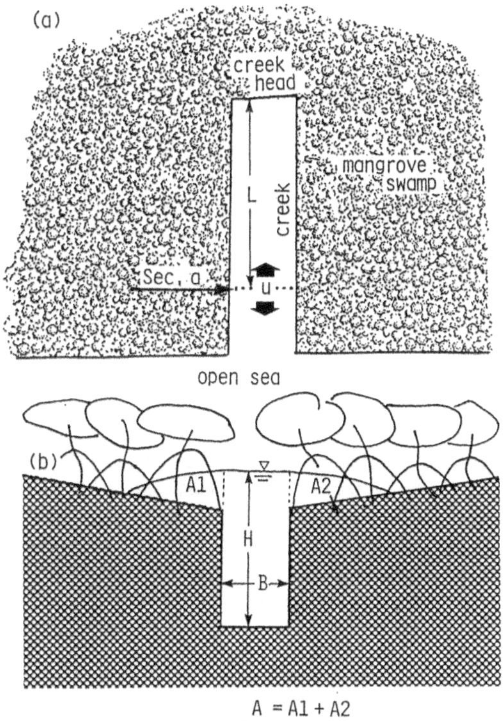

Fig 1 (a) Plan view and (b) cross sectional view of the model creek – mangrove system

Table 1. Ratio of the swamp area to the creek area and the asymmetry of tidal flow observed in various mangrove creeks

	Swamp area/ Creek area	Max. Velocity ($m\ s^{-1}$) Flood	Ebb	
Hinchinbrook Channel (Australia)	2.1	0.5	0.9	(Wolanski *et al.*, 1990)
Tuff Crater (New Zealand)	44.0	0.4	0.6	(Woodroffe, 1985)
Klong Ngao (Thailand)	2.7	0.4	0.8	(Wattayakorn *et al.*, 1990)
Ross Creek (Australia)	***	0.4	0.8	(Larcombe & Ridd, 1992)
Dickson Inlet (Australia)	6.2	0.7	0.8	(Wolanski & Mazda, 1989)
Chwaka Bay (Zanzibar)	***	0.3	0.5	(Wolanski, 1989)
Coral Creek (Australia)	5.5	1.2	1.6	(Wolanski *et al.*, 1980)
Wenlock River (Australia)	***	1.0	2.0	(Wolanski & Ridd, 1986)
Fukido-Gawa (Japan)	12.8	0.5	0.7	(Mazda, unpublished)

Tidal flow asymmetry in the creek

In mangrove creeks, the peak current velocity is usually larger at ebb than at flood tide (Table 1). This result is opposite to the situation in many estuaries without mangroves (Aubrey, 1986). The ebb dominance is believed to help to scour the channel (Wolanski *et al.*, 1992). Friedrichs *et al.* (1992) have studied the flow asymmetry in tidal embayments with inter-tidal flats and salt marshes. However, in such systems, the roughness is much smaller than in densely vegetated mangroves. For instance, the Manning resistance coefficient is about 0.2–0.4 (in MKS units) at Coral Creek, Hinchinbrook Islands (Australia) and 0.2–0.7 at Nakama Gawa, Okinawa (Japan) (Wolanski *et al.*, 1980; Mazda, 1991). A Manning coefficient of 0.4 implies a drag coefficient $\gamma^2 = 4.0$ which is two to three orders of magnitude larger than the value commonly found in many estuaries (Friedrichs *et al.*, 1992).

Basic concept of the model

Here, it is conceptually shown that the mangrove swamp significantly modifies the tidal velocity in the creek.

The model area is composed of a straight creek and fringing vegetated mangrove swamps (Fig. 1a). The creek does not dry out even at low tide. The creek has a flat bottom. The substrate of the mangrove forest slopes gently (Fig. 1b). The creek length is much smaller than the tidal wave length. Creek water inundates the swamp by overland flow and groundwater flow is neglected. The cross sectional mean velocity (u) at Sec.a (Fig. 1) is given by the continuity equation.

$$uHB = \frac{\partial}{\partial t}(HBL + AL) \tag{1}$$

where t is the time, H the water depth in the creek, B the creek width, L the creek length, and A the cross sectional area of the inundated swamp (Fig. 1b). As B and L are constant, Eq. 1 becomes

$$u = \frac{L}{H}\frac{\partial H}{\partial t} + \frac{L}{HB}\frac{\partial A}{\partial t} \tag{2}$$

The first term of the right hand side of Eq. 2 (called u_H, hereafter) is the tidal current in the creek in the

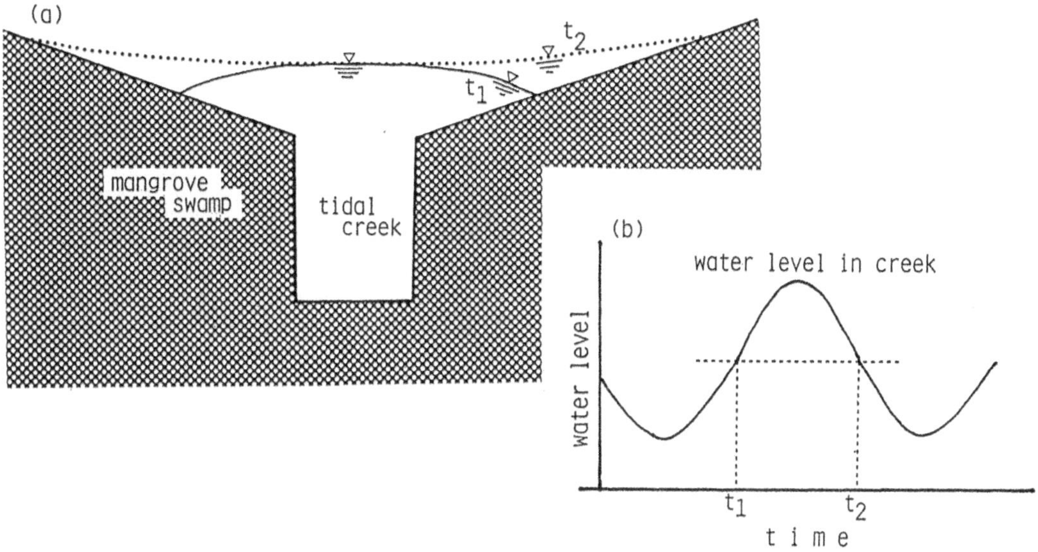

Fig. 2. Sketch of water levels at flood tide (time t_1) and ebb tide (time t_2).

absence of the swamp. The second term (called u_A, hereafter) is due to the swamp.

The water levels in the creek are the same at times t_1 and t_2 (Fig. 2), but are different in the swamp. This suggests that the change of the cross sectional area of the inundated swamp (A) are not tidally symmetric, resulting in the tidal asymmetry of u_A and u.

Numerical model

The depth-averaged momentum and continuity equations are

$$\frac{\partial uh}{\partial t} + \frac{\partial u^2 h}{\partial x} + \frac{\partial vuh}{\partial y} = - gh\frac{\partial \zeta}{\partial x} - \gamma^2 u\sqrt{u^2 + v^2} \quad (3)$$

$$\frac{\partial vh}{\partial t} + \frac{\partial uvh}{\partial x} + \frac{\partial v^2 h}{\partial y} = - gh\frac{\partial \zeta}{\partial y} - \gamma^2 v\sqrt{u^2 + v^2} \quad (4)$$

$$\frac{\partial \zeta}{\partial t} + \frac{\partial uh}{\partial x} + \frac{\partial vh}{\partial y} = 0 \quad (5)$$

where u and v are the vertically-averaged velocity in x- and y-directions, respectively, t is the time, ζ the elevation of water surface, h the depth, g the acceleration due to gravity and γ^2 the drag coefficient. In the creek $\gamma^2 = 0.0026$. In the mangrove swamp γ^2 is about 4.0 for a natural heavy vegetation, as mentioned before. The value of γ^2 varies with vegetation density so that we examined 10 cases with γ^2 varying between 0.0026 and 30 in the swamp. For $\gamma^2 = 0.0026$ in the swamp, there are no mangroves in the swamp.

At the mouth, a semi-diurnal tide with an amplitude of 1.2 m is imposed. The discharge is zero at the head. The swamp starts being inundated when the water is 0.45 m above mean sea level (Fig. 3). The model handles the wetting and drying of the swamp with the tides.

In all cases the tide curve in the creek is found to be independent of γ^2. However the velocity is found to depend strongly on γ^2 (Fig. 4). Two velocity peaks exist at flood tide, and two at ebb tide. Two such peaks of the velocity at flood tide occur for tidal creeks in salt marshes (Bayliss-Smith *et al.*, 1979; Pethick, 1980). Similar effects have been reported from field data by Woodroffe (1985) and Larcombe & Ridd (1992) for tidal creeks with mangrove swamps. The first velocity peak exists even in the absence of the swamp. At flood tide, the second peak decreases with increasing γ^2. At ebb tide, both peaks move closer to each other with increasing γ^2, ultimately merging in one peak. These

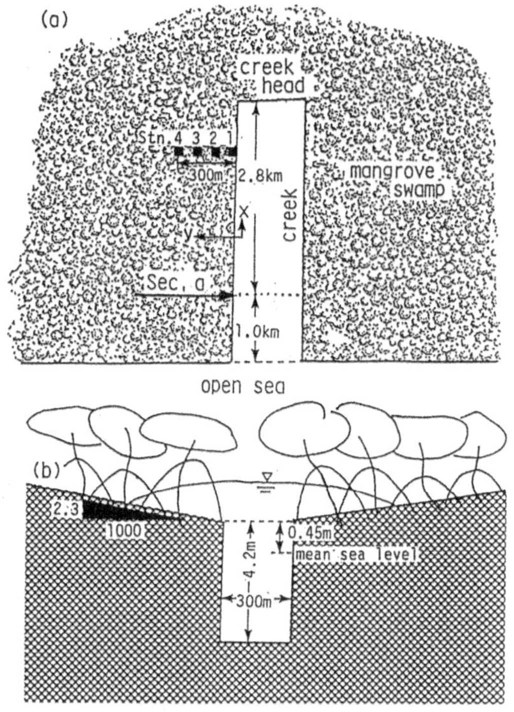

Fig. 3. (a) Plan view and (b) cross-sectional view of the model domain.

features suggest that the second velocity peak is due to the water inundating or draining the swamp.

Time series plot (Fig. 5) of water level, v-component of velocity and volume transport at Stns. 1 to 4 for $\gamma^2 = 4.0$ (see station locations in Fig. 3), shows that the water level in the swamp is quite different from that in the creek. Field data support this finding (Mazda *et al.*, 1992). The asymmetry in the tidal velocity during the filling and the draining the swamp is pronounced. The velocity at ebb tide in the swamp near the creek (Stn. 1) is much larger than at flood tide. The reverse situation prevails far from the creek (Stn. 4). As the flood tide turns to ebb in the creek, the water surface commences to fall near the creek (Stn. 1), while the water is still filling the inner portion (Stn. 4). These features demonstrate that the tidal wave propagation inside the swamp is strongly modified by drag force due to the mangrove vegetation.

Discussion

The maximum velocity in the creek is found to depend strongly on γ^2 in the swamp (Fig. 6a). For large values of γ^2, the maximum velocity is smaller at flood tide (u_{FM}) than at ebb tide (u_{EM}). The tidal asymmetry is parameterised by the ratio, u_{EM}/u_{FM} (called here $r_{E/F}$). $r_{E/F}$ increases with increasing γ^2, has a maximum of 1.24 at $\gamma^2 = 4.0$, and then decreases for increasing γ^2 (Fig. 6b).

The time series plot (Fig. 7) of u, u_H and u_A at Sec.a for $\gamma^2 = 4.0$ shows that $u_H + u_A$ is approximately equal to u, suggesting that the conceptual model (Eq. 2) is justified.

u_H is tidally symmetric but u_A is not. Though u_A is flood dominant, $u_H + u_A$ is ebb dominant. In order to understand this, we focus on the phase relationships between u_A and u_H.

The phase, and also the peak value, of u_A depends greatly on γ^2 (Fig. 8). As shown in Fig. 8, the times of the peak value of u_A both at flood and ebb tides are shifted backward with increasing γ^2 (; phase relationship), also, the peak value of u_A decreases with increasing γ^2 faster at ebb tide than at flood tide (; amplitude relationship). This phase relationship implies that at flood tide the peak value of u_A occurs later than the peak value of u_H for increasing γ^2, while at ebb tide the peak value of u_A occurs closer to that of u_H for increasing γ^2. As a result, the peak value of $u_H + u_A$ decreases at flood tide, and increases at ebb tide, for increasing γ^2. The phase relationship thus causes the ebb dominance in the creek. However, as the ebb flow dominance tends to be cancelled by the peak value reduction with increasing γ^2 (; amplitude relationship), $r_{E/F}$ has a maximum for an intermediate value of $\gamma^2 = 4$ (Fig. 6), being coincidentally a typical value for a mangrove swamp.

Larcombe & Ridd (1992) have argued that in Ross Creek, a mangrove creek in Australia, the ebb flow dominance exports coarse sediments to the open sea and scours a deep tidal channel. Our model suggests that an extremely thick vegetation tends to reduce the ebb flow dominance in the creek, allowing the creek to silt. This finding suggests that in natural mangrove areas, the feed-back between currents, vegetation density and siltation pattern maintains an optimum mangrove ecosystem. A reduction in the size of the swamp, for instance from prawn farming or land reclamation, will result in reducing the tidal asymmetry. This in turn causes the creek to silt and to dry up at low tide, result-

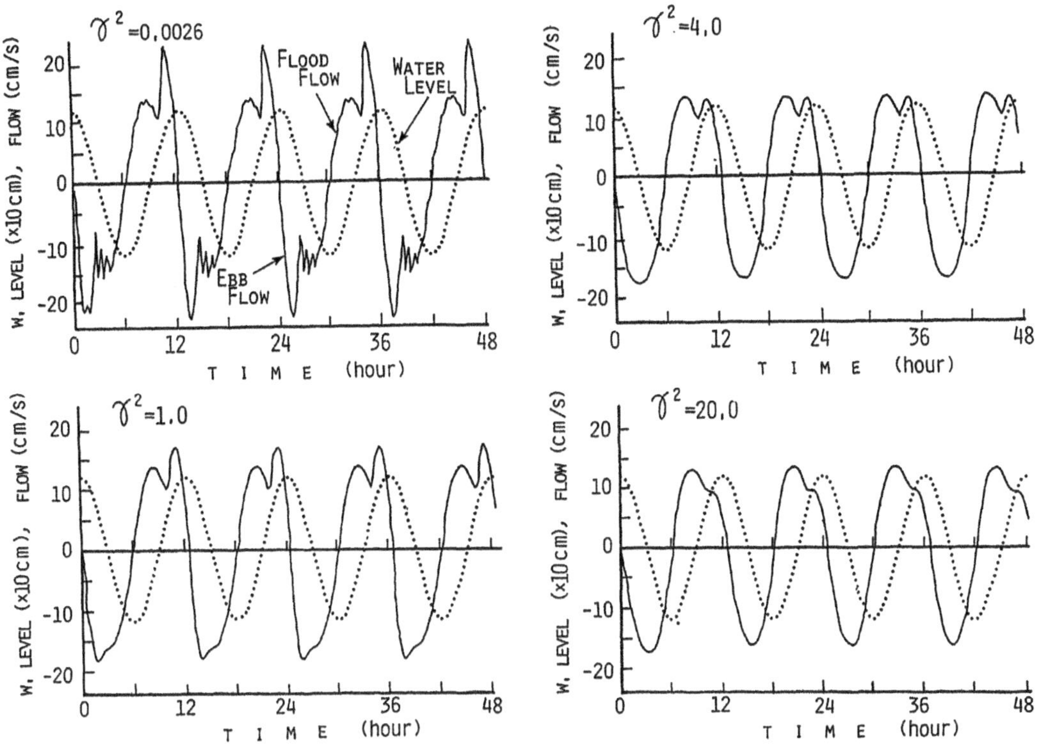

Fig. 4. Time series plot of the predicted velocity and water level at Sec.a for four different values of γ^2 in the swamp.

ing in the loss of the mangroves as a refuge or habitat for marine life.

Conclusion

The current velocity in a mangrove creek is the sum of two components, u_H and u_A. u_H is due to tidal flows in a tidal channel without a flood plain, and is tidal symmetric. u_A is due to the water exchange between the creek and the swamp, and shows a pronounced tidal asymmetry.

Inside the vegetated swamp, the water level and the current velocity are strongly controlled by drag force due to vegetation. The tidal asymmetry is governed by a phase relationship (the time of the peak value of u_A is different from that of u_H) and an amplitude relationship (the peak value of u_A at flood tide is larger than that at ebb tide). Due to the phase relationship, the peak velocity in the creek decreases at flood tide and increases at ebb tide for increasing values of the drag force in the swamp, resulting in the ebb flow dominance in the creek. However, when the drag force is excessive, the ebb flow dominance is reduced due to the amplitude relationship.

These findings suggest that in natural mangrove areas, the feed- back mechanisms among mangrove vegetation, siltation and water flow maintain an optimum mangrove ecosystem.

Acknowledgments

This study was supported by a grant of Nippon Life Insurance Foundation and a grant of Tokai University General Research Organization.

References

Aubrey, D. G., 1986. Hydrodynamic controls on sediment transport in well-mixed bays and estuaries. In J. van de Kreeke

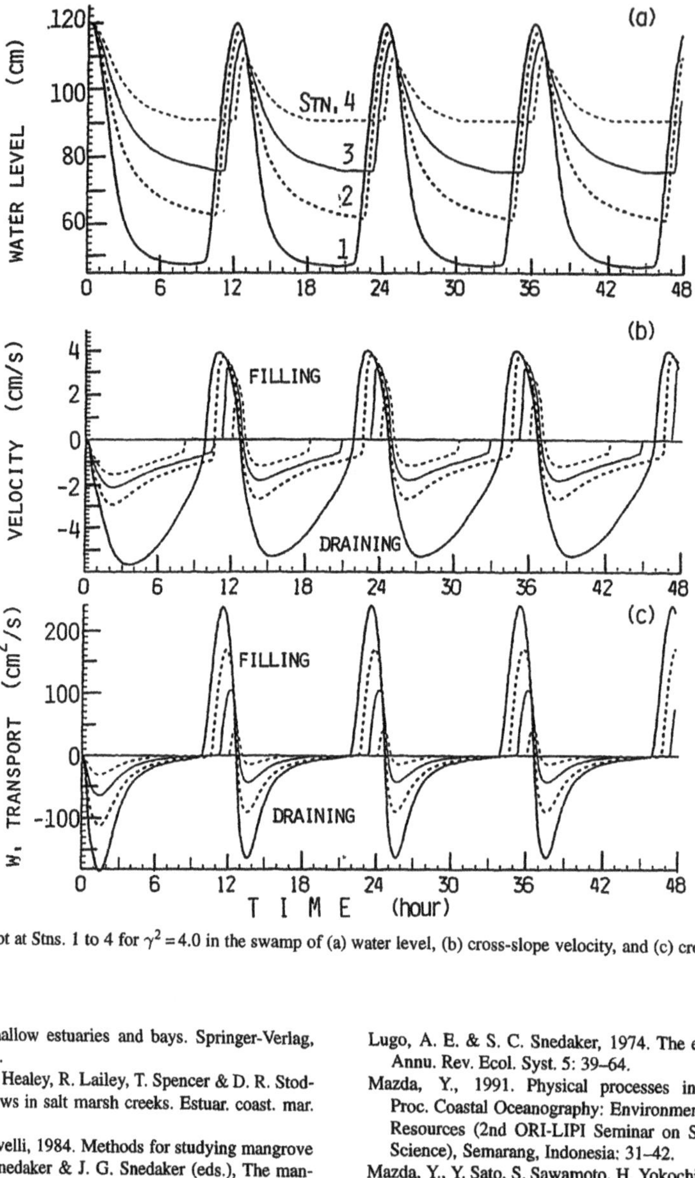

Fig. 5. Time series plot at Stns. 1 to 4 for $\gamma^2 = 4.0$ in the swamp of (a) water level, (b) cross-slope velocity, and (c) cross-slope discharge per unit length.

(ed.), Physics of shallow estuaries and bays. Springer-Verlag, New York: 245–258.

Bayliss-Smith, T. P., R. Healey, R. Lailey, T. Spencer & D. R. Stoddart, 1979. Tidal flows in salt marsh creeks. Estuar. coast. mar. Sci. 9: 235–255.

Cintron, G. & Y. S. Novelli, 1984. Methods for studying mangrove structure. In S. C. Snedaker & J. G. Snedaker (eds.), The mangrove ecosystem: research methods, UNESCO: 91–113.

Friedrichs, C. T., D. R. Lynch & D. G. Aubrey, 1992. Velocity asymmetries in frictionally-dominated tidal embayments: longitudinal and lateral variability. In. D. Prandle (ed.) Coastal and estuarine studies 40: Dynamics and exchanges in estuaries and the coastal zone, Ameriacn Geophysical Union, Washington DC: 277– 312.

Larcombe, P. & P. V. Ridd, 1992. Dry season hydrodynamics and sediment transport in mangrove creeks and some implications for the interpretation of buried mangrove sequences. Proc. Int. Biennial Conf. Physics of Estuaries and Coastal Seas.

Lugo, A. E. & S. C. Snedaker, 1974. The ecology of mangroves. Annu. Rev. Ecol. Syst. 5: 39–64.

Mazda, Y., 1991. Physical processes in mangrove estuaries. Proc. Coastal Oceanography: Environmental, Characteristic and Resources (2nd ORI-LIPI Seminar on Southeast Asia Marine Science), Semarang, Indonesia: 31–42.

Mazda, Y., Y. Sato, S. Sawamoto, H. Yokochi & E. Wolanski, 1990. Links between physical, chemical and biological processes in Bashita-minato, a mangrove swamp in Japan. Estuar. coast. Shelf Sci. 31: 817–833.

Mazda, Y., A. Sase, A. Enda, M. Sawada, H. Hamachi, Y. Maruhashi & S. Yamaguchi, 1992. Environmental survey in the Nakama-Gawa mangrove area, Iriomote Island. Bull. Inst. Oceanic Res. & Develop., Tokai Univ. 13: 1–15.

Pethick, J. S., 1980. Velocity surges and asymmetry in tidal channels. Estuar. coast. mar. Sci. 11: 331–345.

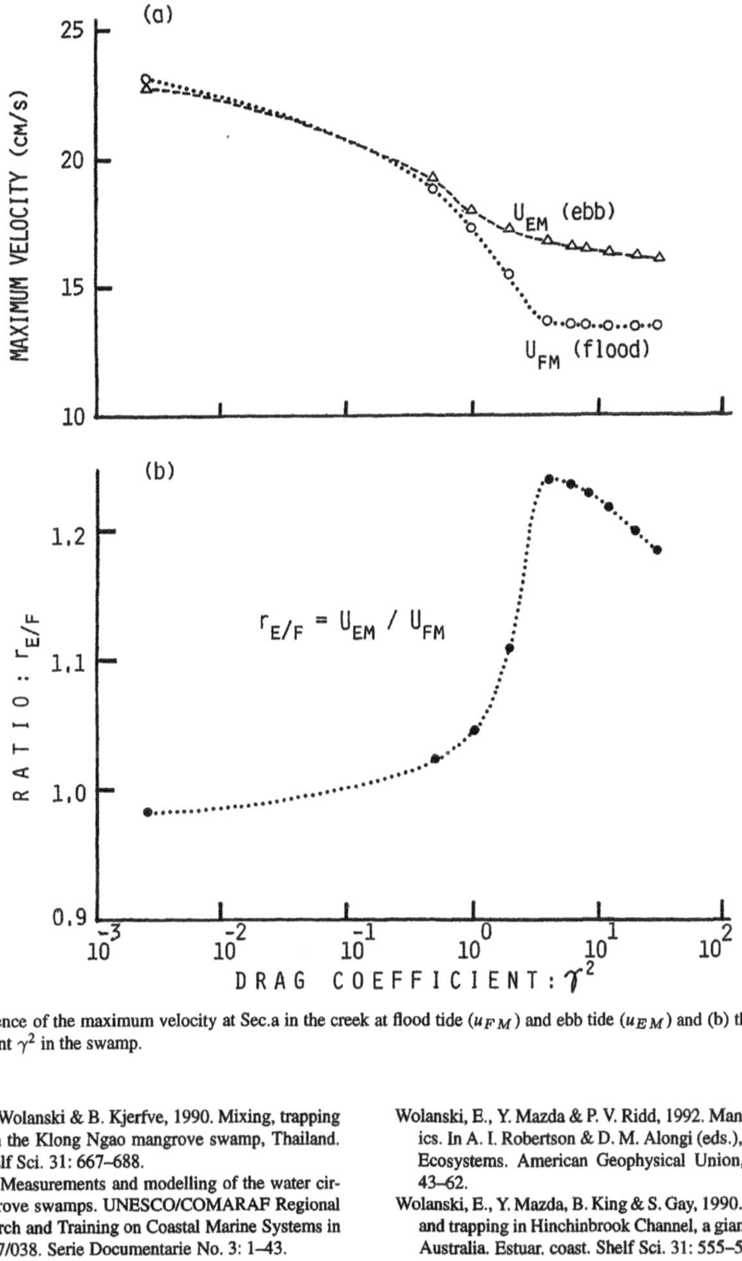

Fig. 6. (a) Dependence of the maximum velocity at Sec.a in the creek at flood tide (u_{FM}) and ebb tide (u_{EM}) and (b) the ratio u_{EM}/u_{FM} on the drag coefficient γ^2 in the swamp.

Wattayakorn, G., E. Wolanski & B. Kjerfve, 1990. Mixing, trapping and outwelling in the Klong Ngao mangrove swamp, Thailand. Estuar. coast. Shelf Sci. 31: 667–688.

Wolanski, E., 1989. Measurements and modelling of the water circulation in mangrove swamps. UNESCO/COMARAF Regional Project for Research and Training on Coastal Marine Systems in Africa — RAF/87/038. Serie Documentarie No. 3: 1–43.

Wolanski, E. & P. V. Ridd, 1986. Tidal mixing and trapping in mangrove swamps. Estuar. coast. Shelf Sci. 23: 759–771.

Wolanski, E. & Y. Mazda, 1989. Dickson Inlet: Preliminary report on the dynamical processes. Australian Institute of Marine Science, Australia.

Wolanski, E., M. Jones & J. S. Bunt, 1980. Hydrodynamics of a tidal creek – mangrove swamp system. Aust. J. mar. Freshwat. Res. 31: 431–450.

Wolanski, E., Y. Mazda & P. V. Ridd, 1992. Mangrove hydrodynamics. In A. I. Robertson & D. M. Alongi (eds.), Tropical Mangrove Ecosystems. American Geophysical Union, Washington, DC: 43–62.

Wolanski, E., Y. Mazda, B. King & S. Gay, 1990. Dynamics, flushing and trapping in Hinchinbrook Channel, a giant mangrove swamp, Australia. Estuar. coast. Shelf Sci. 31: 555–579.

Woodroffe, C. D., 1985. Studies of mangrove basin, Tuff Crater, New Zealand: II. Comparison of volumetric and velocity-area methods of estimating tidal flux. Estuar. coast. Shelf Sci. 20: 431–445.

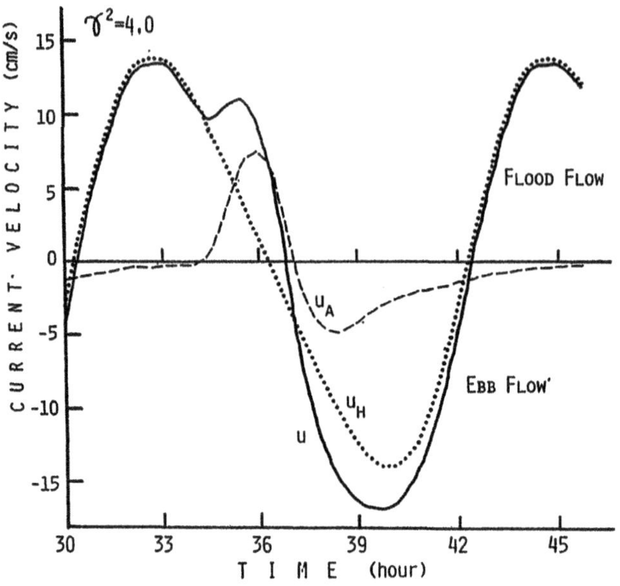

Fig. 7. Time series plot of u_H and u_A (defined by Eq. 2), and u at Sec.a for $\gamma^2 = 4.0$ in the swamp.

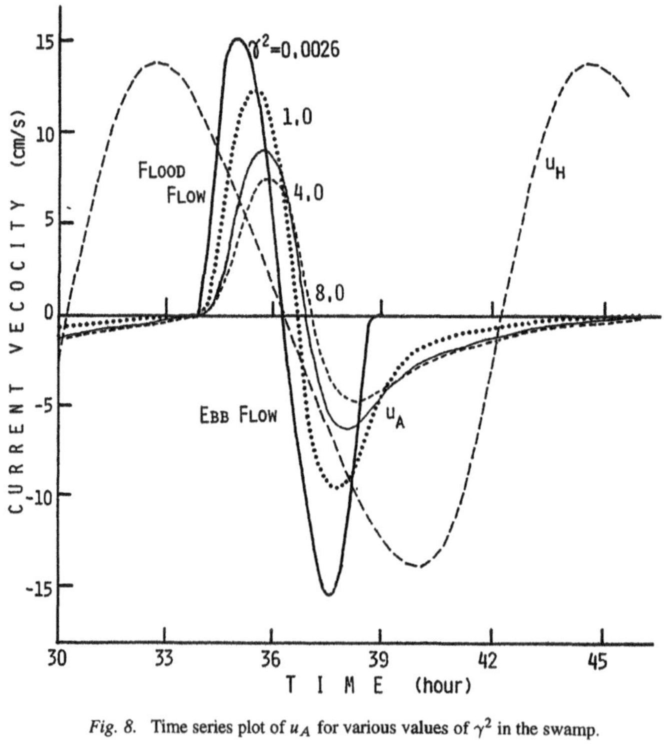

Fig. 8. Time series plot of u_A for various values of γ^2 in the swamp.

Hydrobiologia **295**: 59–65, 1995.
Y. S. Wong & N. F. Y. Tam (eds), Asia-Pacific Symposium on Mangrove Ecosystems.

Ecogeographic variation in *Kandelia candel* from Brunei, Hong Kong and Thailand

Gordon S. Maxwell
Department of Botany, The University of Hong Kong, Pokfulam Road, Hong Kong; (Present address: School of Science & Technology, Open University of Hong Kong, 11/F Trade Department Tower, 700 Nathan Road, Mongkok, Hong Kong)

Key words: Kandelia candel, ecotypes, leaf, propagule attributes, chill tolerance

Abstract

Ecogeographic variation in the widely dispersed but relatively neglected mangrove *Kandelia candel* is examined and described in the geographically isolated populations of this species from Brunei (North Borneo), Hong Kong and Thailand. Morphological attributes of leaf and propagules are compared together with some observations on differential chill tolerance in transplants from Brunei and Thailand growing alongside the wild populations of Hong Kong. Significant differences indicative of ecotypicity were obtained in terms of leaf length and size, propagule length, width and dry weight and chill tolerance of established four year old saplings.

Introduction

Kandelia candel (L.) Druce is a true mangrove (Aksornkoae *et al.*, 1992) species with an interesting geography. It occupies the northern limit of global mangrove distribution: at 31 °23′N, a location at the southern end of Kyu shu Island, Japan (Nakasuga *et al.*, 1974; Chapman, 1975; Hosokawa *et al.*, 1977; Nakagoshi & Nehira, 1986; Sakai & Larche, 1987). Compared to the well known members of the Rhizophoraceae such as *Rhizophora* spp. and *Bruguiera* spp., *Kandelia candel* (a monotypic genus) is relatively poorly known. This may be due to its unusually restricted southern limit for mangrove Rhizophoraceae; a range that appears to be confined to northeast Sumatra and northern Borneo (Tomlinson, 1986). The known distribution of *Kandelia* includes the Ganges Delta (Tomlinson, 1986), western India (Joshi & Bhosale, 1982; Kotmire & Bhosale, 1985; Mulika & Bhosale, 1989). Burma (Davis, 1964), through southeast Asia to south China (Lin & Wei, 1983), the Ryukyu Island and southern Japan (Nakasuga *et al.*, 1974), Taiwan (Li *et al.*, 1975) and Hong Kong (Hodgkiss, 1986).

While *Kandelia* features quite strongly in the mangrove literature of Japan (Nishihira & Urashi, 1976; Walker, 1976; Urashi *et al.*, 1986) and China (Lin & Wei, 1983, Chen *et al.*, 1985 & Gao, 1988), its biogeographic importance in mangrove distribution studies appears to be under emphasised in the broader literature. This may be because relatively few ecological studies of this species have been made. The assessment that *Kandelia* is nowhere abundant (Tomlinson, 1986) would reflect this situation. Recently, it has been reported that *Kandelia* is abundant in the Mai Po marshlands of sub-tropical Hong Kong (Lee, 1989) and locally abundant as a distinctive river edge belt on the Tutong of equatorial Brunei (Maxwell, 1989). The current generalisation by Duke (1992) that the cold tolerance of *Avicennia marina* was seen in no other mangrove species in the world further illustrates that *Kandelia* has been neglected in ecogeographic studies. *Kandelia* is cold tolerant (Sakai, 1974; Nakagoshi & Nehira, 1986) but, to date, such tolerance has been assigned to the *Kandelia* populations of the Japanese islands only. Unlike *Avicennia marina*, in which both tropical and cold tolerant varieties have been described (Duke, 1991) no such information has so far appeared for *Kandelia*.

This study reports on evidence for ecotypicity in ecogeographically isolated populations of *Kandelia* in Brunei, Hong Kong and Thailand. These studies focused on morphological variations in leaf and

Table 1. Morphological attributes used in this study of *Kandelia candel* populations in Brunei, Hong Kong and Thailand.

	Attribute	Description or comment
(a)	Leaf attributes	
1	Length	Length of blade (cm)
2	Width	Widest blade width (cm)
3	Petiole length	In cm
4	Length/width	Ratio of length width; leaf narrowness co-efficient
5	Length × width	Multiple of length and width/2: leaf area (cm^2)
6	General form	Shape descriptors
(b)	Dropper attributes	
7	Length	Length of mature dropper from 'calyx cap' to hypocotyl tip (cm)
8	Width	Maximum hypocotyl width (cm)
9	Dry weights	Oven dry (g)
(c)	Tree attributes	
10	Height	Actual measurements &/or clinometer, or EDM*) In m
11	Dbh	Diameter at breast height or maximum measurable stem diameter; in cm

* EDn = Theodolite with co-axial electronic distance measure

propagule material from these countries. In addition, a study of ecotypic tendencies in transplants from Brunei and Thailand growing in Hong Kong was also made along with some opportunistic observations of differential chill tolerance in these transplants.

Materials and methods

Study sites

There were four study sites, one in each of Brunei and Thailand and two in Hong Kong. The Brunei site was located on the Tutong River (latitude 04 °47′N; longitude 114 °36′E) and consisted of a continuous belt (2 m to 6 m wide and 1.26 km in length) of *Kandelia* trees in pristine condition occupying a pioneer position in front of a mixed mangrove forest dominated by *Rhizophora* spp. and *Avicennia alba*. In Hong Kong, the sites were located at Deep Bay (latitude 22 °29′N; longitude 114 °02′E) and Chek Keng (latitude 22 °25′N; longitude 114 °21′E). *Kandelia* at Deep Bay occurred in dense stands of small trees (*ca* 2.5 m tall) growing in soft mud and mildly saline estuarine waters while those at Chek Keng were squat, dwarf shrubs (*ca* 0.4 m tall) growing on a stony substratum exposed to saline oceanic waters. The Thai site was located on the La-Un River near Ranong in southern Thailand (latitude 10 °11′N; longitude 98 °43′E). Here *Kandelia* grew as a discontinuous riverine belt often in association with *Aegiceras corniculatum*.

Morphological variation

Morphological attributes used are shown in Table 1. They are mainly numeric and are similar to those used by Duke (1990) for the genus *Avicennia* in Australasia. Mature, undamaged (no overt insect herbivory and growth deformities) and uninfected (by fungi) leaves were collected by hand from upper, mid and lower canopy positions on reproductively mature trees in each of the three countries. Leaf maturity was gauged by colour, thickness, condition and the ease at which leaf abscission could be induced by a gentle touch of the petiole. Likewise mature, undamaged propagules were either collected from the forest floor or selected by hand inspection from parent trees. Mature fruit displayed a reddish hue and easily fell following a gentle shaking of a supporting shoot. The parameters of leaf and propagule length and width were measured with Vanier Kanon hand held calipers (pat. No. 946933). Leaf area was measured using a Li-cor Model 3100 meter or approximated using the numeric method of multiplying length × width/2 (Duke, 1990). Wet and dry weights were obtained using Sartorius electronic balances. Wet material was surface dried with tissue paper before weighing. For dry weight determinations, material was oven dried at 80 °C for 3–5 days.

Transplants

Thirty mature propagules from Brunei (Tutong), Hong Kong (Tsim Bei Tsui) and Thailand (La-Un River) were collected fresh and taken to the green house at Kadorrie Agricultural Research Centre, Hong Kong and planted in plastic buckets to *ca* one third of hypocotyl length in fine sand, saturated with harbour seawater (salinity 21 g kg^{-1}). The thirty propagules used from all sources were within ±1.5 cm of the mean lengths of propagule size established from the large propagule samples size on which Table 2 was based. These are described below. Thus these propagules and the six established seedlings on which the

Table 2. Mean measurements with standard error and ranges (in parenthesis below mean) of numeric morphological attributes in populations of *Kandelia candel* from Brunei, Hong Kong and Thailand. Attributes 1–5 and 7–8 in cm; attributes 9 and 10 in g. The samples sizes on which the means are given in the Methods section. The p values on the 1-way ANOVA appear in the Discussion and Results. Attributes are defined in Table 1.

Attribute		*K. candel* populations			
		Brunei (Tutong R.)	Hong Kong Chek Keng	(Mai Po) Deep Bay	Thailand La-Un
(a)	Leaf				
1	Length L	11.8±0.124 (10.00-15 00)	6.69±0.11	9.11±0.131 (7.6-13.6)	15.00±0.11
2	Width W	3.48±0.032 (2.8-4.2)	2.80±0.06	3.9±0.064 (2.9-5.5)	6.23±0.06
3	Petiole length	1.01±0.04	0.71±0.01	1.20±0.016 (0.8-1.5)	1.29±0.03
4	L/W (2.61-5.07)	3.41±0.44	2.40±0.039 (1.97-3.02)	2.45±0.024 (2.06-3.01)	2.63±0.02
5	LxW	20.65±0.332 (12.76-39.00)	9.56±0.341 (2.97-13.80)	17.89±0.570 (11.4-37.4)	50.61±0.692 (33.8-65.36)
6	General form	oblong-elliptic with some marginal curling	oblong-elliptic	oblong-elliptic	Broadly obling-elliptic
(b)	Propagule				
7	Length	31.05±0.30 (18.9-39.5)	19.32±0.236 (10.9-27.0)	22.10±0.291 (14.8-27.95)	46.89±0.602 (40.2-53.6)
8	Width	0.95±0.006 (0.7-1.09)	1.09±0.114 (0.7-1.5)	1.30±0.009 (0.99-1.49)	1.32±0.012 (1.12-1.42)
9	Wet Weight	15.67±0.399 (8.22-20.77)	9.83±0.265 (5.17-15.64)	14.72±0.288 (5.75-21.09)	34.37±1.193 (22.15-47.85)
10	Dry Weight	6.76±0.13 (3.8-10.0)	3.26±0.251 (1.96-5.57)	7.19±0.168 (3.99-10.63)	12.87±1.465 (8.41-13.40)

Table 3 A. Comparative growth (stem height, leaf numbers and size) of Bruneian, Hong Kong (Tsim Bei Tsui) and Thai transplants of *Kandelia candel* at Mai Po after six and twenty-three months. Stem height and leaf size as length by width given in cm; means with ± standard deviation.

	Brunei			Hong Kong (TBT)			Thailand		
Date (growth time)	Stem height (cm)	No.of leaves per plant	leaf size	Stem height (cm)	No. of leaves per plant	leaf size	Stem height (cm)	No. of leaves per plant	leaf size
1.12.89 (six months)	25.25±3.2	21.5±8.5	11.13±1.51 by 2.9±0.37	22.36± 6.3	27.0±9.4	8.76±1.18 by 3.53±0.78	39.2±8.3	11.5±2.69	12.52±1.13 by 4.73±0.76
30.4.91 (twenty-three months)	32.25±8.75	93.5± 5.5	13.2±0.17 by 3.7±0.18	51.25± 19.74	73.5± 20.32	10.6±0.56 by 3.75±0.61	107.25± 39.94	114.0± 26.48	15.0±0.52 by 5.43±0.19

Table 3 B. One way ANOVAS of comparative growth parameters (stem height, leaf numbers and size) in Bruneian, Hong Kong and Thai transplants of *Kandelia candel* at Mai Po after six and twenty-three months December 1989 and April 1991 respectively).

Source of variation (growth parameter)	df	Mean square	*F*	*P*	Significance
A. After six months					
(1) Stem height	17	526.14	12.28	0.0007	***
(2) No leaves per plant	17	426.44	6.62	0.0087	***
(3) Leaf size	17	23.33	13.14	0.005	***
B. After twenty-three months					
(1) Stem height	11	5514.82	7.06	0.0143	**
(2) No leaves per plant	11	2447.18	4.24	0.0503	*
(3) Leaf size	11	24.08	89.02	<0.0001	***

* Marginally significant; ** significant; *** very significant or better

transplant studies were based were deemed representative. Propagules from Brunei and Thailand were air-transported inside plastic bags containing thick paper towels impregnated with seawater. After a four week green house cultivation, six seedlings from each of the three countries were transplanted to a clear, shallow pond edge site near local (Hong Kong) *Kandelia* and *Avicennia marina* trees at Mai Po, Deep Bay, Hong Kong. The number six was a pragmatic response to the planned comparative growth monitoring programme, which became tedious and for which limited travelling and technician time was available. Another restriction was limited planting space available within the Nature Reserve. This site was *ca* 10 × 3 m in size. Although the various droppers were distinctive (Table 2) they were planted in various patterns within a distance of *ca* 30 cm between individual seedlings: droppers from Brunei described a six, while those from Hong Kong a five, and those from Thailand a southern cross pattern. The substratum at this site was a non-anoxic, uniform silty-clay. Growth was monitored in two phases over a four year period. Phase I started on 1 June 1989 and ran until 30 April, 1991 during which time the following growth parameters were recorded at approximately monthly intervals: stem height, length of lateral branches, number of leaves, leaf size, number of plants alive and flowering. During phase 2 from 28 June 1991 to 28 June 1993, the measurements were simplified to assessments of general condition and ability to produce flowers. During the winters of 1990/91, 1991/92 and 1992/93 a special study was made of differential chill tolerance, the monitoring of which focused on foliage dieback, defoliation and tree death. This study involves comparative work with tropical transplants of *Avicennia marina* as well as *Kandelia candel* ecotypes and is the subject of a separate paper currently under preparation. It is not therefore described in detail here but is briefly referred to in the discussion.

Statistical analysis and sample sizes

Attributes 1 to 3, 5, 7, 8 and 10 (Table 1) were analysed using a one-way ANOVA to establish whether the means for the various attributes quantified in Table 2 differed significantly from one another. In addition, post-tests employing the Bonferroni correction were used to handle multiple comparisons (Brunei verses Chek Keng etc.). Transplants were compared in terms of sterm height, number of leaves per plant and leaf size after six and twenty-three months. These parameters were subjected to one-way ANOVA. Calculations were carried out using the IBM compatible GraphPad InStat biostatistical programme.

The leaf sample sizes from Brunei, Chek Keng, Mai Po and La-Un on which the means in Table 2 were based are 105, 59, 84 and 100 respectively. Those of propagule length and width were (in the same order) 155, 145, 100 and 35. For propagule wet weight the respective numbers were: 56, 72, 100 and 27. The corresponding numbers for dry weight were: 56, 15, 71 and 27. In Table 3b, the comparisons made after six months were based on 18 plants, six from each of the three countries represented. Thus the df value is 17, one below the maximum. After twenty-three months, some trees had died and were lost from the sample, hence the df of 11.

Results and discussion

In terms of general leaf form, the three ecotypes of *Kandelia* showed a strong tendency towards a uniform oblong-elliptic shape (attribute 6 in Table 2 and Fig. 1). However, a one-way ANOVA showed that all numeric leaf and propagule differences between the four population attribute means shown in Table 2 were significant ($P = 0.0001$). When subjected to the more statistically stringent Bonferroni corrections to the ANOVA *P* values, all leaf attributes except petiole length in the Mai Po verses La-Un comparison, retained their statistical status. In terms of propagule length, interpopulation differences remained strongly significant following the Bonferroni corrections. With respect to propagule width, the differences between the main populations (Brunei, Mai Po and La-Un) likewise, remained strongly significant. The Chek Keng population was considered to reflect more the conditions of a special habitat (stony substratum and saline, oceanic waters) than the biological potential of *Kandelia* in Hong Kong. Such trends towards smallness (height, leaf size) in very saline and or stony sites have been reported elsewhere, for example in the neotropics (Lopez-Portillo & Ezcurra, 1989; Soto, 1988).

The long, slender propagules from Brunei had a similar dry weight to those from Mai Po which were short and thick: this observation is reflected in the insignificant Bonferroni *P* values ($P = >0.05$) for this attribute.

Ecogeographic differences between the four populations and especially those from Brunei, Hong Kong's Mai Po and Thailand were clearly evident (in the ten numeric attributes studies) and point towards a measure of regional ecotypicity in *Kandelia*. These differences occur in representatives of both actual and transplanted populations (Table 2, 3a and 3b). The differences include important leaf and propagule attributes, as well as tree height.

A gradation in leaf size (length, width and area) from the large Thai leaves through those from Brunei to those of Hong Kong is noted. In this trend the Thai population was outstanding with the size differences between the Bruneian and Hong Kong populations being less marked (Fig. 1). This picture, may also be seen, though less clearly, in petiole length (attribute 3 Table 2).

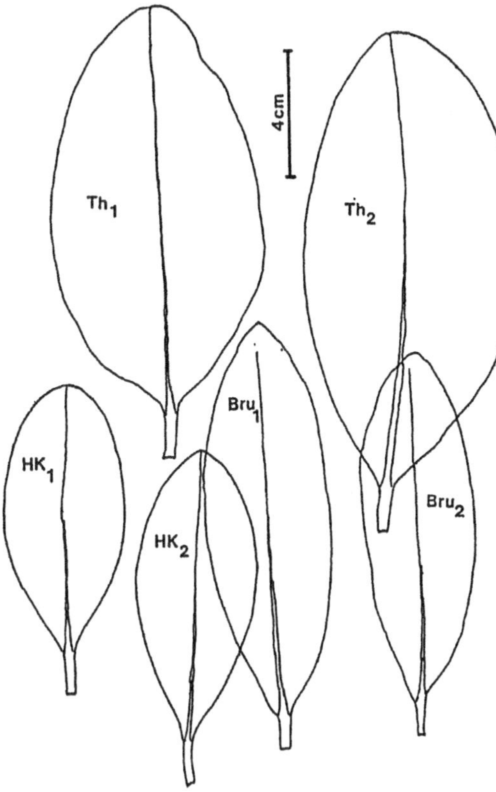

Fig. 1. Variation in leaf size of *Kandelia candel* transplants from Brunei (Bru), Hong Kong (HK) and Thailand (Th) growing at the Mai Po transplant site in Hong Kong.

In terms of propagule attributes several points are noteworthy. With the parameter of length, the trend noted for length width and area, was repeated: the Thai dropper was over twice as long as those from Hong Kong and over one third longer than those from Brunei.

With width (attribute 8 in Table 2) Hong Kong propagules from Mai Po were very similar to those from Thailand and the differences between the Bruneian and Chek Keng Hong Kong populations were minor and insignificant ($P = >0.05$). However, in terms of wet and dry weights, the Thai propagules were as expected from the length and width parameters, over twice as heavy as were those from both Brunei and Hong Kong. In summary, the propagules from Bruneian populations of *Kandelia* were long, slender and relatively light, while those from Thailand were very long and comparatively wide and heavy. In contrast, the Hong Kong propagules were typically short, but also comparatively wide and heavy

In some outstanding members of Rhizophoraceae dropper size mirrors tree size This is especially so with the sometimes massive (20 m + tall) *Rhizophora apiculata* and *R mucronata* trees of the Indo-Malay region, which are also noted for their very large propagules (droppers) (Aksornkoae *et al*, 1992) However, this situation was not clearly evident in the *Kandelia* populations featured in this study The tallest trees were found in Brunei's Tutong River where trees of 5 6 m were *ca* 1 7 m and 2 7 m taller than those from Thailand and Hong Kong populations respectively Interestingly, with the attribute of stem diameter, strong similarities were recorded in all populations except those from Chek Keng The largest trees, in terms of stem diameter, found in Brunei, Hong Kong and Thailand had diameter at breast height (dbh) values of between 10 3 and 10 8 cm It is possible that the stature of *Kandelia* is genetically fixed as that of a small tree, a descriptive title often assigned to *Kandelia* in the review literature of mangrove trees (for example, Chapman, 1975, Tomlinson, 1986, Aksornkoae *et al*, 1992) It is interesting that propagule sizes are relatively large compared to tree size, a situation that pertains even with dwarf *Kandelia* trees (2,8 m or less) of Hong Kong This observation is especially interesting in light of the generalisation that mangroves growing near their climatic limits are characterised by zero reproductive or low reproductive success (Duke, 1992)

Kandelia in Hong Kong copes with a distinctive seasonality (Morton & Morton, 1983) with some cold winters (>5 days around 2–5 °C) (Maxwell, 1993), yet continues to produce propagules in good number and notable size (including wet and dry weight) In evolutionary terms, it appears that *Kandelia* trees, despite their smaller stature, continue to allocate adequate biological resources to propagule formation However, Thai and Brunei transplants while capable of growing in Hong Kong (Table 3a) and flowering, they did not produce fruit and after the severe Hong Kong winter of December 1992/January 1993 only 33% of both Bruneian and Thai transplants survived (Maxwell, 1993) This finding could be considered in two ways Firstly, it tends to support the Duke generalisation of zero reproductive success at the climatic limits of a mangrove species Secondly, taken collectively with the leaf and propagule attributes outlined above, it clearly demonstrates that some ecotypicity exists in ecogeographically isolated Bruneian, Hong Kong and Thai populations of *Kandelia*

Conclusion

Ecogeographic variation in *Kandelia candel* populations from Brunei (North Borneo), Hong Kong and Thailand points to a measure of regional ecotypicity The differences were especially obvious in terms of the leaf attributes of length, width, size and the propagule attributes of length width and weight (both wet and dry) In addition, clear differences were demonstrated in terms of chill tolerance the natural populations of *Kandelia candel* growing at Mai Po in Hong Kong were able to survive an unusually cold winter chill regime of around 1–2 °C on four occasions in January 1993, a tolerance in clear contrast to the tropical transplants from Brunei and Thailand growing at the same site In the latter two only 33% survived

Acknowledgments

I thank Dr John Hodgkiss and Dr Richard Corlett of the University of Hong Kong for sustained advise and help during the five year study period associated with these and related studies I also thank Dr Sanit Aksornkoae and Mr Sonjai Havanond for help with my work on *Kandelia* in Thailand

References

Aksornkoae, S, G S Maxwell, S Havanond & S Panichsuko, 1992 Plants in Mangroves Pub Chalongrat Co Ltd, 99 Tiemruammitr Rd, Huaykhwang, Bangkok 10310, Thailand ISBN 974 89011 7-3 (120 pp)

Chapman, V J, 1975 Mangrove Vegetation Cramer, Lehre

Chen, S, Z Liang & Y Deng, 1985 Guangdong East Mangrove Forest Acta Phytoecologica et Geobotanica Sinica 9 59–63

Davis, J H, 1964 The Forests of Burma Sarracenia 8 1–41

Duke, N C, 1990 Morphological variation in the mangrove genus Avicennia in Australasia systematic and ecological considerations Aust Sys Bot 3 221–39

Duke, N C, 1991 A systematic revision of the mangrove genus Avicennia in Australasia Aust Syst Bot 4 299–324

Duke, N C, 1992 Mangrove floristics and Biogeography, Chap 4 (Pp 63–100) In Tropical Mangrove Ecosystems (Vol 41) A Robertson and D Alongi (eds), American Geophysical Union, Wash D C

Gao, Y Z, 1988 The Chinese Mangroves J Wuhan Botanical Research 6 65–76

Hodgkiss, I J, 1986 Aspects of mangrove ecology in Hong Kong Memoirs of the Hong Kong Natural History Society No 17 107–116

Hosakawa, T, H Tagawa & V J Chapman, 1977 Mangals of Micronesia, Taiwan, Japan, the Philippines and Oceania Chap 14 In Ecosystems of The World, Vol 1 Wet coastal ecosystems Elsevier Scientific, Amsterdam

Joshi, G V & L J Bhosale, 1982 Estuarine ecosystem of India Chapter 2 In Sen, D N & Rajpurohit K S (eds), Tasks for Vegetation Science Vol 2 Dr W Junk Publishers, The Hague, pp 21–33

Kotmire, S Y & L J Bhosale, 1985 A study of mangrove vegetation along Deogad Estuary In Krishna murthy V (ed) marine plants, their biology, chemistry and utilisation, Proc All India Symp mar Plants, Dona Paula, Goa, India, pp 225–230

Lee, S Y, 1989 Litter production and turnover of the mangrove *Kandelia candel* (L) Druce in a Hong Kong tidal Shrimp pond Estuar coast Shelf Sci 29 75–87

Li, H-L , T-S Liu, H Tseng-Chieng, K Tetsuo & C E DeVol, 1975 Flora of Taiwan pp 410–411 & 872–873 Pub Epoch Pub Co Ltd , Taipei, R O C

Lin, P & X-M Wei, 1983 Ecological notes on the mangroves of Fujian, China In Teas, H J (ed), Tasks for Vegetation Science, Vol 8 Dr W Junk Publishers, The Hague 31–36

Lopez-Portillo, J & E Ezcurra, 1989 Response of three mangroves to salinity in two geoforms Funct Ecol 3 355–362

Maxwell, G S , 1989 Kandelia candel & Avicennia marina in Thailand, Hong Kong and Brunei Part I Aspects of the 1989 work Report for Royal Forestry Dept , Bangkhen, Bangkok, Thailand, 22 pp

Maxwell, G S , 1993 Ecogeographic Studies of Avicennia marina and Kandelia candel in Brunei, Hong Kong & Thailand Ph D thesis (unpublished) Botany Department, University of Hong Kong, 394 pp

Morton, B & J Morton, 1983 The Sea Shore Ecology of the Hong Kong Hong Kong University Press

Mulik, N G & L J Bhosale, 1989 Flowering phenology of the west coast of maharshtra (India) J Bombay Nat Hist Soc 86 355–359

Nakagoshi, N & K Nehira, 1986 Growth and mortality of mangrove seedlings transplanted to Hiroshima Hikobia 9 439–449

Nakasuga, T , H Oyama & M Haruki, 1974 Studies of the mangrove community I The distribution of the mangrove community in Japan Jap J Ecol 24 237–246

Nishihira, M & M Urasaki, 1976 Production, settlement and mortality of seedlings A marina, Kandelia candel (L) in Okinawa Abstracts of symposia, Intl Sym on Eco and Man of Trop Shallow Water Communities, Jakarta, July

Sakai, A 1974 Freezing resistance of evergreen and deciduous broad-leaf trees growing on Yakushima island Jap J Ecol 24 35–42

Sakai, A & W Larch, 1987 Frost Survival of Plants Pub Springer-Verlag Berlin

Soto, R , 1988 Geometry, biomass allocation and leaf life-span of Avicennia germinans (L) (Avicenniaceae) along a salinity gradient in Salinas, Puntarenas, Costa Rica Bev biol trop 36 309–324

Tomlinson, P B , 1986 The Botany of Mangroves Camb U P

Urasaki, M , K Nehira & N Nakagoshi, 1986 Dispersal and settlement properties of Kandelia candel (Rhizophoraceae) propagules Plant Species Biol 1 19–26

Walker, E H 1976 Flora of Okinawa and the Southern Ryukyu Islands Smithsonian Institution Press, Washington D C 763–764 & 895

Hydrobiologia **295**: 67–74, 1995.
Y. S. Wong & N. F. Y. Tam (eds), Asia-Pacific Symposium on Mangrove Ecosystems.

Microgeographic genetic structure of the fiddler crab, *Uca arcuata* De Haan (Ocypodidae) in Taiwan

Shong Huang & Jin-Taur Shih
Department of Biology, National Taiwan Normal University, Taipei, 117 Taiwan, Republic of China

Abstract

The genetic structure of *Uca arcuata* in Tanshui mangrove swamp of northern Taiwan was examined. Using as genetic markers, isozymes identified through starch gel electrophoresis indicate that there was moderate genetic differentiation among subpopulations within the population (F_{ST} = .085). Gene flow appeared high when estimated indirectly (Nm = 2.69). The results suggest that the patterns of genetic structure of *Uca arcuata* were influenced by the interaction of local selection due to microhabitat differences and gene flow among fiddler crab colonies in the mangrove swamp.

Introduction

The fiddler crab, *Uca arcuata* (De Haan, 1835) is one of the dominant species of the macrofauna in the estuaries in Taiwan (Crane, 1975; Huang *et al.*, 1989, 1992; Fukui *et al.*, 1989). *Uca arcuata* is distributed in the estuaries either with or without mangrove forest along the west coast of Taiwan. It is a species appropriate for population genetic study because of its broad geographic range (Crane, 1975). Population studies revealed that the population density of *Uca arcuata* and *U. lactea* changes according to season and location (Shih, 1990, 1992; Shih *et al.*, 1991). In addition, *Uca arcuata* is somewhat sensitive to microgeographic differences in physical conditions, because the population size of this species is large in mudflats by river banks and is small in sandy substrates in open areas which are flooded only by springtides (Shih, 1990).

Most nauplii of estuarine crabs are planktonic or interstitial (Anderson, 1982). Swimming larvae are dispersed by tidal action and then molt in a wide range of habitats which have profound effects on their subsequent developing stages (Anderson, 1982; Dittel & Epifanio, 1990; Dittel *et al.*, 1991). The final larval stage, the megalopae, are usually transported by tidal currents to the sediment and then return to the adult habitats randomly, or near to the parental stock area (Jones, 1984; Christy, 1978, 1989; O'Connor, 1991).

Studies on population genetic structure of the genus *Uca* are rare. However, an isozymes study of *Uca rapax* and *U. virens* in Florida showed that the genetic distance (D, Nei, 1978) between populations of *U. rapax* was 0.016 while that between these two species was 0.01–0.035 (Salmon & Kettler, 1987). Hedgecock *et al.* (1982) reviewed genetic variability of four species of *Uca* and found that the mean expected heterozygosity (H_E) are: 0.097 for *U. musica*; 0.028 for *U. princeps*; 0.031 for *U. speciosa* and 0.029 for *U. spinicarpa*. These findings imply that the level of genetic variation and genetic differentiation within and among *Uca* species were low.

Genetic variation can be observed when population subdivision has been affected by environmental patchiness, as areas of favorable habitat intermix with unfavorable ones (Hartl, 1987). Such environmental patchiness may be encountered by the fiddler crabs, because their habitats cover wide ranges in local topography, inundation frequency and food availability (Genoni, 1991). Genetic variation and genetic differentiation within and among populations of the fiddler crab can also be determined by the joint action of three natural forces, namely, the impact of natural selection, gene flow and genetic drift (Ehrlich & Raven, 1969; Endler, 1973; Slatkin, 1987; Wright, 1946, 1948).

This study of genetic structure of *Uca arcuata* attempts to test the hypothesis whether the level of genetic differentiation among subpopulations correlates with habitat variation.

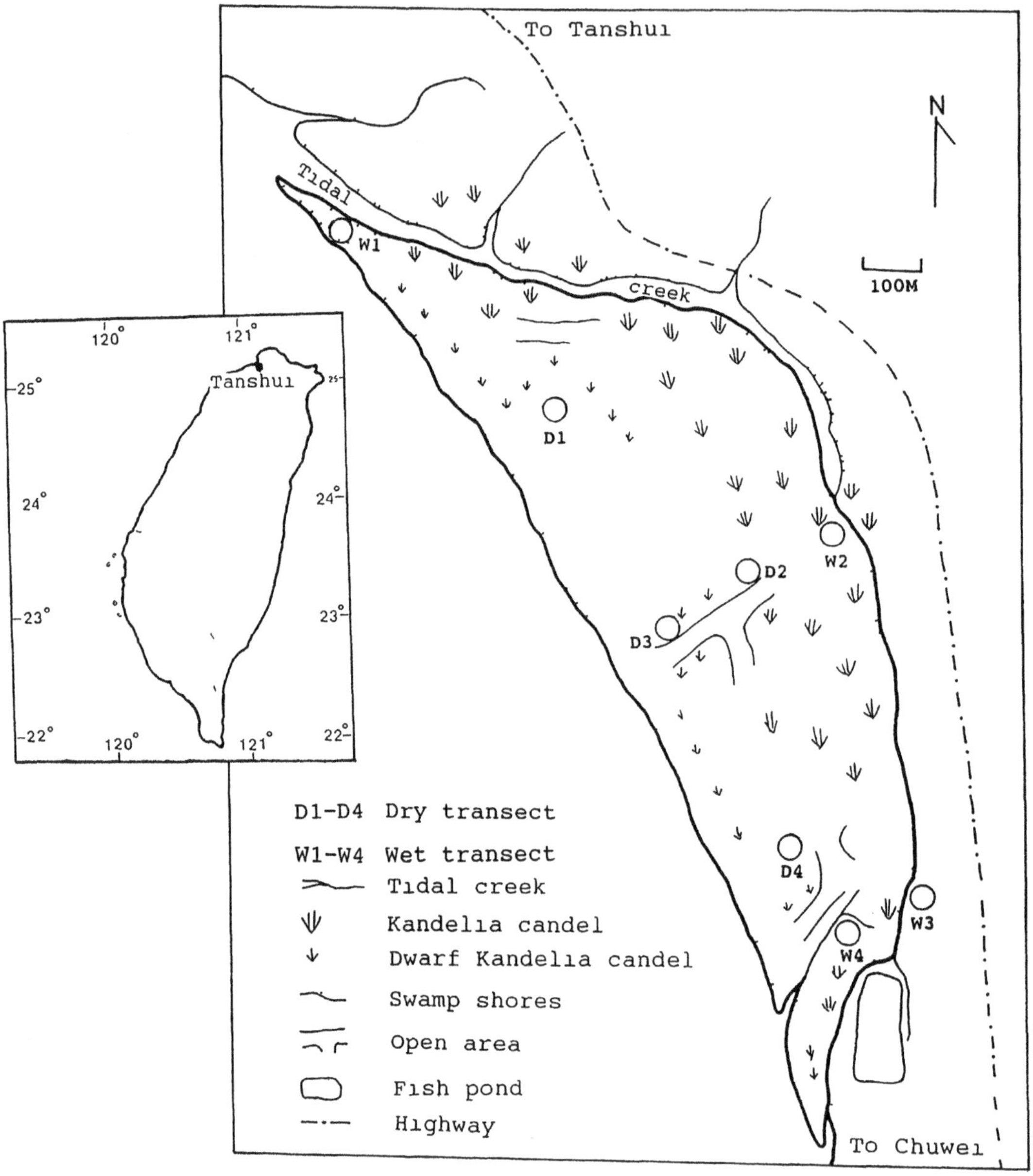

Fig 1 Sampling sites in Tanshui mangrove swamp Inset is the map of Taiwan

Materials and methods

Tanshui mangrove swamp is located at the estuary of the Tanshui River in Taipei County (25°09′N, 121°16′E, Fig. 1). This crescent shape swamp is about 70 hectares and is the largest mangrove swamp in Taiwan. This swamp is subdivided into two areas, namely, sandy riverside and tidal creeks (Chou & Bi, 1990; Shih *et al.*, 1991). The sandy riverside area almost occupies a half of the swamp at westside. Because this area is about three meters above the maximum low springs (MALS, Tidal Level 11, Jones, 1984), it is only flooded during highspring tide. Thus, deforested areas, grassland and dwarf forest, which dominated by *Kandelia candel*, are found in this region. The sandy riverside area was designated as dry transect in this study. On the contrary, the east side of this swamp is low (only two meters above MALS) and flooded most the times through the tidal creek except during neaptide. Tall *Kandelia* forest occurs along tidal creek. This area is designated as wet transect.

Eight subpopulations of male adult *Uca arcuata* were obtained from Tanshui mangrove swamp. Two transects (dry and wet), each including four subpopulations (D-1 to D-4 and W-1 to W-4), were arbitrarily chosen for sampling. Sampling sites were separated at least 100 meters apart. Soil texture was loamy sand along the dry transect while it was silt clay along the wet transect (Chou & Bi, 1990).

Thirty to fourty crabs per site were collected during the Spring of 1993. Hepatopancreas of each crab as ground in 0.1–0.2 ml of phosphate buffer (50 mM, pH 7.5 with 5% sucrose and 0.1% 2-mercaptoethanol). Crude enzymes were absorbed directly onto paper wicks (3 × 8mm), which were then stored in a freezer at −70 °C until electrophoresis. Starch gel electrophoresis and interpretation of allozyme analysis followed the methods of Pasteur & Pasteur (1988) and Kephart (1990). Enzymes were resolved in 11% starch gel (Sigma) using two buffer systems, Borate pH 8.0 and Histidine-HCl pH 7.0 (Soltis *et al.*, 1983).

A total of 6 isozyme loci were examined to be polymorphic at least in one subpopulation, including esterase (EST-1, EST-2), fluorescent esterase (FE-1, FE-2), malic enzyme (ME) and acid phosphatase (ACP). A number of isozymes were monomorphic for the hepatopancreas of *Uca arcuata* (i.e. isocitric dehydrogenase (IDH), phosphoglucomutase (PGM), phosphoglucose isomerase (PGI), leucine aminopeptidase (LAP), superoxide dismutase (SOD) and triosephosphate isomerase (TPI)).

Genetic variation within and among subpopulations, including the percentage polymorphic loci per subpopulation (P), the mean number of alleles per locus (A), mean observed and mean expected heterozygosity (H_O and H_E respectively), were estimated by Nei's formula (Nei, 1973, 1978). Genetic structure was estimated using Wright's F-statistics (Wright, 1943, 1965). These estimations were analyzed using the BIOSYS-1 program of Swofford & Selander (1989). In addition, gene flow rate (Nm) was calculated by using the equation $F_{ST} = 1/(1 + 4Nm)$ (Wright, 1931, 1943).

Results

The ten isozyme systems of this study yielded 12 loci and 25 alleles. Among all enzymes tested, PGM, IDH, SOD, PGI, TPI, LAP are found to be fixed in all *Uca arcuata* subpopulations in Tanshui mangrove swamp. The average number of alleles per locus for all subpopulations is 1.8. The highest number of alleles detected per locus is in EST-1 with five alleles in 4 of the 8 subpopulations. The mean proportion of polymorphic loci (P) within the whole population is 42.7% and may be biased upward because polymorphic loci enzymes were selected for assaying. The mean observed and mean expected heterozygosity across loci for 8 subpopulations were $H_O = 0.189$ and $H_E = 0.216$ respectively (Table 1). Thus, expected heterozygosity is greater than observed heterozygosity ($p < 0.01$).

Deviation from Hardy-Weinberg expectations within subpopulations was measured for each locus. It was found that 33 of the 48 instances (68%) showed no deviation from Hardy-Weinberg equilibrium. However, significant deficiencies were found in 5 of the 6 polymorphic loci when values for all subpopulations within Tanshui mangrove swamp were pooled together (Table 2).

Analysis of genetic structure of 8 subpopulations showed that the inbreeding coefficient (F_{IS}) values ranged from −0.340 at ME locus to 0.270 at EST-1 with a mean of 0.109 (Table 3). Five of the six loci are significantly different from zero for F_{IS} within population. Only the locus ME exhibited a negative value for its F_{IS} indicates that this locus has excess heterozygotes. Meanwhile, 4 of the 6 significant loci were positively different from Hardy-Weinberg equilibrium and indicated that there were excess homozygotes within population. Heterogeneity chi-square analyses for each locus in the overall 8 subpopulations showed similar results to the F_{IS} analysis (Table 3). These phenomena

Table 1. Genetic variability at 12 loci in all *Uca arcuata* subpopulations in Tanshui mangrove swamp. (standard errors in parentheses)

Population	Mean sample size per Locus	Mean no. of alleles per locus	Percentage of loci polymorphic(1)	Mean heterozygosity Directcount	Mean heterozygosity HdyWbg expected(2)
1. WET 1	28.8	1.7	41.7	0.144	0.187
	(0.6)	(0.2)		(0.057)	(0.066)
2. WET 2	28.0	1.8	41.7	0.218	0.227
	(0.9)	(0.3)		(0.081)	(0.083)
3. WET 3	30.4	1.8	41.7	0.233	0.245
	(1.4)	(0.4)		(0.094)	(0.089)
4. WET 4	32.3	1.8	50.0	0.228	0.221
	(0.5)	(0.3)		(0.091)	(0.077)
5. DRY 1	29.8	1.8	41.7	0.215	0.227
	(1.6)	(0.4)		(0.079)	(0.083)
6. DRY 2	26.2	1.6	41.7	0.099	0.204
	(0.8)	(0.2)		(0.046)	(0.073)
7. DRY 3	31.3	1.8	41.7	0.176	0.206
	(0.8)	(0.4)		(0.070)	(0.081)
8. DRY 4	30.9	1.8	41.7	0.202	0.216
	(1.1)	(0.3)		(0.077)	(0.080)
Overall	237.8	2.1	41.7	0.191	0.232
	(6.5)	(0.4)		(0.072)	(0.083)
Mean	29.7	1.8	42.7	0.189	0.216
	(1.9)	(0.7)		(0.047)	(0.018)

(1) A locus is considered polymorphic if the frequency of the most common allele does not exceed 0.95.
(2) Unbiased estimate (see Nei, 1978).

Table 2. Coefficients for heterozygote excess or deficiency (−) of 8 *Uca arcuata* subpopulations in Tanshui mangrove swamp.

Enzyme-Locus	Chuwei population								
	Wet-1	Wet-2	Wet-3	Wet-4	Dry-1	Dry-2	Dry-3	Dry-4	Total
EST-1	−0.554**	0.046*	−0.002	−0.498**	−0.217	−0.739**	−0.264**	−0.195	−0.217
EST-2	−1.00**	—	—	0.045	—	—	—	0.000	−0.242
FST-1	−0.029	0.053	−0.344*	−0.385*	−0.299	−0.449*	−0.442**	−0.208	−0.307**
FST-2	0.328	−0.162	−0.180	0.018	−0.125	−0.261	−0.025	−0.097	−0.175**
ME	0.075	−0.033**	0.476*	0.970*	0.409	−0.148	0.327	0.395	0.278**
ACPH	−0.577**	−0.182	−0.284	0.126	0.122	−1.00**	−0.211	−0.200	−0.332**

Data calculated using Levene's (1949) formula from BIOSYS-1 program.
* Significant at $p < 0.05$; ** significant at $p < 0.01$.

may be resulted from nonrandom mating (e.g. inbreeding) or affected by the Wahlund effect in *Uca arcuata* populations.

The degree of differentiation among subpopulations (F_{ST}) is 0.085 (Table 3). This result indicates that approximately 92% of the detected genetic variation of

Table 3. Contingency Chi-square analysis, F-statistics and amounts of gene flow at all loci in 8 *U. arcuata* subpopulations in Tanshui mangrove swamp.

Locus	Chi-square(1)	F(IS)(2)	F(IT)	F(ST)	Nm(3)
EST-1	125.575**	0.270**	0.330	0.082	2.798
EST-2	26.073*	0.233	0.259	0.034	7.103
FST-1	57.314**	0.266**	0.305	0.053	4.467
FST-2	137.311**	0.052**	0.182	0.137	1.574
ME	71.542**	−0.340**	−0.246	0.071	3.271
ACPH	39.356**	0.263**	0.325	0.084	2.726
Mean		0.109	0.185	0.085	2.691

* $p < 0.05$; ** $p < 0.01$
(1)Levene (1949); (2)Li & Horvitz 1953; (3)$F_{ST} = 1/(1 + 4Nm)$

Table 4. Matrix of genetic distance coefficients of 8 *U. arcuata* subpopulations. Nei (1978) unbiased genetic distance (D).

Population		1	2	3	4	5	6	7	8
1.	WET 1	—							
2.	WET 2	0.033	—						
3.	WET 3	0.048	0.003	—					
4.	WET 4	0.036	0.013	0.022	—				
5.	DRY 1	0.048	0.008	0.012	0.009	—			
6.	DRY 2	0.011	0.024	0.033	0.023	0.021	—		
7.	DRY 3	0.048	0.017	0.027	0.013	0.030	0.032	—	
8.	DRY 4	0.054	0.021	0.026	0.001	0.016	0.031	0.005	—

Uca arcuata resides within subpopulations. Based on Wright's (1978) suggestion, a moderate genetic differentiation of *U. arcuata* exists among subpopulations in Tanshui mangrove swamp. Levels of gene flow (Nm) ranged from 1.574 to 7.103, with a mean of 2.691 (Table 3). This result suggests that the rates of gene flow were relatively high. Mean Nei's genetic distance was 0.024 ± 0.014 (Table 4, Fig. 2). The genetic distance within and between transects revealed no significant correlations with different microhabitats (Table 5).

Discussion

Genetic variation of *Uca arcuata* in Tanshui mangrove swamp as measured at the population level is relatively high (H_T = 0.232). This value is higher than other *Uca* species such as *U. musica* (0.097); *U. princeps* (0.028); *U. speciosa* (0.031), *U. spinicarpa* (0.029) and the group mean of decapods (0.07) (Hedgecock *et al.*, 1982). Also, it is higher than the average heterozygosity of invertebrates in general (0.110, Nevo, 1978); or of other crustaceans such as the coconut crab (0.018, Lavery & Fielder, 1993), penaeid prawns (0.006–0.03, Mulley & Latter, 1980); and of the average for crustaceans (0.073, Hedgecock *et al.*, 1982). High levels of genetic variability are usually caused by abundant gene flow or by random mating within sites (Wright, 1946, 1978).

Five of six polymorphic loci have heterozygotes deviated from Hardy-Weinberg expected proportions in the total population. About 80% of those loci have heterozygote frequencies lower than the H-W genotypic expections. In addition, the expected heterozygosity in total population (H_T) is greater than the

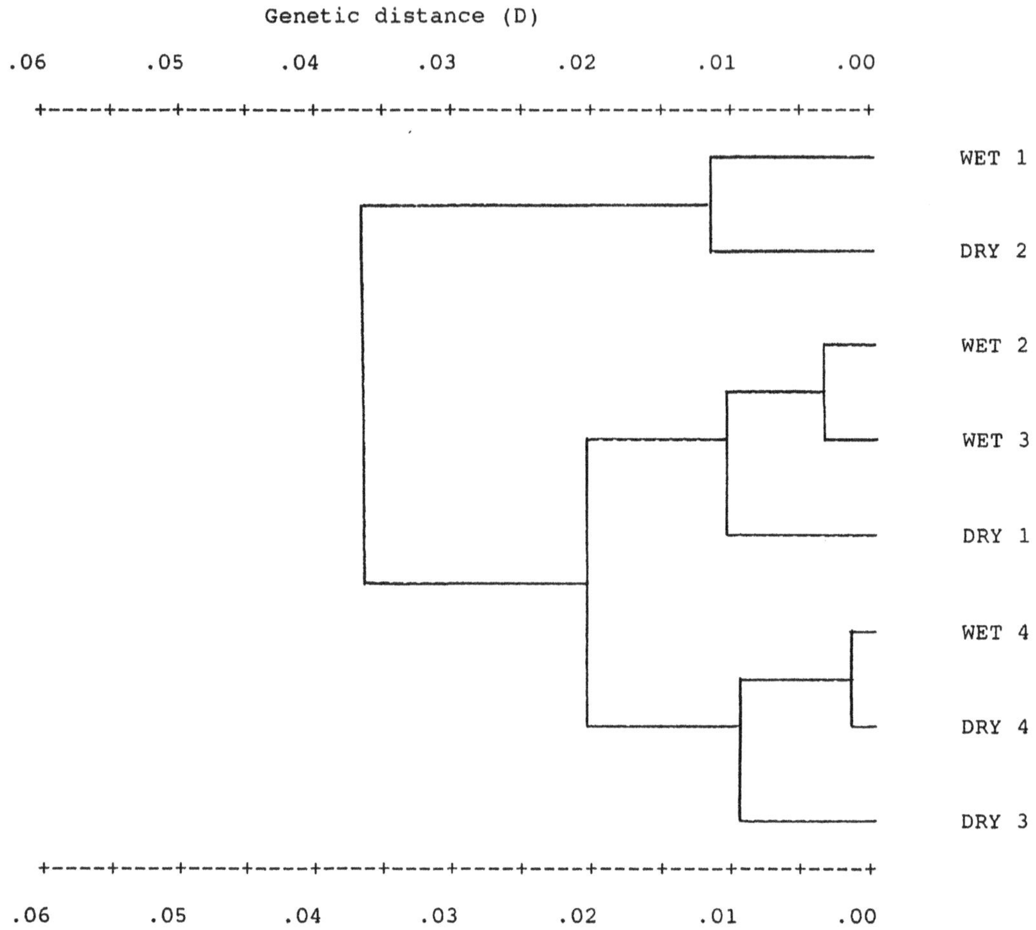

Fig. 2. UPGMA dendrogram showing the genetic relationships among 8 *Uca arcuata* subpopulations.

mean heterozygosity of subpopulations (H_S) (0.232 vs. 0.216). Therefore, significant deficiency of heterozygotes across loci of this study seems to be caused by the Wahlund effect (Li, 1955; Hartl and Clark, 1989). Our results suggest that the *Uca* population may be comprised of more than two unmixed subpopulations. The study of genetic structure of *Homarus americana* also suggested that Wahlund effect may be an important cause for the population subdivision (Hedgecock *et al.*, 1982).

The level of genetic differentiation of *Uca arcuata* in the Tanshui mangrove swamp is about 8.5% (F_{ST} = .085), which is comparable to recorded values for the horseshoe crab (*Limulus*, F_{ST} = 0.076) but lower than that of *Drosophila equinoxialis* (F_{ST} = 0.109)

Table 5. Matrix of genetic distance coefficients averaged by Transect. Coefficient is Nei's (1978) unbiased genetic distance. Range of genetic distances among populations within transect are in parentheses.

TRANSECT	No.of pops.	1	2
1. WET	4	.026 (.003–.048)	
2. DRY	4	.024 (.001–.054)	.023 (.005–.032)

(Nei, 1975). F_{ST} values from 0.05 to 0.15 are considered as indicative of moderate genetic differentiation (Wright, 1978). However, in the present study, because all the crabs were sampled from a mangrove swamp, the degree of genetic differentiation within population (F_{ST} = 0.085) is also considered to be high. Genetic differentiation in a unique community, such as the mangrove swamp, may be caused by microhabitat selection since larval settlement is influenced by two stimuli: a) the presence of adult crab and b) the sediment suitable as adult habitats (O'Connor, 1991). Further analysis of genetic differentiation between dry and wet transects by F-statistics indicates very little genetic differentiation between them (F_{ST} = 0.011). These results suggest that selection forces due to the influence of dry and/or wet habitats, are not strong enough to overcome the high rate of gene flow. In fact, free swimming crab larvae can be transported all over the estuarine habitats during the springtide (Anderson, 1982; Dittel & Epifanio, 1990; O'Connor, 1991). Thus, the genetic variability of each sampled subpopulation in this study, may be the result of the mixture of several initiate subpopulations.

In addition, gene flow, in terms of migration, can occur at different developmental stages of fiddler crab and may have different effects on the migrants. In particular, migration that occurs in the gametic or zygotic stage can cause quite different genetic consequences. Since restriction to free gene exchange of crabs between dry and wet habitats in a mangrove swamp has not yet reported, and since post settlement mortality and pre-settlement habitat selection of *Uca* species are poorly understood, it is difficult to speculate how these factors influence the genetic structure of various subpopulations. Thus, further studies on mating systems of *Uca arcuata* are important in order to evaluate their influence on genetic structure.

Except for the influence of gene flow on the genetic structure of populations, the degree of genetic differentiation in a population is largely determined by neighborhood size. If a population is small (less than 100) and separated from the main group, genetic drift as well as bottleneck effect may occur (Wright, 1969). However, neighborhood size may not affect the results of this study because the population size of *U. arcuata* is relatively large (38.8 ± 13.5 crabs m^{-2}, Shih, 1990) in Tanshui mangrove swamp. Even the male crab's proportion increases from 50 to 90% within population as the study of *U. uraguayensis* (Spivak *et al.*, 1991), the genetic drift may not cause genetic differentiation among subpopulations.

Genetic distance coefficients as well as the UPGMA dendrogram present similar patterns of subpopulational groupings, and show that genetic distances are neither correlated with geographic distance nor with dry and/or wet patterns. Therefore, allelic variation randomly distributed among different microhabitats within a mangrove swamp is most likely due to the interaction between natural selection and gene flow. Microhabitat selection pressure could be evolutionary force which contribute to the observed patterns of genetic differentiation. However, gene flow was found to be abundant among subpopulations within a mangrove swamp, thus, it reduced the impact of natural selection.

In conclusion, the high level of genetic variation within population of *Uca arcuata* is likely due to the high rates of gene flow. The moderate genetic differentiation among *Uca arcuata* subpopulations suggests that the impacts of microhabitat selection within Tanshui mangrove swamp may be important to the genetic differentiation within population. Further investigation on above mentioned factors are necessary.

References

Anderson, D. T., 1982. Embryology. In L. G. Agele (ed.), The Biology of Crustacea, Vol. 2, Embryology and Genetics. Academic Press, New York, 440.

Chou, C. H. & C. C. Bi, 1990. Dynamic distribution of nutrients and variation of environmental factors in Tamshui estuary ecosystem. Proc. Natl. Sci. Conc. B. ROC 14: 131–141.

Christy, J. H., 1978. Adaptive significance of reproductive cycles in the fiddler crab *Uca pugilator*: A hypothesis. Science 199: 453–455.

Christy, J. H., 1989. Rapid development of megalopea of the fiddler crab *Uca pugilator* reared over sediment: Implications for models of larval recruitment. Marine Ecology Progress Series 57(3): 259–266.

Crane, J., 1975. Fiddler crabs of the world (Ocypodidae: Genus *Uca*). Princeton University Press. Princeton, New Jersey.

Dittel, A. I. & C. E. Epifanio, 1990. Seasonal and tidal abundance of crab larvae in a tropical mangrove system, Gulf of Nicoya, Costa Rica.

Dittel, A. I., C. E. Epifanio & O. Lizano, 1991. Flux of crab larvae in a mangrove creek in the gulf of Nicoya Costa Rica. Estuarine Coastal and Shelf Science 32(2): 129–140.

Endler, J. A., 1973. Gene flow and population differentiation. Science 179: 243–250.

Ehrlich, P. R. & P. Raven, 1969. Differentiation of populations. Science 165: 1228–1231.

Fukui, Y., K. Wada, & C. H. Wang, 1989. Ocypodidae, Mictyridae and Grapsidae (Crustacea: Brachyura) from some coasts of Taiwan. J. Taiwan Mus. 42(1): 225–238.

Genoni, G. P., 1991. Increased buddowing by fiddler crabs Uca rapax (Smith) (Decapoda: Ocypodidae) in response to low food supply. J. Exp. Mar. Biol. Ecol. 147: 267–286.

Hartl, D. L., 1987. A primer of population genetics. 2nd ed. Sinauer Associates, Sunderland, Massachusetts.

Hartl, D. L. & A. G. Clark, 1989. Principles of population genetics. Sinauer Associates, Sunderland, Massachusetts.

Hedgecock, D., M. L. Tracey & K. Nelson, 1982. Genetics. In L. G. Abele (ed.), The Biology of Crustacea. Academia Press.

Huang, J. F., H. P. Yu & M. Takeda, 1989. Fiddler crabs (Crustacea: Decapoda: Ocypodidae) of Taiwan. Bull. Inst. Zool., Academia Sinica, R.O.C. 28: 191–209.

Huang, J. F., H. P. Yu, & M. Takeda, 1992. A review of the Ocypodid and Mictyrid crabs (Crustacea: Decapoda: Brachyura) in Taiwan. Bull. Inst. Zool., Academia Sinica, R.O.C. 31: 141–161.

Jones, D. A., 1984. Crabs of the mangal ecosystem. In Por, F. D. & I. Dor (eds.), Hydrobiology of the mangal, Vol. 20. Dr W. Junk Publishers, The Hague. 89–109.

Kephart, S. R., 1990. Starch gel electrophoresis of plant isozymes: a comparative analysis of techniques. Amer. J. Bot. 77: 693–712.

Lavery, S. & D. R. Fielder, 1993. Low allozyme variation in the coconut crab Birgus latro. Comp. Biochem. Physiol. 104B:353–359.

Levene, H., 1949. On a matching problem arising in genetics. Ann. Math. Stat. 20: 91–94.

Li, C. C. & D. G. Horvitz, 1953. Some methods of estimating the inbreeding coefficient. Amer J. Hum. Genet. 5: 107–117.

Li, C. C., 1955. Population genetics. The University of Chicago Press.

Mulley, J. C. & B. D. H. Latter, 1980. Genetic variation and evolutionary relationships within a group of thirteen species of penaeid prawns. Evolution, 34:904–916.

Nei, M., 1973. Analysis of gene diversity in subdivided populations. Proc. Nat. Acad. Sci. USA. 70: 3321–3323.

Nei, M., 1975. Molecular population genetics and evolution. Elsevier, New York.

Nei, M., 1978. Estimation of heterozygosity and genetic distance from a small number of individuals. Genetics 89: 583–590.

Nevo, E., 1978. Genetic variation in natural populations: Patterns and theory. Theoret. Pop. Biol. 13: 121–177.

O'Connor, N.J., 1991. Flexibility in timing of the metamorphic molt by fiddler crab megalopea *Uca pugilator*. Marine Ecology Progress Series 68(3): 243–247.

Pasteur, N. & G. Pasteur, 1988. Practical isozyme genetics. Ellis Horwood Limited.

Salmon, M., & M. K. Kettler, 1987. The importance of behavioral and biochemical differences between fiddler crab taxa, wtih special reference to Uca rapax and U. virens (Salmon and Atsaides). Contributions in Marine Science. 30: 63–76.

Shih, J. T., 1990. Annual estimated densities of Uca arcuata (De Haan, 1835) on Tan-Shui mangrove swamp at Taiwan. Biol. Bull. NTNU 25:1–11. (in Chinese with English abstract).

Shih, J. T., K. Y. Lue & C.H. Wang, 1991. Crab fauna and the activities of ten crab species in Tanshui mangrove swamp of Taiwan. Ann. Taiwan Museum 34: 121–140. (in Chinese with English abstract).

Shih, J.T., 1992. The population variation of *Uca lactea* (De Haan, 1835) in Tanshui mangrove swamp of Taiwan. Ann. Taiwan Museum 35: 67–78. (in Chinese with English abstract).

Slatkin, M., 1987. Gene flow and the geographic structure of natural population. Science 239: 787–792.

Soltis, D. E., C. H. Haufler, D. C. Darrow, & G. J. Gastony, 1983. Starch gel electrophoresis of ferns: A compilation of grinding buffers, gel and electrode buffers, and staining schedules. Amer. Fern. J. 73: 9–27.

Spivak, E. D., M. A. Gavio & C. E. Navarro, 1991. Life history and structure of the world's southernmost *Uca* population: *Uca uruguayensis* (Crustacea, Brachyura) in Mar Chiquita Lagoon (Argentina). Bull. mar. Sci. 48: 679–688.

Swofford, D. L. & R. B. Selander, 1989. Biosys-1. University of Illinois Urbana-Champaign.

Wright, S., 1931. Evolution in Mendelian populations. Genetics 16: 97–159.

Wright, S., 1943. Isolation by distance. Genetics 28: 114–138.

Wright, S., 1946. Isolation by distance under diverse systems of mating. Genetics 31: 39–59.

Wright, S., 1948. On the roles of directed and random changes in gene frequency in the genetics of population. Evolution 2: 279–294.

Wright, S., 1965. The interpretation of population structure by F-statistics with special regard to systems of mating. Evolution 19: 395–420.

Wright, S., 1969. Evolution and the genetic of populations. Vol. 2: The theory of gene frequencies. University of Chicago Press, Chicago.

Wright, S., 1978. Evolution and the genetics of populations. Vol. 4: Variability within and among natural populations. University of Chicago Press, Chicago.

Hydrobiologia **295**: 75–81, 1995.
Y. S. Wong & N. F. Y. Tam (eds), Asia-Pacific Symposium on Mangrove Ecosystems.

Impact of expected climate change on mangroves

C. D. Field
City Polytechnic of Hong Kong, Tat Chee Avenue, Kowloon, Hong Kong

Key words: climate change, mangrove, ecosystems

Abstract

There is a consensus of scientific opinion that the activities of man will cause a significant change in the global climate over the next hundred years. The rising level of carbon dioxide and other industrial gases in the atmosphere may lead to global warming with an accompanying rise in sea-level. Mangrove ecosystems grow in the intertidal zones in tropical and sub-tropical regions and are likely to be early indicators of the effects of climate change. The best estimates of predicted climate change in the literature are presented. It is suggested that a rise in mean sea-level may be the most important factor influencing the future distribution of mangroves but that the effect will vary dramatically depending on the local rate of sea-level rise and the availability of sediment to support re-establishment of the mangroves. The predicted rise in mean air temperature will probably be of little consequence to the development of mangroves in general but it may mean that the presence of mangroves will move further north and south, though this will depend on a number of additional factors. The effect of enhanced atmospheric CO_2 on the growth of mangroves is unknown at this time but that there is some evidence that not all species of mangroves will respond similarly. The socio-economic impacts of the effects of climate change on mangrove ecosystems may include increased risk of flooding, increased erosion of coast lines, saline intrusion and increased storm surges.

Introduction

Mangrove forests are characteristic littoral plant formations of tropical and sub-tropical sheltered coast lines. The plant species that are known as mangroves are derived from a variety of families and they vary in their dependence on littoral habitat. The total world-wide mangrove area is estimated at not less than 170 000 km^2 and there are considered to be some sixty species of trees and shrubs that are exclusive to the mangrove habitat. Mangroves support genetically diverse communities of terrestrial and aquatic fauna and flora that are of direct and indirect environmental, economic and social value to human societies throughout the world. Mangrove ecosystems are being subject to increasing non-sustainable development as economic and population pressures rise in many of the coastal areas of the tropical parts of the world. Mangrove forests are an important source of fuelwood in developing countries and energy for detritus based coastal food-chains, involving fish, molluscs and crustacea of economic value. Mangrove forests also help reduce coastal erosion in some regions of the world by dissipating the force of wave action.

Mangrove ecosystems may be considered dynamic and there is geological and contemporary evidence that they can extend or contract rapidly in response to regional topographical and climatic changes. Such ecosystems are also likely to be affected by stresses related to activities by man and they will exhibit marked spatial and temporal fluctuations as a result of such influences. The problem is to identify changes to mangrove ecosystems due to climate change induced by man's activities from changes that occur naturally. The primary climatic factors to be considered are temperature, atmospheric carbon dioxide concentration and sea level rise.

Climate change

Many reports have been published which address the question of global climate change (GCC) that might arise as a result of the activities of man and examine the possible effects. In particular the following three publications have addressed these issues in great detail: IPCC (Intergovernmental Panel on Climate Change) Climate Change: The IPPC Scientific Assessment. (1990a); IPCC Climate Change: The IPCC Impacts Assessment (1990b); IPCC Climate Change: The IPCC Response Strategies (1991). This review will not attempt to reiterate the extensive discussion that exists on the relatively short term changes to our climate that are likely to result from the activities of man. It will try to summarise the key predictions and then examine what these may imply for the future of mangrove ecosystems.

Global climate has large natural variability at all time and space scales. It is known from geological records that the recurrent variation in the eccentricity of the earth's elliptical orbit around the sun causes 0.2% variation in the amount of solar radiation intercepted by the earth with a period of some 100 000 years. The minimum of this cycle is known to result in the lowering of the global mean temperature and to produce the phenomenon of an ice age. These variations of climate on a geological time scale are not our main concern but a knowledge of the effects may give guidance as to what to expect if significant climate changes occur on a much shorter time scale.

It is known that global warming can be caused by green-house gases. These are gases which can absorb infra-red radiation. The absorption of longwave thermal radiation in the atmosphere, thus preventing the escape of thermal energy, causes the temperature of the atmosphere to rise markedly. The natural presence of these gases causes the earth to be warmer than it would be otherwise. The main greenhouse gases are carbon dioxide, methane, nitrous oxide, and chlorofluorocarbons (CFCs). The commencement of the industrial revolution lead to great amounts of fossil fuel being consumed and the subsequent discharge of large amounts of carbon dioxide to the atmosphere. Since then the activities of man have been continuously elevating the concentration of green-house gases in the atmosphere. Table 1 summarises how the concentrations of these gases are being affected by human activities.

It has been estimated that global average warming has been 0.3 °C and 0.6 °C since the late nineteenth century. The temperature record shows significant differences between the Northern and Southern Hemispheres. It has further been estimated that the rate of increase of global mean temperature during the next century will be 0.3 °C per decade with an uncertainty of 0.2 °C to 0.5 °C per decade. This will result in a likely increase in global mean temperature of about 1 °C above the present value by 2025 and 3 °C before the end of the next century. This estimate is based on the assumption that present conditions continue but makes allowances for population expansion and continued economic growth. It may be considered to be a worst case scenario. It is interesting to note estimates that in the tropics the warming will be both smaller and vary little with season. The reason is that the saturation vapour pressure of water varies non-linearly with temperature, so that at higher temperatures proportionally more of the increase in radiative heating of the surface is used to increase evaporation rather than to increase surface temperature.

The atmospheric carbon dioxide concentration in 1990 was 353 ppmv, which is about 25% greater than the pre-industrial (1750–1800) value of about 280 ppmv, and higher than at any time in the last 160 000 years. Carbon dioxide concentration is currently rising at about 1.8 ppmv (0.5%) per year due to anthropogenic emissions. It has been estimated that atmospheric carbon dioxide concentration will increase to 840 ppmv by the year 2100.

A factor important to mangrove ecosystems is the extent of sea level rise that might accompany an increase in mean global temperature. It is not easy to estimate sea level rise but the estimate is that the average rate of rise in sea level over the last 100 years has been 1.0–2.0 mm y^{-1}. In general the rise appears to be due to thermal expansion of the oceans and to increased melting of mountain glaciers and the margin of the Greenland ice sheet. The worst case scenario discussed above estimates that by the year 2030 global mean sea level will be 8–29 cm higher than today with a best estimate of 18 cm. By the year 2070, the rise in will be 21–71 cm, with a best estimate of 44 cm.

A factor that may be important for the future of mangrove ecosystems is any change in precipitation. It is estimated that there will be a substantially wetter atmosphere as the mean global temperature increases. However, in the tropics though the mean zonal value of precipitation may increase throughout the year there may be also areas of decrease. An accompanying phenomenon is the area of occurrence, frequency and intensity of tropical storms. There is some slight evidence that all of these may increase but it is by no means conclusive.

Table 1. Global man-induced environmental conditions expected at the end of the next century as a result of the IPCC "business-as usual" scenario (IPCC, 1990a).

Atmospheric greenhouse gasses	1990	Current rate of change	2100
Carbon Dioxide	353 ppmv	+0.5 $\%yr^{-1}$	840 ppmv
Methane	1.72 ppmv	+0.9 $\%yr^{-1}$	4 ppmv
CFC-11	280 pptv	+4.0 $\%yr^{-1}$	630 pptv
CFC-12	480 pptv	+4.0 $\%yr^{-1}$	1400 pptv
Nitrous Oxide	310 ppbv	+0.25 $\%yr^{-1}$	420 ppbv
Global mean temperature (land surfaces warm more rapidly than oceans)		+0.3 °C/decade (uncertainty: 0.2-0.5 °C/decade)	+3.0 °C
Global mean sea level (mainly due to thermal expansion and melting of some land ice)		+6.0cm/decade (uncertainty: 3–10 cm/decade)	+60.0 cm

Response of mangrove ecosystems

Sea-level rise

It is difficult to generalise about the effect of climate change on mangrove ecosystems as each system is very much the product of local topographical, climatic and anthropological influences. However, as all mangrove systems occur somewhere between high and low tide marks it is clear that they are likely to be significantly influenced by any changes in sea-level. Different mangrove species appear to have a marked preference for the level of salinity of the surrounding environment and therefore they are to be found at varying distances and elevations from the seaward edge reflecting the degree of mixing of the freshwater input and tidal influx.

Mangrove ecosystems accumulate peat or mud and this gives them the opportunity to adjust to a rising sea level. If the sediment accretion rate equals the rate of rise of sea-level then inundation preferences of the different mangroves species can be maintained. If the rate of sea-level rise exceeds the rate of accretion then some rearrangement of existing vegetation will take place and loss of mangroves will occur if the mean tide level becomes higher than the elevation of the substrate. Ellison & Stoddart (1991) conclude that in the case of low limestone islands where there is little significant allochthonous sediment input, mangrove ecosystems should be able to keep pace with a rising sea-level of 8–9 cm per 100 years, are under stress at between 9 and 12 cm per 100 years and cannot adjust at rates above this level. They believe that the predicted rates of sea-level rise caused by man induced climate change will be too fast for such mangroves to adjust and that there will be a collapse of the mangroves as a viable coastal ecosystem on low islands. The situation with high islands and continental shores will be more dependent on the amount of allochthonous sediment available and that can be highly variable (Woodroffe & Grindrod, 1991). In the case of some macrotidal estuaries, tidal influences can override riverine influences and vertical accretion within the mangrove ecosystem becomes chiefly due to tidally-driven reworking and transport of marine sediments. In northern Australia great mangrove swamps emerged following a dramatic rise in sea-level during the period 8000–6000 BP. It appears from the geological record that previous sea-level fluctuations presented a series of crises and opportunities for mangroves and that they tended to survive or even expand in several refuges, the most likely being continental coastlines with healthy sediment budgets. Woodruff & Grindrod (1991) comment that the factors influencing mangrove establishment at latitudinal limits are complex and incompletely known but that it is likely that previous dramatic temperature perturbations have had less impact than changing sea-levels on mangrove distributions especially on remote oceanic islands.

Temperature rise

Mangrove species show considerable variation in their sensitivity to temperature but the majority of them seem to produce maximal shoot growth when the mean air temperature rises to 25 °C and only *Avicennia marina* continues to produce leaves when the mean air temperature drops below 15 °C (Saenger & Hutchings 1987). It would appear therefore that if the average air temperature increases that the species composition of the mangrove forests may change and the presence of mangroves move further north and south. There is little evidence as to the effect of extremely high temperatures on mangroves but Saenger and Moverly (1985) show that some species demonstrate a declining leaf formation rate at temperatures above 25 °C. The optimum leaf temperatures for photosynthesis in mangroves appears to be 28–32 °C and photosynthetic capacity falls to close to zero at leaf temperatures of 38–40 °C (Clough *et al.*, 1982; Andrews *et al.*, 1984). It is generally accepted that plant development will be accelerated by increased temperature, as long as the temperature reached does not exceed an upper threshold. Very little is known about the effect of changing temperature on metabolic processes in mangroves. Superficially the predicted global warming of between 1.5 °C and 4.5 °C over the next century would seem likely to be of little consequence for the development of various species of mangrove. This impression is reenforced when the expected increase is compared to the diurnal oscillations in temperature, which can be in excess of 20 °C at the limits of mangrove occurrence. However, the temperature increases could become significant when the cumulative effects of temperature on plant development are considered. The elevation of the average temperature of the plant will be a critical factor in terms of growth but how this will be manifest in mangroves remains unknown. Soil warming that will accompany any global temperature rise could escalate the increase of atmospheric CO_2 through stimulation of soil respiration. This is one of several positive feedback mechanisms that may affect global climate.

Rise in atmospheric CO_2 levels

A change in the atmospheric CO_2 level alters the net carbon balance of the plant by changing the substrate resource but development of the plant will be determined principally by the rate modifier temperature and other controlling factors such as enzyme activity and photoperiod. It is therefore difficult to generalise on the effect of changes of atmospheric CO_2 levels on plant development (Rawson, 1992). There is some evidence that elevated CO_2 stimulates plant growth at least in agricultural plant species (Kimball, 1983; Cure & Acock, 1986) where most of the experiments have been carried out with green house grown plants. Eames & Jarvis (1989), in an extensive review, reported some enhancement of growth in juvenile trees and Drake (1992) reported a significant impact of enhanced atmospheric CO_2 on a wetland community of sedge and grasses. Though there is evidence that CO_2 enrichment will enhance growth in seedling tree species there is no equivalent evidence that there will be long term forest growth in response to rising atmospheric CO_2. A growth response to CO_2 may be manifest in below ground processes of forest ecosystems which tend to be nutrient and water limited. At the whole plant level, carbon isotope composition data indicate species variation in regulation of water loss with respect to carbon gain. The limited data suggest that not all species will respond similarly in response to elevated atmospheric CO_2 levels.

In the case of mangroves, Ball & Farquahar (1984a, b) reported that for *Aegiceras corniculatum* and *Avicennia marina* the rate of photosynthesis was limited by stomatal conductance to CO_2 and the internal efficiency of carboxylation involving the enzyme Ru Bp carboxylase. These results suggest that for these mangroves photosynthesis would be enhanced if the ambient CO_2 levels were increased. Contrary to these results, Cheeseman *et al.* (1991) working with *Bruguiera parviflora*, *Bruguiera gymnorrhiza* and *Rhizophora apiculata* suggested that the photosynthetic performance was unlikely to be enhanced by increased levels of ambient CO_2. The effect of CO_2 enrichment on mangrove forests cannot be interpreted within a simple framework as it will depend on complex interactions between several different physiological and environmental factors. Information is needed from long term assessments of growth where high CO_2 concentration, temperature, water stress and nutrient stress are controlled.

Water availability

It is well established that mangroves flourish in warm wet humid conditions where there is plentiful input of fresh water into their normal saline environment. One of the effects of global warming may be to change the pattern of precipitation in the tropics and this could have a profound effect on the growth of mangrove areas. The growth rate of mangroves is critically related to the availability of water to the trees and this is reflected in the soil water content and soil salinity. As most mangroves are tidally inundated, soil water content only becomes a problem when the inundation is occasional and the rainfall very limited. Soil salinity, however, characterises the mangrove habitat and growth of some mangroves has been shown to be maximal under relatively low salinities (Burchett *et al.*, 1984; Clough, 1984). As the salinity of the soil increases the mangroves face the problems of increasing salt levels in the tissues and decreasing availability of water. The increasing salt levels in the tissue may bring about a lessening in the net assimilation rate per unit leaf area (Ball 1988) and therefore reduce growth.

Water availability can also control growth and growth can be expressed as the product of the transpiration rate and the carbon gain per unit water loss (or water use efficiency). As the salinity increases above optimum levels the stomatal conductance declines with an accompanying decrease in transpiration rate, probably reflecting the decline in water potential of the soil. The reduction in stomatal conductance inhibits CO_2 diffusion into the leaf and leads to low assimilation rates. The humidity of the surrounding atmosphere and leaf temperature are also critical factors in these processes.

Mangroves have unusually high water use efficiencies (Ball, 1988) showing adaptation for minimal water use for a given carbon gain, which is reflected in relatively low rates of growth. Clough & Sim (1989) suggest that the water use efficiency of mangroves increases with increasing environmental stress thereby maximising photosynthetic carbon fixation while minimising water loss. Ball & Munns (1992) state that elevated CO_2 can enhance the water use efficiency of mangroves but that this may or may not result in enhanced growth. They also suggest that there may be enhanced growth with elevated CO_2 if growth is limited by water, carbon and nitrogen, but that elevated CO_2 would have little effect on growth when the salinity is too high for a species to maintain water uptake.

If the change in precipitation patterns in mangrove areas is such as to reduce soil salinity then an improvement in growth rates can be expected in some species.

Socio-economic impacts

Throughout much of the tropics the mangrove ecosystems sustain large human populations at subsistence levels. The mangrove ecosystem is valued for the extractable resources it supports, for the nonconsumptive services it provides and its intrinsic ecological value. Mangroves support diverse communities of fauna and flora of direct and indirect economic value and social value to human societies. Fish stocks and exploitable populations of crabs, shrimps and molluscs make up the principal food resource. Timber extraction is also of great importance. Mangrove ecosystems provide a variety of nonconsumptive services including recreational and aesthetic benefits, protection from soil erosion, flood mitigation, filtering of nutrients and protection against saline intrusion.

It is well documented (Saenger *et al.*, 1983; Hamilton & Snedaker, 1984; Field & Dartnell, 1987) that mangroves are under constant development pressure because they are found in coastal and estuarine areas which are also centres of human settlement. Mangrove ecosystems are under extreme pressure from expanding populations and non-sustainable use, such as land reclamation for construction, agriculture and aquaculture. The basic question that arises in the management of a dynamic and complex ecosystem, such as mangroves, is under what conditions should it be maintained and managed for intrinsic value and when should it be reclaimed for alternative purposes. There have been only a few attempts to answer this question (Amarasinghe, 1988; Milliman *et al.*, 1989; Padma Narsey Lal, 1990).

The consequences of possible global climate change have now to be added to an already unstable situation as far as mangrove ecosystems are concerned. It has already been shown that an enriched CO_2 atmosphere and a warmer and wetter climate would on balance favour the growth and expansion of mangroves. The major socio-economic problems are likely to be caused by the effect of rising sea level.

The main consequences will be:

- increased risk of flooding of low lying areas
- increased erosion of vulnerable soft coast
- increased risk of saline intrusion
- possible increase in frequency of storm surges

The loss of economic resources which might occur as a result of sea level rise is extremely difficult to model or predict and will depend heavily on local factors. It must be appreciated that the changes taking place in coastal areas either due to natural causes or human activities are probably already greater than any that might be expected from the predicted change in global climate

Future action

In 1991 a meeting of experts (UNEP-IOC-WMP-IUCN/GCNSMS-11/3 report no 69) recommended that long-term global monitoring systems be established for coral reefs and mangroves to measure the effects of climate change In 1992 a UNEP-UNESCO task team was established to consider the impact of climate change on mangroves The task team has recommended that a long term monitoring and study programme should be implemented at a small number of primary sites for long term and well designed experiments Additional secondary sites should be selected for the gathering of complementary information and intermittent or routine studies covering a wide geographical distribution of sites and types The primary sites should be well documented and include examples of a deltaic system, an arid coast and a low island It was felt that these distinct habitats would provide information that would indicate any effects that might begin to occur as a result of climate change

Such studies would have to be on a long term continuous basis if any trends are to be identified Every effort would have to be made to minimise the effects of human activity and episodic natural events by selecting relatively stable sites

Acknowledgments

I should like to acknowledge discussions with members of the UNEP-UNESCO Task Team on the Effect of Climate Change on Mangrove Ecosystems that have contributed significantly to the development of this paper

References

Amarasinghe M D, 1988 Socio-economic status of the human communities of selected mangrove areas on the West coast of Sri Lanka Mangrove Ecosystems Occasional Paper No 3 UNESCO

Andrews, T J, B F Clough & G J Muller, 1984 Photosynthetic gas exchange properties and carbon isotope ratios of some mangroves in North Queensland In Physiology and Management of Mangroves, Tasks for Vegetation Science 9 (H J Teas, ed), Dr W Junk, The Hague, pp 15–23

Ball, M C, 1988 Ecophysiology of mangroves Trees 2 129–142

Ball, M C & G D Farquhar, 1984a Photosynthetic and stomatal responses of two mangrove species, *Aegiceras corniculatum* and *Avicennia marina*, to long term salinity and humidity conditions Plant Physiol 74 1–6

Ball, M C & G D Farquhar, 1984b Photosynthetic and stomatal responses of the grey mangrove, *Avicennia marina*, to transient salinity conditions Plant Physiol 74 7–11

Ball, M C & R Munns, 1992 Plant responses to salinity under elevated atmospheric concentrations of CO_2 Aust J Bot 40 515–525

Burchett, M D, C D Field & A Pulkownik, 1984 Salinity, growth and root respiration in the grey mangrove *Avicennia marina* Physiol Plant 60 113–118

Cheeseman, J M, B F Clough, D R Carter, C E Lovelock, J E Ong & R G Sim, 1991 The analysis of photosynthetic performance of leaves under field conditions a case study using Bruguiera mangroves Photosynth Res 29 11–22

Clough, B F, T J Andrews & I R Cowan, 1982 Physiological processes in mangroves In Mangrove Ecosystems in Australia Structure, Function and Management (Clough, B F ed) Australian National University Press, Canberra 193–210

Clough, B F, 1984 Growth and salt balance of the mangroves *Avicennia marina* (Forsk) Vierh and *Rhizophora stylosa* griff in relation to salinity Austr J Plant Physiol 11 419–430

Clough, B F & R G Sim, 1989 Changes in gas exchange characteristics and water use efficiency of mangroves in response to salinity and water vapour Oecologia 79 38–44

Cure, J D & B Acock, 1986 Crop responses to carbon dioxide doubling a literature survey Agricult Forest Meteorol 38 127–45

Drake, B G, 1992 A field study of the effects of elevated CO_2 on ecosystem processes in a Chesapeake Bay wetland Aust J Bot 40 579–595

Eamus, D & P G Jarvis, 1989 The direct effects of increase in the global atmospheric CO_2 concentration on natural and commercial temperate trees and forests Adv Ecol Res 19 1–55

Ellison, J E & D R Stoddart, 1991 Mangrove ecosystem collapse during predicted sea level rise holocene analogues and implications J Coast Res 7 151–165

Field, C D & A Dartnall, (eds), 1985 Mangrove ecosystems of Asia and the Pacific status, exploitation and management Australian Institute of Marine Science Queensland Australia, 320 pp

Hamilton, L S & S C Snedaker, 1984 Handbook for mangrove area management East West Center, International Union for the Conservation of Nature and Natural Resources and UNESCO, 123 pp

Hutchings, P & P Saenger, 1987 Ecology of Mangroves University of Queensland Press, Australia, Australia 388 pp

IPCC, 1990a Climate change the scientific assessment Report of working group 1 Cambridge University Press Cambridge, 358 pp

IPCC, 1990b Climate change the impacts assessment Report of working group 11 Australian Government Publishing Service Canberra

IPCC, 1991 Climate change the response strategies Report of working group 111 Island press Washington, D C, 268 pp

Kimball, B A, 1983 Carbon dioxide and agricultural yield an assemblage of 770 prior observations United States Department of Agriculture, Agricultural Research Service, Water Conservation Laboratory Report 14, Phoenix, Arizona, 77 pp

Lal, P. N., 1990. Ecological economic analysis of mangrove conservation: a case study form Fiji. Mangrove Ecosystems Occasional Papers No 6. UNESCO.

Milliman, J. D., J. M. Broadus & F. Gable, 1989. Environmental and economic implications of rising sea level and subsiding deltas: the Nile and Bengal examples. Ambio 18: 340–345.

Rawson, H. M., 1992. Plant responses to temperature under conditions of elevated CO_2 Aust. J. Bot. 40: 473–490.

Saenger, P., E. J. Hegerl & J. D. S. Davie, 1983. Global Status of mangrove ecosystems. IUCN. Comm. Ecological Papers No 3 Gland, Switzerland, 88 pp.

Saenger, P. & J. Moverly, 1985. Vegetative phenology of mangroves along the Queensland coastline. Proc. Ecol. Soc. Austr. 13: 257–265.

Woodroffe, C. D. & J. Grindrod, 1991. Mangrove biogeography: the role of Quaternary environmental and sea-level change. J. Biogeogr. 18: 479–492.

Hydrobiologia **295**: 83–88, 1995.
Y. S. Wong & N. F. Y. Tam (eds), Asia-Pacific Symposium on Mangrove Ecosystems.

The population dynamics of the mangrove *Avicennia marina*; demographic synthesis and predictive modelling

Peter J. Clarke
School of Biological Science, A12, The University of Sydney, 2006, Australia
Now at Botany Department, University of New England, 2351, Australia

Key words: mangroves, *Avicennia marina*, population dynamics, demography, modelling

Abstract

Population dynamics of the widespread mangrove *Avicennia marina* was studied over the complete life-history from zygotes through to adults in southeastern Australia. Zygote survival, propagule dispersal, seedling establishment, seedling recruitment and sapling recruitment were examined by demographic censuses over a range of spatial and temporal scales. Hypotheses about factors regulating survival were tested by manipulative field experiments. Life table statistics for survival and fecundity were used to calculate transition probabilities and their variance for seven stages of life history. These parameters were used as the basis of a stochastic model that predicts population structure after small and large scale perturbations.

Introduction

Modelling the population dynamics of long-lived species, such as mangroves, has been frustrated by problems of complexity related to spatial heterogeneity of populations and the stochastic nature of their dynamics. A useful conceptual model of the spatial and temporal components of a mangrove forest or metapopulation is the 'shifting mosaic' model (Shugart & Urban, 1988). A mangrove forest consists of a mosaic of smaller cohorts, often of differing species, each having a history that depends on episodic recruitment of seedlings (Fig. 1a). The uneven size distributions of mangrove populations in southeastern Australia suggests that episodic mass mortality and/or regeneration occurs at two spatial scales (Fig. 1a). Smaller scale disturbances (type a) result from the death of a few individuals and create gaps of a few crown diameters, whereas larger scale disturbances resulting from storms, and pathogen attack create regeneration sites >0.1 ha.

In many mangrove forests propagules regularly establish because they contain precociously developed embryos, but the recruitment of the established seedlings to older cohorts (saplings) depends on the episodic availability of a 'regeneration niche' (Clarke & Allaway, 1993). Often such opportunities arise after small scale disturbances such as tree fall, which free up sediment and light resources, or larger scale events such as sedimentation or mass mortality (Fig. 1). It is this event driven component of the wider and more complex metapopulation dynamics of mangrove forests containing *Avicennia marina* (Forsk.) Vierh. that this paper addresses.

Life table statistics can be used for demographic modelling of a population, *i.e.* to simulate population change using mathematical algorithms. In particular, matrix population models have been used to project what would happen to the population if demographic attributes remain unchanged, *i.e.* exponential growth (Caswell, 1989). For predictive modelling such assumptions are mostly untenable, but Caswell (1989) argues that they 'reveal something about the present population'. For example, the various indices generated from a deterministic model can be used for interspecific comparison (Harcombe, 1987; Silvertown *et al.*, 1992). While matrix models may be useful for comparative purposes their use as predictors of future population size and structure is limited (e.g. see Burns & Ogden, 1985) because mangrove populations are limited by environmental filtering and competition. Non-linear models can, however, be developed from

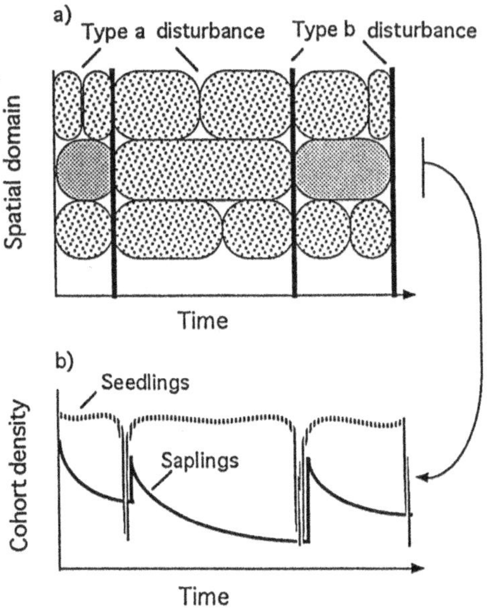

Fig 1 a) Shifting mosaic model with two types of disturbance Small scale disturbance type a, and larger scale disturbance type b b) Sapling recruitment and mortality after disturbance, seedlings constantly present in the understorey

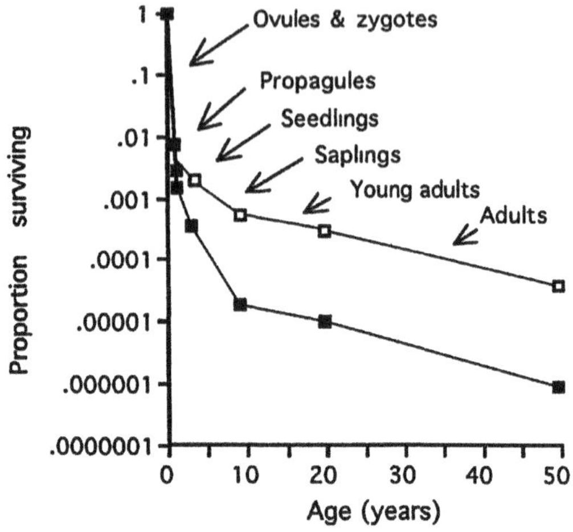

Fig 2 Survivorship throughout the life history Closed symbol type a disturbance, open symbol type b disturbance, e g colonisation Error terms available on request to author

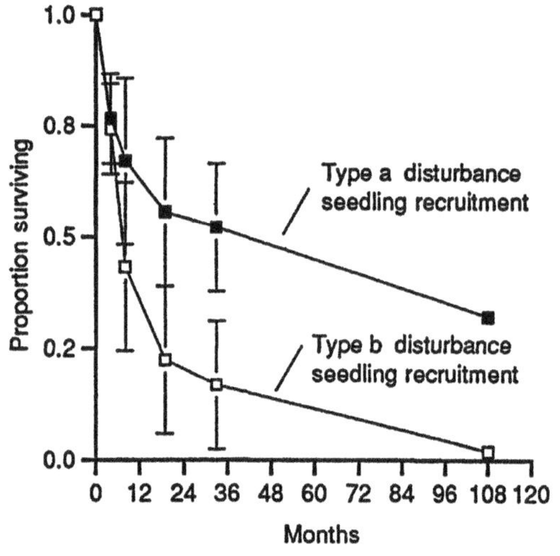

Fig 3 Survival of seedlings and saplings under type a and type b disturbance (See Clarke & Allaway, 1993)

life table statistics together with information about how environmental filtering and competition influence demographic rates

The purpose of this paper is firstly to present a demographic synthesis of the life history of the mangrove *Avicennia marina* var *australasica* Walp Moldenke (Duke, 1991) in south-eastern Australia, and secondly to present a predictive population model for population growth after a regeneration niche has been created

Demographic synthesis

The life cycle

The life cycle of *Avicennia marina* starts with the potential zygote population in the form of female gametophytes. After fertilisation the population consists of precociously developing embryos that are partially dependent on maternal support Once dispersed, the propagule must chance on a favourable place to establish, grow and recruit to older age classes The cycle is complete when offspring are produced by either reproduction forming genets or by growth forming independent clones or ramets

Predispersal mortality

About one third of flower primordia, each containing four ovules, survive to become open flowers but only 3% (0 75% of ovules) survive to become viable fruits some ten months after the initiation of flower buds

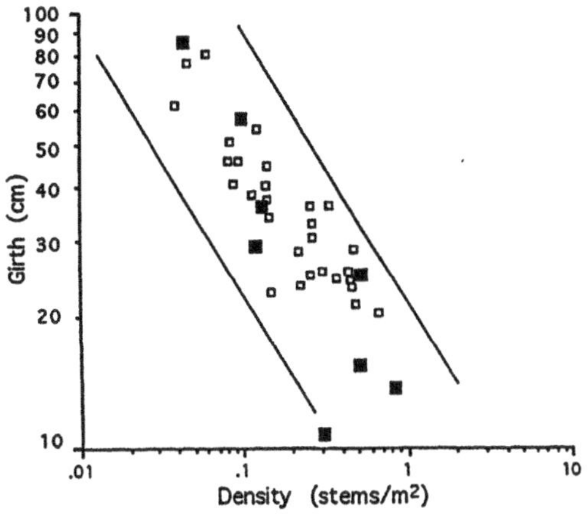

Fig. 4. Correlation of girth and density of trees. Solid symbol *Avicennia marina*. Open symbol data from Jimenez *et al.*, 1985. Thinning slope shown as solid line.

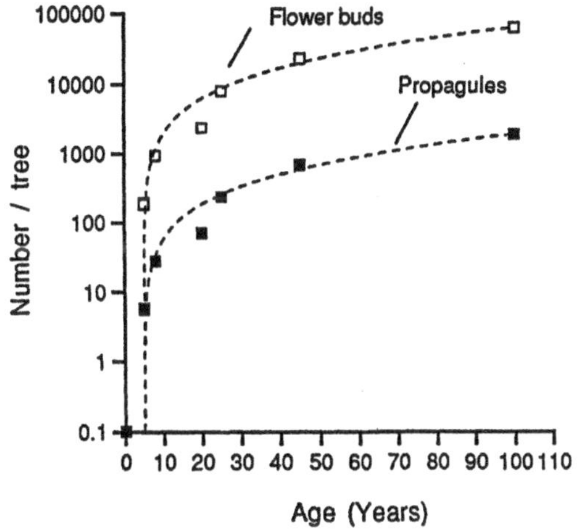

Fig. 5. Fecundity schedule for potential fecundity (flower buds) and realised fecundity (propagules). Details of fecundity errors presented in Clarke & Myerscough, 1991.

(Clarke & Myerscough, 1991) (Fig. 2). Insect herbivores can influence the viability of propagules, but the size of the effect is small in south-eastern Australia (Clarke, 1992). In contrast, rare climatic events, such as hail storms, may reduce predispersal survival close to zero (Clarke, 1992).

Dispersal and establishment

Once released from maternal support propagules have an obligate phase of dispersal about one week, although the dispersal phase may extend over many months (Clarke, 1993). During this period there appears to be little mortality, except for stranding outside the normal habitat range of mangroves (Clarke & Myerscough, 1993). To estimate establishment accurately many thousands of propagules need to be followed over a wide range of conditions. Of these about half of the propagules that strand establish over a wide range of salinity, light and nutrient conditions (Clarke & Myerscough, 1993) (Fig. 2). Some mortality is due to predation (Clarke & Myerscough, 1993), although this mortality is far less than that reported for tropical populations of *Avicennia marina* (Smith, 1987). Apart from predation the main factor inhibiting establishment appears to be tidal and wave buffering.

Seedling growth and survival

Once propagules establish seedling survival is independent of light, nutrient and salinity conditions for during the phase while the cotyledons are attached (Clarke & Myerscough, 1993; Clarke & Allaway, 1993). In contrast, survival of seedlings after the post-cotyledonary phase appears to be largely resource dependent. Seedlings require a combination of light and sediment resources for enhanced growth and survival, *i.e.* recruitment to a reproductive phase (Clarke & Allaway, 1993). About 25% of established seedlings can recruit to a sapling stage (*ca* 10 years) if the regeneration niche is large, *i.e.* after gross canopy and sediment disturbance, whereas after small gap disturbance about 10% recruit to this stage (Clarke & Allaway, 1993) (Fig. 2, Fig. 3).

Sapling and tree survival

In crowded stands populations of saplings appear to thin because of interspecific competition (Fig. 4), but as growth slows the cause of mortality probably shifts towards disturbance effects. One of the problems in predicting the effects of differing environmental conditions of thinning rates is the difficulty in deriving estimates of growth under particular environmental conditions (Landsberg, 1986). Mortality rates change with the age of the tree population (Fig. 2); those in crowded populations have higher natural annual mortality, and this decreases as trees thin and other factors influence mortality. At some stage, however, this rate must increase as older plants appear to be more prone to tree fall and pathogens.

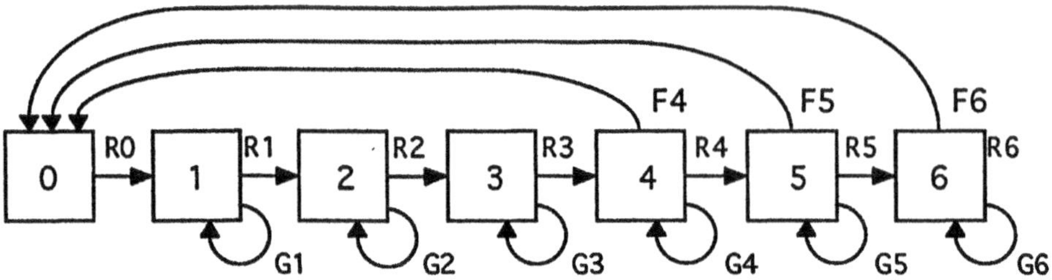

Fig. 6. Life cyde model based on seven life history stages. 0 = propagules, 1 = cotyledonary seedlings, 2 = seedlings, 3 = saplings, 4 = young trees, 5 = trees, 6 = older trees. F = fecundity, R = probability of growth to next stage, G = probability of remaining in stage (mortality = 1-R-G).

Reproduction and fecundity

Plants as young as 5 years can produce viable propagules, but under conditions of crowding and rapid growth most of the plants appear not to be fecund until they become a small tree (*ca* 20y) (Fig. 5) (Clarke, 1992). Fecundity of a tree increases with size and age until maturity where the potential annual fecundity is large (Fig. 5). Mature trees, however, are only fecund every few years, hence averaged over an estimated lifetime of about a century an individual produces about 250 propagules each year (Clarke, 1992).

Modelling

A simple stage-structured model of *A. marina* incorporating the life history outlined above shows seven stages and the probabilities of recruiting to the next stage, the probability of remaining in a stage and at each stage its fecundity (Fig. 6) (Table 1). The transition probabilities differ depending on the type of regeneration niche created and subsequent environmental filtering. The latter effect can, however, be taken into account by incorporating the variance about mean values. The transition probabilities for two types of regeneration are shown in Table 1. Type a is a regeneration sequence after small scale disturbance the size of a few tree crown areas and type b is a sequence following the creation of a new niche (e.g. sedimentation) or the disturbance on the scale of >0.1 ha (e.g. storms, disease, clearing, oil spills). The transition probabilities of recruitment or growth to the next stage, the probability of remaining in a stage and the fecundity rates can be used in Lefkovitch matrices to project population growth (Caswell, 1989).

Clearly, it is unrealistic, in a predictive sense, to assume that the transition probabilities remain unchanged through time such that population growth is exponential *i.e.* deterministic (Fig. 7). Instead, a non-linear component to the model can be introduced so that the transition probabilities decrease as populations become crowded and the canopy closes (Fig. 8). It should be noted that this non-linear projection is stochastic, but for clear presentation of the mean trend the error terms have been excluded from Fig 8. When the maximum density is reached for saplings the transition probabilities for the previous stages are reduced to zero (Table 1). A non-linear modelling projection was made after initialising the population with 1000 propagules that disperse in from surrounding cohorts or sown. The modelled area of the population would be about 100 m^2 *i.e.* 1 propagule per m^2 and the initial transition probabilities used assume regeneration following a large scale disturbance such as an oil spill and smaller scale disturbances such as the formation of a canopy gap. Initial seedling numbers ensure rapid recruitment of saplings, but as they reach maximum density after about a decade further recruitment is restricted as transition probabilities for seedlings reach zero (Fig. 8). Thereafter numbers of saplings decrease as thinning, and recruitment to older stages, occur. After about 20 years very few saplings are left in the understorey and canopy trees stabilise in density at about 0.05–0.1 trees m^{-2}. The predictions the present model makes about the population structure of *Avicennia marina* stands in south-eastern Australia have been verified from stands of known age, but its ability to predict dynamics needs to be tested by observations of sites with a known disturbance history.

Discussion

Much research in mangrove ecology, and plant ecology in general, has been essentially descriptive. Distribu-

Table 1. Annual transition probabilities and fecundity for a stage-based model. Mortality = 1 – Recruitment - Remaining. type a disturbance, small scale; type b disturbance, large scale. Variance values available on request to author.

Rates	Disturbance	Stage 0 Propagules	1 Cotyledonary seedlings	2 Seedlings	3 Saplings	4 Young tree	5 Tree	6 Older tree
Fecundity (F)	Type a	0	0	0	0	0	500	1000
	Type b	0	0	0	0	100	500	1000
Mortality	Type a	0.800	0.250	0.165	0.018	0.029	0.008	0.001
	Type b	0.600	0.104	0.122	0.045	0.029	0.008	0.001
Recruitment (R)	Type a	0.200	0.083	0.010	0.073	0.008	0.012	0.000
	Type b	0.400	0.230	0.045	0.045	0.008	0.012	0.000
Remaining (G)	Type a	0.000	0.666	0.825	0.909	0.963	0.980	0.999
	Type b	0.000	0.666	0.825	0.909	0.963	0.980	0.999

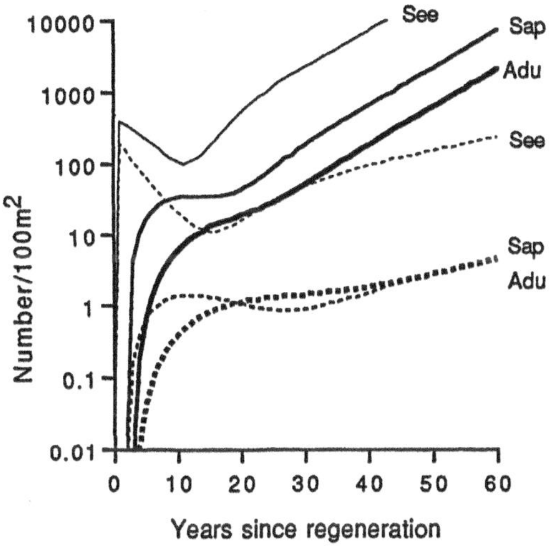

Fig. 7. Deterministic model of population growth after disturbance. Broken line type a disturbance, solid line type b disturbance. Model initialised with 1000 dispersed propagules. See seedlings, Sap saplings, Adu Adults. Confidence limits removed for ease of interpretation.

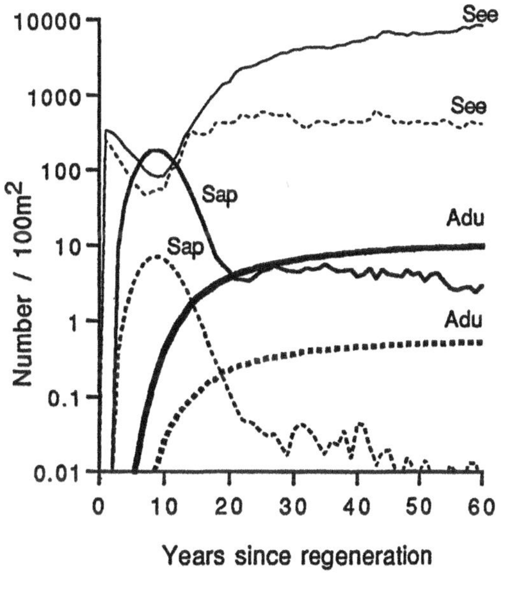

Fig. 8. Non-linear model of population growth after disturbance. Broken line type a disturbance, solid line type b disturbance. Model initialised with 1000 dispersed propagules. See seedlings, Sap saplings, Adu Adults. Confidence limits removed for ease of interpretation.

tions of taxa and patterns of zonations have occupied a plethora of studies and there is now an adequate descriptive background against which hypotheses can be framed. Generality at the landscape scale (geomorphic studies) and precision at the individual scale (ecophysiology) have characterised correlative approaches in attempts to provide explanations and predictions of mangrove distributions, abundance and performance. More recent trends focus on plant populations and experimental manipulation to test specific hypotheses about mangrove processes. These studies seem to provide a conceptual link between the extreme scales of space and time inherent in studies of mangrove vegetation by focussing on the fundamental unit of natural selection

Population studies of mangrove species, and most other long-lived plants, often concentrate on components of the life cycle to explain present distribution and abundance In mangroves, propagule predation (e g Robertson *et al*, 1990), dispersal (e g Rabinowitz, 1978), seedling growth (e g Ball, 1988) and seedling predation (e g Smith, 1987) have been studies to explain distribution and abundance Alone, these explanations may be inadequate because other life history stages may influence the overall population process The present study shows that it is logistically possible to construct life tables for mangrove species and at the same time test hypotheses about regulation of survival, growth and fecundity Clearly, the spatial and temporal variances are large when dealing with field population data, however adequate spatial and temporal replication will increase the precision and generality of the results These data are not easy to collect in an often inaccessible and difficult working environment, nevertheless, they form an important basis for understanding and predicting the gap and regeneration dynamics of mangrove forests (see e g Grant *et al*, 1993) Predictions from these models should be tested either by large scale experimental manipulations or by closer analysis of natural or human induced perturbations rather than stand structure analysis, given the problems of determining the age structure of mangroves Simple stochastic models that incorporate transition probabilities to predict population size may, in the future, be incorporated into more complex shifting mosaic models to assess broader scale effects such as climate change or resource utilisation

References

Ball, M C, 1988 Ecophysiology of mangroves Trees 2 129–142

Burns, B R & J Ogden, 1985 The demography of the temperate mangrove [*Avicennia marina* (Forsk) Vierh] at its southern limit in New Zealand Aust J Ecol 10 125–133

Caswell, H, 1989 Matrix population models Sinauer, Sutherland, Massachusetts, 362 pp

Clarke, P J & W G Allaway, 1993 The regeneration niche of the grey mangrove (*Avicennia marina*) effects of salinity, light and sediment factors on establishment, growth and survival in the field Oecologia 93 548–556

Clarke, P J & P J Myerscough, 1991 Floral Biology and reproductive phenology of *Avicennia marina* in south eastern Australia Aust J Bot 39 283–293

Clarke, P J & P J Myerscough, 1993 The intertidal limits of *Avicennia marina* in southeastern Australia, the effects of physical conditions, interspecific competition and predation on establishment and survival Aust J Ecol 18 307–315

Clarke, P J, 1992 Predispersal mortality and fecundity in the grey mangrove (*Avicennia marina*) in south-eastern Australia Aust J Ecol 17 161–168

Clarke, P J, 1993 Dispersal of grey mangrove (*Avicennia marina*) propagules in south eastern Australia Aquat Bot 45 195–204

Duke, N C, 1991 A systematic revision of the mangrove genus *Avicennia* (Avicenniaceae) in Australasia Aust Syst Bot 4 299–324

Grant, D L, P J Clarke & W G Allaway, 1993 The response of grey mangrove (*Avicennia marina*) seedlings to spills of crude oil J exp mar Biol Ecol 171 273–295

Harcombe, P A, 1987 Tree life tables Bioscience 37 557–568

Jimenez, J A, A E Lugo & G Citron, 1985 Tree mortality in mangrove forests Biotropica 17 35–51

Landsberg, J J, 1986 Physiological ecology of forest production Academic Press, London, 282 pp

Rabinowitz, D, 1978 Dispersal properties of mangrove propagules Biotropica 10 47–57

Robertson, A I, R Giddens & T J Smith, 1990 Seed predation by insects in tropical mangrove forests extent and effects on seed viability and growth of seedlings Oecologia 83 213–219

Shugart, H H & D L Urban, 1988 Factors affecting the relative abundance of forest tree species In P J Grubb & J B Whittaker (ed), Toward a more exact ecology Blackwell Scientific Publications Oxford 249–273

Silvertown, J, M Franco & K McKonway, 1992 A demographic interpretation of Grime's triangle Funct Ecol 6 130–136

Smith, T J, 1987 Effects of seed predators and light level on the distribution of *Avicennia marina* (Forsk) Vierh in tropical, tidal forests Estuar coast Shelf Sci 25 43–51

Hydrobiologia **295**: 89–95, 1995.
Y. S. Wong & N. F. Y. Tam (eds), Asia-Pacific Symposium on Mangrove Ecosystems.

Lower marine fungi (labyrinthulomycetes) and the decay of mangrove leaf litter

G. B. Bremer
School of Biological Sciences, University of Portsmouth, King Henry 1st Street, Portsmouth PO1 2DY, United Kingdom

Key words: fungi, marine, thraustochytrid, mangrove, leaf decay

Abstract

This paper deals with the association of members of the Labyrinthulomycetes (Thraustochytriales and Labyrinthulales) with decaying or decayed leaves at an intertidal mangrove at Morib, Malaysia. Representatives of both orders of these obligately marine unicellular eukaryotes of unresolved taxonomic affinities (Chamberlain & Moss, 1988) were consistently isolated from leaves at all stages of decay from the recently fallen to those in an advanced stage of decay, but not from either green or senescent yellow leaves attached to trees. Baiting experiments using γ-irradiated leaf discs of *Sonneratia* and *Rhizophora* spp. immersed in the aquatic environment of the mangrove, revealed that leaf material was colonised by both labyrinthulids and thraustochytrids within 24 hours of immersion at the test site and these organisms were isolated from the leaf material throughout the 14 day study period. *In vitro* experiments using axenic cultures of three thraustochytrid genera inoculated onto sterile discs of *Sonneratia* leaves and incubated for 14 days caused loss of both biomass and structural integrity of the leaf material. Freeze fracture, followed by scanning electron microscopy of leaves inoculated with a thraustochytrid and a strain of *Labyrinthula*, revealed that penetration of the leaf occurred after 4 days and that the thraustochytrid was associated with localised degradation of internal leaf tissues. Cellulase production by an isolate of *Schizochytrium aggregatum* was detected. The results of all the above investigations are discussed with reference to the role of members of the Labyrinthulomycetes in nutrient cycling in the mangrove.

Introduction

Mangrove leaf litter is an important food source at the heart of food webs which eventually contribute to commercially important coastal and offshore fisheries. The contribution of biomass to the ecosystem is substantial; Lugo & Snedaker (1974) estimated the average annual fall of leaf litter in a forest of red mangrove as 896 g dry weight m^{-2} y^{-1}. Although initially nutritionally poor, the leaf litter is enriched by the decomposition process. Fell and Master (1980) recorded a change in C:N ratios from 120 in senescent *R. mangle* leaves, to 43 in partially decomposed leaves, they also estimated that half of the carbon leaches into the ecosystem and half is dispersed as flocculated debris; the soluble carbon is available to heterotrophs and the flocullate may be directly ingested.

The capability of marine microorganisms particularly fungi to degrade cellulosic materials has received little attention when compared with the volume of literature available on terrestrial species. Heald & Odum (1970) considered that the microflora has a significant role in the recycling of plant material in estuarine and coastal waters and sediments. Observations on marine fungi associated with organic debris from mangroves has centred on the occurrence of ascomycetous fungi growing on wood (Hyde, 1986; Hyde and Borse, 1986; Jones & Tan, 1987) but with relatively little attention having been given to their ability to decay mangrove woods (Mouzouras, 1989).

Although comparable ecological data is available on the lower marine fungi of the mangal (Ulken, 1984; Fell & Master, 1975: Lee & Baker, 1973; Raghu Kumar, 1988) there are also indications of an ability to degrade non-lignified plant tissues (Fell & Master,

1973, 1980). The role of thraustochytrids in particular in the decomposition of mangrove leaves was highlighted by Findlay, Fell, Coleman & Vestal (1986), this study also suggested that the production of the long chain polyunsaturated fatty acids 20:5ω3 and 22:6ω3 by thraustochytrids, and an observed rise in these lipids in decayed mangrove leaves may play a significant role in nutrient enrichment of mangrove leaf debris.

The evidence to implicate thraustochytrids in foliar decay processes in the mangrove is however circumstantial, since unlike some of the mycelial marine fungi which have been shown to possess cellulolytic exoenzymes (Gessner, 1980), there has been no evidence that thraustochytrids produce enzymes capable of degrading complex carbohydrates. This paper is an attempt to resolve some of the unanswered questions surrounding the role of Labyrinthulomycetes in the recycling of particulate carbon in the marine environment.

Study site

The test site for these investigations was an intertidal mangrove at Morib, Peninsular Malaysia, bordering on the Straits of Malacca (2,44N 101.26E). Dominant tree species within the mangrove were *Sonneratia alba*, *Rhizophora apiculata* and *Avicennia marina*, the soil was fine sand, the salinity of the seawater was 22–25‰, the pH 7.8 and the temperature of the seawater during the period of study was 28–30 °C, rising to 37 °C in the shallow pools formed when the tide receded. The location for exposure of the litter bags containing leaf samples was approx. 30 meters from the seaward edge of the mangrove where a pool 30 cm deep was formed at low tide. The studies were carried out in September 1991 and October 1992.

Materials and methods

Isolation studies

The isolation of Labyrinthulomycetes was carried out by direct plating of mangrove leaf material encompassing all stages of decay from recently fallen yellow senescent leaves to the advanced brown fragile leaf material, onto seawater nutrient agar. For thraustochytrid isolation the agar comprised; agar (Oxoid No 3) 10 g; yeast Extract (Oxoid) 1 g; peptone (Oxoid L40) 1 g; natural seawater 1000 ml. For labyrinthulids the agar comprised: agar (Oxoid No 3) 9 g; horse serum (Lab Lemco) 100 ml; natural seawater 1000 ml. Both agars incorporated 500 mg l^{-1} of penicillin G and streptomycin to discourage bacterial growth. Fragments of leaf were cut aseptically and washed in several changes of sterile seawater to remove mould spores and yeasts, then plated out on the agar surfaces. A thin overlay of sterile seawater was then applied to the agar surfaces and the plates incubated at 25 °C and 37 °C in the dark. The plates were examined at 24 hr intervals for Labyrinthulomycete colonies which were than transferred aseptically to fresh agar plates. Several such transfers yielded axenic cultures. The pine pollen baiting method of Gaertner (1968) was used to assess the frequency of occurrence of thraustochytrids in seawater, sediment and detritus samples.

Decay and colonisation of mangrove leaf material in situ

The leaf material used in the study was derived from senescent yellow leaves picked from trees of *Rhizophora* and *Sonneratia* growing at Morib and packed in sterile polythene bags for transport back to the laboratory. 1.5 cm diameter discs were cut from these leaves and placed in sealed polythene bags, the bags and contents were then sterilised by γ irradiation of 28.5 kiloGrays. After sterilisation the bags were coded and the weight of each bag and contents was recorded. For colonisation and decay studies a known wet weight of leaf discs was placed in an nylon mesh bag (mesh size 2 mm) to exclude invertebrate grazers and coded with a numbered tag. The mesh litter bags were then attached to prop roots using plastic cable ties. For studying colonisation by Labyrinthulomycetes, the bags were recovered after 1, 2, 4, 7 and 14 days exposure at the site. For determining weight losses, the bags were recovered at intervals of 1, 2, 4 and 7 weeks. Any sediment attached to the discs was removed by gentle washing and the discs carefully blotted dry. The wet weight of the remaining leaf material was then recorded. Four replicate bags containing leaf discs of *Sonneratia* and four containing leaf discs of *Rhizophora* were exposed and recovered for each time interval.

In vitro *decay studies*

For analysis of the ability of thraustochytrids to decay mangrove leaves, controlled experiments were conducted using axenic cultures. Isolates tested for their ability to cause decay and weight loss were: *Thraustochytrium striatum* Schneider; *Ulkenia pro-*

funda Gaertner, *Schizochytrium aggregatum* Goldstein and Belsky and an isolate of *Labyrinthula*. A culture of *Trichoderma viride* Pers. ex Fr. was used as a reference organism with a well documented cellulolytic decay activity.

To mimic conditions in the mangrove, flasks for inoculation were prepared as follows; 10 g of beach sand was placed in the bottom of a 250 ml Erlenmyer flask and the flask autoclaved. After cooling, 50 ml of pasteurised natural seawater was added to each flask and the flasks incubated for 24 hours at 25 °C and then examined visually to confirm sterility. Weighed amounts of irradiated leaf discs were then added to the flasks and each experimental flask was then inoculated with an actively growing culture of the test organism. Controls were employed consisting of uninoculated flasks. All the flasks were then incubated at 25 °C in the dark in a rotary incubator set at 100 RPM for 14 days. After incubation, each set of leaf discs was removed and washed gently to remove sand particles, excess moisture was removed by careful blotting with paper tissue and each set of discs was then weighed. Weight losses presented are the means of four replicates.

Detection of cellulolytic activity

Cultures of *Schizochytrium aggregatum* (University of Portsmouth culture collection) were grown in a liquid medium comprising: glucose 1 g; yeast extract 0.5 g; peptone 0.5 g; natural seawater 1000 ml. 50 ml of this culture medium in a 250 ml Erlenmeyer flask was inoculated with zoospores of *S. aggregatum* and incubated at 27 °C on a shaker at 200 rpm. Samples of culture fluid were withdrawn daily from day 1 to day 6, centrifuged at 9000 RPM to remove cells and the filtrate assayed for the presence of cellulase using the agar well plate method of Carder (1986). A cellulase standard (Cellulase C2415 Sigma) was employed and uninoculated culture medium was used as a control.

Freeze fracture and scanning electron microscopy of colonised leaves

To provide material for this study, irradiated leaf discs were inoculated with thraustochytrid isolate SLBM and an isolate of *Labyrinthula*, both obtained from decaying mangrove leaves. The isolates were inoculated and grown in the manner described above for decay studies. After 4 days incubation in standing culture, the discs were removed and fixed in 2% glutaraldehyde in cacodylate buffer at pH 7.4 for 8 hours. On removal from the fixative, the discs were immediately plunged into liquid nitrogen for 2 minutes. The frozen leaves were then fractured using coverslip forceps and transferred to a graded series of ethanol for dehydration before being critically point dried. Specimens were sputter-coated with gold and then examined in a JEOL T20 SEM operating at 20 kV.

Results

Isolation studies

Leaves of both *Sonneratia* and *Rhizophora* removed from trees and either in a green photosynthetic state or at the yellow senescent stage, yielded isolates of representatives of a terrestrial mycoflora; *Pestalotia, Trichoderma, Aspergillus, Penicillium* and *Alternaria*. A variety of yeast species was also predominant on all leaf samples. Yellow senescent leaves recovered from shallow pools and estimated as having fallen within two days of recovery, also provided isolates of *Pestalotia*, hyphomycetes and yeasts, in addition, all of the leaf samples at this stage of decay produced colonies of *Labyrinthula* and thraustochytrids. One isolate of a *Phytophthora* sp. was obtained from fallen yellow *Rhizophora* leaves. Leaves of both *Sonneratia* and *Rhizophora* in the more advanced stages of decay when the leaves have become brown and fragile, continued to yield yeast colonies and terrestrial hyphomycetes, however two facultatively marine hyphomycetes, *Dendryphiella arenaria* (Suth.) Pugh et Nicot and *Asteromyces cruciatus* (F. & Mme Moreau) ex Hennebert were also isolated. Large numbers of thraustochytrids of the genera *Thraustochytrium* and *Schizochytrium* were recovered from brown leaves (60% occurrence) but *Labyrinthula* was the most commonly isolated organism from leaf samples at this stage of decay with 80% of the samples positive for this organism.

The use of pine pollen to analyse the frequency of occurrence of thraustochytrids in water, the sandy substratum of the mangrove and the leaf detritus gave results expressed here as the mean percentage of pollen grains colonised by thraustochytrids derived from four separate counts of 100 pollen grains for each of the substrates. The results are: for water 15%; for sediment 46% and for leaf detritus 53%.

Colonisation and decay in situ

Labyrinthula and thraustochytrids were isolated from both *Sonneratia* and *Rhizophora* leaf discs recovered after the initial 24 hour period immersed in the water of the Morib mangrove. Three morphologically differing isolates of *Labyrinthula* were recognised together with several isolates of *Thraustochytrium*, *S. aggregatum* and *Schizochytrium mangrovei* Raghu Kumar. Subsequent recoveries of leaf discs after 2, 7 and 14 days showed similar colonisation by thraustochytrids and labyrinthulids. Throughout the exposure period approximately 40% of the leaf samples of both tree species yielded thraustochytrid isolates, but as the leaf discs decayed, the number of isolates of *Labyrinthula* rose, so that at the end of the 14 day period when the leaf material was soft and brown, over 70% of the samples recovered provided cultures of *Labyrinthula*. Yeast colonies were noted throughout this study and their presence also increased as the decay of the leaf progressed. Fungal mycelium was detected on the culture plates in several instances but identification was not attempted.

The previously weighed sets of leaf discs of both tree species were recovered after 1, 2, 4 and 7 weeks exposure in the mangrove. The change in wet weight of the discs with time is presented in Fig. 1. The rate of weight loss of the discs of *Sonneratia* was initially higher within the first two weeks of the study (48% weight loss) whereas the loss of weight of the *Rhizophora* leaf was slow during the first 4 weeks of the study, but rose rapidly from 16% to 58% between weeks 4 and 7. At the end of the 7 weeks of the study, both sets of discs were dark brown, softened, fragile and badly decayed. Those of *Sonneratia* had lost an average of 78% of their initial weight and the wet weight discs of *Rhizophora* leaves had declined by 58%.

In vitro *decay studies*

The results of this study using axenic cultures to assess the decay of leaves of *Sonneratia* under controlled conditions are presented in Table 1. The three thraustochytrids *Ulkenia visurgensis* Gaertner, *T. striatum* and *S. aggregatum* caused significant decreases in mean wet weight of the leaf discs. *Trichoderma viride* included in the study as a reference organism was responsible for a slightly smaller but comparable decrease in wet weight of the leaf discs. In contrast, the leaf discs inoculated with *Labyrinthula* showed a change in weight not significantly different from the control.

Table 1. Percentage decrease in wet weight of leaf material of *Sonneratia* incubated for 14 days with axenic cultures of fungal isolates.

Isolate	Mean decrease wet weight	Standard deviation
Ulkenia visurgensis	46%	±11.26
Thraustochytrium striatum	43%	±5.43
Schizochytrium aggregatum	40%	±5.87
Trichoderma viride	38%	±14.26
Labyrinthula sp.	14%	±8.93
Control (uninoculated flask)	8%	±3.94

The visual appearance of *Sonneratia* leaf discs after incubation with a culture of *S. aggregatum* for 14 days at 25 °C is shown in Fig. 2A. Compared with the control leaf discs on the left (1), the discs subjected to attack by *S. aggregatum* (2) show considerable loss of biomass and in some cases the discs have become almost skeletonised.

Detection of cellulase activity

The same isolate of *S. aggregatum* used in the above experiments was assayed for production of cellulase in culture. The method adapted from that of Carder (1986) detects clearing zones produced around wells containing culture filtrate in an agar plate containing sodium carboxymethylcellulose. Congo Red was used as a dye to enhance the visualisation of the clearing zones. The result of one assay is presented in Fig. 2B. Clearing zones were not produced by filtrates from 1, 2 and 3 day old cultures of *S. aggregatum* but were produced by filtrates from 4 and 5 day old cultures. Filtrates from 6 day old cultures gave a negative result for cellulase as did the control (C), an uninoculated flask of growth medium. A visual comparison of the amount of cellulase activity present can be made by comparing the clearing zones produced by the culture filtrates with those produced by a cellulase standard (CS) of 5×10^{-3} ml. The consistency of timing of the appearance and disappearance of cellulase in the culture filtrate was confirmed by repeating the experiment 5 times.

Freeze fracture and scanning electron microscopy studies

Leaf discs of *Sonneratia* inoculated with a thraustochytrid and a strain of *Labyrinthula* and cultured for 2 days, were examined in the SEM after fracturing. The inner structure of the leaf was revealed and vegetative cells of both the thraustochytrid and the *Labyrinthula* were detected within the tissues of the leaf. The characteristically fusiform cells of *Labyrinthula* can be seen within the leaf in Fig. 3a, although numerous, the vegetative cells did not appear to be associated with any localised degradation of the plant tissues. The globose sporangia of the thraustochytrid isolate SLBM were present in very large numbers within the leaf, below the cuticle and within the mesophyll (Fig. 3B). At higher magnifications, the sporangia appeared attached to the fabric of the leaf with an indication of localised degradation around the base of the sporangium (Fig. 3C). Circular patches of apparently degraded leaf tissue (Fig. 3D) were detected in some leaf samples. These corresponded in diameter to the base of the thalli of the thraustochytrid and may be areas of attack produced at the interface of the thallus and the leaf surface revealed by the fracturing process when the thalli were pulled away.

Discussion

The mangal is a highly productive system which is not only self sustaining but exports sufficient nutrient to support large trophic webs. The bulk of the energy input is derived from leaf fall and an efficient decay process which according to Subramanian (1988) recycles 60% of the material with only 1–2% persisting as peat.

The recycling of leaf litter in the mangrove is a complex and multi-factorial process which although it involves invertebrate animal species, cannot proceed without decay of the leaves to both reduce particle size and enrich the litter as it decays. The function of enriching nutritionally poor cellulosic plant cell wall material into a protein source can only be carried out by microbial heterotrophs and Heald & Odum (1970) stressed the importance of the microflora in recycling plant material in estuarine and coastal waters and sediments. Fell & Master (1973) documented the sequence of fungal infestation during mangrove leaf decomposition and demonstrated the potential of some species of lower fungi for nutrient enrichment (Fell & Master, 1980). Despite indications that thraustochytrids in particular were associated with decaying cellulosic material in the mangrove (Raghu Kumar, 1988; Ulken, 1984), the only evidence of their physical presence in the leaves was given by Findlay *et al.* (1986). *Labyrinthula*, an obligately marine organism with taxonomic affinities with the thraustochytrids (Moss 1986) has long been known to be associated with a wasting disease of the marine angiosperm *Zostera.* (Renn, 1935). Muehlstein, Porter & Short (1988) were able to confirm that *Labyrinthula* invaded *Zostera marina* L. leaves and reproduced the disease symptoms under laboratory conditions. There is no significant information on the presence or role of *Labyrinthula* in the mangrove.

This study shows that both thraustochytrids and labyrinthulids were associated with mangrove leaf litter in all stages of decay. The indication that the leaves were colonised very soon after falling from the trees was confirmed by the colonisation studies using sterilised leaves placed in the mangrove. Infestation by Labyrinthulomycetes occurred within the first 24 hours of exposure in the mangrove and thereafter they were then present in or upon the leaf material.

The pine pollen baiting of seawater, sediment and detritus indicated a level of inoculum potential, particularly associated with leaf detritus, consistent with such a rapid colonisation of virgin leaf material. The ability of Labyrinthulomycetes to rapidly colonise and invade mangrove leaf tissue, together with their demonstrable ability to penetrate into the internal tissues, suggests they may play an important part in the fragmentation and decomposition of leaves.

In situ decay experiments showed a differential rate of decay of leaf material of *Sonneratia* and *Rhizophora* in the mangrove. *Sonneratia* leaf discs decayed more rapidly, both initially and throughout the study period. It would appear that the *Rhizophora* leaf is more resistant to initial attack but, once decay has started, there is rapid weight loss. This may be due in part to the more robust nature of the *Rhizophora* leaf with its thicker more waxy cuticle which may delay initial attack. The field experiments however showed that decay of leaves of both species is rapid with advanced decay occurring after 7 weeks.

The *in vitro* decay studies using *Sonneratia* leaves, demonstrated the ability of representatives of three genera of thraustochytrids to cause weight losses of between 40% and 46% within 14 days, slightly higher than caused by the isolate of *T. viride* used as a cellulolytic reference organism. In contrast, the *Labyrinthu-*

la isolate caused weight losses not significantly different than those lost from the controls due to leaching processes These results indicate that thraustochytrids have the ability to be potent degraders of cellulosic materials an ability confirmed by the almost skeletonised nature of the leaf discs at the end of the experiment

Despite many reports of thraustochytrids occurring in substantial numbers associated with decaying plant material, (Miller & Jones, 1983, Ulken, 1984, Raghu Kumar, 1985, Findlay *et al*, 1986) however attempts to demonstrate cellulolytic activity (Bahnweg, 1979) proved inconclusive This study provides conclusive evidence that exoenzyme(s) endo β- 1,4 glucanase, capable of degrading carboxymethylcellulose, a soluble form of cellulose is produced by *S aggregatum*, a common and cosmopolitan species of thraustochytrid In nature, thraustochytrids must also be capable of secreting an enzyme able to attack insoluble crystalline cellulose present in cell walls This enzyme, a β-1,4 glucan cellobiohydralase, is often membrane bound and may prove difficult to detect *in vitro*

The results of incubating a thraustochytrid isolate with sterile *Sonneratia* leaves suggest that such an enzyme is present *in vivo*, as after 4 days, there was extensive penetration of the leaf tissue and apparent degradation of cell wall material *Labyrinthula* also penetrated the leaf material rapidly but did not appear to significantly degrade the tissue The *in vitro* decay studies indicate that *Labyrinthula*, although consistently a common member of the mangrove leaf litter microflora, may not be an active decomposer of leaves, causing perhaps only localised penetration and necrosis

In summary, this preliminary study confirms that members of the Labyrinthulomycetes, in particular the thraustochytrids, are strongly implicated in the decay of mangrove leaf litter It is not suggested that they act alone in this respect, but their proven ability to rapidly colonise and penetrate intact leaf material, thus hastening its fragmentation and further decay must afford them a significant role This coupled with the strong indication that leaves decayed by thraustochytrids are enriched with the ω 3 polyunsaturated fatty acids produced by these microorganisms (Findlay *et al*, 1986) suggests an even more important role in introducing these important lipids into the trophic web of the coastal ecosystem

Acknowledgements

The author is grateful to the Browne, Murray and Hill Committee of the Royal Society for providing a travel grant to enable field work at Morib to be carried out Prof Guy Talbot of the Biochemistry Dept University of Laval, Canada, contributed greatly to the work on cellulase activity The author also thanks the Dean of the Institute of Advanced Studies, University of Malaya, Kuala Lumpur for providing laboratory facilities, Prof A J Kutubutheen for support and help during fieldwork visits, Chang May Hing for technical assistance, Colin Derrick for photographic assistance and Prof E B G Jones of the School of Biological Sciences, University of Portsmouth, for his continued support, advice and encouragement without which this study would not have taken place

References

Bahnweg, G , 1979 Studies on the physiology of Thraustochytriales II Carbon Nutrition Veroff Inst Meeresforshung Bremerhaven 17 369–273

Carder, J H , 1986 Detection and condition of cellulase by Congo Red staining of substrates in a cup plate diffusion assay Analyt Biochem 153 75–79

Chamberlain, A H L & S T Moss, 1988 The thraustochytrids a protist group with mixed affinities BioSystems 21 341–348

Fell, J W & I M Master, 1973 Fungi associated with the decay of mangrove (*Rhizophora mangle* L) leaves in South Florida In L H Stevenson & R R Colwell (eds), Estuarine Microbial Ecol ogy University of South Carolina Press Columbia, USA 455–466

Fell, J W & I M Master, 1975 Phycomycete (*Phytophtora* spp nov and *Pythium* sp nov) associated with degrading mangrove leaves Can J Bot 53 2908–2922

Fell, J W & I M Master, 1980 The association and potential role of fungi in mangrove detrital systems Bot Mar 23 257–263

Findlay, R H , J W Fell, N K Coleman & J R Vestal, 1986 In Moss, S T (ed) The Biology of Marine Fungi Cambridge University Press Cambridge UK 91–104

Gaertner, A , 1968 Eine methode des quantitativen Nachweises Niedere, mit pollen koderbare pilze in Meereswater und im sediment Veroff Inst Meeresforschung, Bremerhaven 3 75–92

Gessner, R V, 1980 Degradative enzyme activity by salt marsh fungi Bot mar 23 133–139

Heald, E J & W E Odum, 1970 The contribution of mangrove swamps to Florida fisheries Proc Ann Gulf Carrib Fish Inst 22 130–135

Hyde, K D , 1986 Frequency of occurrence of lignicolous marine fungi in the tropics In S T Moss, (ed), The Biology of Marine Fungi Cambridge University Press Cambridge UK 311–322

Hyde, K D & B D Borse, 1986 Marine fungi from the Seychelles V *Massarina velataspora* a new marine Ascomycete from mangrove wood Mycotaxon 27 161–167

Jones, E B G & T K Tan, 1987 Observations on manglicolous fungi from Malaysia Trans br mycol Soc 89 390–392

Lee, B. K. H. & G. E. Baker, 1973. Fungi associated with the roots of the red mangrove. Mycologia 65: 894–906.

Lugo, A. E. & S. C. Snedaker, 1974. The Ecology of Mangroves. Ann. Rev. Ecol. System. 5: 39–64.

Miller, J. D. & E. B. G. Jones, 1983. Observations on the association of thraustochytrid marine fungi with decaying seaweed. Bot. mar. 26: 345–351.

Moss, S. T., 1986. Biology and phylogeny of the Labyrinthulales and Thraustochytriales. In S. T. Moss (ed.). Cambridge University Press, Cambridge UK: 105–131.

Mouzouras, R., 1989. Decay of mangrove wood by marine fungi. Bot. mar. 32: 65–69.

Muehlstein, L. K., D. Porter & F. T. Short, 1988. *Labyrinthula* sp. a marine slime mold producing the symptoms of wasting disease in eelgrass, *Zostera marina*. Mar. Biol. 99: 465–472.

Raghu Kumar, S., 1985. Enumeration of thraustochytrids (heterotrophic microorganisms) from the Arabian Sea. Mahasagor Bull. Nat. Inst. of Oceanography. 18: 457–465.

Raghu Kumar, S., 1988. *Schizochytrium mangrovei* sp. nov., a thraustochytrid from mangroves in India. Trans. br. mycol. Soc. 90: 627–631.

Renn, C. E., 1935. A mycetozoan parasite of *Zostera marina*. Nature, Lond. 134–416.

Ulken, A., 1984. The fungi of the mangle ecosystem. In F. D. Por & I. Dor. (eds), Hydrobiology of the Mangle. W Junk, The Hague: 27–33.

Hydrobiologia **295**: 97–106, 1995.
Y. S. Wong & N. F. Y. Tam (eds), Asia-Pacific Symposium on Mangrove Ecosystems.

Frequency of occurrence of fungi on wood in Malaysian mangroves

S. A. Alias[1], A. J. Kuthubutheen & E. B. G. Jones[2]
[1] *Department of Botany, University of Malaya, 59100 Kuala Lumpur, Malaysia*
[2] *School of Biological Sciences, University of Portsmouth, King Henry Building, King Henry I Street, Portsmouth, Hampshire PO1 2DY, UK*

Key words: mangrove fungi, frequency of occurrence, marine, tropical

Abstract

Over one hundred fungi have been reported on mangrove wood (Hyde and Jones, 1988; Jones and Kuthubutheen, 1989; Hyde and Jones, unpublished) from tropical and subtropical locations. A variety of factors affect the frequency of occurrence of these fungi, e.g . salinity, length of exposure of substrate, wood species and location within the mangrove. Two aspects are addressed in this paper: a) Comparison of the fungi present at three mangroves in Malaysia (Morib, Kuala Selangor, Port Dickson); and b) The fungi colonising specific mangrove wood, e.g. *Avicennia marina, Bruguiera gymnorrhiza.* Species diversity was greatest at Port Dickson with 63 fungi recorded while species composition varied from site to site. Common fungi at Morib and Kuala Selangor were Halocyphina villosa and Leptosphaeria australiensis along with *Kallichroma tethys* and *Lulworthia grandispora* at the latter site. None of these were common at Port Dickson, the dominant species at this site were *Hypoxylon oceanicum* and *Massarina ramunculicola.* A comparison of the fungi occurring on *Avicennia marina* and *Bruguiera gymnorrhiza* indicated that *Halocyphina villosa* was common on both timbers. Species diversity and abundance were greatest on Avicennia with the following fungi listed as common: *Eutypa* sp., *Kallichroma tethys, Marinosphaera mangrovei, Phoma* sp. and *Julelia avicenniae.*

Introduction

Mangrove vegetation contributes to the primary production to the aquatic environment in the form of leaf and litterfall. Decomposition of this organic material by bacteria and fungi results in protein enriched fragments of detritus. Fungi rather than bacteria have been considered to be the principal source of this increase in nitrogen (Odum and Heald, 1975).

Decomposition of organic material, mainly wood, by marine fungi has recently become a focal point of research (Hyde, 1988b). Prior to the 1980s, little information was available on fungal succession on submerged wood, frequency of occurrence of fungi or their role in the degradation of organic matter in the mangrove ecosystem. Studies were concentrated more on taxonomic and geographical distribution. With the advent of interest in the ecology of these organisms a better understanding of their role and function is now emerging.

Studies of marine fungi occurring on drift and intertidal wood has been undertaken by various workers (Jones & Tan, 1987; Jones *et al.*, 1988; Hyde & Jones, 1988; Hyde, 1989a, b, 1990a, b, 1991; Jones & Kuthubutheen, 1989; Tan *et al.*, 1989; Leong *et al.*, 1991; Tan & Leong, 1990, 1992) and they show that the ascomycetes are common. Decomposition of organic material and wood decay activity of fungi in the mangrove ecosystem has also been demonstrated (Leightley, 1980; Suhirman & Jones, 1984; Mouzouras, 1986, 1989; Vrijmoed *et al.*, 1986a, b).

Tomlinson (1986) lists 54 mangrove trees with 60 associates and to date only 29 species have been investigated for fungi. In Malaysia, the mangrove vegetation in the Indo-Malayan Realm is believed to have reached its optimal development with over 50 species known to exist. Despite the vast area of mangroves in Peninsular Malaysia, only three reports have been published (five mangrove stands) on the occurrence of manglicolous fungi (Jones & Tan, 1987; Jones & Kuthubutheen, 1989; Tan & Leong, 1992).

Thus to gain a better insight of the occurrence of marine fungi in Malaysian mangroves, a study of the frequency of fungi in three mangroves was undertaken. The occurrence of the mycota on specific mangrove trees (*Avicennia marina* var. *resinifera* (Forsk.) Bakh. and *Bruguiera gymnorrhiza* (L.) Lam. was also studied.

Materials and methods

Three mangrove stands were studied for the occurrence of marine fungi (Figure 1). Kuala Selangor has extensive mudflats backed by a small fringe of mangrove forest on the seaward site of the coastal bund. This mangrove has recently been gazetted as a nature park. It is a well established mangrove in an estuary of the Selangor river and brackish (salinity 13–20 ‰). The principal vegetation was *Rhizophora apiculata* Blume, *Avicennia marina* and *Bruguiera gymnorrhiza* and three distinct zones can be observed: *A. marina* (on the river bank), *B. gymnorrhiza* (mid tide) and *R. apiculata* (farther inland). Morib, a large area of mangrove along the coast, the water is brackish (salinity 17-24 ‰), sandy substratum and no zonation was observed. The vegetation was composed of *A. marina*, *Nypa fruticans* (Thunb.) Wurmb., *R. apiculata* and *Sonneratia alba* J. Smith. Port Dickson, two mangrove stands at Blue Lagoon and Cape Rachado with a sandy coastline, rocks, boulders, stone and shingle (salinity 25-30 ‰) and the principal vegetation was: *R. apiculata*, *Rhizophora mucronata* Lam. and *S. alba.*

The study was carried out from December 1991 to August 1992. In the first part of the study, decaying material was collected from the intertidal regions of Kuala Selangor, Morib and Port Dickson mangrove stands. In the second part, collections were made of specific mangrove trees: *A. marina* and *B. gymnorrhiza*. Two types of mangrove subrata were collected: soft tissues (pneumatophores, subterranean roots and young twigs) and lignicolous woody tissue.

At each study site, 200 to 300 samples were collected, placed in clean polythene bags and returned to the laboratory for examination. Surface mud and detritus were washed using sterile sea water. Samples were then incubated in a damp chamber at room temperature for one week and examined for the presence of higher marine fungi. Sterile seawater was added using a fine aerosol spray to prevent the substrata from drying out. Incubated material was periodically examined over a period of three months. Herbarium material and voucher slides of fungi were prepared and are kept in the Department of Botany Herbarium, University of Malaya.

The following data were recorded:

i. List of species and number of collections observed.
ii. Percentage colonization (number of samples supporting fungi × 100 divided by the number of samples examined).
iii. Average number of fungi per sample.
iv. Percentage occurrence of each species (number of collections × 100 divided by the total number of samples examined).

Results

Table 1 lists the fungi and their frequency of occurrence at Kuala Selangor, Morib and Port Dickson mangrove stands.

At Kuala Selangor, a total of 259 samples yielded 51 marine fungi, including 43 Ascomycotina, 2 Basidiomycotina and 6 Deuteromycotina. The most commonly occurring species were *Halocyphina villosa*, *Leptosphaeria australiensis*, *Kallichroma tethys* and *Lulworthia grandispora*. *Marinosphaera mangrovei*, *Eutypa* sp., *Phoma* sp., *Ascocratera manglicola* and *Dactylospora haliotrepha* were frequently encountered. Percentage colonization was high *i.e.* 85.6% and the average number of fungi per sample was 2.2.

For the mangrove stand at Morib, *Leptosphaeria australiensis* and *Halocyphina villosa* were the most common species while *Hypoxylon oceanicum*, *Kallichroma tethys*, *Phoma* sp., *L. grandispora* and *Rhabdospora avicenniae* were frequently encountered. Of the 268 samples examined, 258 samples were found to support fungi (96%) including 43 Ascomycotina, 2 Basidiomycotina and 9 Deuteromycotina. A total of 573 fungal collections were found from 258 samples with an average number of fungi per sample of 2.1.

In comparison with the collections from Kuala Selangor and Morib, Port Dickson showed greater species diversity. The dominant species at this site were *Hypoxylon oceanicum* and *Massarina ramunculicola* while *Leptosphaeria australiensis*, *Kallichroma tethys*, *Halosarphaeia ratnagiriensis*, *Savoryella lignicola*, *Phoma* sp., *Marinosphaera mangrovei*, *Lulworthia grandispora*, *Verruculina enalia* and *D. haliotrepha* were frequent. Sixty-four species were found on 250

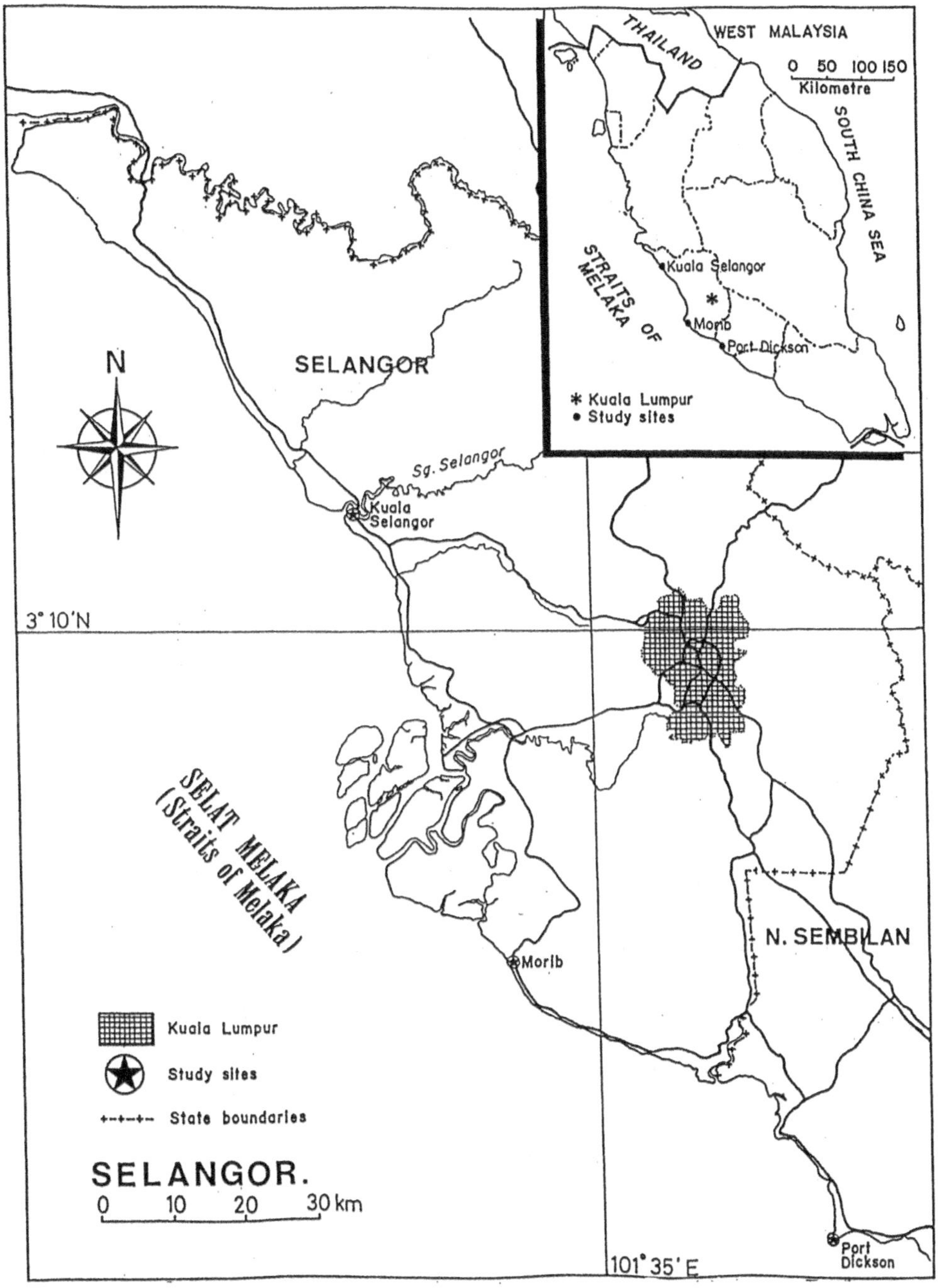

Fig. 1. Map of Malaysia showing study sites (Kuala Selangor, Morib, Port Dickson) (Redrawn by Zainal Isa from Asian Wetland Bureau).

Table 1 Tropical marine fungi collected at 3 test sites in Malaysia

Fungus	Percentage occurrence		
	Kuala Selangor	Morib	Port Dickson
Aegerita sp *	–	1.1	–
Aigialus grandis Kohlm & Schatz	0.4	4.5	3.6
Aigialus mangrovei Borse	0.4	1.1	2.4
Aigialus parvus Schatz & Kohlm	–	0.7	2.4
Aigialus sp 69*	–	0.7	2.4
Aniptodera chesapeakensis Shearer & Miller	0.8	–	–
Aniptodera mangrovei Hyde	–	1.1	–
Antennospora quadricornuta (Cribb & Cribb) T W Johnson	–	1.1	0.4
Antennospora salina (Meyers) Jones*	–	0.4	–
Anthostomella sp 43*	–	1.9	0.8
Anthostomella sp 57*	–	0.4	–
Arthrobotrys oligospora Fres *	–	–	0.4
Ascocratera manglicola Kohlm *	6.9	1.1	3.2
Ascomycete sp 8*	0.4	–	–
Ascomycete 17*	0.4	–	–
Ascomycete sp 21*	0.4	–	–
Ascomycete sp 22*	0.4	1.1	2.0
Ascomycete sp 24*	–	–	4.8
Ascomycete sp 31*	–	–	0.4
Ascomycete sp 33*	–	–	0.4
Ascomycete sp 34*	0.4	–	–
Ascomycete sp 49*	–	–	0.4
Ascomycete sp 53*	–	–	0.4
Ascomycete sp 66*	–	0.8	–
Ascomycete sp 73*	–	–	0.8
Ascomycete sp 77*	–	–	0.4
Ascomycete sp 86*	–	–	0.4
Ascomycete sp 88*	0.8	0.4	4.0
Ascomycete sp 90*	–	–	0.4
Ascomycete sp 91*	–	–	0.4
Ascomycete sp 93*	–	–	0.4
Ascomycete sp 94*	–	0.8	1.2
Biatriospora marina Hyde & Borse	0.4	–	–
Calathella mangrovei Jones & Agerer	3.5	–	–
Cirrenalia basiminuta Raghukumar & Zainal	–	0.4	–
Cirrenalia tropicalis Kohlm	0.4	0.4	0.4
Cirrenalia pygmea Kohlm	0.8	0.4	2.0
Cucullosporella mangrovei (Hyde & Jones) Hyde & Jones	0.4	1.1	0.8
Dactylospora haliotrepha (Kohlm & Kohlm) Hefellner	7.9	3.4	6.8
Dactylospora sp 48*	0.4	–	–
Dictyosporium pelagicum (Linder) G C Hughes	0.4	–	0.4

* New record

Table 1. Continued.

Eutypa sp. 1*	10.3	5.2	–
Eutypa sp.2*	–	1.2	3.6
Eutypa sp. 3*	0.8	–	0.4
Fasciatispora lignicola Alias, Jones & Kultub.*	1.7	0.4	–
Halocyphina villosa Kohlm. & Kohlm.	25.3	16.4	0.4
Halosarpheia abonnis Kohlm.	3.5	4.1	0.8
Halosarpheia cincinnatula Shearer & Crane	0.4	–	–
Halosarpheia fibrosa Kohlm. & Kohlm.	–	0.4	–
Halosarpheia sp.	–	–	0.8
Halosarpheia lotica Shearer	0.4	–	–
Halosarpheia marina (Cribb & Cribb) Hyde	3.1	1.1	–
Halosarpheia minuta Leong	0.8	–	–
Halosarpheia minuta-like	0.4	–	–
Halosarpheia ratnagiriensis Patil & Borse	4.8	4.1	12.4
Halosarpheia retorquens Shearer & Crane	–	–	4.0
Halosarpheia sp.	–	–	0.4
Helicascus kanaloanus Kohlm.	–	–	1.0
Humicola alopallonella Meyers & Moore	–	–	0.8
Hypocrea sp.*	0.4	–	–
Hypoxylon oceanicum Schatz.	4.3	14.9	15.2
Kallichroma tethys (Kohlm. & Kohlm.) Kohlm. & Volkm.-Kohlm.	17.0	15.7	13.6
Kallichroma glabrum (Kohlm.) Kohlm. & Volkm.-Kohlm.*	–	0.4	–
Julella avicenniae (Borse) Hyde*	4.8	7.1	–
Lautospora sp. 85*	–	–	0.4
Leptosphaeria australiensis (Cribb & Cribb) Hughes	17.0	20.5	16.8
Lignincola laevis Höhnk	–	4.1	4.8
Lignincola longirostris (Cribb & Cribb) Kohlm.	5.2	1.5	5.2
Lignincola tropica Kohlm.	0.4	–	0.4
Lineolata rhizophorae (Kohlm. & Kohlm.) Kohlm. & Volkm.-Kohlm.*	–	–	0.4
Linocarpon nypae (P. Henne) Hyde	–	0.4	–
Lophiostoma mangrovei Kohlm.	–	0.4	–
Lophiostoma sp. 26*	–	–	2.8
Lulworthia grandispora Meyers	14.8	7.5	7.2
Marinosphaera mangrovei Hyde	12.2	4.1	7.2
Massarina sp. 4*	0.4	–	–
Massarina ramunculicola Hyde	–	2.2	15.2
Massarina thallasiae Kohlm. & Voklm.-Kohlm.	2.2	2.2	4.8
Massarina velatospora Hyde & Borse	1.3	3.7	5.6
Melaspilea sp.*	–	0.7	0.4
Nais glitra Crane & Shearer	1.3	–	–
Passeriniella savoryellopsis Hyde & Mouzouras	0.4	–	–

* New record

samples and 222 samples supported fungi (53 Ascomycotina, 10 Deuteromycotina, 1 Basidiomycotina). A total of 516 fungal collections were made, the percentage colonization was 88.8% and the average number of fungi per sample was 2.1.

A total of 757 fungal collections from the three mangrove stands yielded 100 species of higher marine manglicolous fungi representing 82 Ascomycotina, 3 Basidiomycotina and 15 Deuteromycotina.

Table 1 Continued

Phaeosphaeria sp 5*	0 4	0 4	0 4
Periconia prolifica Anastasiou	0 4	0 4	–
Phoma sp 1*	9 0	7 8	12 0
Phoma sp 4*	–	0 4	0 4
Phoma sp 41*	–	2 2	7 2
Pyrenographa sp *	–	4 1	1 7
Quintaria lignatilis (Kohlm) Kohlm & Voklm -Kohlm	–	0 7	2 4
Rhabdospora avicenniae Kohlm & Kohlm	–	7 1	1 2
Rhizophila marina Hyde & Jones*	2 6	4 9	4 8
Savoryella lignicola Jones & Eaton	2 6	–	15 2
Savoryella longispora Hyde & Jones*	1 3	4 1	8 0
Savoryella paucispora (Cribb & Cribb) Koch	–	–	0 7
Swampomyces triseptatus Hyde & Nakagiri	–	–	1 6
Trichocladium achrasporum (Meyers & More) Dixon	2 6	–	–
Trichocladium linderi Crane & Shearer (Kohlm)	–	–	0 7
Trichocladium opacum Hughes*	–	–	0 4
Torpedospora radiata Meyers*	–	0 7	–
Verruculina enalia (Kohlm) Kohlm & Voklm -Kohlm	5 2	6 3	6 8
Total	430	483	376
Empty perithecia	63	90	140
Total number of collection	492	573	519
Number of samples examined	259	268	250
Number of sample colonized	204	258	222
% colonization	89 1	96	88
Average number of fungi per sample	2 2	2 1	2 1
Total number of fungi collected	51	54	64
Ascomycotina	45	45	54
Basidiomycotina	2	2	1
Deuteromycotina	5	9	10

* New record

Table 2 presents the results on the occurrence of fungi in the *Avicennia marina* and *Bruguiera gymnorrhiza* zone Of the 95 mangrove samples examined from the *Avicennia* zone, 80 were found to support 30 fungi The number of fungi per sample and percentage colonization were 2 4 and 84% respectively The most common fungi were *Halocyphina villosa* and *Kallichroma tethys Eutypa* sp , *Lulworthia grandispora, M mangrovei, Leptosphaeria australiensis, Phoma* sp and *Julella avicenniae* were frequently encountered The collections in the *Bruguiera* zone were lower in number of samples colonized, percentage colonization and the number of fungi per sample, i e 64, 75 % and 1 2 respectively

Discussion

The mangrove ecosystem is an ideal environment for the growth ard reproduction of fungi, a fact supported by the diversity of fungi encountered (Jones & Hyde, 1988) Hyde and Jones (1988) recognized three ecological niches within the mangrove, *viz* intermittently submerged driftwood, exposed wood of damaged mangroves roots and branches and bark of mangrove roots and branches The percentage occurrence as an expression of the frequency of collections of fungi on these substrata gives an indication of the more common fungi within the mangrove ecosystem (Hyde & Jones, 1988)

Data on the frequency of occurrence of intertidal mangrove fungi in Malaysia were published in Jones

Table 2 Marine fungi collected at *Avicennia* and *Bruguiera* zone

Fungus	Percentage occurrence	
	Avicennia zone	*Bruguiera* zone
Aigialus parvus	1 0*	5 3*
Ascocratera manglicola	5 0	3 2*
Ascomycete sp 8	1 0	–
Ascomycete sp 17	–	1 1
Ascomycete sp 22	4 0	–
Ascomycete sp 34	1 0	–
Ascomycete sp 50	1 0	–
Ascomycete sp 88	2 0	–
Biatrospora marina	1 0*	–
Cirrenalia pseudomacrocephala Kohlm	–	3 1*
Cucullosporella mangrovei	–	2 1*
Dactylospora haliotrepha	4 0	–
Dactylospora sp 48	–	1 1
Eutypa sp	19 0	–
Fasciatispora lignicola	–	1 1
Halocyphina villosa	35 0*	21 1*
Halosarpheia abonnis	5 0	–
Halosarpheia lotica	1 0	–
Halosarpheia ratnagiriensis	12 0*	–
Kallichroma tethys	26 0*	–
Hypoxylon oceanicum	4 0*	7 4*
Julella avicenniae	12 0*	–
Leptosphaeria australiensis	14 0	6 3*
Lignincola longirostris	6 0	3 1*
Lignincola tropica	1 0*	–
Lulworthia grandispora	18 0	4 2*
Marinosphaera mangrovei	15 0*	5 3*
Massarina sp 4	1 0	–
Massarina thallasiae	2 0*	–
Massarina velatospora	–	3 1*
Nais glitra	1 0*	–
Phaeosphaeria sp	–	1 1*
Periconia prolifica	–	4 2
Phoma sp	16 0	6 3
Rhizophilla marina	3 0	5 3*
Savoryella longispora	5 0*	1 1*
Trichocladium opacum	2 0	–
Verruculina enalia	10 0*	3 2*
Total	216	85
Empty fruit body	15	21
Total number of collection	231	106
Number of sample examined	95	85
Number of sample colonized	80	64
% colonization	84	75
Number of fungi per sample	2 4	1 2

* New host record

Table 3 Very frequent and frequent fungi

Mangrove stand	Very frequent fungi	Frequent fungi
Kuala Selangor	*Halocyphina villosa*	*Marinosphaera mangrovei*
	Leptosphaeria australiensis	*Eutypa* sp
	Kallichroma tethys	*Phoma* sp
	Lulworthia grandispora	*Ascocratera manglicola*
		Dactylospora haliotrepha
Morib	*Leptosphaeria australiensis*	*Hypoxylon oceanicum*
	Halocyphina villosa	*Kallichroma tethys*
		Phoma sp
		Lulworthia grandispora
		Rhabdospora avicenniae
Port Dickson	*Hypoxylon oceanicum*	*Savoryella lignicola*
	Massarina ramunculicola	*Leptosphaeria australiensis*
		Kallichroma tethys
		Halosarpheia ratnagiriensis
		Phoma sp
		Marinosphaera mangrovei
		Lignincola longirostris
		Verruculina enalia
		Dactylospora haliotrepha

& Tan (1987), Jones & Kuthubutheen (1989) and Tan & Leong (1992) From these studies, *Halocyphina villosa, Hypoxylon oceanicum, Halosarpheia marina* and *Lulworthia* sp were the most frequently encountered species and *Kallichroma tethys, Lulworthia grandispora, Verruculina enalia, Savoryella lignicola, Lignincola laevis, Dactylospora haliotrepha, Cytospora rhizophorae* Kohlm , *Trichocladium achrasporum* and *Savoryella paucispora* were commonly occurring On the basis of the percentage of occurrence, the fungi observed in the present study can be grouped into very frequent (with 15% and more) and frequent species (6–14%) (Table 3) In comparison with the species encountered in this study and from those reported earlier, *Halocyphina villosa, Kallichroma tethys, L grandispora, Hypoxylon oceanicum, D haliotrepha, Verruculina enalia* and *S lignicola* can be regarded as most commonly occurring in the Malaysian mangrove ecosystem

Hyde & Jones (1988) listed 25 common intertidal mangrove fungi From the present study and the work of Jones and Tan (1987), Borse (1988), Hyde & Jones (1988), Jones *et al* (1988), Hyde (1988a, b, 1989a, b, c, 1990a, b, c), Jones and Kuthubutheen (1989), Tan *et al* (1989) and Tan & Leong (1990, 1992), more than 60 species can now be regarded as commonly occurring in mangrove ecosystems of the West Indo Pacific region Although there are differences in the common species at these study sites, a 'core' group of fungi occurring in the mangrove ecosystem (Hyde, 1990a) can now be recognized

In the present study, the percentage colonization for the three sites were higher than those seen in the Philippines (80%) (Jones *et al* , 1988), Mandai, Singapore (85%) (Tan *et al* , 1989), Pontian Besar, Malaysia (80 4%) (Tan & Leong, 1992) but lower than the 100% obtained at Sungei Geylang Patah (Jones & Tan, 1987) The average numbers of fungi per sample in this study were higher than Sungei Geylang Patah (1 3) (Jones & Tan, 1987) and Pontian Besar (1 5) (Tan & Leong, 1992) It was also higher than in the Seychelles (1 1, 1 5) (Hyde & Jones, 1988) and comparable to Singapore (2 2) (Tan *et al* , 1989)

The differences in the percentage colonization, species diversity and average number of fungi per sample seen in this study, and those reported from the literature can be attributed to several factors Kohlmeyer (1983, 1984), Hyde (1988a, b), Hyde & Jones (1988), Jones *et al* (1988), Jones and Kuthubutheen (1989) and Tan and Leong (1992) discussed some possible factors influencing the occurrence of fungi in a mangrove ecosystem. Amongst the factors are environmental conditions, the nature and origin of substrata examined, the different mangrove tree flora and location of the mangrove sites studied.

Care must be taken in the investigation of the frequency of occurrence of marine fungi from "randomly" collected substrata. For example *Massarina ramunculicola* a was the most common species occurring at Port Dickson mangrove stand and this cou]d be attributed to the large quantity of prop roots of *R. apiculata* collected, while Tan & Leong (1990), accounted for the high percentage of occurrence of *Halosarphaeia retorquens*, *Lignincola laevis* and *Halosarphaeia marina* to the young mangrove twigs which made up the bulk of the material they collected in the four mangroves studied.

Mangrove trees are able to grow at salinities ranging from full sea water (35‰) to fresh water, thus a different fungal flora can be expected within this salinity gradient (Kohlmeyer, 1969). Kohlmeyer (1969) reported on the gradual change from a marine fungal flora to a freshwater flora in the Heeia Swamp, Hawaii: *Kallichroma tethys*, *Lignincola laevis* and *Leptosphaeria australiensis* were found in the area where salinity was 35‰; but farther inland where the salinity ranged from 10‰ to 35‰, *Verruculina enalia* and *Helicascus kanaloanus* were found. In the present work *Halocyphina villosa*, *Kallichroma tethys*, *Eutypa* sp., *Lulworthia grandispora*, *Phoma* sp., *Marinosphaera mangrovei* and *Leptosphaeria australiensis* were found at high percentage frequencies (13–33%) in the *Avicennia* zone where the salinity ranged from 18–20‰ while they are less abundant in the *Bruguiera* zone (salinity 13–14 ‰). *Leptosphaeria australiensis* and *Verruculina enalia* were also frequently found at Morib and Port Dickson where the mangrove stand is situated on the open sea. In the present work, a higher density of species were found in the *Avicennia* zone than in the *Bruguiera* zone. The results of this study give an insight into the pattern of horizontal distribution of the fungi present in the mangrove ecosystem.

Less research has been undertaken on the occurrence of manglicolous fungi on specific mangrove trees (Hyde, 1990a). Hyde (1990a) studied the fungal communities on five trees species namely *A. alba*, *R. apiculata*, *R. mucronata*, *S. alba* and *Xylocarpus granatum*. Some fungi were found to be specific and limited to a single tree species, e.g. *Xylocarpus: Caryospora mangrovei* Kohlm.; *Avicennia: Aigialus grandis* and *Eutypa* sp. The most common fungi colonizing different tree species also varied, e.g. *Xylocarpus: Passeriniella savoryellopsis* and *Sphaerulina cf. oraemaris* Linder; *A. alba, S. alba*: *Hypoxylon oceanicum, Lulworthia* sp., *Halosarpheia ratnagiriensis; Rhizophora* species: *Cirrenalia pygmea, Caryospora rhizophorae* and *Leptosphaeria australiensis*.

Hyde & Jones (1988) listed 90 species of fungi known to inhabit 17 species of mangrove trees which are mostly reported from the Atlantic ocean. With recent publications from the West Indo Pacific region (Hyde, 1988, 1992; Jones & Kuthubutheen, 1989; Jones & Hyde, 1988; Kohlmeyer & Voklmann-Kohlmeyer,1987) more species have been shown to have a wide distribution. For example, *Rhabdospora avicenniae* reported from the Atlantic Ocean, has recently been collected from Malaysia (Jones & Kuthubutheen, 1989; present work).

Hyde (1991) and Tan & Leong (1990) observed that the ascomycetes were the most common taxonomic group in the intertidal mangrove region. In the present study, more than 80% of the collected species were ascomycetes and indicates their importance in the mangrove habitat.

Jones & Kuthubutheen (1989) listed 82 species collected from four mangrove stands in Malaysia. From the present study, and work by other workers (Tan, 1985; Jones & Tan, 1987; Jones & Kuthubutheen, 1989; Tan & Leong, 1992), 169 species of marine fungi have been recorded in Malaysian mangroves while 56 species are new records for Malaysia.

The high percentage of occurrence in the present study reflects the abundance and rich species diversity in Malaysian mangroves. Factors contributing to these, need to be thoroughly investigated before further conclusions on their occurrence and ecology in the ecosystem can be made. Collections from the East-Coast region, which have never been investigated for higher marine fungi, will undoubtly yield more species and further our knowledge of the common mangrove fungi of Malaysia.

Acknowledgements

S. A. Alias would like to thank the University of Malaya for the award of a fellowship and Mr. Kamarudin Mohd. Isa for help with the field studies. Prof. E. B. G . Jones is grateful to the British Council and the University of Malaya for travel grants to visit Malaysia.

References

Borse, B D 1988 Frequency of occurrence of marine fungi from Maharashtra Coast, India Indian J of Marine Sc 17 165–167

Hyde, K D 1988a Observation on the vertical distribution of marine fungi on *Rhizophora* spp at Kampung Danau mangrove, Brunei Asian Mar Biol 5 77–81

Hyde, K D 1988b Studies on the tropical marine fungi of Brunei Bot J Linn Soc 98 135–151

Hyde, K D 1989a Ecology of tropical marine fungi Hydrobiologia 178 199–208

Hyde, K D 1989b Intertidal mangrove fungi from North Sumatra Can J Bot 67 3078–3082

Hyde, K D 1989c Vertical zonation of intertidal mangrove fungi In T Hattori, Y Ishida, Y Maruyama, R Y Morita and A Uchida (eds), Recent Advances in Microbial Ecology Japan Scientific Society Press 302–306

Hyde, K D 1990a A comparison of the intertidal mycota of five mangrove tree species Asian Mar Biol 7 93–107

Hyde, K D 1990b A study of the vertical zonation of intertidal fungi on *Rhizophora apiculata* at Kampong Kapok mangrove, Brunei Aquat Bot 36 255–262

Hyde, K D 1990c *Caryospora mangrovei* sp nov and notes on marine fungi from Thailand Trans mycol Soc Japan 30 333–342

Hyde, K D 1991 Fungal colonization of *Rhizophora apiculata* and *Xylocarpus granatum* poles in Kampung Kapok mangrove, Brunei Sydowia 43 31–38

Hyde, K D 1992 *Julella avicenniae* (Borse) comb nov (Thelenellaceae) from intertidal mangrove wood and miscellaneous fungi from the NE coast of Queensland Mycol Res 96 939–942

Hyde, K D & E B G Jones, 1988 Marine mangrove fungi Mar Ecol (P S Z N I) 9 15–33

Jones, E B G & K D Hyde, 1988 Methods for the study of mangrove fungi In A D Agate, C V Subramanian & M Vanucci (eds), Mangrove Microbiology, Role of Microorganism in Nutrient Cycling of Mangrove Soils and Waters UNDP/UNESCO 9–27

Jones, E B G & Kuthubutheen, A J 1989 Malaysian mangrove fungi Sydowia 41 160–169

Jones, E B G & T K Tan, 1987 Observation on manglicolous fungi from Malaysia Trans Br mycol Soc 89 390–392

Jones, E B G , F R Uyenco, & M P Follosco, 1988 Fungi on driftwood collected in the intertidal zone from the Phillipines Asian Mar Biol 5 103–106

Kohlmeyer, J 1969 Ecological notes on fungi in mangrove forest Trans Br mycol Soc 53 237–250

Kohlmeyer, J 1983 Geography of marine fungi Aust J Bot Supply Ser 10 67–76

Kohlmeyer, J 1984 Tropical marine fungi Mar Ecol (Berlin) 6 320–378

Kohlmeyer, J & Volkmann-Kohlmeyer, B 1987 Marine fungi from Belize with a description of two new genera of ascomycetes Bot Mar 30 195–204

Leightley, L E 1980 Wood decay activities of marine fungi Bot Mar 23 387–396

Leong, W F , T K Tan, & E B G Jones, 1991 Fungal colonization of submerged *Bruguiera cylindrica* and *Rhizophora apiculata* wood Bot Mar 23 387–393

Mouzouras, R 1986 Patterns of timber decay caused by marine fungi In S T Moss (ed), The Biology of Marine Fungi Cambridge University Press, Cambridge 341–353

Mouzouras, R 1989 Decay of mangrove wood by marine fungi Bot Mar 32, 65–69

Odum, W E & Heald, E J 1975 The detritus food web of an estuarine mangrove community In L E Cronin (ed), Estuarine Research Academic Press 265–286

Suhirman & E B G Jones, 1984 Preliminary observations on lignicolous marine fungi in the Java Sea, Indonesia Annls Bogoriens-es 8 35–49

Tan, T K 1985 Observation on marine fungi in Singapore and Penang (Malaysia) Trans Br mycol Soc 85 726–727

Tan, T K & W F Leong, 1990 Mangrove fungi of Singapore and some possible factors influencing their occurrence Trans mycol Soc Japan 31 35–44

Tan, T K & W F Leong, 1992 Lignicolous fungi of tropical mangrove wood Mycol Res 96 413–414

Tan, T K , W F Leong, & E B G Jones, 1989 Succession of fungi on wood of *Avicennia alba* and *A lanata* in Singapore Can J Bot 67 2686–2691

Tomlinson, P B 1986 The botany of mangroves Cambridge Univ Press , Cambridge, 413 pp

Vrijmoed, L L P , I J Hodgkiss, & L B Thrower, 1986a Effects of surface fouling organisms on the occurrence of fungi on submerged pine blocks in Hong Kong coastal water Hydrobiologia 135 123–130

Vrijmoed, L L P , I J Hodgkiss, & L B Thrower, 1986b Occurrence of fungi on submerged pine and teak blocks in Hong Kong coastal waters Hydrobiologia 135 109–122

Hydrobiologia **295**: 107–118, 1995.
Y. S. Wong & N. F. Y. Tam (eds), Asia-Pacific Symposium on Mangrove Ecosystems.

Ecology of mangrove fungi and their role in nutrient cycling: what gaps occur in our knowledge?

K. D. Hyde & S. Y. Lee
Department of Ecology and Biodiversity, The University of Hong Kong, Pokfulam Road, Hong Kong

Key words: mangrove fungi, ecology, nutrient cycling, succession, vertical zonation.

Abstract

Recent investigations have increased our knowledge of the ecology of mangrove fungi. In this paper this information is reviewed with emphasis on biogeography, biodiversity, differences in the tropical and subtropical mycoflora, fungal distribution on mangroves trees, host specificity, vertical zonation and distribution with salinity. Gaps in our knowledge are discussed. There is little knowledge of the role of mangrove fungi in nutrient cycling which is also reviewed. Areas in which knowledge is deficient include quantification techniques for fungal abundance, the nature and activities of fungal extracellular enzymes and fungal modification of mangrove detritus matter, especially the dissolved form.

Introduction

Coastal wetlands are considered to be one of the most productive natural ecosystems on earth (Kohlmeyer & Volkmann-Kohlmeyer 1993a). Both in the mangal and salt marshes the mycota are considered to be extremely important in nutrient cycling. There have been numerous studies on the amount of litter production in mangrove forests (see review by Saenger & Snedaker 1993) and the role of micro-organisms in the decay of leaf litter is well documented (Fell & Master 1973; 1980; Cundell *et al.* 1979; Robertson *et al.* 1992) as is their importance in the mangrove food web (Heald 1971; Odum & Heald 1975; Fell & Master 1980; Tenore *et al.* 1982; Hutchings & Saenger 1987; Robertson *et al.* 1992). Less is known about wood production, but this is also thought to be high (Christensen, 1978; Bunt *et al.* 1979; Boto *et al.* 1984). Studies of the involvement of the fungi in the breakdown of mangrove leaves and wood have also been conducted (e.g., Hyde 1990a). This paper addresses the ecology of these fungi and their role in nutrient cycling, discussing available information and highlighting areas where knowledge is lacking.

Biogeography

Biogeographical distribution maps of marine fungi are published by Hughes (1974), Kohlmeyer (1983) and Jones (1993) and include mangrove fungi. Kohlmeyer & Kohlmeyer (1979) reviewed the literature on higher mangrove fungi and only 42 species were then known to occur on submerged parts of mangroves. Most of these were records of fungi on *Rhizophora mangle* L. in Florida, the Caribbean or central America with few records from other continents (Fig. 1). Collections of mangrove fungi have now been made throughout S.E. Asia (Hyde 1988a, 1989a,b, 1990b,c; Jones & Tan 1987; Jones *et al.* 1988; Jones & Kuthubutheen 1989; Hyde *et al.* 1990, 1993; Vrijmoed *et al.* 1994); in the Indian Ocean (Borse 1988; Hyde 1986; Hyde & Jones 1988, 1989; Jones & Hyde 1990; Chinnaraj 1993); the Pacific Ocean (Hyde, 1990d, 1992e; Kohlmeyer 1984; Kohlmeyer & Volkmann-Kohlmeyer 1987a, 1991, 1993b; Nakagiri 1993) and tropical Atlantic Ocean (Aleem 1980; Kohlmeyer, 1980, 1981; Kohlmeyer & Volkmann-Kohlmeyer 1987b, 1988) and the distribution is now better known (Fig. 1).

In this paper the distributions of four representative mangrove fungi are given, i.e. *Halosarpheia fibrosa* Kohlm. & Kohlm. (Fig. 2); *Halosarpheia marina* (Cribb & Cribb) Kohlm. (Fig. 3); *Lignincola laevis*

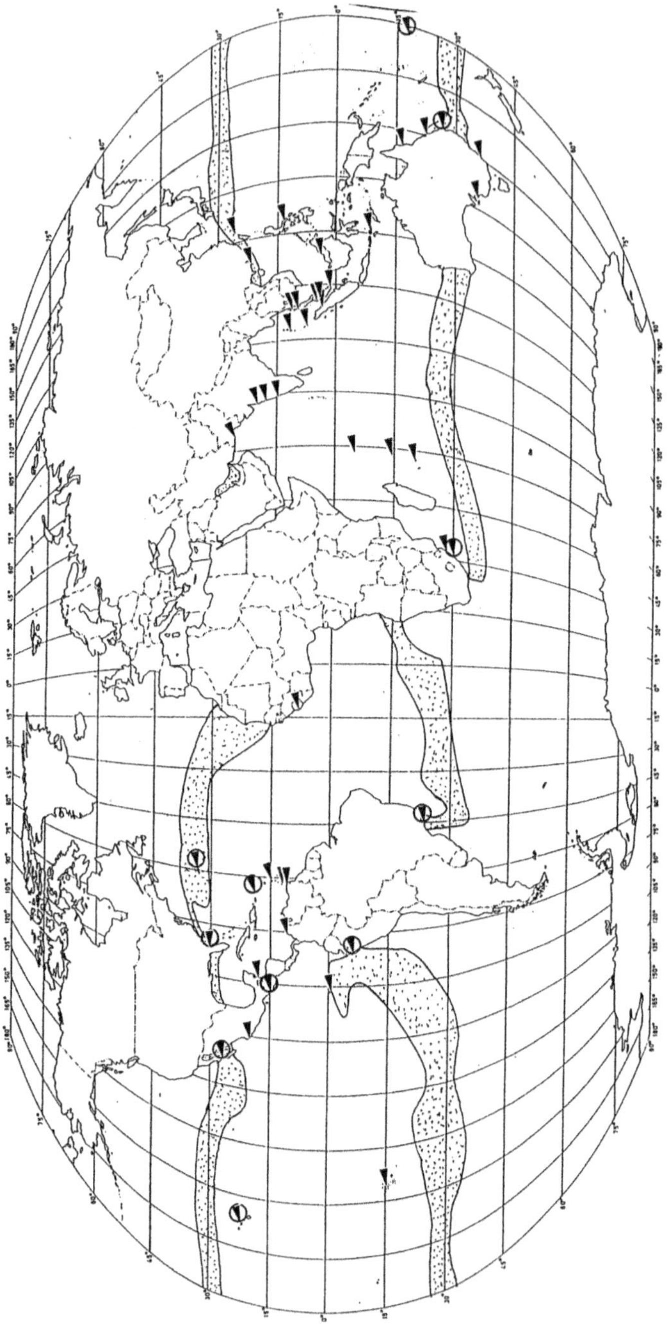

Fig. 1. Locations of reports of mangrove fungi in 1979 (○) and 1993 (◁). Shaded areas are subtropical and non-shaded areas between the shaded areas are tropical.

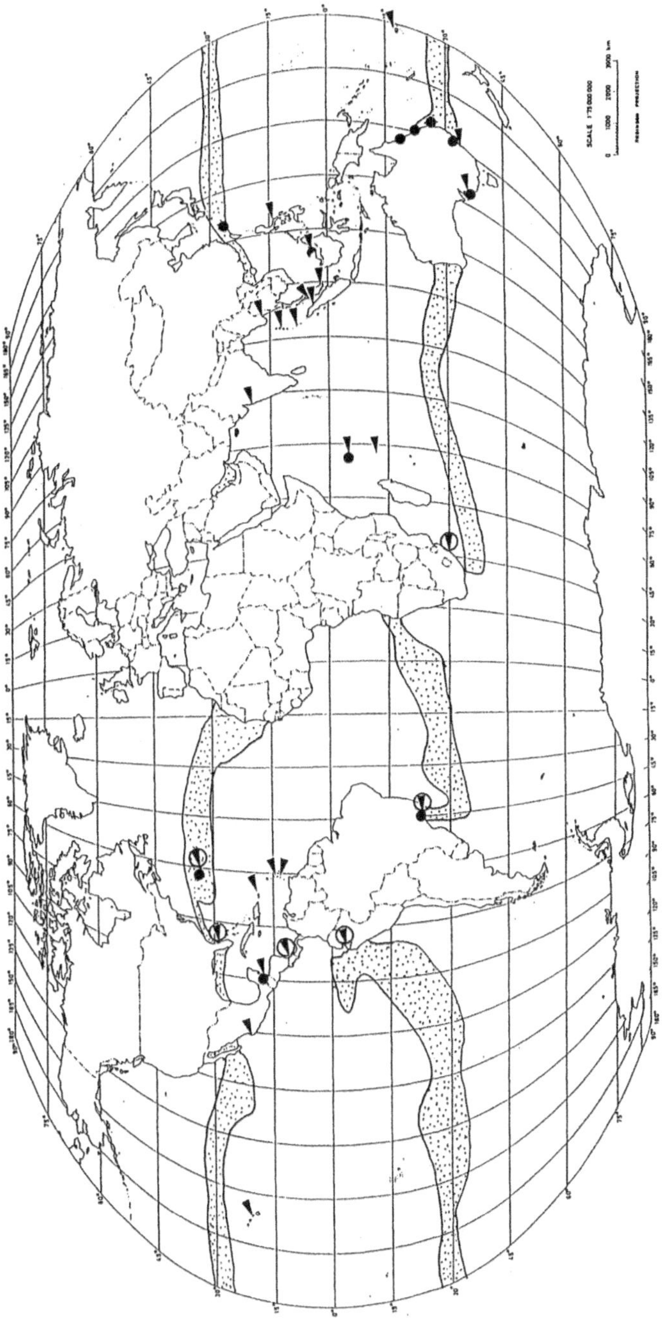

Fig. 2. World distribution of *Lulworthia grandispora* in mangroves in 1979 (○) and 1993 (◁) and *Halosarpheia fibrosa* in 1993 (●).

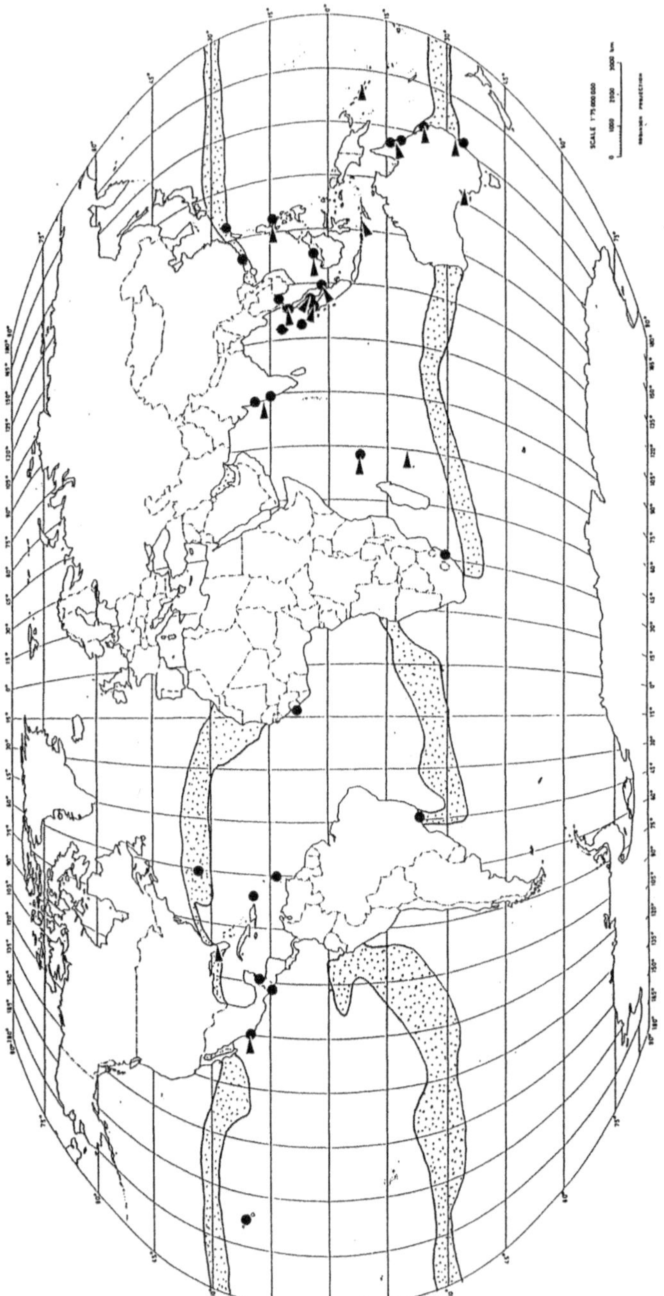

Fig. 3. World distribution of *Lignincola laevis* (•) and *Halosarpheia marina* (▷) in 1993.

Höhnk(Fig. 3) and *Lulworthia grandispora* Meyers (Fig. 2). In all cases the fungi show a pantropical distribution. This is true of most intertidal fungi, since few are host-specific and most of their hosts are pantropical in distribution (Hyde 1990a). Major gaps, however, still occur in our knowledge. There is a dearth of information from Northern and Western Australia and, considering the expanses of Australian shorelines and the potential importance of fungi in nutrient cycling to coastal ecosystems, this is a poor reflection on Australian science. Information is also lacking in the Pacific Ocean, with the exception of the Hawaiian (Oahu) and Society Islands (Moorea). Few studies have been carried out in the mangroves of South America and there is a paucity of information from African mangroves (Fig. 1).

Until recently information on mangrove fungi outside the tropics was not available. However Vrijmoed (1990), Vrijmoed *et al.* (1994) and Hyde (1990d) have published species lists for Hong Kong (subtropical) and Australia (NSW & Victoria: warm temperate) respectively. This enables a comparison with tropical mangrove mycota.

Many mangrove fungi are found in the sub-tropics and the warm temperate regions, but the species diversity in the tropics is much greater. In Brunei 100 species were identified (Hyde 1988a,c), in Hong Kong 60 species (Vrijmoed *et al.* 1994) and in Australia 60 species (Hyde 1990d, 1992a; Kohlmeyer & Volkmann-Kohlmeyer 1991). This may be a reflection of mangrove species richness or the time spent on each study. In Brunei 25 mangrove tree species were examined, in Hong Kong three, while in Australia seven were examined. It may also partly be a reflection of adaptability, with fewer species being able to compete in subtropical conditions. Host specificity does not seem to be the limiting factor, as most mangrove species collected in Brunei do not appear to be host specific (Hyde 1990a). This question could be addressed by growing fungi under controlled temperature regimes.

Notable discrepancies are found in the latitudinal distribution of some mangrove fungi. *Halosarpheia fibrosa*, for example, is a rather illusive taxon described from Bermuda with very few other collections (Fig. 2). The presence of this taxon in warm temperate Australia (Hyde 1990d) may indicate a preference for subtropical as compared to tropical conditions such as those climates found in Bermuda and Southern Australia. Collections in other subtropical locations (e.g. Africa) may answer these questions of temperature selection, although the absence of this taxon from Hong Kong is conspicuous. The collection in tropical Brunei may have been misidentified. Certainly, *H. fibrosa* is rare in the tropics and *Halosarpheia marina* is also common in the subtropics (Fig. 3), as well as in the tropics.

Further notable fungi found in Southern Australia were *Halosphaeria appendiculata* Linder and *Ceriosporopsis halima* Linder. The distribution of these taxa are normally temperate and their presence in warm temperate mangroves is interesting. Some information is now available on the mycota of subtropical mangroves, but further studies are required to establish conclusive differences with tropical mangroves. Growth studies conducted under different temperature regimes may also help us to understand the biogeography of the mangrove fungi.

Biodiversity

Kohlmeyer & Volkmann-Kohlmeyer (1993b) compared the marine mycota of recently introduced *Rhizophora* species in the Pacific Ocean (Oahu and Moorea), with that of established *Rhizophora* stands in the Caribbean (Belize). Only seven and 21 species were identified in Moorea and Oahu respectively, while 43 species are known from *Rhizophora* in Belize. Reasons for the smaller number of taxa in Hawaiian mangroves were given as 1) species being rare or not fruiting on the collected material, and 2) missing species being restricted, in a broad sense, to mangroves and perhaps not yet having reached Hawaii. The even smaller number of fungi in Moorea may be accounted for by 1) the small size of the *Rhizophora* stands and trees, 2) the recent introduction of the mangrove (45 years ago) and 3) the lack of dead stems and roots in the young healthy thriving mangroves (Kohlmeyer & Volkmann-Kohlmeyer, 1993b). The larger number of mangrove fungi on Oahu than on Moorea was also thought to be due to the more diversified terrestrial tree flora, providing an abundant source of driftwood for the development of these fungi. The larger number of fungi in Belize where mangroves have a longer history compared with the Pacific islands, was thought to be representative of the higher mangrove tree diversity; this was extrapolated to the western pacific where a high diversity of the mycota can be expected from a high mangrove diversity. This appears to be true as in Thailand (Hyde *et al.* 1990) recorded 67 mangrove fungi, while Hyde (1988a, c) recorded 100 species in Brunei and Alias *et al.* (1994) have recorded 125 species in Malaysia.

In summary it seems, from the information available, that the diversity of mangrove fungi is dependent on

1. the age of the mangrove stand,
2. the diversity of the mangrove tree flora,
3. the diversity of the terrestrial tree flora, and
4. various micro-habitats in the mangrove (e.g., salinity differences, tidal range).

Host distribution and specificity

Kohlmeyer & Kohlmeyer (1979) discussed host specificity, and reported eight species e.g. *Didymosphaeria rhizophorae* Kohlm. & Kohlm. and *Leptosphaeria avicenniae* Kohlm. & Kohlm., that may be specific to individual host species or genera. Host specificity was also addressed by Hyde & Jones (1988), Hyde (1990a) and Hyde *et al.* (1994). The same conclusions were reached, that there is little evidence for host specificity except for a handful of species, e.g., *Trematosphaeria mangrovis* Kohlm. has not been recorded on hosts other than *Rhizophora racemosa* Meyer, while *Hypophloeda rhizophorae* K.D. Hyde & E.B.G. Jones and *Rhizophila marina* K.D. Hyde & E.B.G. Jones appear to be confined to *Rhizophora* spp. (Hyde 1990a). The question of host specificity is therefore still unresolved and requires further study. It might be possible to grow suspected host specific fungi on different hosts in pure culture, but this would not prove host specificity since the taxa may be able to colonize other timber species in these artificial conditions. The role of endophytes in the mangrove ecosystem should also be addressed. In terrestrial ecosystems endophytic fungi present in living plants often become the saprophytic colonizers of the dead plant material due to their positional advantage (Carroll & Petrini 1983). We do not know if any marine fungi are endophytic in the roots of mangroves and become saprophytic following death. Studies need to be initiated to investigate the endophytic component of mangrove species. Despite this unanswered question the evidence so far indicates that most intertidal mangrove fungi are unspecific in their host requirements.

The question of fungal distribution on host species was investigated by Hyde (1990a). He compared the mycota of 5 intertidal mangrove tree species and found that some fungi were specific and limited to a single tree species, e.g. *Caryospora mangrovei* Kohlm.: *Xylocarpus; Aigialus mangrovis* Borse and *Eutypa* sp.: *Avicennia.* Most were non-specific and developed on more than the tree species (e.g., *Hypoxylon oceanicum* Schatz, *Leptosphaeria australiensis* (Cribb & Cribb) G.C. Hughes and *Savoryella lignicola* E.B.G. Jones & Eaton). However, dominance on each host species differed. *Passeriniella savoryellopsis* K.D. Hyde & Moozouras and *Swampomyces triseptatus* K.D. Hyde and Nakagiri were most common on *Xylocarpus* (hardwood), whereas *Hypoxylon oceanicum, Lulworthia* spp. and *Halosarpheia ratnagiriensis* Borse were most common on *Avicennia alba* Blume and *Sonneratia alba* J. Smith (softwood). On *Rhizophora* spp. (medium), *Cirrenalia pygmea* Kohlm., *Carysporella rhizophorae* and *Leptosphaeria australiensis* were most common. Hyde (1990a) concluded that different taxa may be best adapted for growth on different wood types or that wood quality may provide a stimulatory effect for spore germination. By growing different taxa on different wood species under pure culture and measuring weight losses it would be possible to provide some answers to these questions. It may also be useful to compare germination rates on slivers of different wood species with spores from pure culture or field collections.

One very distinct mangrove habitat is *Nypa* palm and in a series of studies Hyde (199la, 1992b,c, 1993), Hyde & Nakagiri (1989), Hyde & Sutton (1992) have discussed and described fungi from this habitat. The largely distinct mycota found on *Nypa* palm may be accounted for by host specificity, but salinity may also be important as *Nypa* has a somewhat narrow salinity preference (Tomlinson 1986).

Vertical zonation

Papers addressing vertical zonation of fungi in mangrove swamps have been provided by Hyde (1988b, 1989c, 1990c, l991b). In Brunei, Hyde (1988b, 1989c) collected decaying samples of *Rhizophora apiculata* Blume from various tidal levels and found that the fungi were vertically zoned: most were limited either to the upper (e.g. *Hypoxylon oceanicum, Savoryella lignicola*) or lower (e.g. *Antennospora quadricornuta* (Cribb & Cribb) T.W. Johnson, *Thalassogena sphaerica* Kohlm. & Volkm.-Kohlm.) levels, while only two (*Cirrenlia pygmea* and *Lulworthia* sp.) occurred throughout the tidal range. The majority of species were collected above mean tide. In a separate study Hyde (1990c) compared fungal morphology in relation to vertical zonation, and provided the following conclusions:

1. Intertidal mangrove fungi show vertical zonation. Most are confined to a relatively small vertical range, while very few are widely distributed;
2. The greatest diversity of fungi occurs above mean tide where wood dries out superficially during low tides, but is usually inundated daily;
3. Ascomycetes with active spore release are confined to the upper intertidal, while those with passive dispersal mechanisms occur at all levels;
4. Loculoascomycetes are mostly confined to above mean tide;
5. Ascomycetes with carbonaceous ascomata walls are confined to above mean tide, while those with membranaceous walls are common throughout the tidal range;
6. Ascomycetes with superficial ascomata mostly occur above mean tide, while those with immersed ascomata occur throughout the tidal range; and
7. Ascomycetes with coloured and/or ornamented ascospores are confined to above mean tide, while those with hyaline and/or smooth-walled or sheathed ascospores are common throughout the tidal range.

The available data indicate that where there is a small tidal range, e.g., in Florida (0.8 m), this is little evidence for vertical zonation of the intertidal fungi (Kohlmeyer & Kohlmeyer, 1979). In areas where there is a large tidal range, e.g., Brunei (ca 3m) the taxa are vertically distributed. It appears that there are characteristic fungi in the lower intertidal regions where the wood is constantly submerged (e.g. *Antennospora quadricornuta*), while those in the midtidal region mostly have soft structures and passive ascospore release (e.g. *Lulworthia grandispora*). Here, the wood will dry out for short periods, but the fungi will not be subjected to the harsher conditions further up the intertidal gradient. In the high intertidal a third group of fungi is evident, species with protective brittle carbonaceous ascomata and active spore release (e.g. *Hypoxylon oceanicum*). A fuzzy area between the high intertidal region and the terrestrial regions also occurs. In this region many unusual fungi sporulate, but these are unlikely to be true intertidal fungi. *Phellinus* sp, a polypore, is often found in the high intertidal on *Sonneratia grifithii* Korz in Ranong mangrove, Thailand, and Australia (Hyde *et al.* 1994; George & Kenneally 1975). It is unlikely that these fungi can grow in saline conditions, but more likely that the mycelium decays the heartwood which is not affected by salinity and sporulates on the outside surface, perhaps during neap tide periods. K.D. Hyde (unpubl. data) has seen numerous terrestrial fungi sporulating on the mangrove floor in the upper intertidal region during an extreme wet spell in Ranong mangrove, Thailand, in 1988. This occurred during a neap tide period when the mangrove debris was not inundated with sea water. However, as the tidal levels rose and the decaying debris was inundated, the terrestrial fructifications collapsed within a matter of hours despite continuing heavy rainfall.

No information about vertical zonation is available for most parts of the world or most tidal ranges. It would be fascinating to examine the vertical zonation of fungi in areas with a large tidal range (e.g. the Northern Territory, Australia, 7m), although one would suspect the area of supratidal/intertidal interface, supporting "terrestrial taxa", would be much larger.

Distribution with salinity

Very few studies have investigated the effect of salinity on the mycota in mangrove forests. Most reports of intertidal fungi do not record the salinity, and studies of this sort are needed. Perhaps the only three publications that have examined salinity effect superficially are those of Hyde (1992b,c,d). Hyde (1992c) examined the fungi on *Nypa* palm collected in Tg. Api Api Brunei where the salinity is low (14-28‰). *Nypa* is a mangrove palm growing along the fringes of intertidal rivers where it is regularly flushed with river freshwater: the fungi that grow on decaying *Nypa* palm in the intertidal region are mostly different from those in higher salinity mangrove, as described above. However, many of these are unique marine species belonging to genera consistently found on terrestrial palms (Hyde 1988d; Hyde & Nakagiri 1989). Others are marine genera with species unique to *Nypa* (e.g. *Helicascus nypae* K.D. Hyde), while only a handful are found throughout mangrove ecosystems (e.g. *Lignicola laevis*). There are, however, limitations in comparing the mycota of a palm mangrove with that of other dicotyledonous mangroves since they are very different substrates. Hyde (1992d) examined the mycota of decaying intertidal *Kandelia candel* (L.) Druce where the salinity fluctuated between 3-24‰. He found that most of the fungi were typical mangrove species common in mangroves of higher salinity with the exception of a new species *Phomatospora kandeliae* K.D. Hyde. He concluded that the distribution of fungi was probably limited by periods of higher salinity and therefore the mycota was likely to be similar throughout the salinity range of mangroves. However, these are only preliminary findings and further studies should be initiated.

Succession

Only three reports discuss the succession of intertidal fungi on wood and two on the succession of fungi on mangrove leaves and seedlings. Tan *et al.* (1989) examined succession of fungi on wood of *Avicennia alba* and *A. lanata* Ridley, while Leong *et al.* (1991) examined succession on submerged *Bruguiera cylindrica* (L.) Blume and *Rhizophora apiculata*. They submerged young split stems of the tree species in Mandai mangrove and recovered samples periodically. A pattern of succession of fungi developed on the wood with *Lignincola laevis* as an early colonizer, *Verruculina enalia* (Kohlm.) Kohlm. & Volkm.-Kohlm. and *Lulworthia* sp. 1 as an intermediate colonizer and *Aigialus parvus* Schalz & Kohlm. as a late colonizer. Unfortunately this study made only brief mention of the differences in the fungi developing on bark and on the woody parts of the split stems and this should be addressed. A study of the succession of fungi on *Rhizophora apiculata* and *Xylocarpus granatum* König was carried out in Brunei by Hyde (1991b). Poles or small stems of the tree species were attached vertically across the intertidal region and recovered bimonthly. Fungal succession on these timbers differed markedly, but on both a pattern was observable. In the case of *Rhizophora* which remained bark-covered during the duration of the study *Halosarpheia minuta* Leong *et al.* was one of the earliest and most common taxa. *Lulworthia grandispora* then produced ascomata followed by *Manglicola guatemalensis* Kohlm. & Kohlm. and *Anthostomella* sp. (6 months) and *Capillatispora corticola* K.D. Hyde and *Hypophloeda rhizospora* (8 months). With *Xylocarpus granatum*, *Lulworthia* sp. was the first taxon to occur on bark on wood, along with a *Diplodia* sp. and several unidentified coelomycetes. In the later stages several fungi (i.e. *Xylomyces* sp., *Cirrenalia* spp., *Periconia prolifica* Anastasiou, *Leptosphaeria australiensis*, *Lulworthia* spp., *Halosarpheia marina*, *Calathella* sp. and *Hypoxylon oceanicum*) were common on the exposed xylem. The differences in fungi colonizing these two wood species were distinct and illustrate both fungal succession, host specificity, differences in fungi colonizing bark and wood, and resistance against fungal development due to high tannin contents in the bark.

These studies show evidence of fungal succession on submerged wood, but are a measure of the formation of ascomata rather than direct fungi colonization. It may be that some species produce fructifications more speedily than others, even though all taxa may be present in the timber following submergence. This aspect needs further investigation as does the effect of other variables on fungal succession (e.g. salinity, wood species). For succession studies on leaves and propagules the reader should refer to Fell & Master (1973, 1980) and Newell (1986).

The role of fungi in mangrove nutrient cycling

Whereas increasing knowledge is available on the involvement of bacteria in water column and sediment productivity and nutrient cycles, virtually nothing is known about the role played by fungi in such energy pathways (Newell 1984). Even less is known on the actual involvement of fungi in mangrove nutrient cycling. Information on the role of fungi in the decay of mangrove detritus is small compared with that available for salt marsh ecosystems. Despite attention paid to gross decomposition of mangrove leaf litter, most taxonomic studies on mangrove fungi are confined to lignolytic species, which probably play an important role in the turnover of mangrove wood, an emphasis also apparent in studies of fungi-mediated decomposition in terrestrial ecosystems (Stolp 1988). Past studies on mangrove leaf litter decomposition have established that apart from some exceptions (e.g., Robertson 1986), microbial enrichment usually precedes detritivore consumption and enriched detritus is generally more attractive to detritivores (e.g., Bärlocher *et al.* 1989; Graça *et al.* 1993), probably because of the higher availability of nitrogen and the improved digestibility of detrital carbon (Lee 1989; Camilleri 1992; Lee 1993). There are also numerous reports on mangrove leaf, and to a lesser extent, wood decomposition rates. Although fungi have been isolated from decaying mangrove leaves in the past (e.g., Fell & Master 1980; Newell *et al.* 1987; Nakagiri *et al.* 1989), very little is known about the roles played by these fungi in the decomposition process, especially that of leaf litter. Reasons for this deficiency of knowledge may include: a) difficulties in quantifying fungal involvement in the decomposition process; b) lack of knowledge of fungal extracellular enzymes and their activities; c) hitherto emphasis of particulate rather than dissolved components in nutrient cycling studies.

Quantification of fungi abundance

Newell & Statzell-Tallman (1982), Newell *et al.* (1986) and Newell (1992) reviewed the methods used for quantifying fungal presence on decaying herbaceous wetland plant tissues (mainly *Spartina*). Past methods include the estimation of volume or biomass by direct count or their converted values (e.g., Blum *et al.* 1988), but these have been criticized as inaccurate (Newell 1992). Conversion factors available for transforming direct volume counts to biomass are available, the problem being that without a knowledge of which groups of fungi are present, the conversion cannot be given much confidence. Fungi-specific molecules, e.g., glucosamine and ergosterol, have been used for the enumeration of fungi abundance. Newell & Fell (1992) attempted to measure living fungal biomass on decaying mangrove leaves and twigs using ergosterol content, a molecule characteristic of the Eumycota but not produced by vascular plants. The use of these molecules eliminates the problem of contamination from the decomposition substrate. Newell & Fell (1992) indicated that only low levels of ergosterol were detectable in decaying mangrove leaves, reflecting a small contribution from eumycotic species. The involvement of oomycetes may nevertheless be substantial, but because they do not produce ergosterol, their importance could have been overlooked (S.Y. Newell, pers. comm.). Data on fungal quantification is therefore scarce. Before techniques can be established and inter-calibration performed, relating fungal abundance to mangrove decomposition rates quantitatively will still be problematic.

Fungal extracellular enzymes

Both bacteria and fungi process the detritus substrates by the digestive activities of extracellular enzymes. The taxonomy and activities of bacterial extracellular enzymes, unlike those of the fungi, are fairly well known (e.g., Chaloupka & Krumphanzl 1987). Chaloupka & Krumphanzl (1987), for example, make no mention of the type and activities of the enzymes secreted by fungi. The scanty information available on fungal enzymes is largely for terrestrial species. This disparity probably originates from the lack of knowledge on the fungal taxa involved in the mangrove decomposition process. In addition to being a potential indicator of fungal biomass (Newell 1992), extracellular enzyme assays will help understand the kinetics and mechanism of the decomposition process. Recently, Wu (1993) identified 25 genera (42 strains) of fungi from mangroves in the Tansui Estuary near Taipei, Taiwan, and found that most of the ascomycetes were able to secrete a wide range of enzymes potentially capable of decomposing mangrove litter. Other reports on the activities of mangrove fungi in southeast Asia are, however, almost non-existent.

Fungi-mediated production of DOM

It is increasingly recognized that much of the nutrient cycling in aquatic ecosystems takes place in the dissolved form (Wetzel 1984), an aspect largely overlooked in previous trophic studies of mangrove ecosystems (Robertson 1987). Although in some ecosystems some macrodetritivores (e.g., crabs, amphipods) are capable of processing large quantities of mangrove leaf litter per unit time (Robertson 1986; Poovachiranon *et al.* 1986; Lee 1989; Robertson & Daniel 1989), microbes (of which fungi are an important component) are the ubiquitous and constant decomposers of leaf and, in particular, wood litter in the mangal. A large proportion of the end product of fungal processing of mangrove litter is, however, likely to be dissolved rather than particulate. Given the complex nature of the dissolved organic matter in aquatic systems (Wetzel 1984), nothing is known about the type and quantity of DOM which may be derived from fungi-mediated mangrove decomposition.

Another poorly known aspect of the decomposition process is the extent of enrichment possible with fungal colonization of mangrove detritus. Significant increases in the nitrogen content of decaying detritus resulting in decrease in the C:N ratio are common (e.g. Lee 1989). Most detritivores tend to prefer enriched to fresh detritus unless environmental constraints limit the availability, e.g. in tidal forests where mangrove leaf litter export is high (e.g. Robertson 1986). To what extent fungi are responsible for this enrichment is, however, unknown. This enrichment has largely been attributed to the microbial community as a whole and the relative importance of the bacteria and the fungi in the time dimension and the mobilization of different components of the detritus have rarely been investigated.

Acknowledgements

Bonnie To and Sandy Kwok are thanked for typing the manuscript.

References

Aleem, A A , 1980 Distribution and ecology of marine fungi in Sierra Leone (Tropical West Africa) Bot Mar 23 679–688

Alias, A & E B G Jones, 1994 Biodiversity of fungi in tropical mangroves In Fifth International Mycological Congress, Abstracts

Bärlocher, F , S Y Newell & T L Arsuffi, 1989 Digestion of *Spartina alterniflora* Loisel material with and without fungal constituents by the periwinkle *Littorina irrorata* Say (Mollusca Gastropoda) J exp mar Biol Ecol 130 45–53

Blum, L K , A L Mills, J C Zieman & R T Zieman, 1988 Abundance of bacteria and fungi in seagrass and mangrove detritus Mar Ecol Prog Ser 42 73–78

Borse, B D , 1988 Frequency of occurrence of marine fungi from Maharastra Coast, India Indian J Mar Sci 17 165–167

Boto, K G , J S Bunt & J T Wellington, 1984 Variations in mangrove forest productivity in Northern Australia and Papua New Guinea Estuarine, coastal and Shelf Science 19 321–329

Bunt, J S , K G Boto & G Boto, 1979 A survey method for estimating potential levels of mangrove forest primary production Mar Biol 52 123–128

Camilleri, J C 1992 Leaf-litter processing by invertebrates in a mangrove forest in Queensland Mar Biol 114 139–145

Carroll, G C & O Petrini, 1983 Patterns of substrate utilization by some fungal endophytes from coniferous foliage Mycologia 75 53–63

Chaloupka, J & V Krumphanzl, 1987 Extracellular enzymes of microorganisms Plenum Press, New York 216 pp

Chinnaraj, S , 1993 Higher marine fungi from mangroves of Andaman and Nicobar Islands Sydowia 45 109–115

Christensen, B , 1978 Biomass and primary productivity of *Rhizophora apiculata* Bl mangrove in Southern Thailand Aq Bot 4 43–52

Cundell, A M , M S Brown, R Stafford & R Mitchell, 1979 Microbial degradation of *Rhizophora mangle* leaves immersed in the sea Estuarine and Coastal Marine Science 9 281–286

Fell, J W & I M Master, 1973 Fungi associated with the degradation of mangrove (*Rhizophora mangle* L) leaves in South Florida In L H Stevenson & R R Colwell (eds), Estuarine Microbial Ecology University of South Carolina Press, Columbia 455–466

Fell, J W & I M Master, 1980 The association and potential role of fungi in mangrove detrital systems Bot Mar 23 257–263

George, A S & K F Kenneally, 1975 The flora of the Prince Regent river reserve, north-western Kimberley, Western Australia A biological survey of the Prince Regent river reserve, North-West Kimberley, Western Australia in August, 1974 Widl Res Bull West Aust 3 1–116

Graça, M A S , L Maltby & P Calow , 1993 Importance of fungi in the diet of *Gammarus pulex* and *Asellus aquaticus* I feeding strategies Oecologia 93 139–144

Heald, E J 1971 The production of organic detritus in a South Florida estuary University of Miami Sea Grant Technical Bulletin No 6

Hughes, G C 1974 Geographical distribution of the higher marine fungi Veroff Inst Meeresforsch Bremenh Suppl 5 419–441

Hutchings, P & P Saenger, 1987 Ecology of mangroves University of Queensland Press, St Lucia

Hyde, K D , 1986 Frequency of occurrence of lignicolous marine fungi in the tropics In S T Moss (ed), The Biology of Marine Fungi Cambridge University Press, Cambridge 311–322

Hyde, K D , 1988a Studies on the tropical marine fungi of Brunei Bot J Linn Soc 98 135–51

Hyde, K D , 1988b Observations on the vertical distribution of marine fungi on *Rhizophora* spp , at Kampong Danau mangrove, Brunei Asian Mar Biol 5 77–81

Hyde, K D , 1988c Studies on the tropical marine fungi of Brunei II Notes on five interesting species Trans Mycol Soc Japan 29 161–171

Hyde, K D , 1988d The genus *Linocarpon* from the mangrove palm *Nypa fruticans* Trans Mycol Soc Japan 29 339–350

Hyde, K D , 1989a Ecology of tropical marine fungi Hydrobiologia 178 199–208

Hyde, K D , 1989b Intertidal mangrove fungi from North Sumatra Can J Bot 67 3078–3082

Hyde, K D , 1989c Vertical zonation of intertidal mangrove fungi In T Hattori, Y Ishida, R Maryama, R Morita & A Uchida (eds), Recent Advances in Microbial Ecology Japan Scientific Societies Press, Tokyo 302–306

Hyde, K D , 1990a A comparison of the intertidal mycota of five mangrove tree species Asian Mar Biol 7 93–107

Hyde, K D , 1990b *Caryospora mangrovei* sp nov and notes on marine fungi from Thailand Trans Mycol Soc Japan 30 333–342

Hyde, K D , 1990c A study of the vertical zonation of intertidal fungi on *Rhizophora apiculata* at Kampong Kapok Mangrove, Brunei Aq Bot 36 255–62

Hyde, K D , 1990d Intertidal fungi from warm temperate mangroves of Australia, including *Tunicatispora australiensis* gen et sp nov Aust Syst Bot 3 711–718

Hyde, K D , 1991a A new Amphisphaeriaceous fungus from intertidal fronds of *Nypa fruticans* Trans Mycol Soc Japan 31 265–271

Hyde, K D , 1991b Fungal colonization of *Rhizophora apiculata* and *Xylocarpus granatum* poles in Kampong Kapok mangrove, Brunei Sydowia 43 31–38

Hyde, K D , 1992a *Julella avicenniae* Borse comb nov (Thelenellaceae) from intertidal mangrove wood and miscellaneous fungi form the NE coast of Queensland Mycol Res 96 939–942

Hyde, K D , 1992b Fungi from *Nypa fruticans Nipicola carbospora* gen et sp nov (Ascomycotina) Crypt Bot 2 330–332

Hyde, K D , 1992c Fungi from decaying intertidal fronds of *Nypa fruticans*, including three new genera and four new species Bot J Linn Soc 110 95–110

Hyde, K D , 1992d Intertidal fungi from *Kandelia candel* including *Phomatospora kandelii* sp nov Trans Mycol Soc Japan 33 313–316

Hyde, K D , 1992e Intertidal mangrove fungi from the west coast of Mexico, including one new genus and two new species Mycol Res 96 25–30

Hyde, K D , 1993 Fungi from palms V *Phomatospora nypae* and notes on intertidal fungi from *Nypa fruticans* in Malaysia Sydowia 45 199–203

Hyde, K D & E B G Jones, 1988 Marine mangrove fungi Mar Ecol (Berlin) 9 15–33

Hyde, K D & E B G Jones, 1989 Ecological observations on marine fungi from the Seychelles Bot J Linn Soc 100 237–254

Hyde, K D & A Nakagiri, 1989 A new species of *Oxydothis* from the mangrove palm, *Nypa fruticans* Trans Mycol Soc Japan 30 69–76

Hyde, K D & B C Sutton, 1992 *Nypaella frondicola* gen et sp nov , *Plectophomella nypae* sp nov and *Pleurophomopsis nypae* sp nov (Coelomycetes) from intertidal fronds of *Nypa fruticans* Mycol Res 96 210–214

Hyde, K D, A Chalermpongse & T Boonthavikoon, 1990 Ecology of intertidal fungi at Ranong mangrove, Thailand Trans Mycol Soc Japan 31 17–28

Hyde, K D, A Chalermpongse, & T Boonthavikoon, 1993 The distribution of intertidal fungi on *Rhizophora apiculata* In B Morton (ed), The Marine Biology of the South China Sea Proceedings of the First International Conference on the Marine Biology of Hong Kong and the South China Sea, Hong Kong 1990 University of Hong Kong Press, Hong Kong 643–652

Jones, E B G 1993 Tropical marine fungi In S Issac, J C Frankland, R Watling & J S Whatley (eds), Aspects of Tropical Mycology Cambridge University Press, Cambridge 73–89

Jones, E B G & K D Hyde, 1990 Observations on poorly known mangrove fungi and a nomenclatural correction Mycotaxon 37 197–201

Jones, E B G & A J Kuthubutheen, 1989 Malaysian Mangrove Fungi Sydowia 41 160–169

Jones, E B G & T K Tan, 1987 Observations on manglicolous fungi from Malaysia Trans Brit Mycol Soc 89 390–392

Jones, E B G, F R Uyenco & M P Foliosco, 1988 Fungi on driftwood collected in the intertidal zone from the Philippines Asian Mar Biol 5 103–106

Kohlmeyer, J, 1980 Tropical and subtropical filamentous fungi of the Western Atlantic Ocean Bot Mar 23 529–544

Kohlmeyer, J, 1981 Marine fungi from Martinique Can J Bot 59 1314–1321

Kohlmeyer, J, 1983 Geography of marine fungi Aust J Bot Suppl Ser 10 67–76

Kohlmeyer, J, 1984 Tropical marine fungi Mar Ecol (Berlin) 5 329–378

Kohlmeyer, J & E Kohlmeyer, 1979 Marine Mycology The Higher Fungi Academic Press, New York

Kohlmeyer, J & B Volkmann-Kohlmeyer, 1987a Marine fungi from Aldabra, the Galapagos, and other tropical islands Can J Bot 65 571–582

Kohlmeyer, J & B Volkmann-Kohlmeyer, 1987b Marine fungi from Belize with a description of two new genera of ascomycetes Bot Mar 30 195–204

Kohlmeyer, J & B Volkmann-Kohlmeyer, 1988 *Ophiodeira* gen nov (Halosphaeriales) and a survey of higher marine fungi from Saint Croix (Virgin Islands) Can J Bot 66 2062–2067

Kohlmeyer, J & B Volkmann-Kohlmeyer, 1991 Marine fungi from Queensland, Australia Aust J Mar Freshw Res 42 91–99

Kohlmeyer, J & B Volkmann-Kohlmeyer, 1993a Two new genera of Ascomycotina from saltmarsh *Juncus* Systema Ascomycetum 11 95–106

Kohlmeyer, J & B Volkmann-Kohlmeyer, 1993b Biogeographic observations on Pacific marine fungi Mycologia 85 337–346

Lee, S Y, 1989 The importance of Sesarminae crabs *Chiromanthes* spp and inundation frequency on the decomposition of mangrove (*Kandelia candel* (L) Druce) leaf litter in a Hong Kong tidal shrimp pond J exp mar Biol Ecol 131 23–43

Lee, S Y 1993 Leaf choice of the sesarmine crabs *Chiromanthes bidens* and *C maipoensis* in a Hong Kong mangal In B Morton, (ed), Marine Biology of the South China Sea Proceedings of the First International Conference on the Marine Biology of Hong Kong and the South China Sea, Hong Kong, 1990 Hong Kong University Press, Hong Kong 597–604

Leong, W F, T K Tan & E B G Jones, 1991 Fungal colonization of submerged *Bruguiera cylindrica* and *Rhizophora apiculata* wood Bot Mar 34 69–76

Nakagiri, A, 1993 Intertidal mangrove fungi from Iriomote Island IFO Res Commun 16 24–62

Nakagiri, A, S Tokumasu, H Araki *et al* 1989 Succession of fungi in decomposing mangrove leaves in Japan Proc Int Symp Micro Ecol 5 297–301

Newell, S Y 1984 Bacterial and fungal productivity in the marine environment a contrastive overview Colloques Int Cent Natn Rech Scient 331 133–139

Newell, S Y 1986 Mangrove fungi the succession in the mycoflora of red mangrove (*Rhizophora mangle*) seedlings In E B G Jones (ed), Recent Advances in Aquatic Mycology Elek Science 5 1–91

Newell, S Y 1992 Estimating fungal biomass and productivity in decomposing litter In G C Carroll, & D T Wicklow (eds), The fungal community Its organization and role in the ecosystem Marcel Dekker, Inc, New York 521–561

Newell, S Y, R D Fallon & J D Miller, 1986 Measuring fungal-biomass dynamics in standing-dead leaves of a salt-marsh vascular plant In S T Moss, (ed), Biology of Marine Fungi, Cambridge University Press, Cambridge 19–25

Newell, S Y & J W Fell, 1992 Ergosterol content of living and submerged, decaying leaves and twigs of red mangrove Can J Microbiol 38 979–982

Newell, S Y, J D Miller & J W Fell, 1987 Rapid and pervasive occupation of fallen mangrove leaves by a marine zoosporic fungus Appl Environ Microbiol 53 2464–2469

Newell, S Y & A Statzell-Tallman, 1982 Factors for conversion of fungal biovolume to biomass, carbon and nitrogen variation with mycelial ages, growth conditions, and strains of fungi from a salt marsh Oikos 39 261–268

Odum, W E & E J Heald, 1975 The detritus food web of an estuarine mangrove community In L E Cronin (ed), Estuarine Research Vol I Academic Press, New York 265–286

Poovichiranon, S, K G Boto, & N C Duke, 1986 Food preference studies and ingestion rate measurements of the mangrove amphipod *Parhyale hawaiensis* (Dana) J exp mar Biol Ecol 98 129–140

Robertson, A I, 1986 Leaf-burying crabs their influence on energy flow and export from mixed mangrove forests (*Rhizophora* spp) in northeastern Australia J exp mar Biol Ecol 102 237–248

Robertson, A I, 1987 The determination of tropic relationships in mangrove dominated systems areas of darkness In C D Field & A J Dartnall (eds), Mangrove ecosystems of Asia and the Pacific status, exploitation and management Australian Institute of Marine Science, Australia 292–304

Robertson, A I & P A Daniel, 1989 The influence of crabs on litter processing in high intertidal mangrove forests in tropical Australia Oecologia 78 191–198

Robertson, A I, D M Alongi & K G Boto, 1992 Food chains and carbon fluxes In A I Robertson & D M Alongi (eds), Tropical Mangrove Ecosystems American Geophysical Union, Washington D C 293–326

Saenger, P & S C Snedaker, 1993 Pantropical trends in mangrove above-ground biomass and annual litter fall Oecologia 96 293–299

Stolp, H, 1988 Microbial ecology organisms, habitats, activities Cambridge University Press, Cambridge 308 pp

Tan, T K, W F Leong & E B G 1989 Succession of fungi on wood of *Avicennia alba* and *A lanata* in Singapore Can J Bot 67 2686–2691

Tenore, K R, J C Cammen, S E G Findlay & N Phillips, 1982 Perspectives of research of detritus do factors controlling the availability of detritus to macroconsumers depend on this source? J Mar Res 40 473–490

Tomlinson, P B 1986 The botany of mangroves Cambridge University Press, Cambridge 413 pp

Vrijmoed, L. L. P. 1990. Preliminary observations of lignicolous marine fungi from mangroves in Hong Kong. In B. Morton (ed), Proceedings of the Second International Marine Biological Workshop: the marine flora and fauna of Hong Kong and southern China, Hong Kong, 1986, Hong Kong University Press, Hong Kong: 701–706.

Vrijmoed, L. L. P., K. D. Hyde & E. B. G. Jones, 1994. *Diaporthe salsuginea* sp. nov. and *Aniptodera* sp. nov. from Hong Kong and Macau mangroves. Mycol. Res. 96: 699–704.

Wetzel, R. G., 1984. Detrital dissolved and particulate organic carbon functions in aquatic ecosystems. Bull. mar. Sci. 35: 503–509.

Wu, R. Y., 1993. Studies on the microbial ecology of the Tansui Estuary. Bot. Bull. Acad. Sin. 34: 13–30.

Hydrobiologia **295**: 119–126, 1995.
Y. S. Wong & N. F. Y. Tam (eds), Asia-Pacific Symposium on Mangrove Ecosystems.

Observations on vertical distribution of fungi associated with standing senescent *Acanthus ilicifolius* stems at Mai Po Mangrove, Hong Kong

R. B. Sadaba[1], L. L. P. Vrijmoed[2], E. B. G. Jones[3] & I. J. Hodgkiss[1]
[1]*Department of Ecology and Biodiversity, The University of Hong Kong, Pokfulam Road, Hong Kong*
[2]*Department of Biology and Chemistry, City Polytechnic of Hong Kong, Tat Chee Ave., Kowloon, Hong Kong*
[3]*School of Biological Sciences, University of Portsmouth, King Henry I Building, King Henry I Street, Portmouth Hants P01 2DY, UK*

Key words: Acanthus ilicifolius, mangrove, mangrove fungi, vertical distribution

Abstract

The present study was carried out to investigate the higher fungi colonizing the herbaceous mangrove associate *Acanthus ilicifolius*. This paper reports part of an investigation to determine if there is vertical distribution of fungi on the standing plant.

The Mai Po Mangrove, Hong Kong is estuarine with great variations in salinity mainly due to the influence of the Pearl River. Senescent and dead stems of standing *Acanthus ilicifolius* were collected from mangroves in Mai Po from April to December 1992. The maximum tidal range observed was 2.6 m. A stratified sampling strategy was employed to assess the vertical distribution on the standing plant. A total of 44 fungi were collected: 32 Deuteromycotina, 11 Ascomycotina and 1 Basidiomycotina. Very frequent species were *Acremonium* sp.(55%), *Colletotrichum gloeosporioides* cf.(42.5%), *Phoma* sp. (42.5%) *Fusarium* sp.,(25%), *Tubercularia* sp. (24.2%) and *Phialophora* sp. cf. (19.2%). *Agerita* sp., *Corynespora cassiicola*, *Stachybotrys chartarum*, *Trichoderma* sp. and D82 were frequent, while the remaining species were recorded at less than 10%. Vertical zonation of fungi colonizing the standing stems was observed. The apical portions were colonized by typical terrestrial fungi and the basal portions by marine species. This can be attributed to both the nature of the substratum and the degree of exposure due to tidal inundation.

Introduction

There has been a considerable increase in information on mangrove fungi since they were first reported from mangrove roots in Australia by Cribb & Cribb (1955). It is now recognized that mangrove fungi constitute the second largest group of marine fungi (Hyde & Jones, 1988). However, the bulk of the data generated from various areas in the world has concentrated on the description of new species on hosts not previously investigated (Hyde, 1992; Hyde & Nakagiri, 1992; Hyde *et al.*, 1992; Jones & Agerer, 1992; Volkmann-Kohlmeyer & Kohlmeyer, 1993). Only a few systematic ecological studies have been carried out (Hyde & Jones, 1988; Kohlmeyer, 1969; Tan & Leong, 1989). Moreover, these studies were mostly limited to fungi on bark and decorticated wood, roots, pneumatophores (Kohlmeyer & Vittal, 1986; Hyde *et al.*, 1986; Hyde & Jones, 1988), and on leaves either partially immersed or totally submerged in high salinity areas (Anastasiou & Churchland, 1969; Fell & Master, 1973, 1980; Newell, 1976). Little is known of the fungi colonizing herbaceous mangrove plants or their associates (e.g. *Acrostichum speciosum* Willdenow, *Acanthus ilicifolius* L.).

Mangrove fungi on lignocellulose substrata and leaves have been studied recently in Hong Kong (Vrijmoed, 1990; Vrijmoed & Tam, 1990). The present study was undertaken to investigate the higher fungi colonizing *Acanthus ilicifolius* since little is known about fungi colonizing herbaceous substrata in mangroves.

Materials and methods

The study area

The Mai Po Mangrove is situated along the northtwestern coast of the New Territories facing the Pearl river estuary (22°29′N, 114°02′E). This is the largest mangrove community in Hong Kong, covering an area of approximately 0.9 km^2 (Lee, 1990) within the Mai Po marshes along the shoreline of Deep Bay. The dominant mangrove species include *Kandelia candel* (L.) Druce, *Aegiceras corniculatum* (L.) Blanco, *Avicennia marina* (Forsk.) Vierh. and the prickly herbaceous plant *Acanthus ilicifolius.*

The host: Acanthus ilicifolius

Acanthus (Acanthaceae) is the only genus of the family with representatives in mangrove communities. *Acanthus ilicifolius* is a low woody herb up to 2 m high, which occurs on the inland side of channels in mangroves. It is a characteristic associate of mangroves with a known distribution from India to the Western Pacific, tropical Australia, and the Philippines (Tomlinson, 1986). Locally it is found in great abundance at Mai Po. *Acanthus* is a perrenial plant which spreads largely by vegetative means (Tomlinson,1986). The mature stem is largely herbaceous consisting of soft tissue, but the lower portions are woody (soft wood). This is largely due to the growth pattern of *Acanthus*, older portions will be lignified gradually. *Acanthus* therefore offers a range of non-lignified and lignified tissues for colonization, possibly by different ecological groups of fungi.

In Mai Po, *Acanthus* plants grow well above the banks of the major water channel. Thus the lower parts of the plants are submerged only during high tide and are well exposed during low tides. However, the apical portions of tall plants are exposed to the air even at high tide.

Sampling sites within the study area

Stem samples (1.5–2.0 m high) were collected within one to two meters on both sides of the major water channel which runs through the mangrove. The four sampling sites chosen for the determination of selected hydrographical characteristics were just below the wooden boardwalk running along the channel. Water samples were collected in plastic vials and kept in an ice box and brought to the laboratory for pH and salinity measurements. Air and water temperatures were determined *in situ*.

Table 1. Hydrographical data* of the test sites in the Mai Po mangrove, Hong Kong (April - December 1992).

Month	Temperature Air (°C)	Water	Salinity (‰)	pH
April	22.9±1.3	22.2±1.0	2.7±1.3	7.1±0.1
June	24.7±0.1	24.7±0.1	6.3±0.4	7.3±0.0
July	32.3±0.3	28.8±0.5	1.8±1.5	7.2±0.2
August	30.0±2.9	32.8±0.3	5.8±0.8	7.4±0.1
September	29.4±1.4	31.7±1.2	12.3±0.4	7.3±0.0
October	28.3±2.3	26.3±2.2	12.0±0.0	7.6±0.1
November	21.5±0.7	18.2±0.4	20.0±0.0	7.2±0.0
December	19.9±1.5	19.1±0.3	21.0±1.0	7.1±0.0

* Average values from four sites

Collection of samples

To determine whether there is vertical stratification of fungi along the standing stem of *Acanthus*, samples were collected from apical, middle and basal portions of the senescent plant. Apical portions include those portions approximately 1.5–2.0 m from the soil surface while the middle about 0.5–1.0 m. The basal samples were collected directly above the soil surface. No lateral branches were included as they would not provide information relating to vertical stratification of fungal occurrence of the *Acanthus*. All stems collected were approximately 5–10 mm in diameter. Five stems from each level were collected in each month from April to December 1992.

Treatment of samples

Samples were treated following the procedures of Hyde & Jones (1988). The fungi colonizing on the samples were identified on retrieval and after incubation. Voucher slides of all fungi collected were prepared and are held in the first author's collection. Based the frequency of occurrence of the collected fungi, they are classified as 'very frequent' (>20%), 'frequent' (10–20%) and infrequent (<10%) as adopted by Tan and Leong (1989).

Table 2 Frequency of occurrence (%)* of fungi associated with stratified samples of *Acanthus ilicifolius* stems collected in Mai Po between April and December 1992

	Apical	Middle	Basal
ASCOMYCOTINA (11 species)			
Aniptodera chesapeakensis Shearer & Miller	–	5 0	12 5
Gnomonia sp *cf*	2 5	–	–
Halosarpheia retorquens Shearer & Crane	–	–	2 5
Halosarpheia marina cf	–	–	2 5
Leptosphaeria sp	–	2 5	5 0
Lignincola laevis Höhnk	–	2 5	2 5
Marinosphaera mangrovei Hyde	–	2 5	–
Nectria sp 1	5 0	2 5	–
Nectria sp 2	10 0	2 5	–
Ophioceras sp	10 0	–	–
Tirispora unicaudata Jones, Vrijmoed & Read	10 0	2 5	12 5
BASIDIOMYCOTINA (1 Species)			
Agerita sp	5 0	15 0	27 5
DEUTEROMYCOTINA (32 species)			
Acremonium sp	37 5	70 0	90 0
Annellophora sp *cf*	2 5	12 5	5 0
Cirrenalia basiminuta Raghu-kumar, Zainal & Jones	–	5 0	25 0
Clavatospora bulbosa (Anast) Nakagiri & Tubaki	–	–	2 5
Colletotrichum gloeosporioides cf	65 0	25 0	37 5
Coniothyrium sp 1	7 5	15 0	–
Coniothyrium sp 2	2 5	–	–
Corynespora cassiicola (Berk & Curt) Wei	27 5	12 5	–
Fusarium sp	12 5	30 0	35 0
Graphium sp 1	–	10 0	15 0
Graphium sp 2	–	5 0	22 5
Graphium sp 3	–	7 5	7 5
Periconia prolifica Anastasiou	–	10 0	10 0
Phaeoisaria sp	–	7 5	2 5
Phialophora sp *cf*	55 0	2 5	–
Phoma sp 1	30 0	–	55 0
Phoma sp 2	–	–	5 0
Phomopsis sp	2 5	2 5	–
Stachybotrys chartarun (Ehrenb ex Link) Hughes	10 0	22 5	17 5
Stachylidium bicolor Link ex S F Gray	25 0	10 0	–
Trichocladium achrasporum (Meyers & Moore) Dixon in Shearer & Crane	–	–	7 5
Trichoderma sp	15 0	17 5	15 0
Tubercularia sp	40 0	7 5	25 0
Zelosatchmopsis sp *cf*	–	–	5 0

Results

Hydrographical data

Table 1 shows the hydrographical characteristics of the study area as determined at the four sampling points between April and December 1992

Fungi associated with stratified samples of Acanthus ilicifolius stems

Forty-four higher fungi were identified from 120 samples collected from the field and the pooled results are presented in Table 2 In general, the domi-

Table 2 cont..

	Apical	Middle	Basal	
D28 (globose, verrucose, purplish or brown conidia)	10.0	2.5	12.5	
D71 (truncate at one end and rounded on the other, hyaline one-celled conidia)	2.5	–	–	
D82 (fusiform, hyaline, one-celled conidia)	25.0	20.0	17.5	
D36 (filiform, hyaline, one-celled conidia)	–	5.0	2.5	
D6 (ellipsoidal, hyaline, one-celled conidia)	–	5.0	2.5	
D34 (ellipsoidal, 1–3 septate. hyaline conidia)	–	5.0	2.5	
D90 (filiform, septate, hyaline, ends with chambers?, straight or curved)	–	2.5	–	
D43 (cylindrical, hyaline, one-celled conidia)	–	–	2.5	
Summary:	Apical	Middle	Basal	Total
No. of Ascomycotina	4	7	6	11
No. of Basidiomycotina	1	1	1	1
No. of Deuteromycotina	17	24	24	32
Total collections	154	151	194	499
Total no. of species	22	32	31	44
No. of samples examined	40	40	40	120

* Frequency of occurrence (%)= $\frac{\text{No. of collections of a species}}{\text{No. of samples examined}} \times 100$

nant group was the Deuteromycotina with 32 species. The Ascomycotina and Basidiomycotina were represented by 11 and 1 species respectively. The same pattern was observed on the different portions surveyed (Deuteromycotina > > > Ascomycotina > > Basidiomycotina). Based on the overall frequency of occurrence, very frequent species were *Acremonium* sp.(55%), *Colletotrichum gloeosporioides* cf. (42.5%), *Phoma* sp. (42.5%), *Fusarium* sp. (25%), *Tubercularia* sp. (24.2%) and *Phialophora* sp. cf. (19.2%). *Agerita* sp. *Corynespora cassiicola* (Berk. & Curt) Wei, *Stachybotrys chartarum* (Ehrenb. ex Link) Hughes, *Trichoderma* sp., and D82 were frequent (10–20%), while the remaining species were recorded at less than 10%.

Ten species were common to all levels of the stem portions (Table 2). However, their percentage occurrence differed at the three levels. They showed a comparatively higher affinity to a particular level on the stems (Table 2). Thus, *Agerita* sp., *Acremonium* sp., *Fusarium* sp., and D28 had a higher affinity towards the basal portions. *Colletotrichum gloeosporioides* cf:, *Tubercularia* sp. and D82 were more frequently found in the apical portions, whereas *Annellophora* sp. cf. and *Stachybotrys chartarum* were generally restricted to the middle portions. *Trichoderma* sp. was evenly distributed at all three levels.

Thirteen species were collected only once from a particular level (Table 2). These were *Coniothyrium* sp. 2, *Gnomonia* sp. cf., D71 and *Ophioceras* sp. cf. from the apical portion; *Marinosphaera mangrovei* Hyde and D90 from the middle portion; and *Clavatospora bulbosa* (Anast.) Nakagiri & Tubaki, *Halosarpheia retorquens* Shearer & Crane, *Halosarpheia marina* cf., *Phoma* sp. 2, *Zelosatchmopsis* sp. cf., *Trichocladium achrasporum* (Meyers & Moore)Dixon in Shearer & Crane and D43 from the basal portion. The remaining 21 species occurred on more than one level along the stem.

Discussion

In comparing the data generated from this study with those reported by previous researchers, a number of factors require consideration: (1) Nature of the substratum – *Acanthus ilicifolius* comprises both soft and woody tissues and this is expected to affect the ability of the fungi to colonize them; (2) Ecosystem – the Mai Po mangrove is brackish to almost freshwater, depending on season, while other studies involve almost completely saline to brackish water habitats (e.g. Seychelles, Hyde & Jones, 1988). This will significantly affect the fungi recorded on the baits; (3) Vertical distribution – few studies have been undertaken

Table 3 Comparison of the vertical distribution of dominant fungi on *Acanthus ilicifolius* stems and *Rhizophora* spp prop roots (after Hyde, 1988a)

Fungi on *Acanthus ilicifolius* (present study)	Frequency of occurrence[a] (%)	Fungi on *Rhizophora* spp (Hyde, 1988a)	Frequency of occurrence[b]
APICAL (GROUP I)			
Colletotrichum gloeosporioides cf	65	*Leptosphaeria* sp 2	20
Phialophora sp *cf*	55	*Savoryella lignicola* Jones & Eaton	10
Corynespora cassiicola	27	*Halocyphina villosa*	8
Stachylidium bicolor	25	*Cytospora rhizophorae** Kohlm & Kohlm	6
Nectria sp 2	10	*Lignincola tropica** Kohlm	4
Ophioceras sp	10		
MIDDLE (GROUP II)			
Stachybotrys chartarum	22	*Leptosphaeria avicenniae* Kohlm & Kohlm	6
Phaeoisaria sp	8	*Massarina velatospora** Hyde & Borse	10
Marinospaera mangrovei	3	*Dactylospora haliotrepha* (Kohlm & Kohlm) Hafellner	6
D90	3	*Verruculina enalia* (Kohlm) Kohlm & Volkm -Kohlm	6
		*Aigialus parvus** Schatz & Kohlm	4
BASAL (GROUP III)			
Acremonium sp	90	*Trichocladium opacum**	40
Phoma sp 1	55	*Cirrenalia pygmea* Kohlm	20
Fusarium sp	35	*Lulworthia* sp	36
Agerita sp	28	*Humicola alopallonella** Meyers & Moore	13
Cirrenalia basiminuta	25	*Bathyascus grandisporus* Hyde & Jones	6
Graphium sp 2	23	*Antennospora quadricornuta** (Cribb & Cribb) T W Johnson	6
Aniptodera chesapeakensis	13	*Aniptodera* sp	6
Tirispora unicaudata	13		
Trichocladium achrasporum	8		
Zelosatchmopsis sp *cf*	5		
Halosarpheia retorquens	3		
Halosarpheia marina	3		
Clavatospora bulbosa	3		
D43	3		

[a] Frequency of occurrence (%)= $\frac{\textit{No of collections of a species}}{\textit{No of samples examined at a specific level}} \times 100$

[b] No of records

* Found only at this level

on this aspect of mangrove fungal ecology (e.g. Hyde, 1988a).

Effect of substratum

The effect of substratum, in terms of fungal colonization, can be demonstrated by comparing the data presented in this paper with those of Hyde (1988a), Table 3 being the only published information available on stratification of fungal colonization on mangrove wood. There are distinct differences in species composition in these two studies. In the present study, *Acanthus*, (both herbaceous and woody tissues), on the one hand, were colonizcd mainly by Deuteromycotina. However, the woody portions were also colonized by some marine Ascomycotina (e.g. *Halosarpheia retorquens*, *Clavatospora bulbosa*, *Aniptodera chesapeakensis* and *Lignincola laevis*). On the other hand, the woody *Rhizophora* supported more Ascomycotina such as *Hypoxylon oceanicum* Schatz, *Lignincola tropica* Kohlm., *Savoryella lignicola* Jones & Eaton, *Massarina velatospora* Hyde & Borse, *Dactylospora haliotrepha* (Kohlm. & Kohlm.) Hafellner and the common basidiomycete *Halocyphina villosa* Kohlm. & Kohlm. These differences could be attributed to host species and tissue types – *Rhizophora* has little non-lignified tissues. However, the ascomycetes *Aniptodera chesapeakensis*, *Halosarpheia retorquens* and *Lignincola laevis* have been collected from softer tissues like twigs and leaf petioles (Jones *et al.*, 1988).

Ecosystem

The dominant fungi recorded from a brackish site in this study are largely terrestrial Deuteromycotina. Investigations of fungal colonization on woody substrata in similar environments suggested similar results. Shearer (1972) noted that there were more ascomycetes on the balsa blocks submerged in brackish to saline parts of the Patuxent River but with more Deuteromycotina in the freshwater zone. A similar observation was made by Kirk and Brandt (1980), in their study of the fungi in lower Chesapeake Bay using the harder pine and birch wood with longer submergence time. Lastly, Gold (1959) reported no deuteromycetes in Newport River water estuary having a salinity higher than 28.9‰. The difference in species reported in this study could be due to salinity levels (freshwater to brackishwater) and substrata type.

Vertical distribution

In Table 3 the most abundant fungi at each zone or only recorded from one zone are listed. Group I comprises those fungi collected in the apical region of the plant and are typical terrestrial species, while some are known to be parasites e.g. *C. gloeosporioides* cf. Group III comprises largely marine species colonizing the basal part of the plant which was inundated by tidal water at each tidal cycle. Only a few species which colonized in the middle zone were not found at the upper or basal regions or were there in greater abundance. These were largely terrestrial forms.

Some fungi grew well at all levels, e.g. *Acremonium* sp. This may indicate its ability to tolerate varying salinity levels and colonise different tissue types. The isolate was also observed to grow and sporulate on 100% natural sea water corn meal agar (unpublished data). On the other hand, other species of the genus are commonly known to occur in soils which might explain their high frequency on stems close to the mangrove floor. *Fusarium* sp. also occurred on all stem portions of *A. ilicifolius* with increasing frequencies toward the basal portion. As with *Acremonium*, most *Fusarium* species are ubiquitous soil fungi (Domsch *et al.*, 1980) and are also active decomposers of cellulosic plant substrata such as *Spartina alterniflora* Loisel (Meyers *et al.*, 1970) and *Rhizophora mangle* L, seedlings (Newell, 1976).

The high affinity of *Colletotrichum gloeosporioides* cf. towards the apical portion appears to agree with its known distribution, that is, more on herbaceous and 'aerial parts' where it is often associated with the plant disease anthracnose (So, 1991). However, its frequent presence on other portions of the plant also indicates its ability to colonize substrata even in submerged conditions. Moreover, the isolates were observed to grow and sporulate on 100% natural sea water corn meal agar (unpublished data) and Newell (1976) also recorded *Colletotrichum* sp. on *Rhizophora mangle* seedlings. The sporodochial *Tubercularia* sp. was also collected at all levels but with the highest frequencies at the apical portions, indicating its ability to tolerate aquatic conditions with different salinity levels as well as substratum types.

The basal portions of *Acanthus* yielded Ascomycotina and Deuteromycotina commonly reported in mangroves either in brackish or marine waters (Hyde & Jones, 1988; Jones & Tan, 1987) (Table 3). *Halosarpheia retorquens* was originally described from balsa wood in freshwater habitats of the estuaries of Cheapeake Bay (Shearer & Crane, 1980) Since then, it has been reported from other woody mangrove species (Jones *et al*, 1988, Jones & Kuthubutheen, 1989) *Lignincola laevis* Hohnk is cosmopolitan and is known to occur mostly on lignicolous substrata (Kohlmeyer & Kohlmeyer, 1979, Kohlmeyer, 1984) Recently, this was reported on *Pemphis acidula* Forst a dead shoreline tree in India (Chinnaraj, 1993) *Tirispora unicaudata* is a new species to be described by Jones, Vrijmoed, Read and Moss (in press)

Table 3 also lists the vertical distribution of fungi observed on prop roots of *Rhizophora* spp by Hyde (1988) and a comparison is made between his data and those of the present investigation

In conclusion, forty-four species of higher fungi were recorded from 120 samples of standing senescent *Acanthus* stems collected from April to December 1992 The Deuteromycotina was dominant with 32 species, followed by 11 Ascomycotina and 1 Basidiomycotina The very frequent species include *Acremonium* sp, *Colletotrichum gloeosporioides* cf, *Phoma* sp, *Fusarium* sp, *Tubercularia* sp and *Phialophora* sp cf These very frequent species collected in this study are different from other mangrove fungi reported on lignocellulose substrata A vertical zonation of fungi on the standing senescent *Acanthus* stems was also observed This investigation demonstrates that both the tissue type (lignicolous versus herbaceous) and varying degrees of exposure due to tidal inundation are important factors governing species distribution colonizing a vertically-orientated substratum in the mangroves

Acknowledgments

The first author would like to express his thanks to the Department of Biology and Chemistry, City Polytechnic of Hong Kong for the use of their laboratory facilities and the Caltex Green Fund for supporting his participation in the Asia Pacific Symposium on Mangrove Ecosystems in Hong Kong Prof E B G Jones is grateful to the British Council and City Polytechnic of Hong Kong for financial support to work in Hong Kong The authors are grateful to Dr Brian Sutton of IMI for identification of certain species This study was made possible through the UPGC Grant No 904-044

References

Anastasiou, C J & M Churchland, 1969 Fungi on decaying leaves in marine habitats Can J Bot 47 251–257

Chinnaraj, S, 1992 Higher Marine Fungi of Lakshadweep Islands and a Note on *Quintaria lignatilis* Cryptogamie Mycol 13 312–319

Chinnaraj, S 1993 Manglicolous fungi from atolls of Maldives, Indian Ocean Ind J mar Sci 22 141–142

Cribb, A B & J W Cribb, 1955 Marine fungi from Queensland – I Univ Queensl Pap Dep Bot 3 77–81

Domsch, K L, W Gams & T H Anderson, 1980 Compendium of soil fungi Academic Press, London, 859 pp

Fell, J W & I M Master, 1973 Fungi associated with the degradation of mangrove (*Rhizophora mangle* L) leaves in South Florida In L H Stevenson & R R Colwell (eds), Estuarine Microbial Ecology University of South Carolina Press, Columbia 455–465

Fell, J W & I M Master, 1980 The association and role of fungi in mangrove detrital systems Bot mar 23 257–263

Gold, H, 1959 Distribution of some lignicolous ascomycetes and fungi imperfecti in an estuary J Elisha Mitchell Sci Soc 75 25–28

Hyde, K, 1988a Observations on the vertical distribution of marine fungi on *Rhizophora* spp, at Kampong Danau Mangrove Brunei Asian Marine Biology 5 77–81

Hyde, K, 1988b *Phomatospora acrostichi sp nov* a marine fungus of pinnae of *Acrostichum speciosum* Trans br mycol Soc 90 135–138

Hyde, K, 1989 *Caryospora mangrovei* sp nov and notes on marine fungi from Thailand Trans mycol Soc Japan 30 333–341

Hyde, K, 1992 *Aigialus striatispora sp nov* from intertidal mangrove wood Mycol Res 96 1044–1046

Hyde, K, C A Farrant & E B G Jones, 1986 Marine fungi from Seychelles III *Antiptodera mangrovii sp nov* from mangrove wood Can J Bot 64 2989–2992

Hyde, K & E B G Jones, 1988 Marine Mangrove Fungi Mar Ecol (PSZNI) 9 15–33

Hyde, K & A Nakagiri, 1992 Intertidal fungi from Australia The genus *Swampomyces* including *S triseptatus* sp nov Sydowia 44 122–130

Hyde, K & L L P Vrijmoed, S Chinnaraj & E B G Jones, 1992 *Massarina armatispora sp nov* a New Intertidal Ascomycete from mangroves Bot mar 35 325–328

Jones, E B G & R Agerer, 1992 *Calathella mangrovei sp nov* and observations on the mangrove fungus *Halocyphina villosa* Bot mar 35 259–265

Jones, E B G & A J Kuthubutheen, 1989 Malaysian Mangrove Fungi Sydowia 41 160–169

Jones E B G & T K Tan, 1987 Observations on Manglicolous Fungi from Malaysia Trans br mycol Soc 89 390–391

Jones E B G, R Uyenco & P Follosco 1988 Fungi on driftwood collected in the intertidal zone from the Philippines Asian Mar Biol 5 103–106

Jones, E B G, L L P Vrijmoed, S J Read and S T Moss *Tirispora*, a new Ascomycetous genus in the Halosphaeriaceae Can J Bot (in press)

Kirk, P W J & J M Brandt, 1980 Seasonal distribution of Lignicolous marine fungi in the Lower Chesapeake Bay Bot mar 13 657–668

Kohlmeyer, J, 1969 Ecological Notes on Fungi on Mangrove Forests Trans br mycol Soc 53 237–250

Kohlmeyer, J , 1984 Tropical Marine Fungi P S Z N I Marine Ecology, 329–378

Kohlmeyer, J & B P R Vittal, 1986 *Lophiostoma mangrovis*, A new marine ascomycete from the tropics Mycologia 78 485–489

Kohlmeyer, J & E Kohlmeyer, 1979 Marine Mycology, the Higher Fungi Academic Press, New York, 690 pp

Lee, S Y, 1990 Primary productivity and particulate organic matter flow in an estuarine mangrove-wetland in Hong Kong Mar Biol 106 453–463

Leong, W F, T K Tan, K D Hyde & E B G Jones, 1990 *Payosphaeria minuta gen et sp nov*, an Ascomycete on Mangrove Wood Bot mar 33 511–514

Meyers, S P, M E Nicholson, P Miles, J S Rhee & D G Ahearn, 1970 Mycological studies in Barataria Bay, Louisiana and biodegradation of oyster grass, *Spartina alterniflora*, Louisiana State Univ Coastal Bull Louisiana State Univ Coast Bull 5 111–124

Newell, S Y, 1976 Mangrove fungi The succession in the mycoflora of mangrove (*Rhizophora mangle* L) seedlings In E B G Jones (ed), Recent Advances in Aquatic Mycology Elek Science, London 51–91

Shearer, C , 1972 Fungi of the Chesapeake Bay and its tributaries III The distribution of wood-inhabiting Ascomycetes and Fungi Imperfecti of the Patuxent River Am J Bot 59 961–69

Shearer, C & J L Crane, 1980 Fungi of the Chesapeake Bay and its tributaries VIII Ascomycetes with unfurling appendages Bot mar 23 607–615

So, M L , 1991 Hong Kong Plant Diseases Urban Council Hong Kong, 121 pp

Tan, T K & W F Leong, 1989 Succession of fungi on wood of *Avicennia alba* and *A lanata* in Singapore Can J Bot 67 2686–2691

Tomlinson, P B , 1986 The Botany of Mangroves Cambridge University Press, London, 413 pp

Volkmann-Kohlmeyer, B & J Kohlmeyer, 1993 Biogeographic observations on Pacific marine fungi Mycologia 85 337–346

Vrijmoed, L L P, 1990 Preliminary observations of lignicolous marine fungi from mangroves in Hong Kong In B Morton (ed), Proceedings of the Second International Marine Biological Workshop The Marine Flora and Fauna of Hong Kong and Southern China Hong Kong University Press, Hong Kong 707–706

Vrijmoed, L L P & N F Y Tam, 1990 Fungi associated with leaves of *Kandelia kandel* (L) Druce in litterbags on the mangrove floor of a small subtropical mangrove community in Hong Kong Bull mar Sci 47 261

Hydrobiologia **295**: 127–134, 1995.
Y. S. Wong & N. F. Y. Tam (eds), Asia-Pacific Symposium on Mangrove Ecosystems.

Substrate type and microbial interactions as factors affecting ascocarp formation by mangrove fungi

T. K. Tan[1], C. L. Teng[1] & E. B. G. Jones[2]
[1]*Department of Botany, National University of Singapore, Kent Ridge, Singapore 0511*
[2]*School of Biological Sciences, University of Portsmouth, King Henry I Street, Portsmouth PO1 2DY, United Kingdom*

Key words: mangrove fungi, ascocarp formation, colonisation

Abstract

The growth and ascocarp formation of *Aigialus parvus, Lignincola laevis* and *Verruculina enalia* in single and mixed cultures on wood of *Avicennia alba, Bruguiera cylindrica* and *Rhizophora apiculata* was studied. In pure cultures, these fungi grew well on all three species of wood. Except for *Aigialus parvus* on *B. cylindrica*, all three fungi also formed abundant ascocarps. The time needed for ascocarp formation after inoculation ranged from six to eight weeks for *L. laevis* on all wood species, and for *V. enalia* on *Avicennia alba*; to ten weeks for *V. enalia* on *B. cylindrica* and *R. apiculata*; and 12 weeks for *Aigialus parvus* on *Avicennia alba* and *R. apiculata*.

Mixed cultures involving two or three of the test fungi delayed the onset of sporulation and affected the abundance of ascocarps formed. Sporulation by *Aigialus parvus* on *Avicennia alba* and *R. apiculata* was markedly reduced by *L. laevis* alone or in combination with *V. enalia*. Likewise, sporulation by *L. laevis* was suppressed by *Aigialus parvus* and/or *V. enalia*. In contrast, the presence of *L. laevis* enhanced ascocarp formation by *V. enalia* on all types of wood. These observations suggest the presence of interference competition among the test fungi.

Introduction

Wood submergence experiments had been carried out to establish the pattern of colonisation of marine fungi, as well as their ability to utilize different wood substrata. Previous studies by Tan *et al.* (1989) on fungal colonisation of *Avicennia alba* Blume and *Avicennia lanata* Ridley, and by Leong *et al.* (1991) using wood of *Bruguiera cylindrica* (L.) Blume and *Rhizophora apiculata* Blume showed a certain pattern of succession. Early colonisers such as *Lignincola laevis* Hohnk and *Verruculina enalia* (Kohlm.) Kohlm. & Volkm.-Kohlm. (formerly *Didymosphaeria enalia* Kohlm.) were species appearing on wood blocks after six to eighteen weeks of submergence. Intermediate colonisers such as *Aigialus parvus* Schatz & Kohlm. were species that appeared after 18 to 32 weeks of submergence.

Various factors such as salinity, temperature, type and condition of substrata, tidal exposure and microbial interactions could influence colonisation of wood by marine and mangrove fungi. It is only recently that the possible role of microbial interactions on fungal succession in the marine environment has been investigated. Miller *et al.* (1985) showed that the presence of *Ceriosporopsis halima* or *Amylocarpus encephaloides* decreased the number of *Lulworthia* perithecia in submerged wood panels, suggesting the presence of interference competition. Strongman *et al.* (1987) examined the pair-wise interaction of 27 marine fungi and found that certain species had high antagonistic activities. In our study, the preference of *Aigialus parvus, Lignincola laevis* and *Verruculina enalia* for different mangrove wood, and microbial interactions between the three fungi – factors which could possibly influence their ability to colonise the wood – were investigated.

Materials and methods

Aigialus parvus, *Lignincola laevis* and *Verruculina enalia* obtained from the School of Biological Sciences, University of Portsmouth (UK) were maintained on yeast glucose seawater agar (YGA) containing per litre seawater: D-glucose 10 g, yeast extract 3 g, Oxoid agar powder 15 g.

Wood from the trunks of *Avicennia alba*, *Bruguiera cylindrica* and *Rhizophora apiculata* was obtained from the Mandai mangroves, sawn into 5 × 1 × 1 cm blocks and sterilised by autoclaving. These blocks were then placed, five replicate blocks of each wood species per flask, in sterile 250 ml Erlenmeyer flasks each containing 50 ml filtered and sterilised seawater. The wood blocks were not submerged but kept above the surface of the seawater by a mesh support placed in each flask.

The wood blocks of each tree species were inoculated with the three fungi in single pure cultures (controls) as well as in combinations of two or three species. The inoculum consisted of mycelial discs (1 cm diameter) of the fungus grown on YGA. Two mycelial discs were aseptically placed on the surface of each wood block. The flasks were then incubated on a 150 rpm rotary shaker set to operate at 12 h cycles so that the wood blocks were alternately bathed in seawater and exposed to dry at 12 hourly intervals. The seawater in the flasks was changed for fresh seawater weekly. The wood blocks were examined biweekly after the initial incubation period of four weeks, for 18 weeks. Details on mycelial growth on the four broad surfaces of each block were noted and the number of ascocarps counted.

The data obtained on ascocarp formation in single and mixed cultures were analysed using STATGRAPHICS. For each species, one-way analyses of variance (ANOVA) and a 95% Confidence Interval Multiple Range Test were performed.

Results

Growth and sporulation of Aigialus parvus

Growth on the surface of *Avicennia alba* and *R. apiculata* blocks was good, appearing as thin mycelial mats and aerial hyphae. However, there was no apparent surface growth on the blocks of *B. cylindrica*. Sporulation on *Avicennia alba* and *R. apiculata* was observed from the 12th week (Figs 1a & 1c). This increased rapidly until the 18th week when the average number of *Aigialus parvus* ascocarps per block was higher on *Avicennia alba* (234±56 ascocarps) than on *R. apiculata* (45±10 ascocarps). The ascocarps were randomly distributed on the wood surfaces of both mangrove timbers. There was no ascocarp formation on *B. cylindrica* blocks during the 18 week period (Fig. 1b).

The number of ascocarps of *Aigialus parvus* when co-cultured with *L. laevis* was significantly reduced on both *Avicennia alba* (Fig. 1a) and *R. apiculata* (Fig. 1c) blocks. Results of the statistical tests are summarised in Table 1. Sporulation was also observed to occur two weeks later than the Controls on *Avicennia alba* (Fig. 1a) and four weeks later on *R. apiculata* (Fig. 1c). Ascocarps were produced when *Aigialus parvus* was co-cultured with *L. laevis* on *B. cylindrica* but the number was not significant compared to the absence of sporulation in the Control (Fig. 1b, Table 1). It was not possible to differentiate between the mycelia of the two fungi on the wood surfaces.

There was a significant and total suppression of ascocarp formation in the presence of *V. enalia* alone, and *V. enalia* together with *L laevis* (Figs 1a–1c, Table 1). In particular, the mycelium of *Aigialus parvus* was observed to have either failed to establish from the start or to be eliminated in the later part of the incubation period from the 6th to 8th week.

Growth and sporulation of Lignincola laevis

Mycelial growth on the surfaces of *Avicennia alba* blocks was good, but less so on those of *B. cylindrica* and *R. apiculata* on which the mycelial mat was scanty. As in the case of *Aigialus parvus*, the abundance of ascocarps formed corresponded with the abundance of surface mycelial growth of the fungus. Sporulation was seen to occur from the 6th week on *Avicennia alba* (Fig. 2a) and *R. apiculata* (Fig. 2c) and the 8th week on *B. cylindrica*. In all cases, the ascocarps first appeared near the inoculation point and later, randomly over the wood surfaces. The number of ascocarps formed continued to rise with incubation and by the 18th week, there were more *L. laevis* ascocarps per block on *Avicennia alba* (1456±360 ascocarps) than on *B. cylindrica* (156±28 ascocarps) and *R. apiculata* (83±27 ascocarps) (Figs 2a–2c).

The formation of ascocarps by *L. laevis* was also significantly reduced when it was co-cultured with the other two fungi (Table 1). In the presence of *Aigialus parvus*, *L. laevis* did not form ascocarps on *Avicennia alba* blocks (Fig. 2a). On *B. cylindrica* blocks, only 36

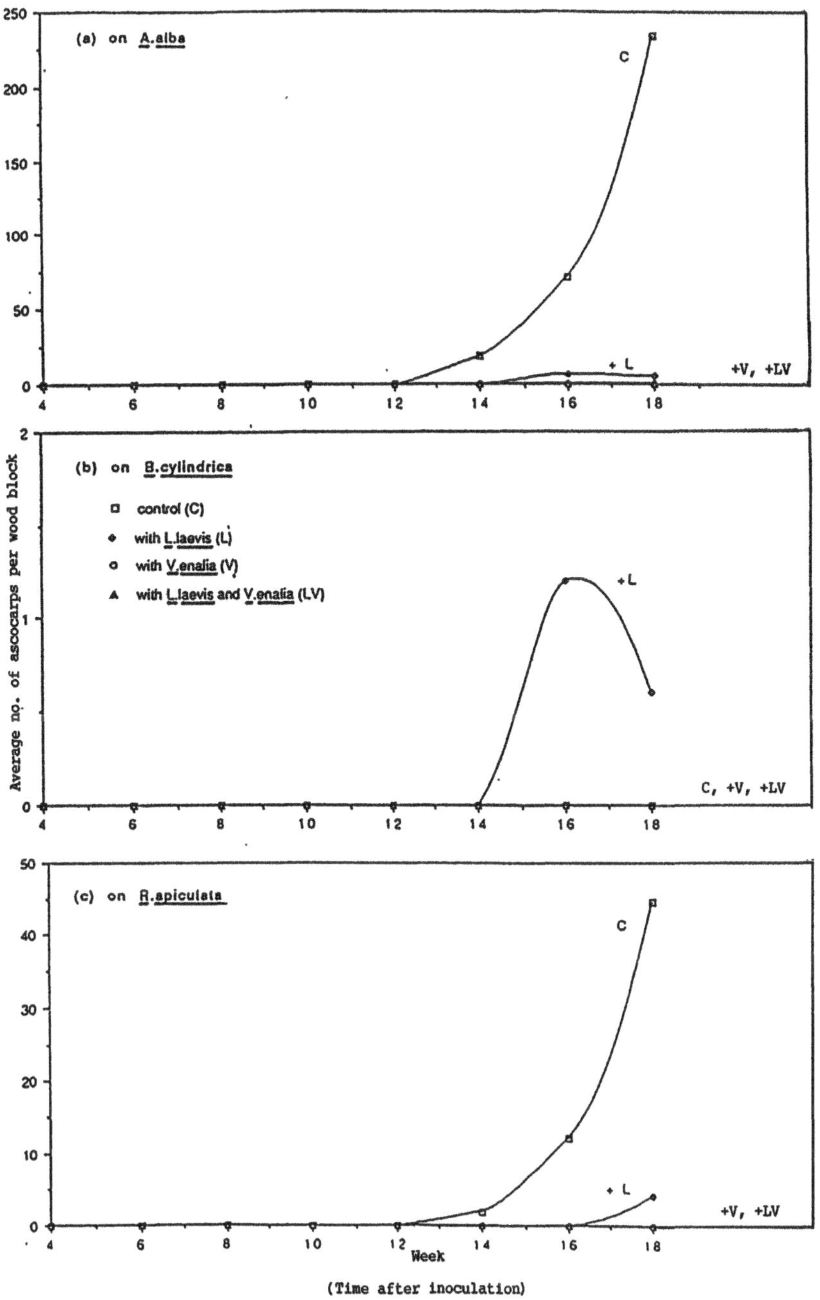

Fig. 1. Number of *Aigialus parvus* ascocarps formed in single and mixed cultures.

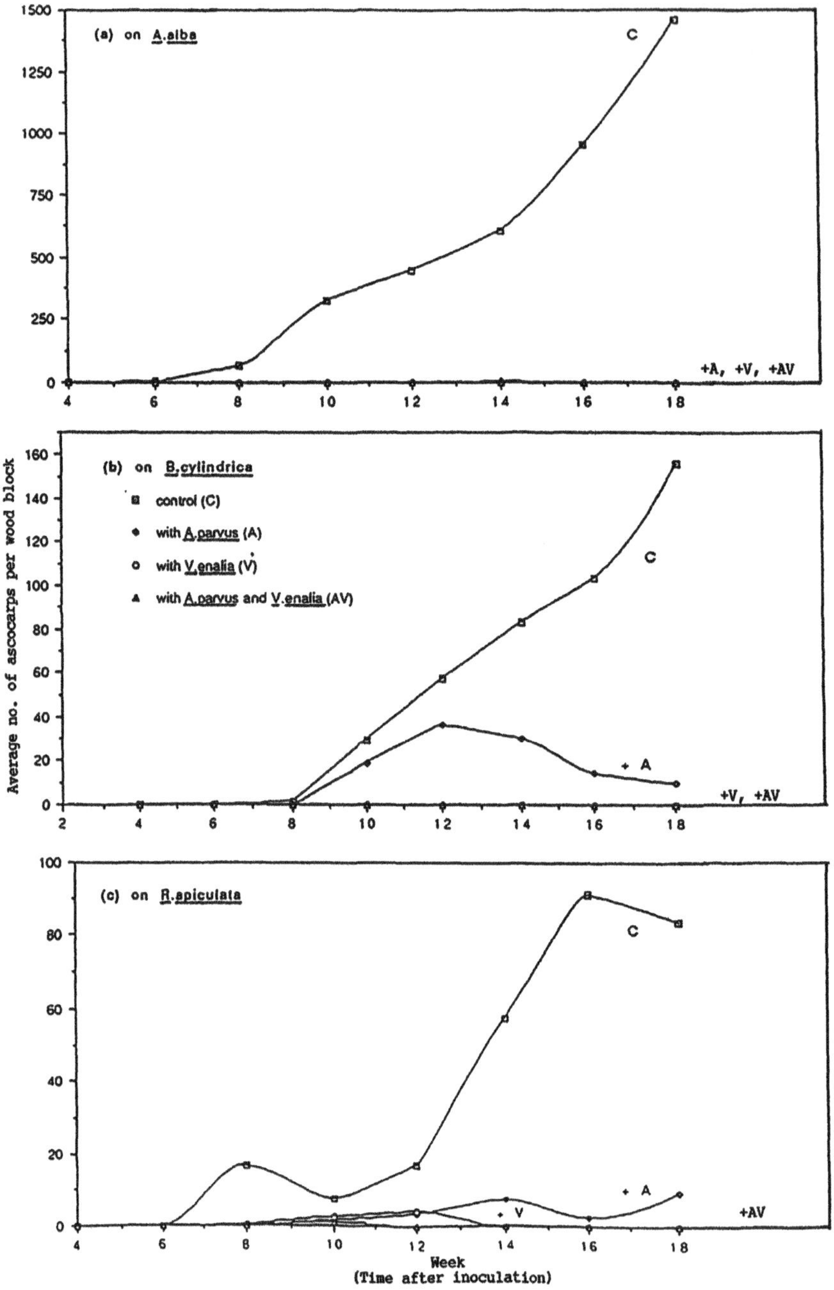

Fig. 2. Number of *Lignincola laevis* ascocarps formed in single and mixed cultures.

Table 1. 95% Confidence Interval Multiple Range Test of number of ascocarps formed by three fungi in single and mixed cultures after 18 weeks using log (number of ascocarps)

	Avicennia alba				*Bruguiera cylindrica*				*Rhizophora apiculata*			
Aigialus parvus (A)	both L & V	*			control	*			with L & V	*		
	with V	*			both L & V	*			with V	*		
	with L		*		with V	*			with L		*	
	control			*	with L	*			control			*
Lignincola laevis (L)	both A & V	*			both A & V	*			both A & V	*		
	with V	*			with V	*			with V	*		
	with A	*			with A	*			with A		*	
	control		*		control		*		control			*
Verruculina enalia (V)	control	*			with A	*			with A	*		
	with A	*			both A & L	*			control	*		
	with L		*		control		*		with L	*	*	
	both A & L		*		with L			*	both A & L		*	

Note
1. Single and mixed culture treatments are arranged in ascending order of log (number of ascocarps)
2. Aligned asterisks within fungal species tested: no significant difference at $p = 0.05$
3. Unaligned asterisks within fungal species tested: significant difference at $p = 0.05$

(±9) ascocarps were observed by the 12th week but these declined to 9 ascocarps by week 18 (Fig. 2b). Likewise, only a few ascocarps (9) were formed on *R. apiculata* blocks (Fig. 2c). It was not possible to differentiate between the mycelia of the three fungi.

In the presence of *V. enalia*, ascocarp production by *L. laevis* on *Avicennia alba* and *B. cylindrica* was significantly suppressed (Figs 2a & 2b, Table 1). Only about four ascocarps were produced on *R. apiculata* blocks from the 8th to the 12th week before these disappeared by the 14th week (Fig. 2c). The mycelium of *L. laevis* was observed to be encroached upon by *V. enalia* on *Avicennia alba* blocks and appeared eliminated from all the three wood species by the 8th to 10th week.

Growth and sporulation of Verruculina enalia

Surface growth of the fungus on *Avicennia alba* blocks was good with the mycelium forming a thick mat, but poor on *B. cylindrica* and *R. apiculata* on which the mycelium appeared dispersed. Sporulation began from the 6th week on *Avicennia alba* (Fig. 3a) but was much later on *B. cylindrica* and *R. apiculata* (week 10) (Figs 3b & 3c). The number of ascocarps increased with time and by the 18th week there were more *V. enalia* ascocarps per block on *Avicennia alba* (742±180 ascocarps) than on *B. cylindrica* (30±8 ascocarps) and *R. apiculata* (53±14 ascocarps) (Figs 3a–3c).

On *Avicennia alba* and *R. apiculata* the presence of *Aigialus parvus* did not significantly affect the number of *V. enalia* ascocarps formed compared to that of the Control (Fig. 3a, Table 1). However, on *B. cylindrica* blocks, ascocarps of *V. enalia* were not formed at all when it was co-cultured with *Aigialus parvus*. Mycelial growth of *V. enalia* was good on *Avicennia alba*, moderate on *R. apiculata* but poor on *B. cylindrica*. Mycelium was thin and highly dispersed on blocks of *B. cylindrica*.

Blocks with *V. enalia* co-cultured with *L. laevis* showed a marked increase in ascocarp formation on all three wood species. The number of fruit bodies formed by the 18th week on *Avicennia alba* (1572±360) and *B. cylindrica* (205±58) was significantly different from those of the Controls (742±180 and 30±8 respectively) (Figs 3a & 3b, Table 1). Co-culturing with both *Aigialus parvus* and *L. laevis* resulted in the formation of large numbers of ascocarps on *Avicennia alba* (1965±150) and *R. apiculata* (280±64). This was significantly higher than the number formed when it was co-cultured with *L. laevis* alone (Figs 3a & 3c, Table 1). However, ascocarps of *V. enalia* were not

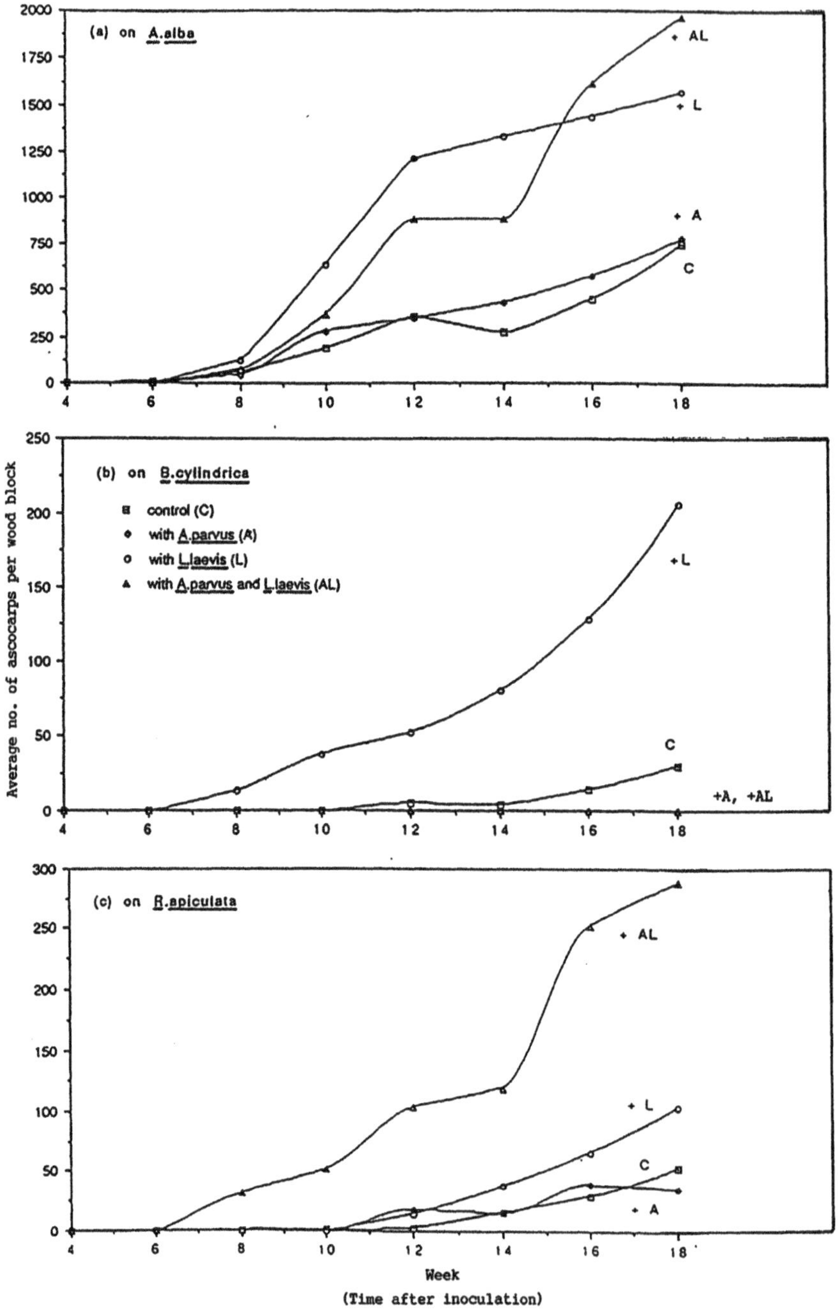

Fig. 3. Number of *Verruculina enalia* ascocarps formed in single and mixed cultures.

formed on *B. cylindrica* when it was co-cultured with the other two fungi (Fig. 3b).

Discussion

Leightley (1980) showed in submergence experiments that weight losses in the wood of *Bruguiera gymnorhiza* and *Rhizophora stylosa* were lower than that of *Avicennia marina* var. *resinifera*. The wood of *Rhizophora* sp. is known to be more lignified than that of *Avicennia* (Jones & Hyde, 1988). These observations suggest that the wood of *B. cylindrica* and *R. apiculata* could be more durable than that of *Avicennia alba*, thereby reducing the ability of *Aigialus parvus*, *L. laevis* and *V. enalia* to grow and sporulate on *B. cylindrica* and *R. apiculata*. In contrast, the less lignified and softer nature of *Avicennia alba* wood could have facilitated mycelial growth and sporulation of the three fungi.

The shorter incubation time taken by *Aigialus parvus*, *L. laevis* and *V. enalia* to form ascocarps on *Avicennia alba* than on *B. cylindrica* and *R. apiculata* agreed with Tan *et al.*'s (1989) observation that colonisation of *Avicennia alba* (based on ascocarp formation) was rapid: nine of the total of 18 species that they observed on *Avicennia alba* appeared between 18 and 32 weeks of submersion. In comparison, the ascocarps of the majority of fungal species were observed on *B. cylindrica* and *R. apiculata* only after a longer submersion of 22 to 54 weeks (Leong *et al.*, 1991). This further demonstrates that *Avicennia alba*, among the three mangrove wood species, constitutes a substrate that readily promotes sporulation of marine fungi, irrespective of whether these were in pure cultures as in this current study or in mixed (natural) cultures as in Tan *et al.*'s (1989) study. This characteristic of *Avicennia alba* could be related to a lower tannin content. According to Allen *et al.* (1974), the wood of *B. cylindrica* and *R. apiculata* have relatively higher levels of tannin in their tissues than that of other mangrove trees, including *Avicennia alba*.

Tan *et al.* (1989) and Leong *et al.* (1991) reported that *Aigialus parvus* was fairly common on submerged *Avicennia alba*, *B. cylindrica* and *R. apiculata* wood blocks with percentage occurrences ranging from 26.5% to 38.3%. It was considered an 'intermediate' coloniser whose ascocarps appeared on *R. apiculata* by the 18th week, and on *Avicennia alba* and *B. cylindrica* by the 22nd week. However, in this current study *Aigialus parvus* sporulated 10 weeks earlier on *Avicennia alba* and six weeks earlier on *R. apiculata*. This earlier sporulation showed that the presence of mixed fungal species on the wood blocks of the two earlier studies (Tan *et al.*, 1989; Leong *et al.*, 1991) could have delayed sporulation of *Aigialus parvus*, suggesting the presence of microbial interactions.

The present study also shows that *L. laevis* and *V. enalia* were able to colonise and sporulate early on all three wood species. *L. laevis* was consistently an 'early coloniser' in the studies of Tan *et al.* (1989) and Leong *et al.* (1991), forming ascocarps on the wood blocks from the 6th week onwards and persisting to as late as the 50th week on *Avicennia alba* and *B. cylindrica*. Likewise in this study, ascocarps of *L. laevis* were observed about 6 weeks after inoculation. However, the highest number of ascocarps were formed on *Avicennia alba*, in contrast to its higher incidence on *B. cylindrica* and *R. apiculata* as reported by Leong *et al.* (1991).

Verruculina enalia was reported by Leong *et al.* (1991) to be very common on all three wood species with percentage occurrences ranging from 63.3% to 70.9%. It was also an 'early' coloniser, forming ascocarps from the 11th week onwards until the 60th week on all three wood species (Leong *et al.*, 1991). In this study, *V. enalia* took about the same period of time (10 weeks) to start ascocarp formation on *B. cylindrica* and *R. apiculata*, but not on *Avicennia alba* which occurred 5 weeks earlier.

The results of this study, although complementing what was earlier observed by Tan *et al.* (1989) and Leong *et al.* (1991) on the sequence of sporulation by *Aigialus parvus*, *L. laevis* and *V. enalia* on mangrove wood in nature, showed that the time taken for ascocarps to be formed and their relative abundance could be different under pure or single culture condition. This suggested the role of microbial interactions which were demonstrated in this study through the use of mixed cultures.

The effect of mixed cultures on the number of ascocarps formed could be attributed to the relative ability of the different species to colonise and grow on the substrata, competition for nutrients and interference competition. The latter could be a major factor in decreasing or suppressing ascocarp formation when *Aigialus parvus* and *L. laevis* were co-cultured. Both fungi were shown by Teng (1993), using the approach of Strongman *et al.* (1987), to exhibit mutual inhibition of mycelia in agar cultures. This could lead to limited growth of both types of mycelia and subsequently, limited sporulation or absence of sporulation compared to the respective single cultures. Likewise, interference competition could have existed in the mixed cultures of *V enalia* and the other two species of fungi – except that *V enalia* was the most 'aggressive' of the three species This is based on our observation that the mycelium of *V enalia* generally tend to encroach upon the mycelia of the other species on the wood surfaces, and Teng's (1993) finding that *V enalia* had the highest index of antagonism (36) compared to *Aigialus* (18) and *L laevis* (15) An interesting feature of the interaction between *V enalia* and *L laevis* in our study is the marked increase in ascocarp formation by *V enalia* in the presence of *L laevis* While fungal reproduction is generally known to be enhanced by nutrient stress, this is an unlikely factor for *V enalia* since *L laevis* grew poorly and was not a strong competitor in the presence of *V enalia* However, it is possible that sporulation of *V enalia* was enhanced by metabolites secreted by *L laevis*

References

Allen, W E , H M Grimshaw, J A Parkinson & C Quarmby, 1974 Chemical Analysis of Ecological Materials Blackwell Scientific Publications, Oxford

Jones, E B G & K D Hyde, 1988 Methods for the study of mangrove marine fungi In A D Agate, C V Subramanian & M Vannucci (eds), Mangrove Microbiology, Role of Microorganisms in Nutrient Cycling of Mangrove Soils and Waters UNDP/UNESCO Publication 9–27

Leightley, L E , 1980 Wood decay activities of marine fungi Bot mar 23 387–395

Leong, W F , T K Tan & E B G Jones, 1991 Fungal colonisation of submerged *Bruguiera cylindrica* and *Rhizophora apiculata* wood Bot mar 34 69–76

Miller, J D , J A Findlay, E B G Jones & Y E Moharir, 1985 Colonisation of wood blocks by marine fungi in Langstone Harbor Bot mar 28 251–257

Strongman, D B , L Calhoun, J A Findlay, J D Miller & N J Whitney, 1987 The biochemical basis for interference competition among some lignicolous marine fungi Bot mar 30 21–26

Tan, T K , W F Leong & E B G Jones, 1989 Succession of fungi on wood of *Avicennia alba* and *A lanata* in Singapore Can J Bot 67 2686–2691

Teng, C L , 1993 Substrate preferences and interactions affecting fungal colonisation of mangrove wood B Sc Hons thesis, Botany Department, National University of Singapore

Hydrobiologia **295**: 135–140, 1995.
Y. S. Wong & N. F. Y. Tam (eds), Asia-Pacific Symposium on Mangrove Ecosystems.

Continental scale patterns in mangrove litter fall

John S. Bunt
212 Lower Plateau Road, Bilgola, NSW 2107, Australia

Key words: mangroves, litterfall, Australia, climate

Abstract

Litter fall was monitored in stands of the mangrove species *Rhizophora stylosa* Griff., *Ceriops tagal* (Perr.) C. B. Robinson and *Avicennia marina* (Forsk.), Vierh. at approximately monthly intervals over a single annual cycle at selected locations around the coastline of Australia and throughout the distribution of each species. Concurrent data were obtained from a single location near Port Moresby in Papua New Guinea. The materials recovered in sub-canopy catchers were sorted into major categories and dried and weighed as leaves, petiolar stipules, twigs and other woody tissues, reproductive parts (flowers, flower buds, fruit and propagules) and residual detritus. This paper considers the principal findings of the study among which it may be reported that the highest total annual litter recoveries at individual catchers were 1598 g dry wt m^{-2} for *A. marina*, 2369 g dry wt m^{-2} for *R. stylosa* and 1290 g dry wt m^{-2} for *C. tagal*. Significant regional differences in litter fall emerged when data from major climatic zones were compared. The outcome of this analysis is detailed in the body of the paper.

Introduction

During the exploratory phases of a program on the productivity of tropical coastline wetlands, a group at the Australian Institute of Marine Science undertook a detailed study of litter fall in the mangroves of Hinchinbrook Island in northern Queensland (Bunt, 1982). Later, during 1982/3, in an endeavour to set this work in a larger context, collaboration with a number of interested individuals and organizations made it possible to arrange the collection of litter concurrently over a 12–14 month period at mangrove sites widely distributed around the Australian coastline and extending to one location in Papua New Guinea. Attention was directed at three of the most commonly occurring species viz.; *Avicennia marina* (Forsk.) Vierh., *Rhizophora stylosa* Griff. and *Ceriops tagal* (Perr.) C. B. Robinson. The purpose of this report is to present and consider the principal findings of that undertaking. Details of the data are held at the Australian Institute of Marine Science. Phenological trends in the data for *A. marina* have been examined independently by Duke (1990).

Materials and methods

Study areas

The project encompassed virtually the entire distribution of the mangroves around Australia over a considerable range of environmental conditions. At the same time, it will be appreciated that the density of sampling sites could not be high for reasons of cost, logistic challenges and limits on the considerable effort required in the processing of litter samples at monthly intervals over an annual cycle. Nonetheless, 27 study sites, including a number at remote locations, were established as indicated in Fig. 1 and Table 1.

Sampling procedures

Litter traps were fabricated to a standard design (Bunt, 1982) using 1 mm grade Sarlon shade cloth attached to PVC tubing to provide a 1 m^2 entry. The litter catchers were hung from their PVC frames as inverted cones and suspended in the field with stout cords to convenient mangrove trunks and branches high enough above the substrate to avoid inundation by tides. Draw string

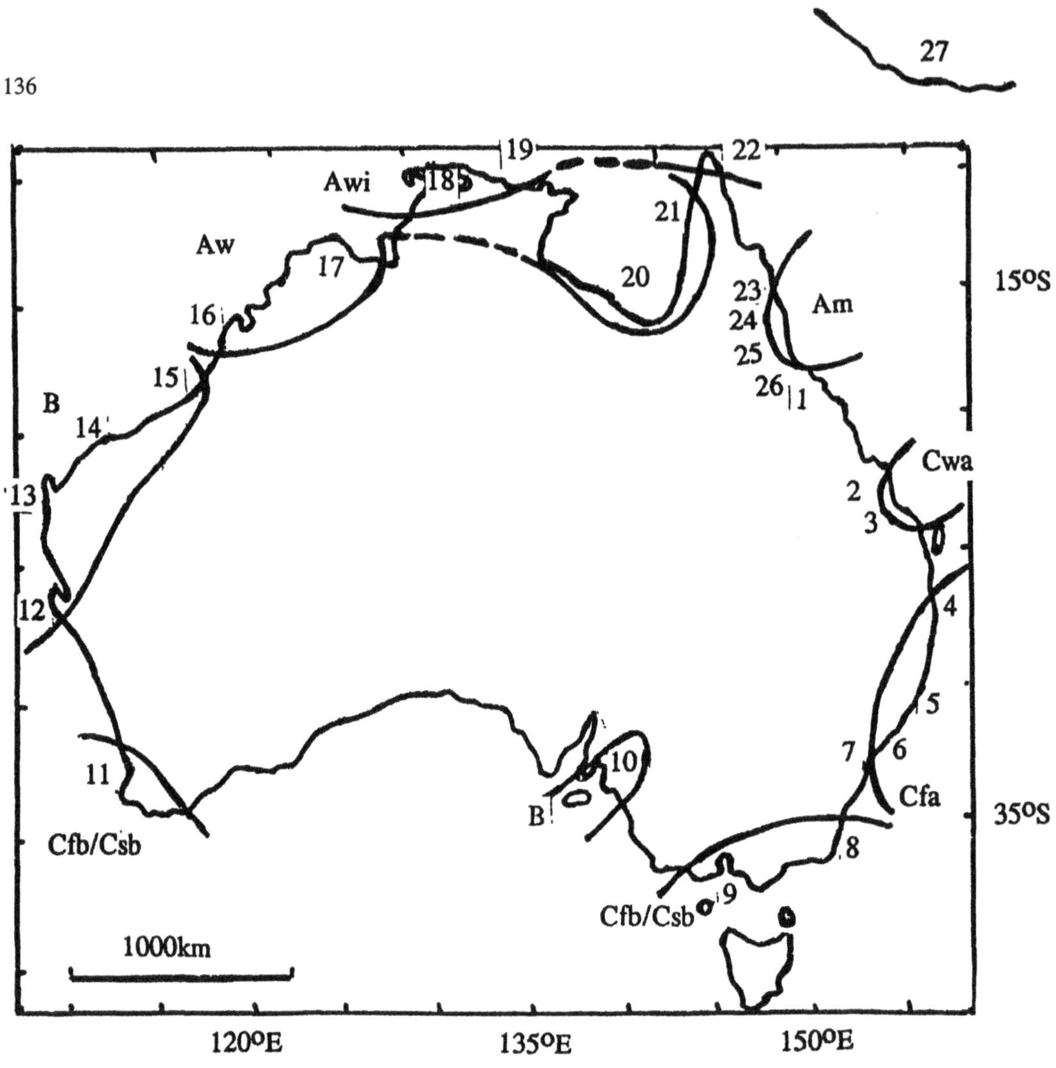

Fig. 1. Australia (south coast of Papua New Guinea offset) with sampling locations (1–27) and climatic zones. Dotted lines indicate zones with segments which are geographically separated.

closures at the base of each trap allowed for ready recovery of accumulated litter.

All sites at locations shown in Fig. 1 were selected in the field, choosing sections of mangrove and individual trap positions as far as possible representative of each location. With the exception of those in New South Wales, all catchers were installed by the author or by N. C. Duke, who was also responsible for receiving and processing the litter collections.

With well over 30 mangrove species occurring in Australia at lower latitudes, attention was restricted to *A. marina*, *R. stylosa* and *C. tagal*. *A. marina* occurs throughout the distribution of the Australian mangroves. *R. stylosa* and *C. tagal* are limited to the tropics and sub-tropics. All are major species. Two varieties of *C. tagal* are widespread and Duke (1990) has now clearly established the existence of three varieties of *A. marina*. In this account, however, it has not been possbible to pay detailed attention to litter fall differences beyound the level of species.

Two and, in some cases, three traps were installed in essentially monospecific stands of every species available at the chosen locations. Of a total of 27 such locations, 24 were sampled for *Avicennia*, 16 for *Rhizophora* and 13 for *Ceriops*. The traps were closed by the collaborators listed by Duke (1990) on or within a

Table 1. List of locations and species sampled (see Fig. 1)

Geographic locations			*A. marina*	*R. stylosa*	*C. tagal*
1.	Chunda Bay	19°21′S 147°14′W	x	x	x
2.	Curtis Is.	23°36′S 151°09′W		x	x
3.	Callipe River	24°02′S 150°57′W			x
4.	N. Stradbroke Is.	27°33′S 153°28′W	x		
5.	Nambucca River	30°37′S 153°01′W	x		
6.	Port Stephens	32°42′S 152°10′W	x		
7.	Botany Bay	33°59′S 151°12′W	x		
8.	Merimbula	36°52′S 149°55′W	x		
9.	Westernport Bay	38°25′S 145°12′W	x		
10.	Port Gawler	34°55′S 138°36′W	x		
11.	Bunbury	33°20′S 115°39′W	x		
12.	Carnarvon	24°53′S 113°40′W	x		
13.	Exmouth	21°55′S 114°07′W	x	x	
14.	Dampier	20°40′S 116°42′W	x	x	
15.	Port Hedland	20°18′S 118°35′W	x	x	
16.	Broome	17°57′S 122°13′W	x	x	
17.	Wyndham	15°27′S 128°06′W	x		
18.	Darwin	12°24′S 130°52′W	x	x	x
19.	Maningrida	12°03′S 134°16′W		x	x
20.	Mornington Is.	16°32′S 139°23′W	x	x	x
21.	Weipa	12°38′S 141°53′W	x	x	x
22.	Jacky Jacky Creek	10°50′S 142°35′W	x	x	x
23.	Cooktown	15°28′S 145°15′W	x	x	x
24.	Daintree River	16°06′S 145°09′W	x	x	x
25.	Trinity Inlet	16°54′S 145°45′W	x	x	x
26.	Hinchinbrook Is.	18°21′S 146°14′W	x	x	x
27	Port Moresby (Papua New Guinea)	09°30′S 147°10′W	x	x	x

day of July 26, 1982 and emptied as nearly as possible at monthly recorded intervals for 12–14 months. In most instances, this schedule was maintained successfully and the recovered materials regularly mailed to the Australian Institue of Marine Science for processing.

Sample processing

All litter materials were sorted manually on receipt into the following categories: leaves, petiolar stipules; wood; reproductive parts; and remaining debris. The sorted materials were then dried to constant weight at 70 °C.

Results and discussion

Reducing the accumulated data to its limits yields the information shown in Table 2. Overall, the most productive species in litter fall terms was *R. stylosa* with a mean annual yield of 965 g m^{-2} followed by *C. tagal* with 675 g m^{-2} and, only a little further behind, by *A. marina* with 620 g m^{-2}. At the same time, litter production from location to location and even from catcher to catcher was highly variable. Coefficients of variation in litter yield from individual catchers ranged from 0.35 in *Ceriops* to 0.62 in *Avicennia*. It is worth noting that the ranges of yields are comparable with those compiled from the literature by Snedaker & Brown (1982), principally, although not entirely from studies in Florida and Puerto Rico and, of course, at those locations, relevant to other species. The overall mean

Table 2. Total litter fall (g dw m^{-2} yr^{-1}); summer data

Species	Mean	Range	Coefficient of variation
A. marina	620	110–1598	0.62
R. stylosa	965	284–2369	0.47
C. tagal	675	311–1290	0.35

of 738 g m^{-2} from the Australian data compares quite well with the mean of 812 g m^{-2} derivable from the Snedaker & Brown (1982) compilation.

While mean values for mangrove litter fall over a major continental coastline are undoubtedly of interest in global terms, they cannot, of course, reveal patterns at regional or more local scales. To that end, examination of the data for possible correlations with several likely environmental controls taken singly was generally disappointing. On the other hand, a one-way analysis of variance of the data for annual litter fall grouped according to source from major climatic zones within the Koppen system as applied to Australia (e.g. see Gentilli, 1986) proved quite encouraging. The zones of interest have been marked in Fig. 1, their geographic boundaries having been set as a reasonable compromise between a number of atlas sources. Locations included in each zone, some, it should be noted, at or close to zonal boundaries, together with their principal climatic characteristics, are as follows:

Awi: hot; small annual temperature variation; dry winter
locations 18, 19 and 22.

Aw: hot; dry winter
locations 16, 17, 20, 21, 1

Am: hot; generally short dry season (on average)
locations 23, 24, 25, 26

B (mainly BWh): hot; arid
locations 10, 12, 13, 14, 15

Cwa: hot summer, dry winter
locations 2, 3

Cfa: hot summer; uniform rain
locations 4, 5, 6, 7

Cfa/Csb: long mild summer; cool winter
locations 8, 9, 11

Port Moresby (Papua New Guinea): similar to Australian zone Am but hotter and wetter.

The mean annual data for total litter fall and its principal components for each species by climatic zone are shown in Table 3 along with levels of significance in the differences between zones. With the exception of woody materials in *R. stylosa*, the differences in yield for this species and for *A. marina* were found to be significant at or better than the 1% level. Differences in yield of *C. tagal* were less variable and achieved significance at the 5% level only in terms of total litter fall and with respect of interpetoliar stipules, a minor constituent.

The highest mean yield of total litter in *A. marina* (1049 g m^{-2} yr^{-1}) was recorded in zone Am with a single catcher in the same zone (location 23; Endeavour River) yielding a maximum for the species of 1598 g m^{-2} yr^{-1}). For *R. stylosa* and *C. tagal*, maximum mean yields of 1877 and 945 g m^{-2} yr^{-1} respectively were recorded at Port Moresby where the climate is similar to zone Am. In Australia, on the other hand, the highest mean yield for *R. stylosa* of 1192 g m^{-2} yr^{-1} was recorded in zone B although yields up to 2173 g m^{-2} yr^{-1} were recorded at single catchers in zone Am. Although differences in yield between climatic zones for *C. tagal* are not significant, the highest mean yield of litter in Australia for this species (855 g m^{-2} yr^{-1}) and the highest individual total (1290 g m^{-2} yr^{-1} at location 20; Mornington Island) were both obtained in zone Aw. It should be noted that the higher Papua New Guinea mean yield for this species included exceptionally high contributions of reproductive tissues.

It is interesting that the highest mean litter yield for *R. stylosa* in Australia was recorded in the arid zone B, whereas individually higher yields were obtained under much wetter conditions in zone Am and in the rather similar climate prevailing near Port Moresby. Climate, of course, is not the only determinant of mangrove productivity. Indeed, the data presented here make it plain that a great deal of variability persists when litter fall data are grouped in terms of long term climatic patterns. Clearly, the capacity for litter production at individual sites must depend heavily on controls such as salinity regime, nutrient inputs and topographic characteristics of the substrate operating at relatively local scales. In those terms, it seems reasonable to conclude that the most favourable climatic conditions are likely to be found, not necessarily where the highest mean yields occur, but in the zone where the highest individual litter fall yield is recorded. In other words, it seems likely that full response to any favourable climate becomes possible only where other local conditions are not limiting to growth and production. On that basis, zone Am would be identified as least limiting climatically for *A. marina*, although only in respect of the variety *eucalyptifolia* (Zipp. ex

Table 3. Litter fall (totals and components) in a set of climatic zones together with significance measures of differences between zones by analysis of variance.

A. marina					
Climate zone	Litter fall (g dw m^{-2} yr^{-1}) +/- std deviation				
	Total	Leaves	Wood & debris	Reprod. parts	
Cfa	307(183)	236(154)	49(33)	22(19)	
Cfb/Csb	436(148)	310(174)	69(51)	58(58)	
B	879(168)	520(98)	138(33)	219(120)	
Aw	430(211)	305(143)	81(51)	43(37)	
Am	1049(457)	523(249)	268(150)	241(171)	
Awi	234	199	27	10	
PNG*	829(33)				
* Port Moresby					
Significance level	<1%	1%	1%	<1%	
R. stylosa					
	Total	Leaves	Wood/debris	Reprd. parts	Stipules
B	1172(354)	668(178)	125(51)	244(215)	134(51)
Aw	987(243)	624(191)	117(77)	154(124)	91(26)
Am	752(364)	539(316)	68(37)	67(44)	77(51)
Awi	589(332)	420(197)	54(73)	61(55)	53(26)
Cwa	690(395)	294(127)	22(4)	310(223)	64(40)
PNG	1877(697)	1048(136)	157(62)	515(412)	158(58)
Sig.	<1%	1%	5%	1%	<1%
C. tagal					
Aw	855(237)	485(131)	106(33)	243(117)	21(4)
Am	678(225)	450(152)	75(33)	119(122)	34(1)
Awi	487(186)	309(104)	61(29)	103(73)	15(7)
Cwa	568(43)	386(95)	61(40)	86(51)	35(26)
PNG	945	437	18	467	22
Sig.	5%	>5%	>5%	>5%	>5%

Miq.) N. C. Duke, since this is the only one occurring in that zone (see Duke, 1991). The data available suggests that *A. marina* var *marina*, which is found only in Western Australia, is most favoured by zone B and that *A. marina* var *australasica* (Walp.) Moldenke, limited to parts of eastern and southeastern Australia, finds optimal climatic conditions in zone Cfa.

Like *A. marina* var *eucalyptifolia*, *R. stylosa* would also be most favoured by zone Am including the rather similar climate near Port Moresby. Using the same reasoning, zone Aw might be expected most favourable climatically for *C. tagal*.

For regional and other larger scale estimates of mangrove litter fall, it is clearly important that both zonal climate and controls operating at more local levels be taken into account. Where estimates of litter fall are required for extensively distributed species or species associations, isolated observations may provide misleadingly high or low statistics. For example, considerable variability is evident in the growing body of information now available for mangrove litter fall in S. E. Asia. To illustrate the point, recorded annual yields of litter for mixed stands of *Rhizophora mucronata* Lamk. and *R. apiculata* Bl. in Indonesia range from 494–1290 g m^{-2} (Sukardjo, 1989 and Sukardjo, unpubl. respectively). At one location in South Sumatra, the Salch River, albeit in stands of different floristic composition, Soerianegara *et al.* (1985)

report annual litter yields ranging from 622–1255 g m^{-2} Notwithstanding similarities between these fairly typical S E Asian data and those here reported for Australia, for neither region is there yet sufficient data to draw critical comparisons

Acknowledgements

This study originated within and was supported by the Australian Institute of Marine Science The many individuals and organizations who lent their support to the undertaking have been listed by Duke (1990) Their interest and willing assistance is here gratefully recognized The author also wishes to express his thanks to his wife, Eleanor, for her considerable enthusiasm in the field assisting with the installation of litter catchers He also wishes to recognize N C Duke's commitment to organising and undertaking the sorting of litter and assembling the data that accumulated over the period of the study

References

Bunt, J S , 1982 Studies of mangrove litter fall in tropical Australia In B F Clough (ed) pp 223–237 Mangrove Ecosystems in Australia A N U Press, Canberra 223–237

Duke, N C , 1990 Phenological trends with latitude in the mangrove tree *Avicennia marina* J Ecol 78 113–133

Duke, N C , 1991 A systematic revision of the mangrove genus *Avicennia* (Avicenniaceae) in Australasia Aust Syst Bot 4 299–324

Gentilli, J , 1986 Climate In D N Jeans (ed) Australia, a Geography, Vol 1 Sydney Univ Press, Sydney 14–48

Snedaker, S C & M S Brown, 1982 Primary productivity of mangroves In A Mitsui and C C Black Jr (eds) Handbook of Biosolar Resources, Vol 1, Part 2, CRC Press, Florida 477–485

Soerianegara, I , Z Coto, T B Suselo, P H J Nainggolan, Supriyanto, K Sumawidjaja, S Rahardjo, Dj Purwanto & E A Adiwilage, 1985 Ekologi produksi edosistem mangrove sebagai salah satu dasar penentuan pola pemanfaatan daerah pantai Kerjasma antara Pusat Penelitian dan Pengembangan Biologi Tropika, Institut Pertanian Bogor dengan Kantor Menteri Negara Kependudukan dan Lingkungan Hidup

Sukardjo, S , 1989 Litter fall production and turnover in the man grove forests in Muara Angke-Kapuk, Jakarta In I Soerianegara, P M Zamora, K Kartawinata, R C Umaly, S Tjitrosomo, D M Sitompul and U Rosalina (eds) Mangrove management its ecological and economic considerations Biotrop Special Publication 37 129–143

Hydrobiologia **295**: 141–148, 1995.
Y. S. Wong & N. F. Y. Tam (eds), Asia-Pacific Symposium on Mangrove Ecosystems.

The growth performances of two mangrove crabs, *Chiromanthes bidens* and *Parasesarma plicata* under different leaf litter diets

P. W. Kwok & S. Y. Lee
The Swire Marine Laboratory, University of Hong Kong, Cape d'Aguilar, Shek O, Hong Kong

Key words: grapsid crabs, detritivory, growth, survival, assimilation efficiency

Abstract

The growth performance of adult individuals of the detritivorous mangrove grapsid crabs *Chiromanthes bidens* and *Parasesarma plicata* common in the Mai Po Marshes, Hong Kong, was followed in a long-term laboratory rearing experiment (10 months). Individual crabs' moulting frequency, growth increment and mortality when fed four kinds of leaf litter available in their natural habitats, viz. brown (two weeks of decomposition) and yellow (fresh litter) *Kandelia candel*, and brown and yellow *Avicennia marina* were followed. The survival period was strongly related to litter type when the data from both crab species were pooled, being longest for crabs fed with brown *Avicennia marina*, followed by brown *Kandelia candel* > yellow *Avicennia marina* > yellow *Kandelia candel*. A higher proportion of crabs moulted twice when yellow *Avicennia marina* was supplied, but more crabs moulted for a third time when fed with brown *Avicennia marina*. The growth increment of the two species of crabs after moulting was found to be significantly related to their pre-moult size (ANOVA, $P<0.05$), but effects of the four treatments were non-significant. Increment of *Chiromanthes bidens* was significantly greater than that of *Parasesarma plicata* under the brown *Kandelia candel* treatment. Litter treatment has no effect on the time taken for the crabs to moult, only in the case of the yellow *Avicennia marina* treatment did *Parasesarma plicata* take a longer time to moult a second time.

Introduction

Brachyuran crabs, especially Grapsidae and Ocypodidae, form a major component of the macrofauna in the Indo-Pacific mangrove ecosystems. Although both groups of crabs burrow, the major food source of the Grapsidae is mangrove leaf litter whereas the ocypodids are deposite-feeders, and the distributions of these crabs do not seem to be restricted by the particle size of the mud (Macnae, 1968). Much work has been done to investigate the relationship between distribution and habitat preferences of the macrofauna and the mangrove environments they inhabit (Warner, 1967; Sasekumar, 1973; Hartnoll, 1975; Seiple, 1979; Macintosh, 1984). The Grapsidae are important not only because of their burrowing activities which can affect nutrient cycling and forest productivity (Smith *et al.*, 1991), or the topography of mangrove swamps (Warren & Underwood, 1986); but also their role as a primary link in the food web in mangrove forests (Camilleri, 1989). Through their feeding activities, mangrove leaves are broken down into fine particulate organic matter (FPOM) of different sizes and composition (Camilleri, 1992), affecting food availability to detritivores that feed on the mud surface. Leaf-consuming crabs can speed up the decomposition of mangrove leaves and so affect energy flow within the mangal (Robertson, 1986). Consumption experiments have shown that the crabs are more effective in assimilating decomposing than senescent leaves (Giddins *et al.*, 1986). The nutritional value, which is often expressed as the carbon to nitrogen ratio, and the tannin content of different mangrove leaf types have been found to be important in affecting the crabs' choice of leaf diet (Camilleri, 1989; Lee, 1993).

Because of the different nutritional qualities of the leaf litter available, crabs that inhabit different mangrove forests and thus feeding on different qualities of leaves may have different growth performance and survivorship. The two dominant grapsid crabs commonly found in the Mai Po Marshes, Hong Kong, namely *Chiromanthes bidens* and *Parasesarma plicata*, are found in two different parts of the mangrove forest respectively dominated by *Kandelia candel* and *Avicennia marina*. In order to study the effect of litter quality and its decomposition status on the secondary production of these two species of crabs, a long-term rearing experiment (10 months) was carried out. Four different leaf litter types, viz. brown and yellow *Kandelia candel* (BK, YK) and brown and yellow *Avicennia marina* (BA, YA) were supplied *ad libidum* to the crabs. Each crab was assigned to a particular litter treatment and remained so throughout the experimental period. Individual crab's moulting frequency, growth and survivorship were followed throughout the experimental period and compared among the four treatments.

Materials and methods

Adult individuals of *C. bidens* and *P. plicata* of carapace widths ranging from 15 mm to 21 mm were collected from the mid-intertidal *Kandelia candel* forest in the Mai Po Marshes, northwest Hong Kong. Twenty-six individuals of each crab species were used in the experiments. Only those individuals with a hard carapace (*i.e.* in the intermoult stage), bearing no eggs or mutilations were used. Their sexes were recorded and carapace width measured to the nearest mm by vernier calipers. These fifty-two crabs were kept individually in aquaria and were assigned randomly to the four litter treatments. Members of males and females of each crab species was nearly equal (3 or 4), making a total of thirteen crabs in each treatment. The initial carapace width of the crabs were found to be not significantly different between the two species of crabs. (mean carapace width of CB = 17.6 mm and PP = 17.3 mm) or with respect to the four treatments assigned to them. (ANOVA: F = 2.657, DF = 3 and F = 0.063, DF = 1, both $P > 0.05$).

Rearing experiment

The crabs were kept individually in small plastic aquaria made by joining two sections of PVC pipes (12 cm long for the upper section, 6.5 cm long for the lower section and 10.5 cm diameter), a piece of plastic mesh (mesh size of 7 × 3 mm) is inserted in between the two sections to allow faeces to pass through and be collected in the lower chamber. Four holes were made in the lower chamber to allow water to go into the aquaria. Crabs were kept in these aquaria with 1 cm depth of 50% sea water (15 ppt) above the mesh, from August 1992 to June 1993. These aquaria were kept in two tanks with running, filtered and aerated water with temperature maintained at 20 °C. Water in the tanks was changed every two days and faeces removed from each aquarium.

Leaf litter diets

Four litter treatments: brown and yellow *Kandelia candel* and brown and yellow *Avicennia marina*, each with 13 replicates, were used. Yellow (senescent) leaves which could be easily abscissed were respectively collected from monospecific stands of *Kandelia candel* and *Avicennia marina* in the Mai Po Marshes. Part of the collection of the yellow leaves of both species was put into two plastic litter bags of 2.25 mm^2 mesh size, and left on the respective forest floors for two weeks before they were used in the experiments. This process allowed the leaves to undergo decay and microbial enrichment with minimal attack from small detritivores, and at the same exposure time standardized the decomposition stage of the leaves offered to the crabs. The C/N ratio (carbon to nitrogen ratio) of the leaves was measured by a CHN autoanalyser and were found to be 18.4 (± 0.35) for BK, 49.1 (± 0.48) for YK, 22.2 (± 0.15) for BA and 27.4 (± 0.53) for YA; all the S.E. were calculated from three replicates of homogenized materials from large leaf samples. Preliminary trials showed the crabs would not consume more than 2 g (wet weight) of leaf litter each day. Excess amounts (about 4 g wet weight per crab per day) of the four types of litter were supplied to the crabs in all the experiments.

Growth increment and intermoult period

The crabs were examined every day and all individuals which moulted were recorded. The intermoult period (*i.e.* the no. of days taken to moult) was counted from the first day of the experiment until the day the crabs moulted. If the crabs moulted again during the experimental period, the second intermoult period was counted from the day when their carapace became hard after the first moult until the day the crabs moulted again. As soon as the carapace of the moulted individuals became hardened, the new carapace width was measured and the percentage increase in carapace width was then calculated by comparing the new and the old values.

Mortality and survival period

During the routine daily examination of the crabs, the occurrence of any dead individuals was also recorded, those died during moulting were counted as moulting failures. The survival period was then calculated from the first day of the experiment to the day the crab died.

Assimilation efficiency

Assimilation efficiency (AE), was calculated using the equation of Conover (1966):

$$AE = (L - F)/(1 - F) \times L,$$

where L is the ratio of ash-free dry weight (AFDW) to dry wt. of litter and F is the same ratio for faeces.

Results

Assimilation efficiency

Since the amount of faeces collected from each crab species was small, materials from the two species of crabs had to be pooled together for analysis. The mean assimilation efficiency for the two species of crabs is 32.19% (S.E. = 1.75, n = 3) for BK, 9.20% (S.E. = 0.98, n = 3) for YK, 31.0% (S.E. = 1.84, n = 5) for BA and 13.04% (S.E. = 1.05, n = 4) for YA. ANOVA of the arcsine-transformed data suggest that the assimilation efficiencies of the four litter types are significantly different (F = 52.19, DF = 3 & 13, P<0.005), with those of BK and BA being significantly higher than those of YA and YK. There was, however, no significant difference in the assimilation efficiency in between the two species of leaves when materials of the same decomposition stages were compared (Tukey test).

Moulting period

The number of days taken by the two species of crabs, *Chiromanthes bidens* and *Parasesarma plicata*, to moult for the first time was found to be independent of the litter treatments, crab species or sex, when the pre-moult size of the crabs was included as a covariate (3-ways ANOVA: F(litter) = 1.169, DF = 3; F(species) = 0.004, DF = 1; F(sex) = 0.116, DF = 1, P>0.05).

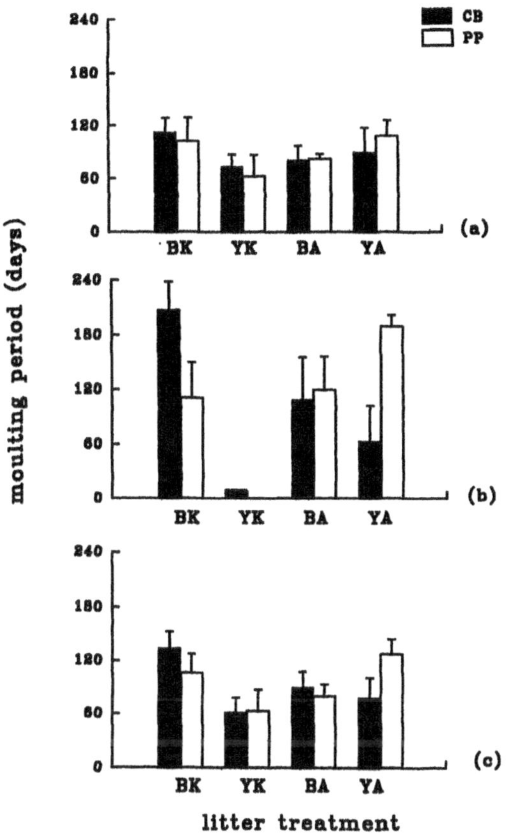

Fig. 1. The no. of days taken (± S.E.) by the two species of crabs, CB (*Chiromanthes bidens*) and PP (*Parasesarma plicata*), to moult (a) for the first time, (b) for the second time and (c) the mean moulting period for all moults, under the four litter treatments: BK (brown *Kandelia candel*), YK (yellow *Kandelia candel*), BA (brown *Avicennia marina*) and YA (yellow *Avicennia marina*).

As there was only one individual of *Chiromanthes bidens* that moulted twice under the YK treatment, this treatment was not included in the analysis of the crabs' second intermoult period. Again, days taken to moult was not significantly related to the litter treatments, crab species or sex. The moulting period of the two species of crabs between the four litter treatments were not significantly different. Although the period taken for the first moult did not differ significantly with that for the second moult, the mean time taken to moult the second time was longer than that for the first time (Fig. 1).

When the number of days to achieve the first and second moults were pooled together, the results also showed that moulting period was independent of the types of litter fed to the crabs or the species of crabs (3-ways ANOVA: $F(\text{litter}) = 2.444$, DF = 3; $F(\text{species}) = 0.036$, DF = 1; $F(\text{sex}) = 0.74$, DF = 1, $P > 0.05$).

Size increment

Similar to moulting period, increment in carapace width after the first moult was also analysed separately from the increment after the second moult. The data were expressed as the percentage increase in carapace width and were arcsine-transformed for a three-way ANOVA of litter treatments, crab species and sexes with pre-moult size as covariate. For the first moult, effects of litter treatment, crab species and sexes were found to be non-significant. Among the four litter treatments, only BK caused a significantly larger mean percentage increase in *Chiromanthes bidens* (ANOVA: $F = 11.772$, Df = 3 & 1, $P < 0.05$) than in *Parasesarma plicata*. No significant regression of log percentage increase in carapace width against pre-moult size was found in all the first moult cases.

Increments in the second moult were analysed in the same way as those in the first moult. The results, again, suggested there is no significant effect of litter, crab species or crab sexes.

If the percentage increase from the first and second moults were pooled together, only in the case of *Chiromanthes bidens* did litter treatment cause a significant difference (ANOVA: $F = 3.396$ Df = 3,27, $P < 0.05$). Growth in size was largest when (*Chiromanthes bidens* was fed BK, followed by YK and BA and YA. Only YA was significantly different from the BK treatment (Student-Newman-Keuls ranges test, $F = 3.396$, DF = 3 & 27, $P < 0.05$). Under the BK treatment, *Chiromanthes bidens* was also found to attain a significantly (ANOVA: $F = 4.616$ Df = 1, $P < 0.05$) larger increase after moulting than *Parasesarma plicata* (Fig. 2). Regression analysis of percentage increase in carapace width with pre-moult size suggested that for *Chiromanthes bidens*, pre-moult size related significantly with the percentage increase of the carapace width in the *Avicennia* treatment (for BA, $r^2 = 0.387$ $F = 5.675$ Df = 1&9, $P < 0.05$; for YA, $r^2 = 0.707$ $F = 9.646$ Df = 1 & 4, $P < 0.05$) (Fig. 3)

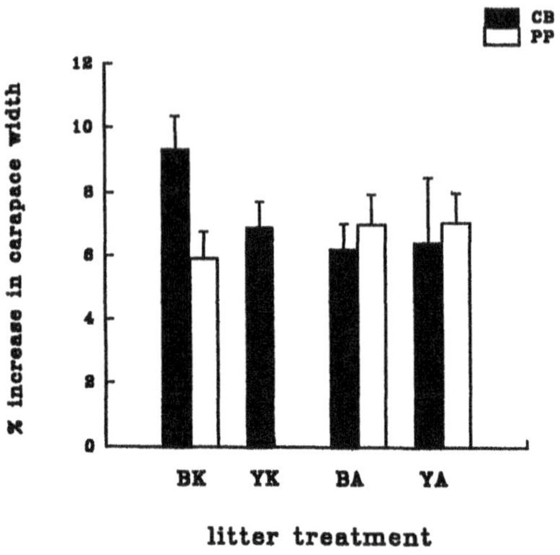

Fig. 2. The mean percentage increase (± S.E.) in carapace width after all moults of the two species of crabs: CB (*Chiromanthes bidens*) and PP (*Parasesarma plicata*), under the four litter treatments.

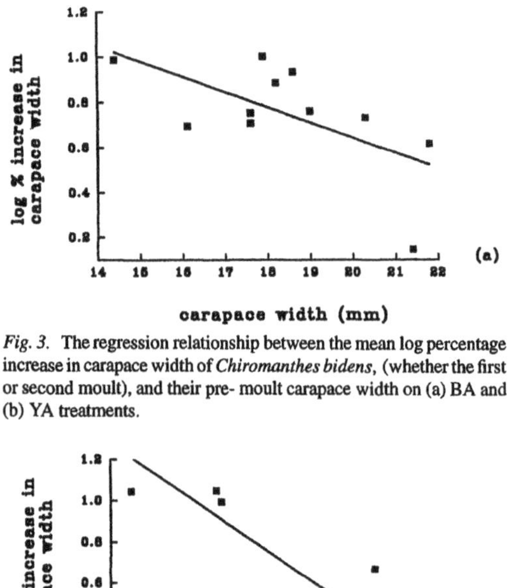

Fig. 3. The regression relationship between the mean log percentage increase in carapace width of *Chiromanthes bidens*, (whether the first or second moult), and their pre- moult carapace width on (a) BA and (b) YA treatments.

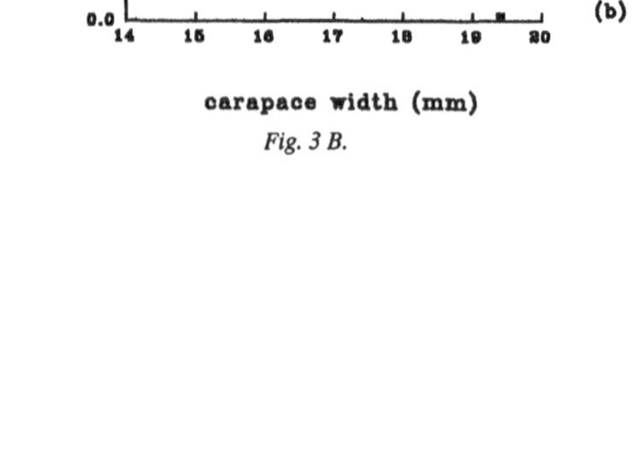

Fig. 3 B.

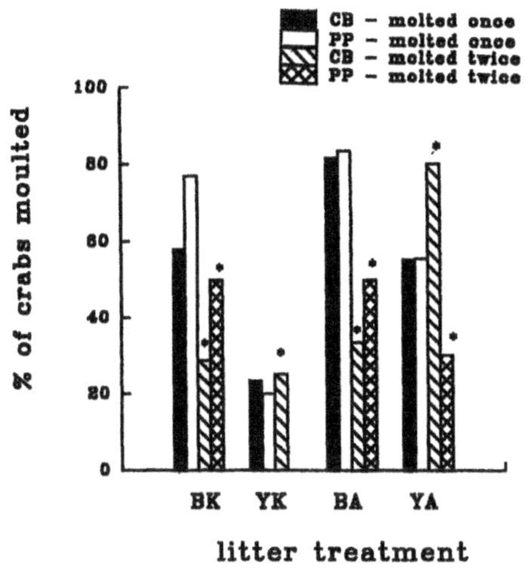

Fig. 4. The percentage of the two species of crabs (CB and PP) that have moulted once or twice when fed with different litter types. * denotes significant differences ($P<0.01$, χ^2 test) between the two species of crabs in the four litter treatments.

Table 1. Percentage of crabs that moulted once and twice under the four litter treatments. CB - *Chiromanthes bidens*; PP - *Parasesarma plicata*

	CB		PP	
	1st moult	2nd moult	1st moult	2nd moult
BK	58.3	28.6	76.9	50.0
YK	23.5	25.0	20.0	0.0
BA	81.8	33.3	83.3	50.0
YA	55.6	80.0	55.6	30.0
	moulted once		moulted twice	
BK	68.0		41.2	
YK	21.6		12.5	
BA	82.6		42.1	
YA	55.6		46.7	
1st moult	$\chi^2 = 2.485$ df = 3, $P<0.5$			
2nd moult	$\chi^2 = 53.005$ df= 3, $P<0.001$			
moulted once or twice	$\chi^2 = 3.544$ df= 3, $P<0.5$			

Moulting frequency

The number of times a crab moulted during the experimental period was independent of the litter treatments. (ANOVA $F=1.05$ Df=3,1, NS).

For the first moult, differences in the percentage of crabs moulted in particular treatments was found to be non-significant for the two species of crabs and the four litter treatments. In the second moult, the percentage of crabs moulted twice was calculated as the number of crabs that moulted a second time in each particular litter treatment. The percentages of the two species of crabs which moulted a second time are significantly dependent on the litter treatment. (Chi-square test, $\chi^2=53.005$ Df=3, $P<0.001$, Fig. 4).

When the two species of crabs were analysed together, it was found that the percentage of crabs that could moult once or twice was not significantly related to the four litter treatments (Table 1).

Mortality

Litter treatment significantly affected the survival periods of the crabs (ANOVA: $F=2.806$ Df=3,1, $P<0.05$). The survival period was the longest for BA with the descending order of BK>YA>YK (LSD test, $P<0.05$) (Fig. 5 & Table 2). Litter treatment effects were, however, consistent for the two crabs. The survival period of the two species of crabs under the four treatments was not significantly different from each other.

Table 2. Survival period of the two species of crabs under the four litter treatments and the occurrence of moulting-related mortality.

	CB	PP	moulting failure (%)	
	survival days	survival days	1st moult	2nd moult
BK	143.5	209.3	23.5	14.3
YK	126.6	119.8	12.5	100.0
BA	219.7	188.3	5.3	0.0
YA	177.7	173.3	20.0	0.0

ANOVA: on survival period (days)
F (litter) = 2.806 df = 3, $P<0.05$
χ^2 test on moulting failure for 1st and 2nd moult: χ^2 = 86.83 df = 3, $P<0.001$

The percentage of crabs that died during moulting in the first and second moults was found to be significantly dependent on the litter treatments (Chi-square test, $\chi^2=86.83$ Df=3, $P<0.001$). (Fig. 5 & Table 2) Yellow *Kandelia candel* leaves was found to be least conducive to successful moulting, whereas *Avicennia*

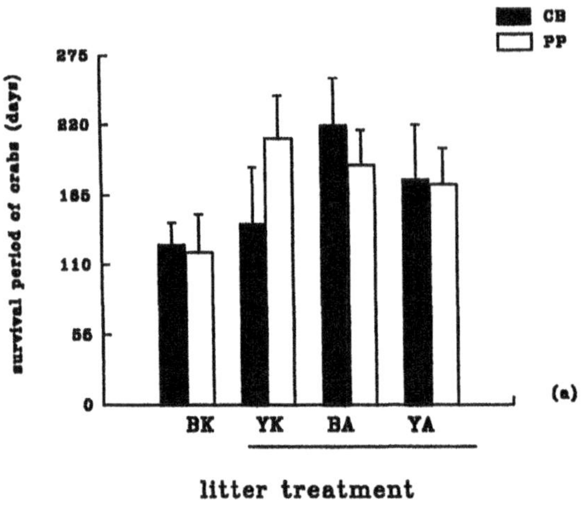

Fig. 5 A. The no. of days the two species of crabs survived under the four litter treatments. Each bar shows the actual number of survival days ± S.E. (treatments joined by the same line are not significantly different).

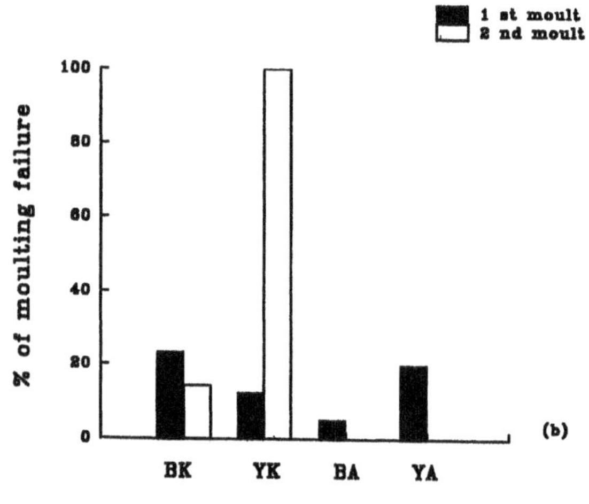

Fig. 5 B. The percentage of moulting mortality for the litter treatments during the first and second moults.

marina leaves, whether brown or yellow, seemed to enhance moulting success.

Discussion

The intermoult period is an important parameter influencing growth rate in Crustacea. The interval can be markedly affected by extrinsic factors like food quality or temperature (Hartnoll, 1983). In general, intermoult period tends to increase with size consistently (Hartnoll, 1983), meaning that the crab moults less frequently as it grows bigger. By regressing the log intermoult period against size, the slope of the regression line can reflect the rate of lengthening of the intermoult period as size increases. Different crab species will have different slopes, depending on the growth strategies of the crabs (Hartnoll, 1983).

Because of the uncertainty of the exact moulting condition of the experimental crabs when they were collected (although they were all in the intermoult period), time taken to have the first moult cannot accurately reflect the actual moulting interval of the crabs under the different diet treatments. The moulting period of the second moult could, however, be accurately followed. The first moulting period was not significantly different from the second one in all the four treatments (t-test $t=2.69$, $P>0.05$), suggesting that the four litter treatments had an insignificant effect on the moulting period of *C. bidens* and *P. plicata*, or the interval may be overwhelmingly dictated by intrinsic factors like hormone secretion (Hopkins, 1992).

Moulting frequency, similar to moulting period, varies with the size of the crabs. Indeed, the nutritional values of the litter treatments may also be important in providing enough energy for the crabs to moult and grow. *Avicennia marina* leaves have a low tannin and high nitrogen content and are thus more preferred by crabs (Camilleri, 1989; Lee, 1993).Tannins have been reported to interfere with digestion (Mattson, 1980) and nitrogen assimilation of high tannin foods can be 30% lower than that of animal diet (Wolcott & Wolcott, 1987). In a southern African mangrove swamp, the large detritivorous mangrove grapsid, *Sesarma meinerti*, has a very high (82.44%) assimilation efficiency for six week old *Avicennia marina* leaves (Emmerson & McGwynne, 1992). In this experiment, significantly higher assimilation efficiency were also attained by *C. bidens* and *P. plicata* for decaying mangrove litter. Although the crabs had a higher assimilation efficiency for BK than BA in the present study, BK did not seem to facilitate moulting. Therefore, assimilation efficiency and nutrient availability are not the only factors affecting the crabs' ability to moult. But, in the case of YK, the crabs' inability to moult for the second time is probably due to the poor assimilation efficiency (12.51%) for this litter and most crabs actually did not live long enough to moult again.

Some previous works on crabs have suggested that food quantity and quality may affect growth performance and reproductive output but not survival (Micheli, 1993). In this study, none of the sexually mature females (carapace width>15 mm) produced eggs although they were kept individually (sperm is usually stored up in the spermatheca). This may again be due to malnutrition and it suggests that a pure litter diet probably cannot provide enough energy or nutrients for the crabs to reproduce. It is not rare to see crabs feeding on the mud surface in the field, which contain high abundance of invertebrates and microbe-enriched fine detritus.

The size increment after each moult, which is another important component of overall growth rate in Crustacea, usually decreases with increasing size (Hartnoll, 1983). Very few *C. bidens* and *P. plicata* could moult successfully for a second time in the experiment and have an increase in their carapace width, the majority of them either had a failure in the second moult and resulted in mortality, or they had 0% increase during the experimental period. Brown leaves (in this case, two weeks old) is found to be favoured by crabs in the mangrove swamp because of the decreased carbon to nitrogen ratio and lower level of unpalatable materials like tannins through leaching (Camilleri & Ribi, 1986; Giddins *et al.*, 1986; Neilson *et al.*, 1986; Robertson, 1988; Lee, 1989). The C/N ratio of brown *Kandelia candel* used in the present experiment is 18.4 with nitrogen at 2.45%, while for brown *Avicennia marina*, the C/N ratio is 22.1 with 1.85% nitrogen. So it seems that brown *Kandelia candel* is a better quality food than brown *Avicennia marina*. However, only *C. bidens* responded more significantly in their carapace width increment to this litter treatment. This result matches well with the field observation that *C. bidens* is more abundant in the mangrove forest dominated by *Kandelia candel* than that dominated by *Avicennia marina*. However, differences in the growth performances of the crabs on different litter diets (which are actually available in the field) is probably not the only reason for the observed abundance differences. Other factors like consumption rate on different litter types, predation risk, the crabs' feeding habits and other environmental parameters in the two forests are still subject to further investigation.

The brown leaves are shown to be better quality food again because they gave the longest survival periods. While brown *Kandelia candel* is found to be important in giving faster growth, brown *Avicennia marina* is shown to be the better food to maintain survivorship. YK is the poorest food for which the crabs can only have a 9.20% assimilation efficiency. It is therefore not surprising to find that the survivorship under this treatment is significantly lower than the other three.

Conclusion

From this experiment, it was found that the decaying leaves of *Kandelia candel* and *Avicennia marina* are good quality food for crabs' survival. However, the crabs' moulting period and moulting increment do not seem to be much affected by the litter type offered. Further, *Kandelia candel* leaves is good for the crabs' moulting increment while *Avicennia marina* leaves may enhance moulting frequency. Thus, crabs that inhabit mangrove forests dominated by different tree species may have different growth and survival rates. Given the reported effects of grapsid crabs on mangrove community structure and dynamics (e.g. Smith *et al.*, 1991), it therefore appears that the interaction between this important faunal component and its host mangroves is a two-sided one.

Acknowledgments

This study was supported by a Postgraduate Studentship awarded to P. W. Kwok by the University and Polytechnic Grant Committee (UPGC) and a grant from the Research Grants Council to S. Y. Lee.

References

Camilleri, J. C., 1989. Leaf choice by crustaceans in a mangrove forest in Queensland. Mar. Biol. 102: 453–459.

Camilleri, J. C., 1992. Leaf-litter processing by invertebrates in a mangrove forest in Queensland. Mar. Biol. 114: 139–145.

Camilleri, J. C. & G. Ribi, 1986. Leaching of dissolved organic carbon (DOC) from dead leaves, formation of flakes from DOC, and feeding on flakes by crustaceans in mangroves. Mar. Biol. 91: 337-344.

Conover, R. J., 1966. Assimilation of organic matter by zooplankton. Limnol. Oecanogr. 11: 338–354.

Emmerson, W. D. & L. E. McGwynne, 1992. Feeding and assimilation of mangrove leaves by the crab *Sesarma meinerti* de Man in relation to leaf-litter production in Mgazana, a warm-temperate southern African mangrove swamp. J. exp. mar. Biol. Ecol. 157: 41–53.

Giddins, R. L., J. S. Lucas & M. J. Neilson, 1986. Feeding ecology of the mangrove crab *Neosamartium smithii* (Crustacea:Decapoda:Sesarmidae). Mar. Ecol. Prog. Ser. 33: 147–155.

Hartnoll, R. G., 1975. The grapsidae and ocypodidae (Decapoda-Brachyura) of Tanzania. J. zool. Soc. Lond. 177: 305–328.

Hartnoll, R. G., 1983. Strategies of crustacean growth. Aust. Museum Memoir. 18: 121–131.

Hopkins, P. M., 1992. Hormonal control of the molt cycle in the Fiddler Crab *Uca pugilator*. Am. Zool. 32: 450–458.

Lee, S. Y., 1989. The importance of sesarminae crabs *Chiromanthes* spp. and inundation frequency on mangrove (*Kandelia candel* (L.) Druce) leaf litter turnover in a Hong Kong tidal shrimp pond. J. Exp. Mar. Biol. Ecol. 131: 23–44.

Lee, S. Y., 1993. Leaf choice of the sesarmid crabs *Chiromanthes bidens* and *C. maipoensis* in a Hong Kong mangal. In Morton, B. (ed.), Marine Biology of the South China Sea, Hong Kong University Press, Hong Kong, pp.

Macintosh, D. J., 1984. Ecology & productivity of Malaysian mangrove crab populations. Symp. Mangr. Env. 354–377.

Macnae, W., 1968. A general account of the fauna and flora of mangrove swamps and forests in the Indo-West-Pacific region. Adv. Mar. Biol. 6: 73–270.

Mattson, W. J. Jr., 1980. Herbivory in relation to plant nitrogen content. Annu. Rev. Ecol. Syst. 11: 119–161.

Micheli, F., 1993. Effects of mangrove litter species and availability on survival, moulting and reproduction of the mangrove crab *Sesarma messa*. J. exp. mar. Biol. Ecol. 171: 149–163.

Neilson, M. J., R. L. Giddins & G. N. Richards, 1986. Effect of tannins on the palatability of mangrove leave. Mar. Ecol. prog. ser. 34: 185–186.

Robertson, A. I., 1986. Leaf-burying crabs: their influence on energy flow and export from mixed mangrove forests (*Rhizophora* spp.) in a northeastern forest. J. exp. mar. Biol. Ecol. 102: 237–248.

Robertson, A. I., 1988. Decomposition of mangrove leaf litter in tropical Australia. J. exp. mar. Biol. Ecol. 116: 235–247.

Sasekumar, A., 1973. Distribution of macrofauna on a malayan mangrove shore. J. anim. Ecol. 43: 51–69.

Seiple, W. H., 1979. Distribution, habitat preferences & breeding periods in the Crustaceans *Sesarma cinereum* and *S. reticulatum* (Brachyura:Decapoda:Grapsidae). Mar. Biol. 52: 77–86.

Smith III, T. J., K. G. Boto., D. F. Stewart & R. L. Giddins, 1991. Keystone species & mangrove forest dynamics: the influence of burrowing by crabs on soil nutrient status and forest productivity. Estuar. coast. Shelf. Sci. 33: 419–432.

Warner, G. F., 1967. Life history of the mangrove tree crab, *Aratus pisoni*. J. zool. Soc. Lond. 153: 321–335.

Warren, J. H. & A. J. Underwood, 1986. Effects of burrowing crabs on the topography of mangrove swamps in New South Wales. J. exp. mar. Biol. Ecol. 102: 223–235.

Wolcott, D. L. & T. G. Wolcott, 1987. Nitrogen limitation in the herbivorous land crab *Cardisoma guanhum*. Physiol. Zool. 60: 262–268.

Hydrobiologia 295: 149–158, 1995.
Y. S. Wong & N. F. Y. Tam (eds), Asia-Pacific Symposium on Mangrove Ecosystems.

Nutrients and heavy metal contamination of plants and sediments in Futian mangrove forest

N. F. Y. Tam[1], S. H. Li[2], C. Y. Lan[2], G. Z. Chen[2], M. S. Li[2] & Y. S. Wong[3*]
[1]*Department of Biology and Chemistry, City Polytechnic of Hong Kong, Tat Chee Avenue, Kowloon, Hong Kong*
[2]*Research Institute of Environmental Sciences, Zhongshan University, PRC*
[3]*Research Centre/Biology Department, Hong Kong University of Science and Technology, Clear Water Bay, Kowloon, Hong Kong*
(Corresponding author)*

Abstract

An ecological survey was carried out to determine the levels of nutrients and heavy metals in the sediments and leaf tissues of two dominant mangrove plant species, *Kandelia candel* and *Aegiceras corniculatum*, in Futian mangrove forest, Shenzhen, the People's Republic of China. The spatial and seasonal variations of these elements were also investigated. The results show that there was no major difference between two sampling sites 150 m apart. In both sites, the sediment concentrations of total and NH_4^+- N, total and extractable P, total and extractable K, total organic carbon were consistently higher in the landward locations and decreased gradually towards the sea. The sediment samples collected at the seaward edge of the mangrove plant community had the lowest levels of nutrient and organic matter. The vertical variations (from the land to the sea) of sediment heavy metals were less obvious and no particular trend could be identified. Extremely high contents of Cu, Cd, Pb, Cr and Zn were found at certain locations, suggesting the occurrence of some local contamination. The mean total metal concentrations in sediments decreased in the order Mn > Zn > Cu > Cr = Pb > Cd for the sample locations. Most of the heavy metals were not in a bioavailable form as the concentrations of extractable metals were relatively low (< 1% of total metals). Pb, Cr and Cd were not detected in leaf samples. Leaf C, N, P and K contents were similar between the two species and no significant difference was found among locations, although *A. corniculatum* seemed to have lower Mn concentrations than *K. candel*. With reference to temporal variations, no significant difference in sediment concentrations of some nutrients and metals was found between the spring and autumn seasons.

Introduction

Futian mangrove forest (114°03′E, 22°32′N) is situated within a Nature Reserve of Shenzhen Special Economic Zone, the People's Republic of China, neighbouring the Mai Po Nature Reserve of Hong Kong. Geographically, this mangrove area is located towards the northern limits of mangrove development. Its flora and fauna, the productivity and environmental factors are different from other mangrove stands located in tropical regions. Very little information is available on mangrove ecosystems of the sub-tropical regions and most of these studies were focused on species composition, community structure and zonation. Until now, no research data have been reported on the organic matter, nutrients and heavy metals in sediments and plants of mangrove forests in Southeast coast of China, in spite of their significance in determining mangrove function and productivity. These baseline data are also important in designing the long-term management and conservation policies of a mangrove forest.

Moreover, sediments are important sinks for land-derived contaminants especially heavy metals, and have been widely used to identify sources of pollution, to evaluate its extent and to diagnose the environmental quality of aquatic systems. Mangrove ecosystems, being extremely rugged, have considerable capacity to withstand domestic, poultry and even industrial discharges; their sediments and plants are capable of retaining pollutants (Clough & Attiwill, 1982; Dwivedi & Padmakumar, 1983; Por, 1984). The levels of pollutants immobilized in the sediment depend not only on the physical, chemical and biological properties of the sediment, but also on the background levels of the pollutants. It is thus essential to understand the background concentrations if the mangrove forest is to be used as effluent discharge and treatment facility. An ecological survey was therefore conducted (1) to investigate the levels of organic matter, nutrients and heavy metals of sediments and dominant plant species in Futian mangrove ecosystem; and (2) to evaluate the spatial and temporal variations of these elements in sediments and plant leaves.

Materials and methods

The ecological survey was carried out in Futian Nature Reserve located in the Northeast coast of Shenzhen Gulf, the PRC. The reserve stretches 11 km from east to west and covers an area of 304 hectares. Along the muddy shore of this reserve, mangrove forests with an area of about 112 hectares are dominated by *Aegiceras corniculatum* (L.) Blanco, *Kandelia candel* (L.) Druce and *Avicennia marina* (Forsk.) Vierh. The plants are about 4–5 m high and 20 years old. The mangrove ecosystem is inundated by incoming tides twice a day and the largest tidal range during spring is about 2.8 m.

Two sites (A and B, each had an area 180 m × 10 m), of about 150 m apart within the Futian mangrove forest, were selected. Surface sediment samples (0–20 cm) were taken at regular intervals from the landward to seaward regions of each site. Triplicate samples were collected, air-dried, passed through a 2-mm sieve and then analyzed for organic carbon (Walkley & Black, 1934), pH and electrical conductivity (1:5 soil:water extract), NH_4^+-, NO_2^- and NO_3^--N, (extracted with 2N KCl at 1:4 ratio followed by steam distillation), total Kjeldahl nitrogen (Micro-Kjeldahl digestion followed by steam distillation), available-P (Olsen *et al.*, 1954) and total P (Murphy & Riley, 1962). Total metals, including K, Cu, Zn, Mn, Cr, Cd and Pb were determined by digesting the sediment samples in conc. HNO_3 at 160 °C, then analyzed by flame atomic absorption spectrophotometry. The available metals were measured by extracting the soil samples with 1M ammonium acetate (pH 4) at 1:5 soil:water ratio and analyzed by atomic absorption spectrophotometry. Two ecological surveys were carried out during spring (April) and autumn (September) periods in order to explore the seasonal variations of element levels in sediments. During the April sampling, at the same interval where sediment samples were taken, mature leaves of *A. corniculatum* and *K. candel* were collected and analyzed for nutrients (organic matter, N, P and K) and heavy metals (Cu, Zn, Mn, Cr, Cd and Pb) using the same techniques as described above. The mean, standard deviation, and range of the nutrient and heavy metal concentrations were calculated. All data were expressed in terms of 105 °C oven-dried weight basis. The spatial and temporal variations of sediment nutrient and heavy metal concentrations were determined by the 3-ways analysis of variance (ANOVA): differences between Sites A and B; between April and September sampling time, and between different sites along the transects (tidal gradient) were evaluated.

Results and discussion

Sediment pH and electrical conductivity

Sediment pH. The mangrove sediments in Futian Nature Reserve were very clayey, consisting of 54.7% clay, 32.4% silt and a relatively low percentage of sand (12.9%). The high silt and clay content suggested that deposition of alluvium continued to take place and the mangrove was completely inundated during high tides. The sediment also had high cation exchange capacity (28.56 meq 100 g^{-1}). Figure 1A shows that all the soils had neutral to acidic pH varying from 7.0 to 4.2. The most acidic sediments were found in the middle part of Site A in both spring and autumn. More alkaline pH values were recorded in sediment samples collected from the seaward regions (Fig. 1A). It has been reported that acidic pH of mangrove sediments was partly due to the oxidation of FeS_2 and FeS to H_2SO_4, and partly resulted from the decomposition of mangrove litter and the hydrolysis of tannin in mangrove plants which released various kinds of organic acids (Liao, 1990). The lack of mangrove litter in the foreshore region explained why a higher pH was often found in the seaward locations. The pH differences along the tidal gradient could also be a result in the frequency of inundation by seawater which has a strong buffering capacity. The pH recorded in the present study was more acidic than those reported in Sai Keng (Hong Kong) and Jiulong Jiang Estuary (Fujian); the pH of these two mangrove areas were 6.7 and 6.9, respectively (Lin *et al.*, 1987; Tam & Wong, 1993). However, very acidic pH (2.62–6.10) was also found in sediments of Hainan Mangroves (Liao, 1990).

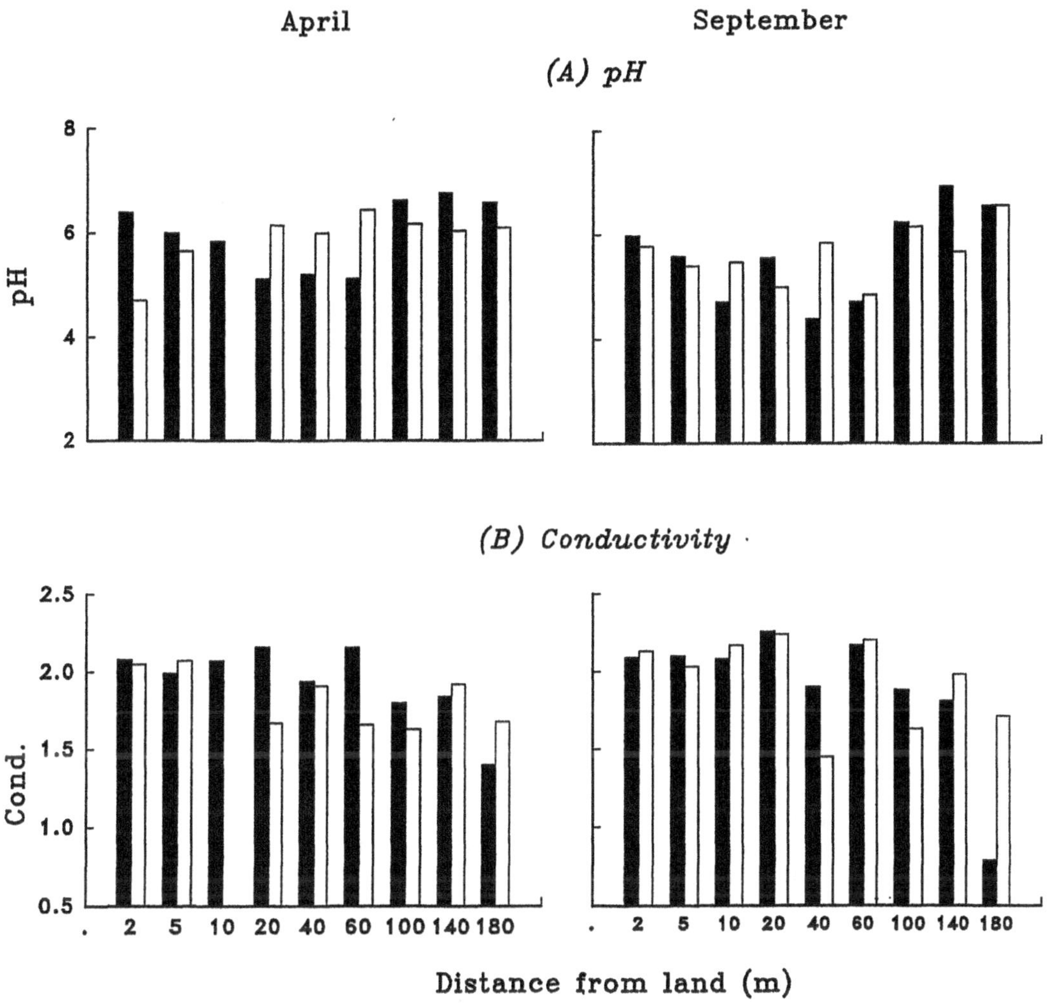

Fig. 1. The profile of sediment pH and electrical conductivity (mS cm^{-1}) along the vertical positions of Futian Mangrove Ecosystem. (■: Site A; □: Site B).

Sediment conductivity. Most sediment samples had electrical conductivity at the range of 1.4–2.2 mS cm^{-1} and the variations between Sites A and B were small (Fig. 1B). No specific trend was identified along the vertical positions of the shore except the locations closer to sea seemed to have lower conductivity values. In general, samples collected from spring period (April) had similar soluble salt content as those from autumn season (September).

Sediment nutrient concentrations

Sediment K. The total K concentrations of sediments collected in Spring time (April) had fairly uniform values from the landward to seaward regions of the mangrove, with an average value of 1.6% (dry weight basis), and there was no major difference between two sites (Fig. 2A). In general, September samples showed a significantly lower K content than April samples, with highest values recorded in the most landward region and the lowest values found in the seaward

Table 1. Results of ANOVA tests showing the differences between Sites A and B; between April and September sampling time; and between sampling sites along the tidal gradient.

Soil Parameters	Sampling site		Sampling Time		Tidal gradient	
	F value	Sig. of F	F value	Sig. of F	F value	Sig. of F
Total K	3.09	0.083	132.16	0.000	6.61	0.000
Organic C	22.73	0.000	5.18	0.026	99.21	0.000
Total P	1.04	0.310	12.55	0.001	14.21	0.000
Ext. P	1.76	0.189	0.13	0.722	3.19	0.004
Total N	5.74	0.019	2.34	0.130	40.49	0.000
NH_4^+-N	12.20	0.001	0.45	0.506	5.21	0.000
NO_3^--N	4.29	0.042	19.69	0.000	3.73	0.001
Total Cu	0.67	0.415	4.54	0.036	0.57	0.798
Total Zn	0.15	0.696	0.59	0.442	1.32	0.246
Total Mn	0.05	0.829	7.50	0.008	6.32	0.000
Total Cd	13.94	0.000	2.62	0.110	0.91	0.510
Total Cr	1.99	0.162	0.04	0.836	0.79	0.610
Total Pb	0.05	0.829	0.22	0.630	1.06	0.403
Ext. Cu	4.27	0.043	0.86	0.356	3.46	0.002
Ext. Zn	0.22	0.640	18.94	0.000	7.39	0.000
Ext. Mn	6.55	0.013	50.52	0.000	16.63	0.000
Ext. Cd	23.69	0.000	48.54	0.000	4.37	0.000
Ext. Cr	0.30	0.585	135.83	0.000	0.84	0.570
Ext. Pb	0.62	0.434	10.23	0.002	1.73	0.108

side (Table 1). This might be related to the fact that K, a very mobile and soluble element, was washed away by heavy rains during the summer season. In the September samples, Site B seemed to have higher K levels than Site A (Fig. 2A). The K content recorded in this study was significantly lower than the Fujian Mangroves which had total K of 2.07% (Lin *et al.*, 1987) but higher than that of the Hainan Mangroves (0.42–1.19% K) (Liao, 1990).

Sediment organic C. Organic carbon content was relatively high in the landward regions (around 4.5%) and decreased gradually towards the sea, with the lowest value of 0.4% at the location closest to sea (Fig. 2B). The trend and magnitude of organic carbon was comparable to the Sai Keng mangrove forest in Hong Kong (Tam *et al.*, 1993) but significantly lower than that of Malacca mangroves in west coast of Malaysia (20% organic matter; Thong *et al.*, 1993). The organic matter is derived from two sources. First, mangrove stem and leaf litter may be incorporated into the soil surface, whilst death of roots adds organic matter to the soil at varying depths. Second, particulate organic matter derived from wastewater discharges and tidal waters is trapped into the soil. Moreover, the landward regions were only flooded with tidal water during the high spring tide, i.e., two tides a month. Therefore, organic matter from plant litter and its decomposition could accumulate on the ground surface for a longer period of time resulting for higher percentage of organic C. In general, Site A had significantly higher content of organic carbon than Site B, and the differences between April and September samples were relatively small when compared with the spatial variations (Table 1).

Sediment P. The total P concentrations decreased along the tidal gradient (landward station: 0.18%; mid-shore: 0.14%; low-shore: 0.03%) (Fig. 2C). Sites A and B had similar P content, and September samples had significantly lower P values than the April samples in both sites (Table 1). This mangrove soil had higher P concentrations than other mangroves in Southeast coast of China, such as Hong Kong Sai Keng man-

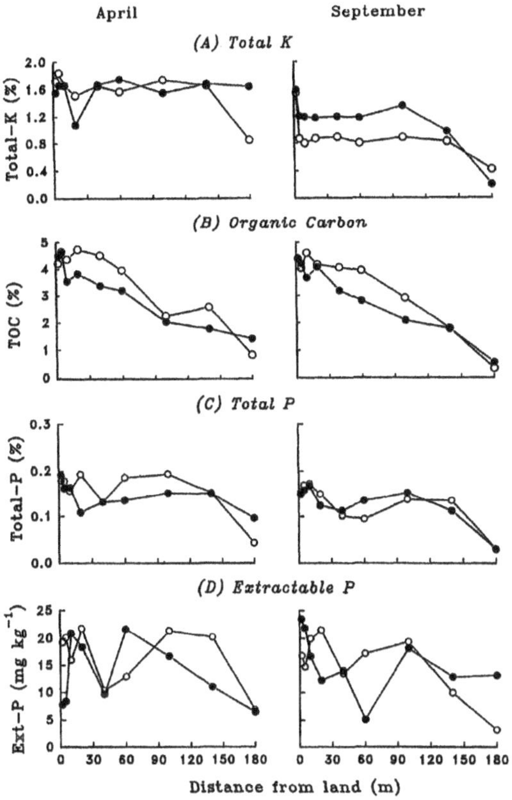

Fig. 2. The profile of sediment total K, organic carbon, total and extractable P along the vertical positions of Futian Mangrove Ecosystem. (o: Site A; •: Site B).

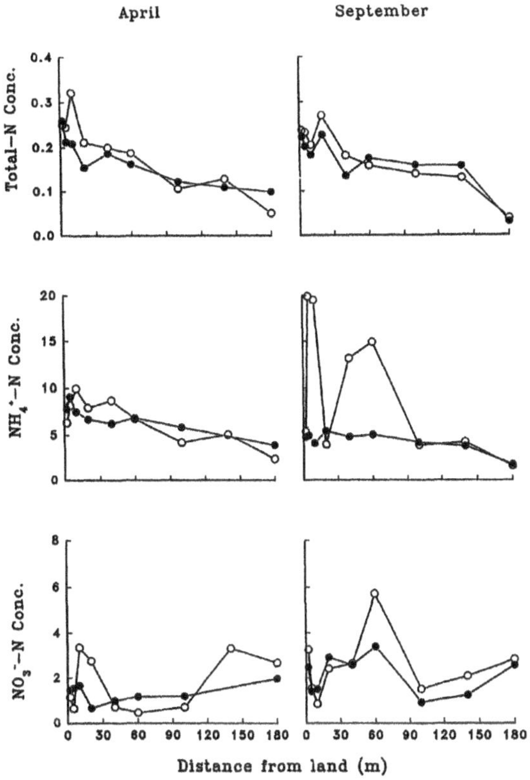

Fig. 3. The profile of sediment total N (%), $NH_4{}^+$-N (mg kg^{-1}) and $NO_3{}^-$-N (mg kg^{-1}) along the vertical positions of Futian Mangrove Ecosystem. (o: Site A; •: Site B).

groves (0.01–0.1 %; Tam *et al.*, 1993) and Fujian Jiulong mangrove (0.05%; Lin & Lin, 1985). The high P levels, especially at the back of the shore might be related to P contamination due to discharge of municipal effluents and other human activities such as leaching of fertilizers used in agriculture. It is also related to the fact that landward region was less affected by tidal water as this region was only flooded during high spring tide period. Most phosphorus in the sediment was not bio-available and less than 5% of the total P could be extracted by sodium bicarbonate. The extractable P content was at the range of 5–24 mg kg^{-1} (Fig. 2D), with wide variation along the shore. The $PO_4{}^{3-}$-P levels fell within the range recorded in Australian Brisbane mangrove area (MacKey *et al.*, 1992) and northern Australian mangroves (Boto & Wellington, 1983). Highest concentrations of extractable P were also found at regions close to land, suggesting the major source of P was from inland water, human discharges or litter decomposition.

Sediment N. The trend of changes in total N along the vertical positions of the mangrove forest was similar to that of organic carbon, with higher total N concentrations at the more inland locations (0.25%) declining gradually towards the sea; the lowest value (0.04%) was recorded at the seaward region (Fig. 3A). The N difference between Sites A and B was significant but between the two sampling periods was small (Table 1). The N concentrations observed in this study fell within the range reported in Sai Keng, Hong Kong mangroves (Tam *et al.*, 1993), in mangroves of Goa, India (Jagtap, 1987) but slightly lower than that in Fujian mangroves (Lin & Lin, 1985). A large proportion existed as

Table 2. The mean concentrations and range values (in bracket) of total and extractable heavy metals in Futian mangrove soil.

Heavy metal	Total concentration (mg kg^{-1})	Extractable concentration (mg kg^{-1})	Ext/Tot (%)
Cu	41.1 (15.8–308.1)	0.97 (0.52–1.07)	2.36
Zn	146.1 (53.1–423.1)	5.5 (10.6–14.1)	3.79
Mn	438.8 (124.5–789.2)	189.7 (11.3–462.1)	43.23
Cd	2.96 (0.27–7.94)	0.34 (0.04–0.67)	11.49
Cr	34.2 (6.8–56.7)	0.032 (5.67–76.97)	0.09
Pb	35.4 (0.1–63.0)	0.32 (0.02–1.28)	0.89

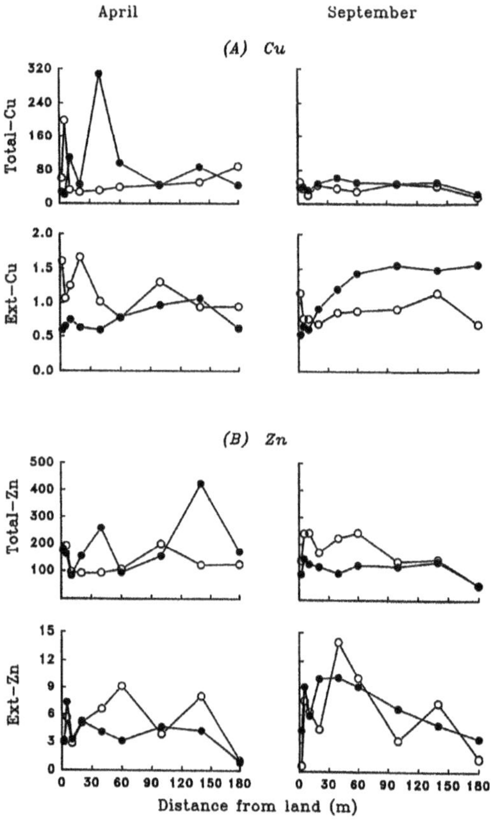

Fig. 4. The profile of sediment total and extractable Cu and Zn content (mg kg^{-1}) along the vertical positions of Futian Mangrove Ecosystem. (o: Site A; •: Site B).

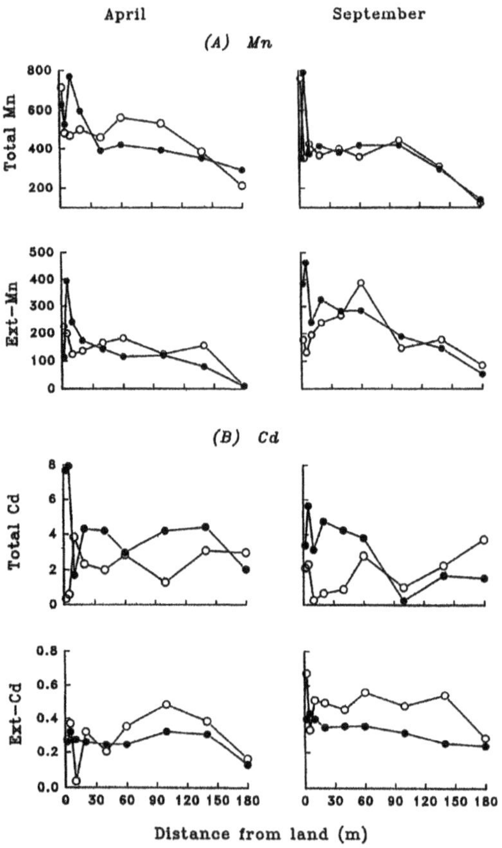

Fig. 5. The profile of sediment total and extractable Mn and Cd content (mg kg^{-1}) along the vertical positions of Futian Mangrove Ecosystem. (o: Site A; •: Site B).

organic N, and the total amount of inorganic N (NH_4^+ + NO_2^- + NO_3^-) was less than 1% of total N. In general, nitrite was not detected in the mangrove sed-

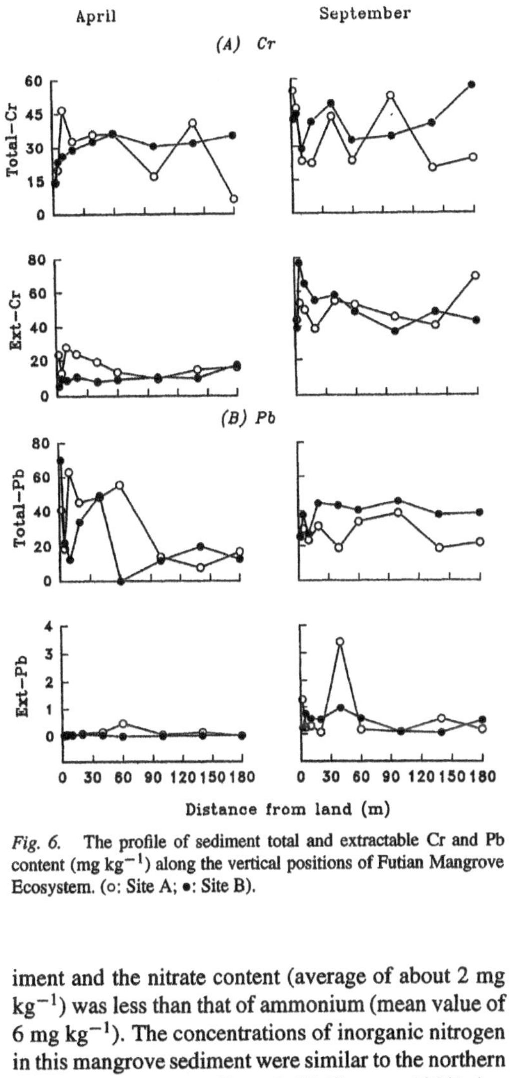

Fig. 6. The profile of sediment total and extractable Cr and Pb content (mg kg^{-1}) along the vertical positions of Futian Mangrove Ecosystem. (o: Site A; •: Site B).

iment and the nitrate content (average of about 2 mg kg^{-1}) was less than that of ammonium (mean value of 6 mg kg^{-1}). The concentrations of inorganic nitrogen in this mangrove sediment were similar to the northern Australian mangrove (Boto & Wellington, 1983), but lower than that reported in the Brisbane Mangroves (MacKey *et al.*, 1992). Variations of ammonium and nitrate along the vertical positions of the mangrove forest did not show any specific trend although ammonium appeared to decrease slightly towards the sea (Fig. 3). The spatial and temporal variations in inorganic N concentrations found in this study (Table 1) might be related to the fact that some samples collected in September had extraordinary high NH_4^+ content. This suggested that local contamination due to human activities might have occurred. Very few comparable data are available from other mangrove forests distributed in same geographical location, thus it is difficult to discuss the nutrient status of the mangrove forest under study.

Sediment heavy metal content

Total heavy metals. The heavy metals, except Mn, did not exhibit any specific trend of variations along the vertical positions of the mangrove forest (Figs. 4–6; Table 1). Highest concentrations of Mn were recorded at the landward region then gradually dropped and reached a very low level towards the sea (Fig. 5). In general, the two sampling sites and the two sampling times did not show any significant difference in terms of total heavy metal concentrations (Table 1). The mean and range of heavy metal content are summarized in Table 2. The mean metal concentrations were found to decrease in the order of Mn > Zn $\geq$ Cu, Cr, Pb > Cd. This pattern is similar to the River Tees estuary sediments reported by Davies *et al.* (1991). Along the shore positions, some samples tended to have exceptionally high heavy metal content, suggesting that "hot spots" occurred from time to time, probably due to local contamination from anthropogenic inputs.

The Cu, Zn and Cd contents were higher than those obtained in the Brisbane study although Pb and Cr were within the range of Brisbane sediment (MacKey *et al.*, 1992). When compare with Sai Keng, Hong Kong mangrove forest in same geographical location (Tam *et al.*, 1993), the total heavy metal concentrations recorded in present study, except Pb, seemed to be higher. As no other data are available from previous study of mangrove sediments of Southeast China, and the background levels of this region were unknown, it is difficult to make a relevant comparison or meaningful conclusion on heavy metal contamination. Nevertheless, the total heavy metal concentrations of this mangrove forest, except Cd, did not exceed the reference values which had been used as the basis for estimating the pollution level, and all figures are well below the signal values (the upper limits) for standard soil containing 25% clay and 10% organic matter (Japenga *et al.*, 1990). This indicates that the heavy metal concentrations recorded in Futian mangroves would not cause harmful effects on organisms. This comparison also demonstrated that the sediments of Futian mangrove forest were not seriously polluted by heavy metals. It is true that Shenzhen has only been developed in recent years and very little industrial activities had been carried out in the townships surrounding this

Table 3. Concentrations of nutrients (%) and heavy metals (mg kg^{-1}) in leaf tissues of *A. corniculatum* and *K. candel.*

Mangroves	N	P	K	Cu	Zn	Mn	Sources
A. corniculatum							
Futian, PRC	1.32	0.14	0.53	4.12	85.2	166	The present study
	(0.01)	(0.04)	(0.09)	(0.76)	(28.6)	(89)	
Queensland, Australia	0.85	0.10	0.48	5.8	15.5	158	Spain and Holt (1980)
	(0.06)	(0.01)	(0.08)	(1.1)	(3.0)	(67)	
Fujian, PRC	1.22	0.12	0.87	ND	ND	ND	Lin and Lin (1985)
	(0.01)	(0.00)	(ND)				
K. candel							
Futian, PRC	1.39	0.13	0.59	4.05	69.7	1048	The present study
	(0.01)	(0.01)	(0.12)	(1.12)	(25.2)	(269)	
Fujian, PRC	1.88	0.15	0.89	ND	ND	ND	Lin and Lin (1985)
	(0.20)	(0.01)	(ND)				
Tanshui, Taiwan	1.90	0.15	ND	ND	ND	ND	Chen (1982)

Values of mean and standard deviation (in bracket) were shown; ND: not determined.

mangrove forest, thus heavy metal pollution should not be a problem in this area.

Extractable heavy metals. Most heavy metals except Mn were not in a bioavailable form and most extractable metals were less than 4% of total metals (Table 2). The proportion of extractable to total concentration differed from metal to metal and the order was: Mn > Cd > Zn > Cu > Pb > Cr. The variation of extractable heavy metals along the vertical positions of the shore were similar to that of total heavy metals (Figs. 4–6). Seasonal variations were not high although September samples appeared to have higher extractable metals, particularly Cr, Cd and Zn, than that of April samples (Table 1). The sediment Mn, behaved markedly different from that of the other heavy metals, declining towards the sea with more than 40% being extractable. Similar results had been reported by Real *et al.* (1993). They suggested that increased salinity impedes the oxidation process and Mn is able to cross the oxic layer and return to the water column, leading to lower Mn concentration in sediments at seaward locations.

Nutrient concentrations in plant leaves. The concentrations of carbon, phosphorus, nitrogen and potassium in leaves of *K. candel* and *A. corniculatum* were relatively constant along the vertical positions of the mangrove forest and did not change with sample locations (Fig. 7). The content of these elements in leaf samples collected in Site A did not differ from those of Site B. There was also no significant variations between the two plant species in terms of leaf nutrient and organic carbon content, indicating different plant species in the same mangrove area had similar nutrient levels (Table 3). The P and N values of this study were comparable to that found in Fujian Jiulong Jiang mangrove but K content of the present study was significantly lower (Lin & Lin, 1985; Lin *et al.*, 1987). However, the N, P and K content of *A. corniculatum* leaves were slightly higher than that of mangroves in Queensland (Spain & Holt, 1980). As the nutrient concentrations of leaf tissues reflected the soil nutrient concentrations which differed significantly geographically, the leaf N, P and K content of same plant species would vary from mangroves to mangroves.

Heavy metal levels in vegetation

Heavy metals such as Cr, Pb and Cd were not detected in all leaf samples and the variations of leaf Cu, Zn and Mn content along the shore were shown in Figure 8. It seemed to be a general trend that leaves collected from landward regions had higher Cu, Zn and Mn than those from the seaward side, although plants collected from some locations had exceptionally high heavy metals.

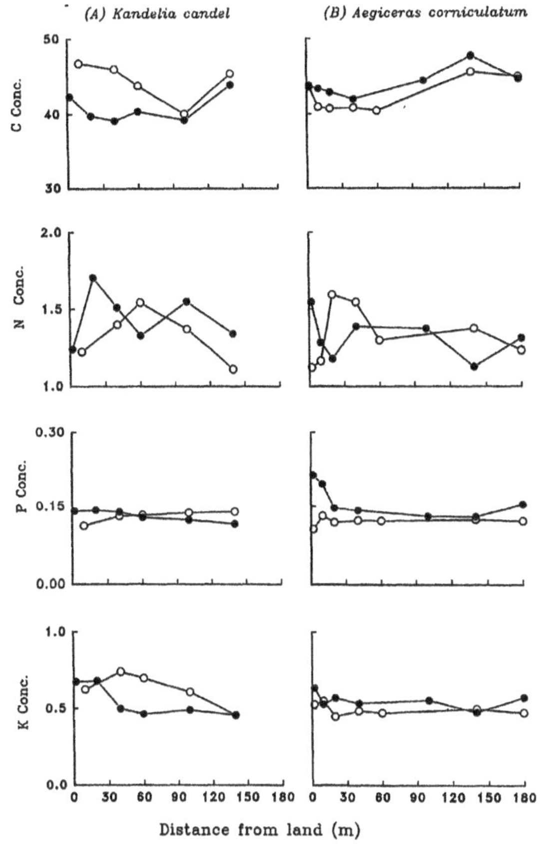

Fig. 7. Variations of leaf organic carbon, N, P and K concentrations (%) in *K. candel* and *A. corniculatum* along the vertical positions of Futian Mangrove Ecosystem. (o: Site A; •: Site B).

Two plant species had accumulated similar amounts of Cu and Zn in their leaves but *K. candel* had significantly higher leaf Mn concentration than *A. corniculatum*. The average Cu and Mn concentrations of *A. corniculatum* recorded in this study were comparable to that of Queensland mangrove (Table 3). As no previous study had been reported on the heavy metal content of mangrove plants especially in Southern part of China, it is impossible to discuss whether the plants in this mangrove forest had been contaminated. Nevertheless, Allen *et al.* (1974) stated that the Cu, Zn and Mn concentrations in plant materials fell in the range of 2.5–25, 15–100 and 50–1000 $\mu g\ g^{-1}$, respectively. This suggested that the two mangrove plants did not have excessive accumulation of these three heavy met-

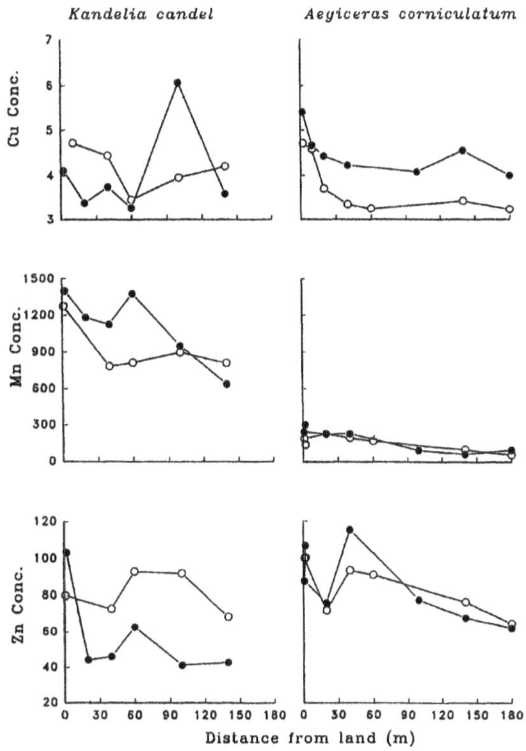

Fig. 8. Variations of leaf Cu, Mn and Zn concentrations (mg kg^{-1}) in *K. candel* and *A. corniculatum* along the vertical positions of Futian Mangrove Ecosystem. (o: Site A; •: Site B).

als in their leaves, and heavy metal pollution was not a problem in Futian mangrove forest.

Conclusion

The present ecological survey provides the background information on the nutrient and heavy metal status of sediments and leaves of *A. corniculatum* and *K. candel* in a subtropical mangrove located in Southeast coast of China. The results show that nutrients, N, P, K and organic C in this mangrove sediment were highest at the landward region and declined gradually towards the sea. This indicates the significance of tidal inundation, the importance of mangrove litter and its decomposition on elemental composition of the sediment. The seasonal variations of sediment element composition were generally insignificant. The heavy metal concentrations in sediments did not show any specific trend of changes along the vertical positions of the shore. In general, their values were low except that some localities might contain extraordinary high concentrations of heavy metals which were probably due to human contamination. Moreover, the extractable fractions of heavy metals were very small, indicating that most heavy metals were not bioavailable. This was reflected by the low heavy metal content in plant leaves. The leaf element content of *A. corniculatum* was comparable to that of *K. candel* except Mn; the former plant species had lower concentrations.

Acknowledgements

The authors would like to thank the officers and technicians in Futian Nature Reserve, Shenzhen Special Economic Zone, the PRC for their assistance in field sampling. We would also like to express our gratitude to Research Grant Council of Hong Kong and the Croudier Foundation for grant support.

References

Allen, S. E., H. M. Grimshaw, J. A. Parkinson & C. Quarmby, 1974. Chemical Analysis of Ecological Materials. Blackwell Sci. Publ., Oxford.

Boto, K. G. & J. T. Wellington, 1983. Nitrogen and phosphorus nutritional status of a northern Australian mangrove forest. Mar. Ecol. Press Ser. 11: 63–69.

Chen, M. Y., 1982. Ecological study of the mangroves of Tan-Shui estuary. Quart. J. Chinese Forestry 15: 17–25.

Clough, B. F. & P. M. Attiwill, 1982. Primary productivity of mangroves. In H.J. Teas (ed), Biology and Ecology of Mangroves. Tasks for Vegetation Science Series Vol. 8, Dr. W. Junk Publishers, Lancaster: 213–222.

Davies, C. A., K. Tomlinson & T. Stephenson, 1991. Heavy metals in River Tees estuary sediments. Environ. Technol. 12: 961–972.

Dwivedi, S. N. & K. G. Padmakumar, 1983. Ecology of a mangrove swamp near Juhu Beach, Bombay with reference to sewage pollution. In H. J. Teas (ed.), Biology and Ecology of Mangroves. Tasks for Vegetation Science Series Vol. 8, Dr. W. Junk Publishers, Lancaster: 163–170.

Jagtap, T. G., 1987. Seasonal distribution of organic matter in mangrove environment of Goa. Indian J. Mar. Sci. 16: 103–106.

Japenga, J., K. H. Zschuppe, A. J. DeGroot & W. Salomons, 1990. Heavy metals and organic micropollutants in flood plain of the River Waal, a distributary of the River Rhine, 1958–1981. Netherlands J. Agric. Sci. 38: 381–397.

Liao, J. F., 1990. The chemical properties of the mangrove Solonchak in the northeast part of Hainan Island. The Acta Scientiarum Naturalium Universititis Sunyatseni (Suppl.) 9(4): 67–72.

Lin, P. & G. H. Lin, 1985. Studies on the mangrove ecosystems of the Jiulong Jiang River Estuary in China. IV. the accumulation and biological cycle of nitrogen and phosphorus elements in the *Kandelia candel* community. Acta Phytoecologica et Geobotanica Sinica 9(1): 21–31.

Lin, P., L. Su & Q. Y. Lin, 1987. Studies on the mangrove ecosystems of the Jiulong River Estuary in China. II. Accumulation and biological cycle of potassium and sodium elements in the *Kandelia candel* community. Acta Ecologica Sinica 7(2): 102–110.

MacKey, A. P., M. Hodgkinson & R. Nardella, 1992. Nutrient levels and heavy metals in mangrove sediments from the Brisbane River, Australia. Mar. Pollut. Bull. 24(8): 418–420.

Murphy, J. & J. P. Riley, 1962. A modified single solution method for the determination of phosphate in natural wastes. Anal. Chem. Acta 27: 31–36.

Olsen, S. R., C. V. Cole, F. S. Watanabe & L. A. Dean, 1954. Estimation of available phosphorus in soils by extraction with sodium bicarbonate. U.S. Dept. Agric. Circ. 939: 1–19.

Por, F. D., 1984. The ecosystem of the mangrove forests. In F. D. Por & I. Dor (eds.), Hydrobiology of the Mangal. Dr. W. Junk Publishers, The Hague: 1–14.

Real, C., R. Barreiro & A. Carballeira, 1993. Heavy metal mixing behaviour in estuarine sediments in the Ria de Arousa (NW Spain). Differences between metals. The Sci. Tot. Environ. 128: 51–67.

Spain, A. V. & J. A. Holt, 1980. The Elemental Status of the Foliage and Branchwood of Seven Mangrove Species from Northern Queensland. Division of Soils, Divisional Report No. 49, CSIRO, Australia.

Tam, N. F. Y., L. L. P. Vrijmoed, S. H. Li & Y. S. Wong, 1993. The chemical properties of soil and its association with litter and plant production in a sub-tropical mangrove community in Hong Kong. Proc. Int. Conf. on Marine Biology of Hong Kong and the South China Sea. Hong Kong, October 1990 (in press).

Tam, N. F. Y. & Y. S. Wong, 1993. Mangrove soils as sinks for wastewater-borne pollutants. Proc. Asian-Pacific Symp. on Mangrove Ecosystems. Hong Kong, September 1993 (in press).

Thong, K. L., A. Sasekumar & N. Marshall, 1993. Nitrogen concentrations in a mangrove creek with a large tidal range, Peninsular Malaysia. Hydrobiologia 254: 125–132.

Walkley, Y. A. & I. A. Black, 1934. An examination of the Detjareff method for determining soil organic matter and a proposed modification of the chromic acid titration method. Soil Sci. 37: 29–38.

Hydrobiologia **295**: 159–166, 1995.
Y. S. Wong & N. F. Y. Tam (eds), Asia-Pacific Symposium on Mangrove Ecosystems.

Forest structure and biomass of mangroves in the Mgeni estuary, South Africa

T. D. Steinke, C. J. Ward & A. Rajh
Marine Science Unit, University of Durban-Westville, Durban, South Africa

Key words: biomass, estuary, forest structure, mangrove

Abstract

Forest structure and biomass were determined in a mangrove stand dominated by *Bruguiera gymnorrhiza* (L.) Lam. Trees in 5 m^2 sample plots were harvested at ground level and then further cut into 1 m strata for separation into living wood, dead wood, leaves, reproductive material and pneumatophores. Mean above-ground living biomass was calculated at 94.49±7.83 t dry matter ha^{-1}, while dead wood contributed a mean mass of 7.63±0.89 t dry matter ha^{-1}. Excavations of roots yielded a below-ground biomass of 9.67 t dry matter ha^{-1} which represented only 9.8% of the above-ground value. There was a mean density of 4700 living stems ha^{-1} with plant heights ranging from 0.57 m to 5.80 m. Mean LAI was 4.95±0.80. As a basis for estimating standing biomass, regression lines were fitted to biomass values from individual trees of *B. gymnorrhiza* and *Avicennia marina* (Forssk.) Vierh. of various sizes. A comparison of these relationships with methods used by previous workers for estimating biomass suggests that most other methods cannot be applied without modification for local stands of mangroves.

Introduction

Mangroves on the African continent reach their southernmost limits in the subtropical to warm temperate estuaries along the eastern coastline of South Africa. Although the total area of mangroves is relatively small (approximately 1058 ha) (Ward & Steinke, 1982), these communities have a significant role to play in the conservation and ecology of the estuarine ecosystems, and in the provision of poles for the construction of corrals, fish traps, etc. (Moll *et al.*, 1971; Steinke & Charles, 1986; Ward *et al.*, 1986; Steinke & Ward, 1988, 1990). It is therefore essential to have an acceptable method for the assessment of the biomass of our mangrove communities if management of this resource for a sustainable future use is to be successful. This study was undertaken to provide a method for such assessments.

Study site

The study was conducted in the Beachwood mangrove swamp at the mouth of the Mgeni River (29 °48′S, 31 °03′E). The Beachwood area, which is long and narrow (approximately 2.60 km × 0.35 km), lies roughly in a NE-SW direction, parallel to the coast. The mangroves occur along the creek separating a range of dunes on the sea front from scrub and urban development on the landward side. The main mangrove species are *Bruguiera gymnorrhiza* (L.) Lam. and *Avicennia marina* (Forrsk.) Vierh., although *Rhizophora mucronata* Lam. is also present. While the total area of mangroves in the Mgeni estuary was calculated from aerial photographs to be 44 ha (Ward & Steinke, 1982), floods in 1987 reduced the stand slightly. Two stands dominated by *B. gymnorrhiza*, with scattered trees of *A. marina*, were chosen for this study. In these stands the mangroves varied from saplings to tall trees (*B. gymnorrhiza* 0.5–14.0 m, *A. marina* 1.0–8.0 m). The stands comprised canopy trees with an increasing number of saplings towards the margins. The DBH (diameter of the stem at breast height) of the canopy members varied, an indication of variation in age of members of the stand. Tidal coverage in the stands occurred only at high spring tides.

Materials and methods

Stand biomass

Four 5 m^2 plots were chosen at random in a mangrove stand dominated by *B. gymnorrhiza* and, using a chain saw, the plants in each plot were removed at ground level. The above-ground parts were cut into 1 m strata which were further separated into living wood, dead wood, leaves, reproductive material and, in the case of the few *A. marina* trees, into pneumatophores. With the exception of random samples of leaves which were taken for leaf area determinations on a Li-Cor L1-3100 Area Meter, the plant material was oven-dried in a forced-draught oven at 80 °C until it could be weighed to constant mass. The leaf samples were also oven-dried after leaf area determinations, area/mass ratios determined and total leaf area calculated for each stratum using linear regression. Roots were excavated to a depth of approximately 500 mm in only two of the above plots to give an estimate of below-ground biomass. The excavated material was washed carefully under a jet of water to remove adhering mud, a sieve being used to retain fine roots. Root material was also oven-dried and weighed as above.

Light measurements were taken randomly at 1 m intervals from the top of the canopy to ground level, using a Li-Cor L1-188 Integrating Quantum Photometer. Mean values were obtained for each stratum.

Individual tree biomass

For biomass determinations of individuals, thirteen trees of *B. gymnorrhiza* (3 cm to 14 cm DBH) and five trees of *A. marina* (5 cm to 10 cm DBH) were chosen at random outside the four plots sampled for stand biomass. The individual trees were harvested at ground level and cut into 1 m strata, followed by further separation into the components mentioned above, before being oven-dried and weighed. Accurate determinations of biomass of pneumatophores (*A. marina*) and knee roots (*B. gymnorrhiza*) were not possible without extensive excavation to trace the roots of individual trees, nor was below-ground biomass obtained for this reason.

Statistical analyses

Results were statistically treated either by fitting curves to the biomass values from individual trees or by calculation of standard errors (Rayner, 1969).

Results

To avoid seasonal effects, all harvests were carried out at the same time of the year, *i.e.* October to December.

Stand biomass

As the mean heights of mangroves varied, it was not possible to combine the data and analyse the results statistically for each stratum. Consequently, the results of only one plot were presented in Fig. 1. This plot was chosen because it provided representative results with approximately mean values. The distribution of leaves and reproductive material increased towards the crown, while living and dead wood increased towards the base of the trees. For the first three metres from the base there was negligible or no reproductive material; this was present only in the upper parts of the canopy which received better illumination (Table 1). Conversely, dead wood yielded higher values in the more heavily shaded strata. Leaf data down to three metres revealed area/mass ratios showing a steady increase which coincided with the increase in percentage light interception (Table 1). Below this level the high ratios were well correlated with high light interception. The LAI (leaf area index) for this plot was 4.11.

The mean total values of biomass for the different components of the four stands and mean stand characteristics are presented in Table 2. In both species, stems represented the greatest percentage of the biomass (86.0% and 91.8% for *B. gymnorrhiza* and *A. marina* respectively), while leaves made up only 13.2% and 6.2% respectively. Mean total above-ground biomass amounted to 94.49±7.83 t dry matter ha^{-1}, while below-ground biomass yielded only 9.67 t dry matter ha^{-1}, which represented 9.8% of the above-ground value. Dead wood contributed a mean mass of 7.63±0.89 t dry matter ha^{-1}. Trees of *B. gymnorrhiza* were almost ten times more numerous than *A. marina* because of the high number of saplings of the former species. Consequently, *B. gymnorrhiza* was represented by trees with a lesser DBH and mean height (2.66±0.16 m and 4.56±0.53 m for *B. gymnorrhiza* and *A. marina*, respectively). The height of the canopy trees was the same in both species (5–6 m), with trees ranging from 0.57 m to 5.75 m and from 0.97 m to 5.80 m in *B. gymnorrhiza* and *A. marina* respectively. The mean LAI of four stands was 4.95±0.80, which is higher than that obtained for apparently similar stands

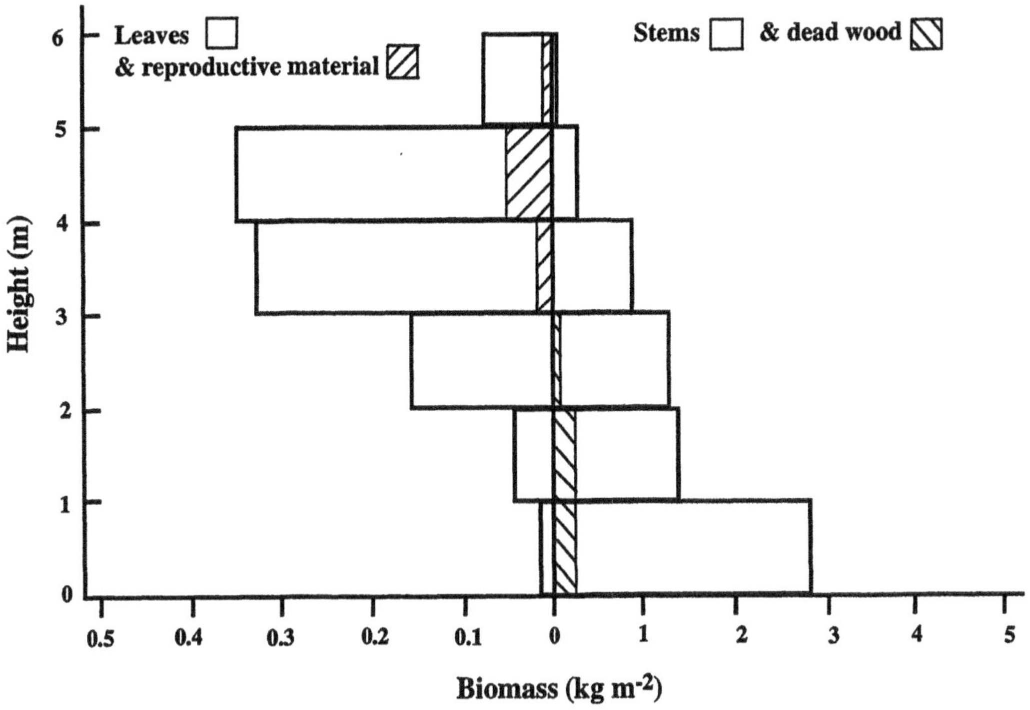

Fig. 1. Stratal distribution of above-ground biomass and dead wood in a representative plot of mangroves.

Table 1. Stratal distribution of leaf components and light interception in a representative plot of mangroves.

Strata (m)	Leaf mass (g)	Leaf area (dm^2)	Area/mass ratio	Percentage light interception
5–6	430.51	14721.50	34.20	26
4–5	1749.82	63409.86	36.24	58
3–4	1669.36	66322.98	39.73	80
2–3	780.33	41205.10	52.80	95
1–2	225.33	13335.50	59.18	96
0–1	104.15	6717.95	64.50	98

of *B. gymnorrhiza* in southern Japan and Indonesia (Suzuki & Tagawa, 1983; Ogino & Chihara, 1988).

Individual tree biomass

Using the data obtained from individual trees (Table 3), it was found that the power curve, $y = ax^b$, where y = above-ground biomass (kg) and x = DBH (cm), gave a very good estimate of biomass for *B. gymnorrhiza* ($r = 0.99$, $P < 0.01$ for stem, leaves and total biomass). In the case of *A. marina*, the above equation once again provided the best estimate, although with not such a good fit ($r = 0.95$, $P < 0.01$ for stems and total biomass, $r = 0.97$, $P < 0.01$ for leaf biomass) (Fig. 2).

Stratal distribution of biomass components from individual trees revealed similar patterns to those for stands (Fig. 1), although there were no living or dead

Table 2. Mean distribution of biomass among different components and mean characteristics of a closed mangrove stand ($x \pm$ SE).

Species	Stems (t ha^{-1})	Leaves (t ha^{-1})	Reproductive material (t ha^{-1})	Pneumatophores (t ha^{-1})	Total above-ground biomass (t ha^{-1})	Roots (t ha^{-1})	Dead wood (t ha^{-1})	DBH (cm)	Stems ha^{-1}
Bruguiera	64.22	9.84	0.61	–	74.67		5.88	2.31	42500
gymnorrhiza	±13.72	±2.20	±0.21			9.67	±0.99		
Avicennia	18.20	1.22	–	0.40	19.82		1.75	4.19	4500
marina	±11.75	±0.79		±0.28			±1.03		
Total	82.42	11.06	0.61	0.40	94.49	9.67	7.63	–	47000
					±7.83		±0.89		

Table 3. Mean distribution of above-ground biomass and dead wood among different components of individual trees.

Species	Height (m)	DBH (cm)	Stems (kg)	Leaves (kg)	Reproductive material (kg)	Total above-ground biomass (kg)	Dead wood (kg)
Bruguiera	4.67	3.44	2.76	0.47	0.002	3.23	1.45
gymnorrhiza	5.55	3.88	3.80	0.41	0.002	4.21	0.38
	5.84	5.79	8.13	1.28	0.084	9.49	0.71
	6.22	5.92	11.14	0.79	0.174	12.10	0.99
	6.72	6.94	13.49	1.39	0.085	14.96	0.94
	7.62	7.32	17.88	1.58	0.220	19.68	1.15
	6.23	7.32	–*	1.26	0.099	–*	1.56
	8.05	7.45	18.98	1.74	0.463	21.18	1.83
	6.77	8.28	22.59	2.12	0.091	24.80	0.68
	7.72	8.66	22.84	2.03	0.097	24.97	1.92
	10.16	11.14	49.77	2.90	0.106	52.78	3.66
	7.68	11.58	47.41	3.53	0.208	51.15	3.17
	13.50	13.50	101.13	5.67	0.488	107.29	3.22
Avicennia	4.92	5.47	4.94	0.37	–	5.31	1.13
marina	7.72	6.05	10.49	0.53	–	11.02	0.48
	7.63	6.24	11.71	0.63	–	12.34	0.57
	8.15	7.16	–*	0.86	–	–*	1.55
	7.51	9.99	30.51	1.40	–	31.91	1.62

* This component lost in a fire

branches at the lower levels. Except in the case of *B. gymnorrhiza* with a DBH less than 5 cm, stems represented at least 90% of the above-ground biomass (Table 4). In this species the ratio of leaves to stems decreased with DBH and therefore probably with age of trees (Whittaker, 1975). The density of the stems (wood only) was 805±14 ($n=31$) and 780±13 ($n=21$) kg dry mass m^{-3} fresh volume for *B. gymnorrhiza* and *A. marina* respectively.

Discussion

Both sets of data confirmed the high proportion of biomass in stems. Variations in leaf biomass with age

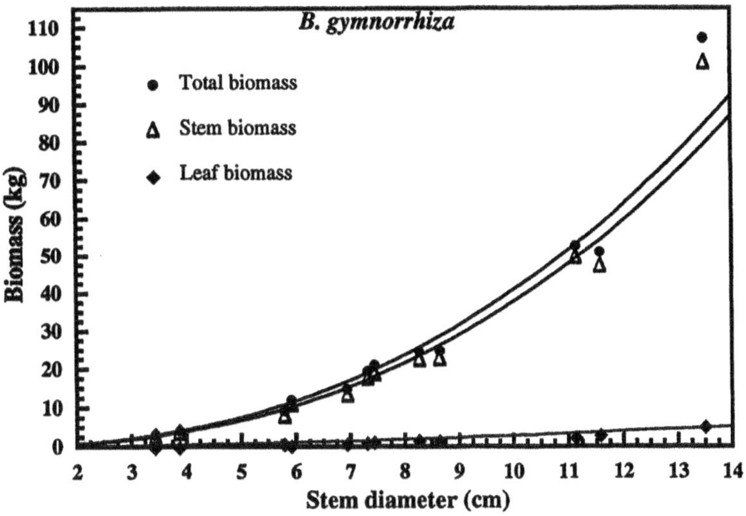

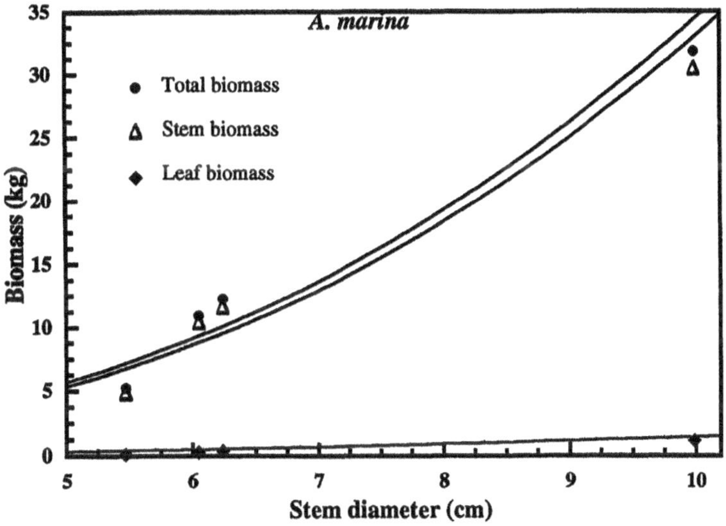

Fig. 2. Relationships between DBH and biomass for different components of above-ground biomass.

of trees could not be attributed to seasonal effects as all harvests were carried out in late spring to early summer (October to December) when there was normally no peak in leaf fall (Steinke & Charles, 1986). Light measurements through the canopy revealed that leaves in the top three metres intercepted most of the light. This was due to the high proportion of leaves near the top of the canopy with a taper towards the base of the stand (Christensen, 1978). It is significant that production of reproductive material was restricted virtually to these top layers. This suggests that only near the top of the canopy was the light intensity sufficiently high to sustain a level of photosynthesis to allow for the production of propagules. This supports field observations with a portable infrared gas analyser (Steinke and Rajh, unpublished work) that shade leaves of *B. gymnor-*

Table 4. Mean percentage distribution of above-ground biomass among different components for trees of specified size classes ($x \pm SE$).

Species	DBH (cm)	Stems	Leaves	Reproductive material
Bruguiera	<5	87.86±2.38	12.15±2.38	0.05±0.01
gymnorrhiza	5–10	90.11±0.81	8.89±0.83	0.99±0.25
	10–15	93.75±0.53	5.90±0.51	0.35±0.08
Avicennia marina	5–10	94.68±0.56	5.32±0.56	–

rhiza are usually below the light compensation point. Although the stand was short in comparison with those in some tropical regions, the leaves from the canopy to the lowest stratum showed similar adaptations to the light and temperature environment. Area/mass ratios varied, with thinner leaves of larger surface area at the base and thicker leaves of smaller area at the top. Leaf orientation also changed from nearly vertical at the top to the lowest leaves which are held normal to the incoming radiation (Andrews *et al.*, 1984).

The results from large individual trees revealed that leaves and dead wood were not present in the lower strata. In the closed stand, however, the presence of these components at these levels was probably an indication of juveniles. The small mean size of *B. gymnorrhiza* and also the relatively high LAI were also due to a large number of younger plants. For this reason the approximate annual increment in biomass was calculated from data for individual trees. From early aerial photographs (Ward, personal observation) it was deduced that mangroves became established in that particular area approximately 32 years before harvesting. Assuming the larger trees to have been among the early colonisers, an annual increment of 0.78 kg (*B. gymnorrhiza*) and 1.00 kg (*A. marina*) was estimated, which is lower than the values obtained in the tropics (Christensen, 1978).

Relatively little information is available on below-ground biomass in mangroves and the figures that have been produced show top/root ratios that vary over a wide range. The ratio of 10.20:1 obtained in this study was very different from that of Clough & Attiwill (1975, 0.59:1), Briggs (1977, 0.71–0.98:1), Golley *et al.* (1962, 2.27:1) and Tamai *et al.* (1986, 2.26:1), although closer to that of Silva *et al.* (1991, 4.00:1). There are a number of reasons, including those put forward by Komiyama *et al.* (1987) and Silva (1991), which could account for the high ratios. The temperate mangroves, for example, appear to have the lowest ratios. The majority of estimates of root biomass have been gained through sample cores which varied in depth from 350 mm to (in one case) 1500 mm. While excavation to only 500 mm could have resulted in the loss of some information in this work, Silva *et al.* (1991) found the highest biomass value in the top 100 mm (89.9 g m^{-2}), decreasing steadily with depth to less than 5 g m^{-2} at 600 mm. The high percentage of fine roots found by Komiyama *et al.* (1987) suggests that it is this component that might have been underestimated in our study. Clearly, the importance of mangrove roots for the ecosystem requires that this area should receive further study.

For the estimation of above-ground biomass various methods developed by previous researchers were applied to determine which provided the closest relationship under our conditions (Briggs, 1977; Suzuki & Tagawa, 1983; Putz & Chan, 1986; Tamai *et al.*, 1986; Clough & Scott, 1989; Silva *et al.*, 1991). Comparative estimates from specific trees harvested in this study, covering the most abundant local size classes, indicated that our equation over-estimated slightly the biomass of smaller trees, but provided a good estimate in the case of larger trees of both species (Table 5). Furthermore, with the exception of the formula of Suzuki and Tagawa (1983) for small trees of *B. gymnorrhiza*, our equation provided the closest estimate. The low estimate from the method of Putz and Chan (1986) and Silva *et al.* (1991) are perhaps understandable as they were developed for trees of *Rhizophora apiculata* and *R. mangle*, respectively. The variation with a similar technique used by Clough and Scott (1989) could well be due to differences in population density which these authors have suggested may account for differences. It would appear that Briggs (1977) applied his formula

Table 5 Comparative estimates of total above-ground biomass (kg) for trees of *B gymnorrhiza* and *A marina*

Species	DBH (cm)	Briggs	Suzuki & Tagawa	Putz & Chan	Tamai *et al*	Clough & Scott	Silva *et al*	This study	Actual biomass
Bruguiera	5 79	–	9 4	14 2	10 8	10 7	8 0	10 7	9 5
gymnorrhiza	11 14	–	40 5	73 6	56 5	48 2	39 9	52 8	52 8
Avicennia	5 47	4 5	–	–	–	–	–	7 2	5 3
marina	9 99	23 0	–	–	–	–	–	34 3	31 9

to *A marina* either widely spaced or on the edge of a swamp, whereas in this study the trees were in a closed, dense stand

Not only did the power curve give the 'best fit' to the data for both species, but this technique was also found to be the most convenient to apply Attempts to use height as a parameter were not successful A close relationship between height and biomass was not found in our mangroves (Table 3), damage to the growing point was found frequently, and this caused branching of the main stem and consequently shorter trees Height of individual trees was also difficult to measure accurately in a closed, extensive canopy, even that of the relatively low-growing trees which formed the major component of our local communities These differences endorse the recommendation by Clough & Scott (1989) that local checks must be conducted to ensure that a particular relationship is valid

Although old trees of *B gymnorrhiza* have been recorded with a height of approximately 20 metres in South Africa (Macnae, 1963), this is unusual as the majority are of the size measured in this study or only slightly taller Our mangroves are also of limited extent, man's influence on these communities in the past has been intense, and, particularly in the southern areas where they are nearing the limit of their distribution, growth is slower than under more favourable tropical conditions It is hoped that future exploitation of the mangroves in this country will be preceded by environmental impact assessments which will include estimates of biomass using the methods elaborated in this paper In view of the key role which mangroves play in many of our estuarine ecosystems, it is essential that their production potential should be assessed and maintained at a high level

Acknowledgments

The authors wish to express their appreciation to all who assisted with field work, in particular S Kasavan Mrs R Bunsee is thanked for typing the article To the University of Durban-Westville for their financial contribution, and the Natal Parks Board, thanks are also due

References

Andrews, T J , B F Clough & G J Muller, 1984 Photosynthetic gas exchange properties and carbon isotope ratios of some mangroves in North Queensland In H J Teas (ed) Tasks for Vegetation Science Physiology and management of mangroves Dr W Junk Publishers, The Hague, 106 pp

Briggs, S V, 1977 Estimates of biomass in a temperate mangrove community Austr J Ecol 2 369–373

Christensen, B , 1978 Biomass and primary production of *Rhizophora apiculata* Bl in a mangrove in southern Thailand Aquat Bot 4 43–52

Clough, B F & P M Attiwill, 1975 Nutrient cycling in a community of *Avicennia marina* in a temperate region of Australia In G E Walsh, S C Snedaker & H J Teas (eds), Univ Gainesville, Florida Proc Internat Symp Biol Mgmt Mangroves, Hawaii 1 137–146

Clough, B F & K Scott, 1989 Allometric relationships for estimating above-ground biomass in six mangrove species For Ecol Mgmt 27 117–127

Golley, F B , H T Odum & R F Wilson, 1962 The structure and metabolism of a Puerto Rican red mangrove forest in May Ecology 43 10–19

Komiyama, A , K Ogino, S Adsornkoe & S Sabhasri, 1987 Root biomass of a mangrove forest in southern Thailand I Estimation by the trench method and the zonal structure of root biomass J Trop Ecol 3 97–108

Macnae, W , 1963 Mangrove swamps in South Africa J Ecol 51 1–25

Moll, E J , C J Ward, T D Steinke & K H Cooper, 1971 Our mangroves threatened Afr Wildlife 25 103–107

Ogino, K & M Chihara, 1988 Biological system of mangroves A report of East Indonesian Mangrove Expedition, 1986 Ehime Univ

Putz, F. E. & H. T. Chan, 1986. Tree growth, dynamics and productivity in a mature mangrove forest in Malaysia. For. Ecol. Mgmt 17: 211–230.

Rayner, A. A., 1969. A first course in biometry for agriculture students. Univ. Natal Press, Pietermaritzburg, 626 pp.

Silva, C. A. R., L. D. Lacerda, L. F. F. Silva & C. E. Rezenda, 1991. Forest structure and biomass distribution in a red mangrove stand in Sepetiba Bay, Rio de Janeiro. Revta brasil. Bot. 14: 21–25.

Steinke, T. D. & L. M. Charles, 1986. Litter production by mangroves. I: Mgeni Estuary. S. Afr. J. Bot. 52: 552–558.

Steinke, T. D. & C. J. Ward, 1988. Litter production by mangroves. II. St Lucia and Richards Bay. S. Afr. J. Bot. 54: 445–454.

Steinke, T. D. & C. J. Ward, 1990. Litter production by mangroves. III. Wavecrest (Transkei) with predictions for other Transkei estuaries. S. Afr. J. Bot. 56: 514–519.

Suzuki, E. & H. Tagawa, 1983. Biomass of a mangrove forest and a sedge marsh on Ishigaki Island, South Japan. Jap. J. Ecol. 33: 231–234.

Tamai, S., T. Nakasuga, R. Tabuchi & K. Ogino, 1986. Standing biomass of mangrove forests in southern Thailand. J. Jpn. For. Soc. 68: 384–388.

Ward, C. J. & T. D. Steinke, 1982. A note on the distribution and approximate areas of mangroves in South Africa. S. Afr. J. Bot. 1: 51–53.

Ward, C. J., T. D. Steinke & M. C. Ward, 1986. Mangroves of the Kosi System, South Africa: Their re-establishment since a mass mortality in 1965/66. S. Afr. J. Bot. 52: 501–512.

Whittaker, R. H., 1975. Communities and ecosystems. Current Concepts in Biology Series. MacMillan, New York, 158 pp.

Hydrobiologia **295**: 167–181, 1995.
Y. S. Wong & N. F. Y. Tam (eds), Asia-Pacific Symposium on Mangrove Ecosystems.

Genetic diversity, distributional barriers and rafting continents – more thoughts on the evolution of mangroves

Norman C. Duke*
Smithsonian Tropical Research Institute, PO Box 2072, Balboa, Republic of Panama
* *Current address: 22-24 Victoria Street, Townsville Queensland 4810, Australia*

Key words: mangrove, *Avicennia*, evolution, fossils, pollen

Abstract

Without continental drift, the diversity and distribution of many species, including mangrove plants, would be very different today. First, there would be fewer pantropic genera and many more endemics. Second, their characteristics would not be as common and widespread as some are today. Continental drift has brought about the massive mixing and dispersal of genes in geologically recent times, greatly enhancing the evolutionary process; particularly for flowering plants - the angiosperms, which evolved during this period.

Mangrove plants are comprised of approximately 70 species from 20 quite different angiosperm families. Most taxa are characterized by special physiological abilities and structural forms, enabling them to live in both seasonally fluctuating saline conditions, and water-saturated soils. Their occurrence is mostly tropical, perhaps because of harsh physiological conditions of intertidal habitats; but distributions of specific taxa do not fully concur with the idea of a completely tropical evolution, at least for some important species.

At least one genus of mangrove tree, *Avicennia*, occurs around the world, chiefly in tropical estuarine habitats, although they also range into temperate latitudes, especially in the south. Around the world, there are no more than ten species of *Avicennia* recognised today, but their diagnostic determinants were inadequate prior to recent studies using both numerical analyses of morphological parameters and isozymes. Such analyses significantly reduced the number of apparent species, notably around Australia, and provided a basis for the revision of distributional records throughout the Indo West Pacific region. One species, *A. marina*, was found to be widespread and morphologically variable with genes divided into characteristic groupings of at least three geographic areas in the region. Based on these findings, there are several novel inferences to be made regarding the evolution of this genus. A western Gondwanan origin is proposed, with subsequent radiation of several taxa facilitated via the tectonic dispersal of southern continental fragments.

Introduction

There are two important paradigms associated with the evolution and biogeography of mangrove plants. One is the idea of a common centre-of-origin for all taxonomic entities and species. The second is the idea of a centre-of-diversity linked with a centre-of-origin. These are quite old concepts, and they still form the basis of our present understanding of the evolution of mangrove plants. When applied to specific taxa however, they do not explain present-day distributions (Duke, 1992; Tomlinson, 1986), leaving fundamental questions un-answered. For instance, why do mangroves have such remarkable distributional disjunctions and discontinuities? And why do they occur so widely around the world, especially in view of these discontinuities and their overall genetic uniformity? It appears that hypotheses based on centre-of-origin and centre-of-diversity concepts are chiefly based on a precept of exaggerated long-distance dispersal to explain the presence of certain taxa in widely disparate regions. In view of these doubts, and some recent findings on intra-specific variation, it is time for a re-appraisal of evolutionary models of mangroves.

The genetic composition of todays mangrove flora, while subject to present-day climatic and geographical conditions (Duke, 1992), is largely relict. Species are presumably where they are today because of past events and circumstances. And, these were influenced by respective physiology, ecology, dispersal success, geological circumstances, evolutionary rate, and origin of each taxon. These concepts are summarized in eight generalized statements concerning the biogeography and evolution of mangroves, based on Tomlinson (1986) and expanded here:

1. *Ecological and Climatic Conditions.* Mangroves are generally restricted to sheltered tropical coastlines where mean monthly seawater temperatures rarely drop below 20 °C. Some mangrove taxa are further restricted to areas of higher rainfall and to larger riverine estuaries.
2. *Topographic Height.* Mangroves are generally restricted to a very small topographic range from around mean sea level to high water of spring tides.
3. *Distribution by Sea.* Mangroves have water-buoyant propagules which are dispersed by sea and estuarine currents, driven by wind, waves, tides and ocean circulation. Note: sea-drift, and its possible deposition as fossil beds, may extend far beyond the range of growing plants.
4. *Distributional Change.* Where climatic conditions and sea levels change, the distributional ranges of mangroves expand or contract.
5. *Appreciable Geological Age.* Mangroves have long fossil records extending back to the Cretaceous era, at least to around 100 million years ago. This often applies to specific genera, demonstrating their genetically conservative nature and relatively slow rates of genetic change.
6. *Diversity and Systematic Groupings.* Some extant mangrove taxa extend around the world, while others have very restricted ranges. The concentration of most species in the Indo-Malesian area defines the chief centre of diversity.
7. *Disjunctions of Distribution.* There are several major distributional disjunctions for mangroves around the world. Two chief examples are: the disjunction between the Atlantic East Pacific and Indo West Pacific populations; and, the disjunction to the north and west of New Guinea.
8. *Continental Drift Theory.* The evolution of mangroves, and angiosperms generally, took place at the same time as the southern super-continent of Gondwana broke apart. Continental fragments were scattered around the world, carrying their respective complements of plants and animals. Some parts of the former southern continent are now in contact with the old northern land mass of Laurasia.

The relative importance and influence of each statement above is species-dependent. The first two identify the predominant and major physiological constraints which particularly influence within-site species distributions. They can be considered primary factors, chiefly based on temperature, moisture and tides. The third concerns dispersal of propagules. This has been greatly over-rated, and I suggest that propagules do not primarily provide for long-distance dispersal, but rather for enhanced re-establishment locally. Briefly, there are two major problems encountered in assessments of the dispersal capabilities of mangroves: one is the assumed uniformity of this ability in all taxa; and the other is the often exaggerated claim of long-distance dispersal capabilities, despite contradictions encountered in records of extant distributional disjunctions.

Consider the global disjunction which separates conspecifics within either the Indo West Pacific or Atlantic East Pacific regions. These are joined at present by the Pacific Ocean which forms a major biogeographic barrier to most mangrove species, with one exception, a species of 'New World' (Atlantic East Pacific) *Rhizophora* in the south-western Pacific islands of Fiji to New Caledonia. This distribution is discontinuous with American mangroves, and the trans-Pacific distribution is interrupted by the absence of this species from intervening islands with suitable habitats. Clearly, this demonstrates present limitations of long-distance dispersal for Rhizophora. The reason for such a situation must involve a change in current environmental conditions which restrict present-day dispersal. This is covered in the remaining statements, 4 to 8, which all relate to historical change in geological and climatic conditions, including changes in genetic make-up and speciation.

McCoy & Heck (1976) investigated similarities between mangroves and two other tropical shallow water habitats, finding they shared global patterns of occurrence, species diversity, and fossil transitions, despite their different genetic make-up and dispersal capabilities. For example, compare the non-buoyant propagules of seagrasses with the dispersal specialist taxa in corals and in mangroves. In addition, there is considerable variation of dispersal capabilities within these groups; further challenging the importance of long-distance dispersal. But, other evidence describes markedly similar floral elements from widely disparate regions around the world; noting five major examples described by Hutchinson (1973): eastern North America and eastern Asia, western Africa and eastern Brazil, Mascarene and southern India, southern Africa and western Australia, and New Zealand and southwestern South America. There must be an explanation which encompases all these observations, based on both a better understanding of changing distributional constraints, and the dispersal capabilities for each taxon.

Mangrove species are typically sea-water dispersed by buoyant propagules. They are expected therefore to be less influenced by sea barriers which restrict most other plants. But, even so, their ability to disperse across water has definable limits. These are notably species-specific, and often far less than expected; for example, *Avicennia* propagules apparently travel less than 200 nautical miles before they perish, based on personal observations and studies of buoyant longevity by Steinke (1975, 1986). This contrasts dramatically with *Rhizophora* propagules which survive several months at sea (Rabinowitz, 1978), and are capable of traveling much greater distances. It is therefore unreasonable to lump all mangrove genera in one combined evolutionary model based on extant distributions alone.

Given all the different ancestral sources of mangrove plants and their present-day wide distributions, their co-evolution appears more circumstantial since they almost certainly originated at different locations and at different times. The fact that distributions overlap so much attests to some greater influence, and I suggest this was the direct result of movements in continental fragments described in the theory of continental drift. To unravel the major processes involved, each taxon must be dealt with separately, tracing both their extant populations and their putative fossil remains. The following discussion deals with this at three levels, notably habitats, selected families, and one specific genus, the *Avicennia*.

Origin of mangroves and ancestral groups

It is convenient first to investigate the origin of mangrove taxa, making up the habitat. Were they ancestral to land plants, or were they derived from land plants? The latter seems more plausible since the majority of mangrove plants are angiosperms and these first appeared in the fossil record as sparse montane shrubs during the Early Cretaceous (Takhtajan, 1980). Furthermore, while mangrove orders and families are diverse, none include appreciably primitive characteristics. For this reason, it is assumed that mangrove species were derived from either freshwater swamp plants or halophytes, or both. These observations are implicit in any discussion of the origin and evolution of the various taxa, and the habitat they form.

The intertidal habitat is inhospitable to most plants, it has obvious spatial limitations, and similar 'mangrove' capabilities and characteristics have evolved in only a few members of several very different plant groups (Duke, 1992). In view of these comments, it is important to ask whether mangrove ancestors would have dispersed into intertidal habitats without some dynamic physical processes 'directing' genetic change. One such physical process might have been extended periods of progressive sea level rise. In these circumstances, coastal environments would have been subject to constant encroachment of marine conditions, leaving selection of marine-tolerant offspring as the only possibility for survival. However, sea level changes are notably dynamic and they generally fluctuate widely in geological time. Another process might meet the constant encroachment model described above, and it occurs irrespective of sea level change. This is the formation of new intertidal habitat when continental land masses break apart over millions of years; for example, consider the break-up of the Gondwanan supercontinent since Cretaceous times.

In Fig. 1, world maps show four geological eras, based on Lambert equal-area projections with both poles shown at the same time, used by Briggs (1987). Black shapes depict the present day world map, while shaded ones represent postulated maps for Early Cretaceous, Late Cretaceous and Eocene eras. By presenting these as overlays, the relative movement of land masses is shown through time. Note the separation of the Gondwanan supercontinent parts, South America, Africa, Antarctica, India and Australia. Their dispersal took place at different times and rates. In this depiction, Antarctica has moved very little while others have moved considerably. For example, note the rapid northward movement of India. Also note that all, except Antarctica, are now in relatively close contact with portions of the old northern supercontinent, and all are disconnected from each other.

These tectonic events and the evolution of angiosperms, including mangroves, all took place over the same period. I suggest that mangrove species, and their habitat, evolved and diversified because of the breakup of Gondwanaland. This would account for the

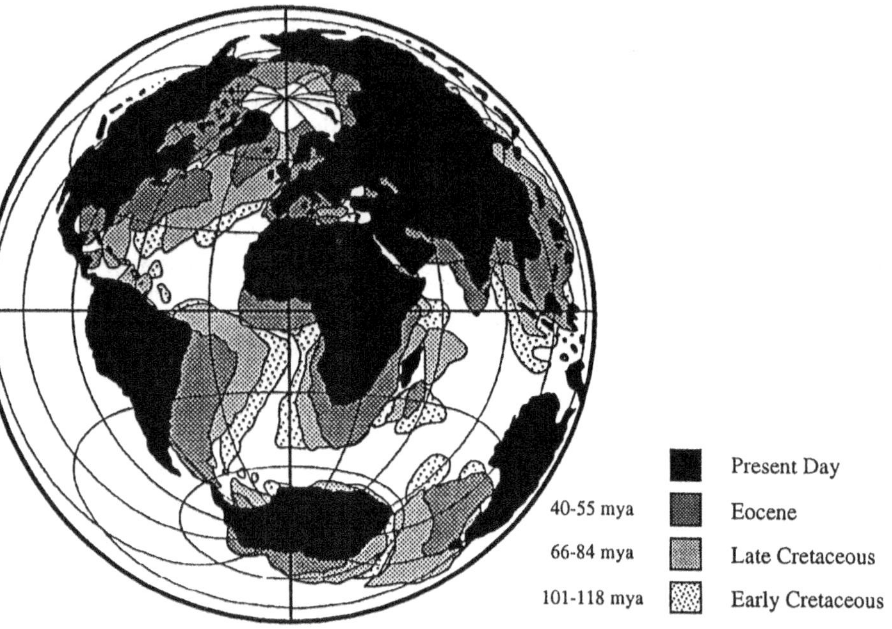

Fig 1 World maps of four geological eras, from early Cretaceous to present day These are overlaid to depict the movement of continents during the latter stages of the breakup of the Gondwanan supercontinent, as described in the theory of continental drift Maps are presented in Lambert equal area projection based on Briggs (1974) Note 'mya' is an abbreviation for the unit, 'million years ago'

wide dispersal and distribution of species around the earth (Fig 1)

The next question concerns speciation, notably in the creation of genetically unique plant species This process is implicit in an evolutionary model, and there are at least three ways a new species might form (Takhtajan, 1980) First, where species disperse from a centre-of-origin, those at the centre might multiply by subtle mutation and genetic drift The pattern created is of younger more recently evolved taxa toward the centre and older conservative ones around the margins In this way, the centre-of-diversity marks the centre-of-origin A second mechanism is where the process of diversification takes place where a continuous range of one taxon is divided and each sub-group becomes genetically isolated The sub-groups become different over time in this vicariant process as each is subject to different biological selection pressures and the influence of different mutations A third mechanism is where new species evolve by 'budding' in a process also called the 'founder' effect In this process, small satellite populations become genetically isolated either by range contraction of the larger population, or by exceptional long distance dispersal events In each instance, genetic diversification from the original population could occur quite dramatically since genes of the founder population might comprise only a limited subset of the original larger one Furthermore, any mutant genes would have a proportionally greater influence (Takhtajan, 1980), permitting rapid change in the new population

The circumstances of continental drift might have greatly enhanced these evolutionary processes Land masses breaking apart could create vicariant populations with sibling species and varieties, while land masses moving closer together might form founder populations, also resulting in new taxa

There are problems however, with inferring evolutionary processes from extant distributions The major difficulty is in discerning those taxa which have retained ancestral characters from those that are derived or new This may be resolved in a number of ways using anatomical and other morphological comparisons, or from genetic comparisons, or from direct observation of fossils. Unfortunately, the latter more direct approach is not possible since mangrove fossils are generally scarce. Nevertheless, inferences from comparative studies do provide important insights, provided phenotypic and genetic differences of morphological characters have been clearly distinguished. Note that mangrove trees often include a wide range of growth forms, and degrees of growth expression, correlated directly with environmental parameters. These include salinity, temperature, and inundation frequency (Soto & Corrales, 1987; Duke, 1990a), as well as available light; for example, there is significant morphological variation within single individuals between leaves from the upper and lower canopy (Duke, 1990a). Using comparative anatomical, morphological and genetic information, it is possible to construct phylogenies for comparison with the limited fossil evidence. This would test various hypotheses and provide an evolutionary model for on-going evaluation of clearly defined taxa.

Hypotheses on the evolution of mangroves

Several hypotheses proposed explanations for extant distributions of mangroves. In one, van Steenis (1962) suggested a primary radiation in the Malesian area prior to dispersal chiefly eastward across the Pacific. While advocating long-distance dispersal, he expressed reservations about suitability of apparently drier habitats along the ancient Tethys coastline leading to the Atlantic. This route, however, was later advocated by Chapman (1976, 1977) in a second hypothesis which accepted the Malesian centre-of-origin and subsequent radiation in the Late Cretaceous. In a third hypothesis, McCoy & Heck (1976) proposed the centre-of-diversity differed from the centre-of-origin, which, they suggested, occurred along the ancient Tethys coastline. Fourthly, Specht (1981) accepted an Australasian centre-of-origin, based chiefly on pollen observations by Churchill (1973), and postulating Early Cretaceous (or even earlier) origins. Mepham (1983), in a fifth hypothesis, accepted a Late Cretaceous time of origin with radiation from the Gondwanan-east Tethyian area; a migration northward would have been achieved by tectonic movements of India and Australia, making it possible for dispersal westward through the Tethys.

These hypotheses have essential differences, but they are generally similar, reflecting available evidence at respective dates of writing. In most cases, a centre-of origin is proposed, and this was altered when fossil records apparently contradicted earlier proposals matching centre-of-diversity and centre-of-origin concepts. There are, however, two views on subsequent radiations. Most hypotheses suggest that particular radiation routes led to extant taxa en route. The hypothesis of McCoy & Heck however, proposed that radiation was initially uniform before contracting and leaving disjunct refuge populations that evolved into extant taxa. But in either case, Tomlinson (1986) pointed out that extant distributions were not explained by any hypotheses, and he suggested that they generally used greater time spans than were likely. A greater criticism is that none of these hypotheses fully considered individual taxa, taking into account their major differences in morphology and life history processes, including dispersal ability and habitat preference. It is worth noting that downstream estuarine mangroves might simply drift along a continuous coastal niche, while upstream forms might 'estuary hop'; the latter being reminiscent of terrestrial plants which 'island hop'. Presumably, those less dependent on particular estuarine conditions, preferring more coastal conditions rather than upstream lower salinities, would have the greatest chance of wider dispersal.

In addition, Tomlinson (1986) pointed out that mangrove taxa were geologically old and included some remarkably conservative and stable genera. These observations were based chiefly on the detailed review of fossil pollens provided by Muller (1981). This review is summarized in the schematic of Fig. 2, where all families with mangrove genera are depicted, showing oldest records of orders, families and genera since the earliest angiosperm pollen. Muller recorded this earliest pollen occurred around the Early Cretaceous, approximately 110 million years ago in the Aptian period. The first mangrove orders and families appeared during the Santonian in the Late Cretaceous, around 80 million years ago. The first mangrove genus recognized was *Nypa* in the Maestrichtian, around 69 million years ago. *Rhizophora* pollen appeared in the Late Eocene. *Avicennia* and *Sonneratia* in the Early Miocene. These observations however appear remarkably conservative compared with limited macro-fossil data. For example, *Nypa* fruit (Whitmore, in Heywood, 1978) in Mid Cretaceous, fruit and leaves of *Avicennia* (Berry, 1916, 1936) in Eocene, and the earliest angiosperm fossil dating from 135 million years ago

(Takhtajan, 1980). There is also the disputed (Muller, 1981) claim by Churchill (1973) of *Avicennia* pollen from the Eocene. The time lines of mangrove evolution (Fig. 2), based on the review of palynological evidence by Muller (1981), therefore appear relatively conservative.

Connections between Indo West Pacific and Atlantic East Pacific regions

Possible dispersal routes of mangrove genera occurring in both world regions may have followed three possible paths, and in either direction. Each was approximately the same distance, considering past conditions, but two further factors must be considered. One is the proximity of sites for growth and further dispersal, and the other is the suitability of climatic conditions along the way.

The first route, via the ancient Tethys Sea, was favoured in one hypothesis of mangrove evolution (McCoy & Heck, 1976). This route was tropical and it was apparently available until the Miocene. Potential problems for dispersal of mangroves were suggested by van Steenis (1962) however, who noted that dry climatic conditions would have severely restricted the range of suitable mangrove habitat.

The second route, across the east Pacific, was suggested but dismissed because of its need for 'island hopping' and long distance dispersal. Current phanerozoic maps (Fig. 1) however, show an almost continual coastline between eastern Asia and western North America during Late Cretaceous and Paleocene times. The tropics during that time were reportedly around 52 ° latitude North and South, and climate was subtropical at cooler sections of the route. This route therefore cannot be dismissed so easily although there is no fossil evidence to support it. There is evidence however, suggesting the limited western migration across the southern Pacific (Duke, 1992) of one mangrove species, *Rhizophora samoensis*, which is indistinguishable morphologically from *R. mangle* (Tomlinson, 1986). The presence of unoccupied suitable habitats within their range suggests a Late Cretaceous connection between western Pacific and Atlantic East Pacific, possibly crossing to South America via an old island archipelago present during the early formation of the Pacific Plate (Schlanger & Premoli-Silva, 1981; Schlanger *et al.*, 1981). The migration appears to have been one-way, and distances between the putative ancient islands was large enough to restrict all mangroves except the dispersal specialist, *Rhizophora*.

A third path around Antarctica or southern Africa has never been discussed, although evidence of *Avicennia* in Gondwana was accepted by Specht (1981) and Mepham (1983). Clearly, if *Avicennia marina* occurred there, it could have migrated around the subtropical-warm temperate shores of Antarctica, connecting between eastern South America and Australia at some time prior to the Eocene. There are no fossil records of *Avicennia* to support this idea, other than those disputed (Muller, 1981) ones from Western Australia (Churchill, 1973). Migration between southern Africa and Australia was also possible apparently via an island archipelago which included the Indian subcontinent during the Late Cretaceous (Kemp & Harris, 1974; Norton & Molnar, 1977). In any case, a southern connection appears to be a plausible alternative, comparable with other routes.

Evolution of *Avicennia*

Phylogenies from morphological and genetic evidence

The large polymorphic genus, *Avicennia*, is well-known ecologically, systematically and genetically (Duke, 1988, 1990a, 1990b, 1991, 1992) in comparison with other mangrove taxa. Accordingly, this genus has the best information for the development of a specific evolutionary model.

All *Avicennia* taxa in the world (Duke, 1991, 1992; Tomlinson, 1986) were evaluated in multivariate analyses using diagnostic morphological characters. The results show four major groupings of species (Fig. 3), including (1) *A. marina* and *A. alba*; (2) *A. officinalis* and *A. integra*; (3) *A. rumphiana*; and, (4) *A. germinans*, *A. schaueriana* and *A. bicolor*. The last group comprises all Atlantic East Pacific species. In this analysis, three groups were arranged around *A. rumphiana*, which could be interpreted as suggesting a central or intermediate phylogenetic role of this species. However, a detailed appraisal of anatomical features of Indo West Pacific species (Tan & Keng, 1965), suggested that *A. officinalis* was more primitive. The groups were also divided by flower-size, making two important associations within *Avicennia*, notably groups 1 and 3 with small flowers, and groups 2 and 4 with larger flowers.

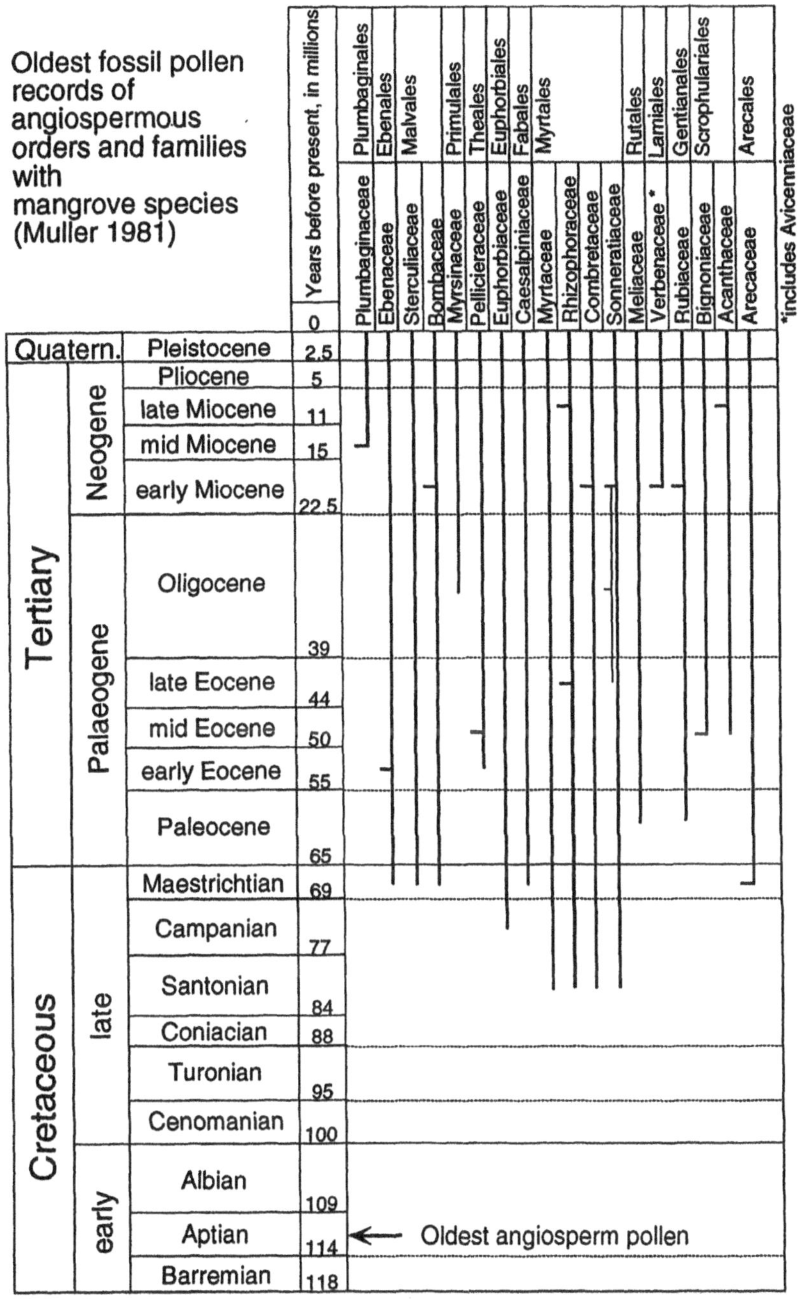

Fig. 2. Timelines of fossil pollen observations for all orders and families of angiosperm mangrove plants reviewed by Muller (1981). The oldest record of angiosperm pollen, accepted by Muller, is included for reference. Note: solid lines depict oldest records of mangrove families; bar flags depict oldest records of mangrove genera; and, dashed lines depict oldest records of mangrove orders.

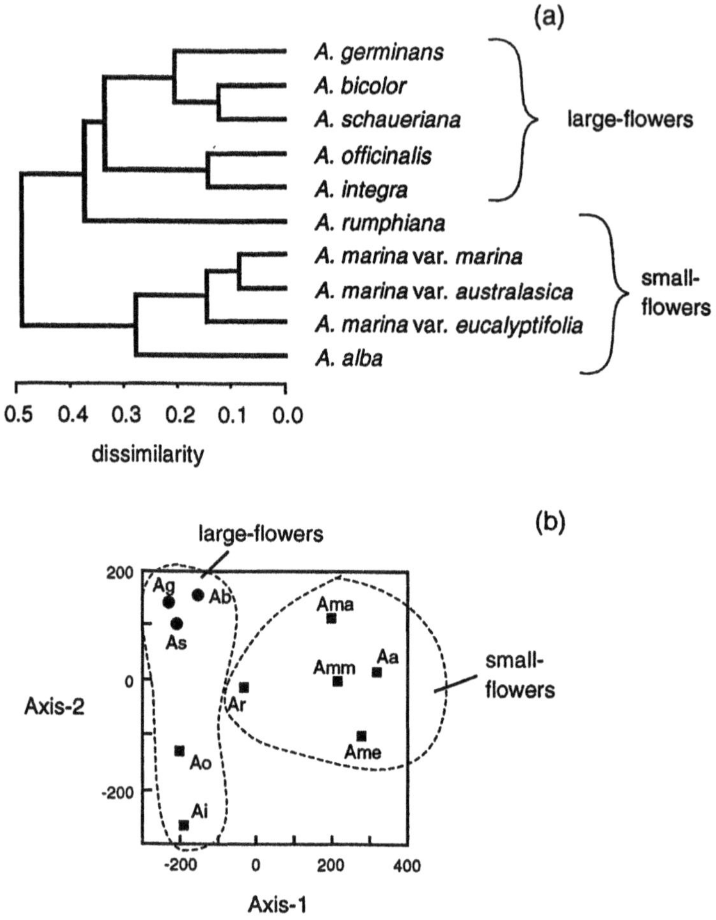

Fig. 3. (a) Dendrogram showing fusion sequence using morphological characters for major *Avicennia* taxa in the world. Data consisted of ordered multistate attributes of major morphological characters (Duke, 1988). The cluster analysis used the UPGMA method on dissimilarity measures derived from Gower's algorithm. Note, taxa are also grouped as small and large-flower forms. (b) Plot of principal coordinate analysis of morphological characters for major *Avicennia* taxa in the world, as listed in Fig. 3a. The analysis used dissimilarity measures derived from Gower's algorithm. Note, circles depict Atlantic East Pacific taxa, squares depict Indo West Pacific taxa, and all are also grouped as small and large-flower forms.

Detailed studies of the widespread small-flower species, *A. marina*, revealed three varieties based on morphology (Duke, 1990a), electrophoretic patterns and carbohydrate composition (Duke, 1988). Measures of genetic identity, determined by electrophoresis suggest an order of phylogenetic derivation of *A. marina* varieties from var. *marina*, to var. *australasica*, to var. *eucalyptifolia*. This was deduced from the first and second criteria for recent progenitor-derivatives which have less variation than progenitors (Gottlieb, 1973; Crawford, 1983). Morphological differences also suggested an intermediary role of *A. marina* var. *marina* to other varieties (Duke, 1990a).

Limitations of dispersal and growth for Avicennia

The chief mode of reproduction in *Avicennia* is the sexual production of water-borne propagules. These apparently withhold root development for around four days, depending on salinity and temperature, after which they sink (Steinke, 1975, 1986). This would limit dispersal to within 100–200 nautical miles in average sea conditions, influenced by currents and wind-blown drift. Dispersal is also limited by the inability of adults to reproduce in the colder climates of higher latitudes (Duke, 1990b). Growth is limited further by salinity conditions (Burchett *et al.*, 1984; Clough 1984), and each of these characters is expected to differ for each species. For example, *A. marina* has a wide estuarine range upriver from the mouth, while *A. integra* has a much smaller range midway in mostly hyposaline conditions (Duke, 1992). This has the effect of limiting the latter species to estuaries with continuous, although variable, freshwater input. All these factors are subject to present-day geological and climatic conditions; and the distributional disjunctions observed now between estuaries, might be the result of differing conditions in the past.

Genetic variation is also an important consideration, although it is believed to be relatively stable. Evidence of genetic stability is apparently shown in two examples where diagnostic characteristics were maintained in populations believed to have been isolated for approximately 40 million years, notably: *A. germinans* in North America and west Africa (Tomlinson, 1986); and, *A. marina* var. *australasica* in south-eastern Australia and New Zealand (Duke, 1991).

Disjunctions and discontinuities of Avicennia distribution

Evidence of disjunctions and discontinuities in distributions are found for all levels of species occurrence; notably global, regional, and within regions. The division of *Avicennia* species within two different hemispheres of the world represents a global disjunction where no species are shared (Duke, 1992). They are isolated by natural barriers of both land and sea, preventing exchange of genetic material. Reasons for this occurrence might be explained by historical considerations providing the necessary genetic and physical links in the past. In this way, the two global regions are also thought to be centres of secondary radiation with lesser differences between taxa and other distributional disjunctions; based on smaller changes to both genes and distributional barriers. All this combines to create additional levels of complexity in the mosaic of present-day distributional anomalies.

Examples of disjunctions and discontinuities within regions are shown by the occurrence of *Avicennia marina* varieties in Australasia, *A. officinalis* in New Guinea, and *A. integra* in Australia (Duke, 1991, 1992). In general, historical changes in the Indo West Pacific are more complex than those in the Atlantic East Pacific (Briggs, 1987). This is reflected chiefly in respective numbers of species and is used as evidence for the centre-of-origin in hypotheses of mangrove evolution, specifically suggesting either locations in Indo-Malesia (van Steenis, 1962; Chapman, 1976, 1977), or Australasia (Specht, 1981; Mepham, 1983). The duality of these putative centres is reflected in equal numbers of species, and a high proportion (ca. 20%) of species with localized affinities and endemism (Duke, 1992). For *Avicennia*, this situation was partially shown with *A. alba* and *A. officinalis* predominantly found in Indo-Malesia, *A. integra* and *A. marina* var. *australasica* and *A. marina* var. *eucalyptifolia* in Australasia.

Avicennia marina var. *australasica* occurs in northern New Zealand and south-eastern Australia (Duke, 1991). These countries have not been sufficiently close for crossmigrations for at least 40 million years, and maybe even 65 million years, back to the Late Cretaceous. This all sounds rather unlikely considering the reported Early Cretaceous beginnings of the evolution of angiosperms, but some scenario like this does match the southward disposition of this taxon today. It also concurs with fossil *Avicennia*-like pollen recorded by Churchill (1973) from Late Eocene deposits in southwestern Australia. The strength of this latter evidence, as already noted, was weakened by doubts expressed by Muller (1981).

The richer bio-diversity of the Indo West Pacific region is also generally explained by the tectonic dispersal of continental fragments (Briggs, 1987). For example, the major zone of faunal and floral overlap observed between Malesia and Australasia presumably resulted from the relatively recent contact between these areas in the late Miocene. This zone included many closely related (sibling) species of mangroves with wider distributions in their respective areas (Duke, 1992).

This important discontinuity and zone of phylogenetic duality, however, is not particularly clear for species of *Avicennia*. The most common species, *A. marina*, occurs widely through both regions, and other species have overlapping and different Malesian or Australasian affinities. This is particularly evident around New Guinea where many mangrove species are distributed along either northern or southern coastlines according to their respective floral affinities (Duke, 1992). Therefore, while a species like *A. alba* (Indo Malesia to northern New Guinea only) might be considered part of the Indo Malesian flora, *A. officinalis*

(Indo Malesia to southern New Guinea only) is difficult to categorize. Similarly, the occurrence of *A. rumphiana* along both coastlines cannot be classified either. The reasons may be related to different stages of phylogenetic development and dispersal, possibly shown by ever greater respective eastern limits of occurrence, where greater ranges were observed in putatively younger ancestral forms, *A. officinalis*, *A. rumphiana*, and *A. alba*, respectively.

There are two major species groupings for Indo West Pacific species, represented by *A. marina* with small flowers, and *A. officinalis* with large flowers (Fig. 3). The latter group includes Atlantic East Pacific species, and I suggest that a common progenitor of this large-flower group migrated between regions. However, small-flower taxa are more widely distributed in the Indo West Pacific today, and they exclusively occupy all putative dispersal routes to extant, large-flower Atlantic East Pacific habitats. If this phylogenetic inference is correct, the range of large-flower forms must have decreased while that of the small-flower group increased. In any case, an evolutionary model for *Avicennia* must account for this distributional discontinuity.

The direction of dispersal between regions is another problem, depending on where *Avicennia* evolved first. There is little evidence to suggest where this might be, although the genus appears to have developed early in the evolution of angiosperms (Barlow, 1981), suggesting a South American, or western Gondwanan origin. In support of this idea, it is suggested that the breakup of the continents would have created vast inter-continental estuaries, slowly changing from riverine swamps and flood plains to more marine conditions. The largest example was the division of Africa and South America (Fig. 1), and considering the combined river catchments involved, it is difficult to imagine a better environment and circumstances for the evolution of mangrove characteristics. This would have taken place over millions of years, providing the opportunity for a wide range of riverine and terrestrial taxa to develop the specialized characters we see in mangroves today.

The earliest fossil evidence of Avicennia

The early evolution of flowering plants is currently unresolved, although there are some implications of both the area and the time of origin (Takhtajan, 1980; Barlow, 1981). The story is complicated however by possible polyphyletic beginnings with at least three major groups arising at different times and places (Krassilov, 1977), although this concept was dismissed by Cronquist (1981). In any case, taxa with tricolpate pollen (note: *Avicennia* pollen is tricolpate) first appeared in western Gondwana (Brenner, 1976) in the Early Cretaceous, and by the mid-Cretaceous they were dispersed widely (Barlow, 1981). Their radiation included Australia where tricolpate pollen first appeared in the latter part of the Early Cretaceous, around 115 million years ago (Dettman, 1981). This arrival in Australia was characterized by plants already showing wide ecological adaptation (Raven & Axelrod, 1974). There were two routes available to plants at that time. Firstly, a tropical one, from Africa via India and an island archipelago (Kemp & Harris, 1974), which reportedly remained open until the Late Cretaceous, around 65–70 million years ago. Second, there was a polar route from South America via Antarctica which remained open until the Oligocene, around 35 million years ago (Raven, 1979). The first route was limited by access between South America and Africa which ended much earlier, and the latter route was restricted by its mostly, at best, warm subtropical climate.

The scant older fossil record for *Avicennia*, summarized in Table 1, indicates a wide distribution extending from the Atlantic East Pacific to Indo West Pacific regions by the Eocene, around 40 million years ago. Observations of pollen by Muller (1964,, 1981) suggest a relatively late arrival in Malesia during the middle Miocene, around 20 million years ago. However, these records are likely to be remarkably conservative; for *Avicennia*, this might be a function of the active collection of pollen by insects, thus reducing its concentration in sediments and future fossil beds. In any case, fossil records for *Avicennia* are very limited, and some are considered doubtful. The remarkable report by Churchill (1973) of *Avicennia* pollen in south-western Australian Eocene deposits, was questioned by Muller (1981); such doubts must be resolved by both a re-assessment of the relevant samples, and the collection of new pollen material and macro fossils from more sites around the world.

A model for the evolution of Avicennia *taxa*

The model draws on the points already raised, notably the phylogenetic affinities of the different taxa, based on morphological and genetic evidence, the fossil record, and the most recent records of extant distribution. It is also described in the context of past geolog-

Table 1. Oldest fossil records of *Avicennia* and '-like' forms, noting: age (mya = million years ago), location and source reference.

Taxa	Component	Greatest age	mya	Site	Authority
A. nitidaformis	leaves	Eocene	38–55	Mississippi	Berry (1916)
A. eocenica	fruit	Eocene	38–55	Tennessee	Berry (1916)
Avicennia-like*	pollen	late Eocene	40	SW Australia	Churchill (1973), * doubted by Muller (1981)
Avicennia	pollen	early Miocene	20	Marshall Islands, west Pacific	Leopold (1969) in Muller (1981)
Avicennia	pollen	mid Miocene	14	Borneo	Muller (1964)
Avicennia	pollen	late Miocene	10	N South America	van Steenis (1969)
Avicennia	pollen	late Miocene	10	Nigeria	van Steenis (1969)
Avicennia	pollen	late Miocene	10	NW Borneo	Muller (1981)
A. miocenica	leaves	Miocene	6–22	Columbia	Berry (1936)
A. lanceolata	leaves	Tertiary	3–65	Columbia	Moldenke (1960)
Avicennia	pollen	Pliocene	4	Guyana	Wymstra (1971) in Muller (1981)
A. germinans	leaves	Pleistocene	<2.5	Trinidad	Moldenke (1960)

ical and climatic conditions as presently known (e.g. Smith *et al.*, 1981; Briggs, 1974, 1987). This evaluation is also taken in appreciation of the evolution of other plant and animal groups, as described by Briggs (1987). Clearly, there is great benefit in comparing other biota whose fossil records are more complete and informative. This is especially so for biota having similar dispersal limitations, and similarly demonstrating the biological effectiveness of putative dispersal routes or barriers.

A western Gondwanan origin of Avicennia

I suggest that the earliest *Avicennia* ancestors appeared during the Early Cretaceous, along the north-western coast of Gondwanaland. This is based chiefly on two pieces of evidence. First, this was the site of the earliest angiosperm fossils. Second, this was the future site of the slow-expanding super-estuary between South America and Africa. This condition, occurring over a prolonged period, I believe to be essential for the selection and development of the specialized traits which characterize mangrove plants today. Based on the occurrence of large-flower forms in both regions today, it is assumed this form also had large-flowers. It might also have had glabrous inner surfaces of the petal lobes, a fully hairy radical on the propagule, and ovate to lanceolate leaves with rounded apices; possibly linking *A. schaueriana* in the Atlantic East Pacific and *A. officinalis* in the Indo West Pacific. It is of interest that these forms cannot be easily distinguished on present botanical descriptions. I propose that these represent western and eastern derivatives of an ancestral pan-Tethyian equatorial distribution during the mid Cretaceous (Fig. 4). It was apparently restricted to the northern shores of the southern continent. The preferred habitat was possibly in larger estuaries and in mid- to upstream locations like their present day relatives. *Avicennia bicolor* and *A. integra* are in this group, and their occurrences on the American Pacific coast and the north Australian coast respectively, perhaps represent outlier populations at the extremes of this ancient range. The events which would isolate these populations and allow their eventual speciation probably took place at different times. The first step occurred when the continents of Africa and South America separated during the Late Cretaceous, around 80–90 million years ago. Around this time the formation of an earlier isthmus between North and South America might also have isolated the future *A. bicolor* populations. The last of this group, *A. integra*, might have been isolated from *A. officinalis* in relatively recent times. This will be discussed later. Meanwhile, as the Gondwanan break-up continued into the Late Cretaceous, this further isolated the ancestral *Avicennia* populations. Those in South America stayed much as they are today, but those in north-western Africa, north-eastern Africa, India and Australia were subjected to major changes, leading them to change and speciate further. The populations in Africa were presumably divided in the north (and possibly the south) by very dry climates.

Populations in north-western Africa eventually came within dispersal range of the spreading fragments of the northern super-continent, Laurasia, during the Late Cretaceous (Fig. 4, also Fig. 1) and the Paleocene, around 60–80 million years ago. It is likely these were restricted to the western hemisphere of the northern continent by a channel connecting with the Arctic Sea (Fig. 4). In view of the warmer climatic conditions and the connection between Europe and North America at the time, it is proposed that Avicennias spread west to North America via southern Europe. By this time, the character of the ancestral species had changed to that of *A. germinans*. This might have occurred during the original crossing of the narrowing Tethys Sea when small founder populations flourished on the northern continent once they crossed from Africa. Or it could have occurred earlier when South American and African populations first became separated. In any case, by the late Eocene, around 50 million years ago, the wide- ranging *A. germinans* was split by the formation of the North Atlantic Ocean and the slow contraction of tropical climates, which had reached 50-60 ° N. This forced species toward the more equatorial distributions of today. During this time the older American isthmus had gone, but it was reformed in the late Pliocene, around 2–3 million years ago. This resulted in the isolation of *A. germinans* on Atlantic and Pacific coasts of North and Central America.

In the east, the situation became more complex as the breakup of the ancient supercontinents continued, and warmer climates contracted. I suggest a model that took place in two stages, based on the development and dispersal of large and small-flower forms. The large-flower, *A. officinalis*, ancestor probably began its crossing to the northern continent and Malesia during the Late Cretaceous, around 65–100 million years ago, as the Indian sub-continent brushed past the north-eastern part of Africa (Norton & Molnar, 1977). Apparently, this enabled such a taxon with relatively poor dispersal abilities and restricted habitat requirements to invade the northern super-continent, notably in the area of Indo Malesia. When this took place it was probably isolated from the *A. germinans* progenitor to the west, initially by a physically divided east and west sides of Africa, and later by dry climatic conditions (van Steenis, 1962). Presumably, the species became extinct in east Africa as conditions dried. Once reaching south-eastern Laurasia in the Eocene, around 38-54 million years ago, it spread relatively quickly east along the wet tropical southern shores of the old Sund Peninsula. Later, with the approach of the Australian continent in the Miocene, around 10-27 million years ago, it probably spread there via New Guinea. The retraction of this range later on, presumably left the populations in northern Australia which became *A. integra*.

The appearance of small-flower taxa

Meanwhile, I suggest that the evolution of small-flower taxa took place around the south-eastern shores of the ancient Tethys Sea, during the mid Cretaceous. At this time, the eastward spread of large-flower forms had reached north-eastern Africa. Evolution of the first small-flower form, *A. marina*, possibly took place in one of two very different ways however, depending on the progenitor. In each case, it was a large flower-form, but it could have been either *A. officinalis* or *A. schaueriana*, depending on whether it migrated around the northern or southern shores of Africa, respectively. The northern route provides a simple model with only one dispersal route, but the southern route would help explain apparent comparable cold tolerances in *A. schaueriana* and *A. marina*, shown by their extreme southern distributions in respective hemispheres (Duke, 1991, 1992). In the first case, founder populations possibly formed on offshore islands, via Madacascar and India. In the second case, founder populations might have been established around the southern coast of Africa.

In either case, it seems likely that the taxon reached the Indian subcontinent prior to contact with north-east Africa. After its contact with Africa, India was also colonized by the large-flower form, *A. officinalis*. This being the case, *A. marina* would then have followed two migration routes, one north to Malesia, and the other south-east to Australia and New Zealand. This southern migration probably took place across an old archipelago which crossed the developing Indian Ocean (Kemp & Harris, 1974) linking Australia and Africa until the Late Cretaceous. After *A. marina* arrived in Australia, the continent apparently became fragmented as sea levels rose in the mid to Late Cretaceous, causing further isolation of these populations; presumably leading to the formation of the two other varieties in northern Australia, and south-eastern Australia and New Zealand. The New Zealand population has been isolated ever since. Meanwhile those in Australia were re-united preventing further genetic separation, while retaining their sub-specific identities. At

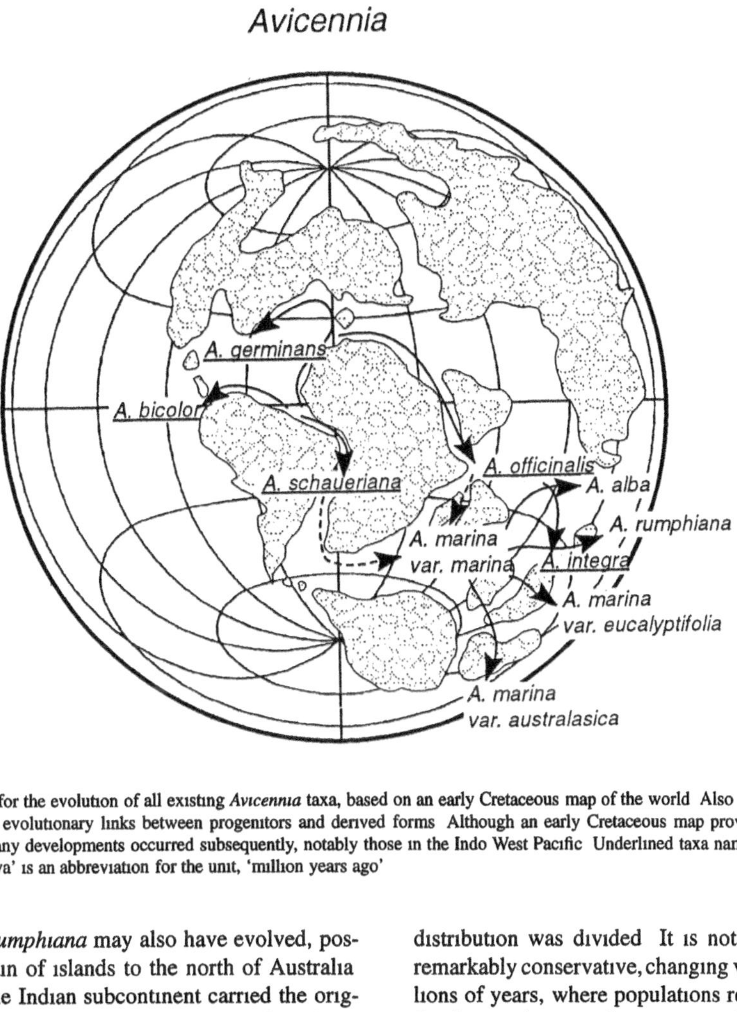

Fig 4 A model for the evolution of all existing *Avicennia* taxa, based on an early Cretaceous map of the world Also refer to Fig 1 Arrows depict spatial and evolutionary links between progenitors and derived forms Although an early Cretaceous map provides the basis for this representation, many developments occurred subsequently, notably those in the Indo West Pacific Underlined taxa names depict large-flower species Note 'mya' is an abbreviation for the unit, 'million years ago'

this time, *A rumphiana* may also have evolved, possibly on a chain of islands to the north of Australia Meanwhile, the Indian subcontinent carried the original A marina variety together with *A officinalis* to Malesia and southern Asia During this crossing, the most recent species of *Avicennia*, *A alba*, possibly developed from founder populations of *A marina* as the Asian continent drew nearer

Conclusion

In developing this model, the chief principle was to have new species and varieties evolve only whilst ancestral species invaded new sites, or, after a wider distribution was divided It is notable that taxa are remarkably conservative, changing very little over millions of years, where populations remained large and fixed in one location In any case, where small populations became genetically isolated, either by founder or vicariant events, this may have allowed greater expression of their mutant genes (Takhtajan, 1980) These were normally present in larger populations, but their low frequency precluded significant expression Of course, this process was driven by both the changing climate and by the breakup of the Gondwanan supercontinent This may also be the reason why the mangrove flora of today is so rich in species and diversity of morphological characters In the first instance, the expanding super-estuary provided the geographic circumstances for the evolution of major mangrove characteristics Second, the dispersal of continental fragments made sure those new forms and their derivatives were distributed around the world

Mangrove forests are as complex and variable in makeup and evolutionary processes as any forest habitat Plant species comprising these ecosystems have come from a variety of ancestral groups, and their co-occurrence in the present time should not be taken as a measure of common evolution or origin Clearly, changes taking place in this habitat were greatly influenced by the massive displacement of continents during the last 100 million years Mangrove evolution, diversification and dispersal apparently were accelerated by continental drift Although perhaps not all taxa were involved in these changes, there appears to be no other explanation for the extant distributional ranges and disjunctions of most, especially polyspecific, genera This being the case, we may find that some groups will extend back further than our present fossil records indicate, and in different places than we may at first expect

Acknowledgements

During the preparation of this article, the author was supported by the Marine Spill Response Corporation, USA, and the Smithsonian Tropical Research Institute, Panama My thanks also to the Hong Kong University of Science and Technology, particularly Dr Yuk-Shan Wong, and Dr S Y Lee of the University of Hong Kong, for their encouragement and support

References

Barlow, B A, 1981 The Australian flora its origin and evolution In, Flora of Australia, Australian Government Public Service, Canberra

Berry, E W, 1916 The lower Eocene floras of south-eastern North America U S Geological Survey Professional Paper 91 347–348, pl 104, 107, Fig 4, 6

Berry, E W, 1936 Miocene plants Bull Torrey Botanical Club 63 65, pl 3, Fig 1 3 Brenner, G J, 1976 Middle Cretaceous floral provinces and early migrations of angiosperms In C B Beck (ed), Origin and Early Evolution of Angiosperms, Columbia University Press, New York

Briggs, J C, 1974 Marine Zoogeography, McGraw-Hill, New York

Briggs, J C, 1987 Biogeography and plate tectonics In, Developments in Palaeontology and Stratigraphy, 10 Elsevier, Amsterdam

Burchett, M D, C D Field & A Pulkownik, 1984 Salinity, growth and root respiration in the grey mangrove, *Avicennia marina* Physiol Pl 60 113–118

Chapman, V J, 1976 Mangrove Vegetation Cramer, Vaduz

Chapman, V J, 1977 Introduction In Wet Coastal Ecosystems Ecosystems of the World, 1 Elsevier, Amsterdam

Churchill, D M, 1973 The ecological significance of tropical mangroves in the early Tertiary floras of southern Australia Geological Society of Australia Special Publication 4 79–86

Clough, B F, 1984 Growth and salt balance of the mangroves *Avicennia marina* (Forsk) Vierh and *Rhizophora stylosa* Griff in relation to salinity Australian J Pl Physiol 11 419–430

Crawford, D J, 1983 Phylogenetic and systematic inferences from electrophoretic studies In S D Tanksley & T J Orton (eds), Isozymes in Plant Genetics and Breeding, Elsevier, Amsterdam, Part A 257–288

Cronquist, A, 1981 An Integrated System of Classification of Flowering Plants Columbia University Press, New York

Dettman, M E, 1981 The Cretaceous flora In A Keast (ed), Ecological Biogeography of Australia W Junk, The Hague

Duke, N C, 1988 The mangrove genus *Avicennia* (*Avicenniaceae*) in Australasia, Ph D Thesis Botany Dept, James Cook University of North Queensland, 195 pp

Duke, N C, 1990a Morphological variation in the mangrove genus *Avicennia* in Australasia systematic and ecological considerations Australian Systematic Bot 3 221–239

Duke, N C, 1990b Phenological trends with latitude in the mangrove tree *Avicennia marina* J Ecol 78 113–133

Duke, N C, 1991 A systematic revision of the mangrove genus *Avicennia* (*Avicenniaceae*) in Australasia Australian Systematic Bot 4 299–324

Duke, N C, 1992 Mangrove Floristics and Biogeography In A I Robertson & D M Alongi (eds), Tropical Mangrove Ecosystems, Coastal and Estuarine Studies Series, American Geophysical Union, Washington, D C, 41 63–100

Gottlieb, L D, 1973 Genetic differentiation, sympatric speciation, and the origin of a diploid species of *Stephanomeria* Am J Bot 60 545–553

Heywood, V H, 1978 Flowering Plants of the World Mayflower Books, New York

Hutchinson, J, 1973 The Families of Flowering Plants, 3rd edn Oxford University Press, London

Kemp, E M & W K Harris, 1974 The vegetation of Tertiary islands on the Ninetyeast Ridge Nature 258 303–307

Krassilov, V A, 1977 The origin of angiosperms Bot Rev 43 143–176

Mepham, R H, 1983 Mangrove floras of the southern continents, Part 1, The geographical origin of Indo-Pacific mangrove genera and the development and present status of Australian mangroves South African J Bot 2 1–8

McCoy, E D & K L Heck Jr, 1976 Biogeography of corals, seagrasses, and mangroves an alternative to the center of origin concept Syst Zool 25 201–210

Moldenke, H N, 1960 Material towards a monograph of the genus *Avicennia* L I and II Phytologia 7 123–168, 179–252, 259–263

Muller, J, 1964 A palynological contribution to the history of the mangrove vegetation in Borneo In L M Cranwell (ed), Ancient Pacific Floras University of Hawaii Press, Honolulu

Muller, J, 1981 Fossil pollen records of extant angiosperms Bot Rev 47 1–142

Norton, I O & P Molnar, 1977 Implications of a revised fit between Australia and Antartica for the evolution of the Eastern Indian Ocean Nature 267 338–340

Rabinowitz, D , 1978 Dispersal properties of mangrove propagules Biotropica 10 47–57

Raven, P H , 1979 Plate tectonics and southern hemisphere biogeography In K Larsen & L B Holm-Nielsen (eds), Tropical Botany Academic Press, London

Raven, P H & D I Axelrod, 1974 Angiosperm biogeography and past continental movements Ann Missouri Bot Gardens 61 539–673

Schlanger, S O , H C Jenkyns & I Premoli-Silva, 1981 Volcanism and Vertical Tectonics in the Pacific Basin Related to Global Cretaceous Transgressions Earth and Planetary Science Letters 52 435–449

Schlanger, S O & I Premoli-Silva, 1981 Tectonic, Volcanic and Paleogeographic Implications of Redeposited Reef Faunas of Late Cretaceous and Tertiary Age from the Nauru Basin and Line Islands In, Initial Reports of the Deep Sea Drilling Project 61 817–827

Smith, A G , A M Hurley & J C Briden, 1981 Phanerozoic paleocontinental world maps In, Cambridge Earth Science Series Cambridge University Press, Cambridge

Soto, R & L F Corrales, 1987 Variacion de algunas caracteristicas foliares de *Avicennia germinans* (L) L en un gradiente climatico y de salinidad Revista de Biologia Tropical 35 245–256

Specht, R L , 1981 Biogeography of halophytic angiosperms (saltmarsh, mangrove and sea-grass) In A Keast (ed), Ecological Biogeography of Australia W Junk, The Hague

Steenis, C G G J van, 1962 The distribution of mangrove plant genera and its significance for palaeogeography Proc Kon Net Amsterdam, Series C 65 164–169

Steenis, C G G J van, 1969 Plant speciation in Malesia with special reference to the theory of non-adaptive saltatory evolution Biol J Linn Soc 1 110

Steinke, T D , 1975 Some factors affecting dispersal and establishment of propagules of *Avicennia marina* (Forsk) Vierh In G Walsh, S Snedaker & H Teas (eds), Proceedings of the International Symposium in Biology and Management of Mangroves 2 402–414

Steinke, T D , 1986 A preliminary study of buoyancy behaviour in *Avicennia marina* propagules South African J Bot 52 559–565

Tan, A S & H Keng, 1965 Comparative morphological studies of four species of *Avicennia* (Verbenaceae) in Singapore J Singapore Nat Acad Scien 1 8–29

Takhtajan, A , 1980 Outline of the Classification of Flowering Plants (Magnoliophyta) Bot Rev 46 225–359

Tomlinson, P B , 1986 The Botany of Mangroves Cambridge University Press, Cambridge, 413 pp

Hydrobiologia **295**: 183–191, 1995.
Y. S. Wong & N. F. Y. Tam (eds), Asia-Pacific Symposium on Mangrove Ecosystems.

Temporal distribution and abundance of shrimp postlarvae and juveniles in the mangroves of Muthupet, Tamilnadu, India

R. Mohan[1], V. Selvam[2] & J. Azariah[2]
[1]*Sultan Qaboos University, Department of Fisheries Science & Technology, P.O. Box 34, Al-Khod 123, Muscat, Sultanate of Oman*
[2]*Department of Zoology, University of Madras, Guindy Campus, Madras 600 025, India*

Key words: penaeid shrimp, postlarvae, juvenile, abundance, mangroves, India

Abstract

The temporal distribution patterns of the predominantly occurring postlarvae and juvenile shrimps in the mangrove and associated habitats of Muthupet, India were investigated for two years from February 1984 to January 1986. Among the eight commercially important species recorded, *Penaeus indicus* H. Milne Edwards, *P. merguiensis* De Man, *P. monodon* Fabricus and *Metapenaeus dobsoni* (Miers) were predominant. The postlarval recruitment size varied with species: *P. indicus* and *P. merguiensis* recruited at the size of 9–11 mm total length (TL), *P. monodon* at 12–14 mm TL and *M. dobsoni* at 4–6 mm TL. The species *P. indicus*, *P. merguiensis* and *M. dobsoni* were observed continuously throughout the study period with maximum abundance occurring from July to September in 1984–85 and from August through October in 1985–86. *P. monodon* occurred seasonally from November to January in both years. Postlarvae and juvenile catches were low during low salinity and high salinity periods and a higher density was observed in the months of moderate water salinity. Large numbers of *P. indicus*, *P. merguiensis* and *M. dobsoni* clearly showed the preference to the detritus rich muddy substrate, whereas *P. monodon* did not show any preference and was equally abundant over different substrate types.

Introduction

India, a major shrimp supplier in the Indo-Pacific region, generates considerable revenue through export of shrimp to Japan, USA, Europe and other countries. The annual production of penaeid shrimps in India was estimated to be 117 000 tonnes in the year 1985. The value of Indian shrimp exported to United States alone was estimated to be US $ 56.9 million (U.S. Department of Commerce, 1988).

Most of the penaeid shrimp species in India have a similar life history pattern. Adults spawn in deep waters, and post larvae and juveniles use inshore estuarine and coastal waters as nursery habitats for their early development. The use of mangroves as nursery grounds for penaeid prawns in India (Suseelan & Kathirvel, 1980; Chakroborthy *et al.*, 1982; Sambandam *et al.*, 1982; Babu & Babu, 1986; Silas, 1986; Subramanian, 1987; Rao, 1990) and in other parts of the world (de Freitas, 1986; Stoner & Zimmerman, 1988; Vance *et al.*, 1990) has been reasonably well studied. A positive statistical correlation between commercial yields of shrimp and extent of mangrove forests has been well established. (MacNae, 1974; Martosubroto & Naamin, 1977; Turner, 1977; Staples *et al.*, 1985; Silas, 1986; Chong *et al.*, 1990). Destruction of mangroves was considered as one of the major factors contributing to the decline in prawn catches in El Salvador (Daugherty, 1975). High abundance of juveniles in tropical estuaries and success of shrimp fishery in the tropical inshore areas has been attributed to high concentrations of mangrove-derived detritus in the nearshore ecosystems (Stoner & Zimmerman, 1988; Longhurst & Pauly, 1987).

In India, 256 000 hectares (Blasco, 1976) of mangrove swamps present along the east and west coasts play a vital role as nursery grounds for penaeid shrimps. Surveys of shrimp fry resources in mangrove estuaries on the east coast of India indicated catch rates of up to 4600 and 6200 fry of *P. monodon* and *P. indi-*

cus respectively per man hour of operation (ICAR, 1978). Postlarvae and juveniles have been traditionally exploited by artisanal fisheries and as wild seeds for aquaculture (Macintosh, 1982; Silas, 1986). Fishing at this stage reduces the recruitment of the next fishery, ultimately affecting the spawning potential to the extent that postlarval recruitment could be diminished (Garcia, 1988). Despite many studies of prawn seed resources on estuarine mangroves along Indian coastline (Macintosh, 1982), very little information is available on the mangrove lagoon of Muthupet at the Palk Straight along the east coast.

The present paper is a study describing the species composition and quantitative monthly variation of postlarval and early juvenile shrimps in mangrove estuarine areas of Muthupet, Tamil Nadu, India.

Description of the study sites

The mangrove swamp of Muthupet (lat. 10 ° 25′N; long. 79 ° 30′E) is located at the southern end of the Cauvery delta, 73 km from Thanjavur in the State of Tamil Nadu, India (Fig. 1). The drainage tributaries of the river Cauvery enter the Palk strait through a lagoon in this region. An extensive mangrove swamp occurs in the deltaic interface area between the drainage tributaries, lagoon and sea. The swamp is composed of many islands and islets that are traversed by numerous interconnecting channels, creeks, inlets and rivulets. In the river Koreiyar, which is the main source of freshwater to Muthupet mangrove lagoon, a weir was installed at about 11 km upstream to control and augment water for irrigation in the upland fields near the rivers.

Materials and Methods

The two-year study from February 1984 to January 1986 was conducted at three sampling stations (Fig. 1) established along the river Koreiyar and lagoon, surrounded by pure stands of *Avicennia marina* (Forsk.) Vierh. The water depth at the sampling sites varied from 0.5 to 1.5 m. Station I was in a bayou connected to the river Koreiyar and characterized by muddy substratum covered by a thick stratum of detrital peat material. Pneumatophores of mangroves were present and algal vegetation was not observed. Station II was in the river Koreiyar, 600 m downstream from Station I. The substratum was muddy and stratum containing very thin layer of detritus. Station III was located in the lagoon where the sandy substratum was densely covered by algal vegetation dominated by *Gracilaria* and *Enteromorpha* spp.

A rectangular pocket seine net having a mesh size of 2 mm, 2 m in length and 1 m deep was used for the postlarval and juvenile collections. For each sampling, the net was dragged twice by two people, starting from shallow waters and continuing to the adjacent shore, covering a total area of 10 m^2. The collections were carried out for a period of two years on a fortnightly basis. The collected samples were preserved in 5% formalin for further analysis in the laboratory. The postlarvae and early juveniles less than 20 mm total length (TL) were sorted out and identified to species by their color, location and distribution of the chromatophore on the various parts of the body, length of the rostrum, presence or absence of rostral spines, spinnules on dorsal carina of the 6th abdominal segment and spine on the carapace. The descriptive guidelines outlined by Muthu (1978) and Motoh & Buri (1980) were followed.

Surface water temperature and salinity were also measured during sampling with a stem thermometer and temperature compensated refractometer. Rainfall data recorded at Meteorological Department, Adirampattinam were used. Analysis of Variance (ANOVA) was carried out by SAS General Linear Models (GLM) procedures to examine the effects of year, month, station and interactions on temperature, salinity and numerical abundance of shrimps (SAS, 1985). Following ANOVA results, Tukeys' Studentized Range Test was also conducted to compare the specific variability of shrimp density among the stations (SAS, 1985).

Results

Annual precipitation was 1412 mm in 1984–85 and 1285 mm in 1985–86 (Fig. 2). The major percentage of rainfall in this area was during the monsoon months between October and December. However, unprecedented rainfall was recorded in February (284 mm), March (160 mm) and July (230 mm) during the year 1984–85.

The seasonal variation in mean surface water temperature (Fig. 2) ranged from 24.2 to 32.6 °C in 1984-85, a difference of 8.2 °C, whereas in 1985–86, it ranged between 28.6 and 31.4 °C, a difference of only 2.8 °C. In both years, annual lows occurred in February and highs in May. ANOVA test (Table 1) showed highly significant ($P<0.01$) interannual and month-

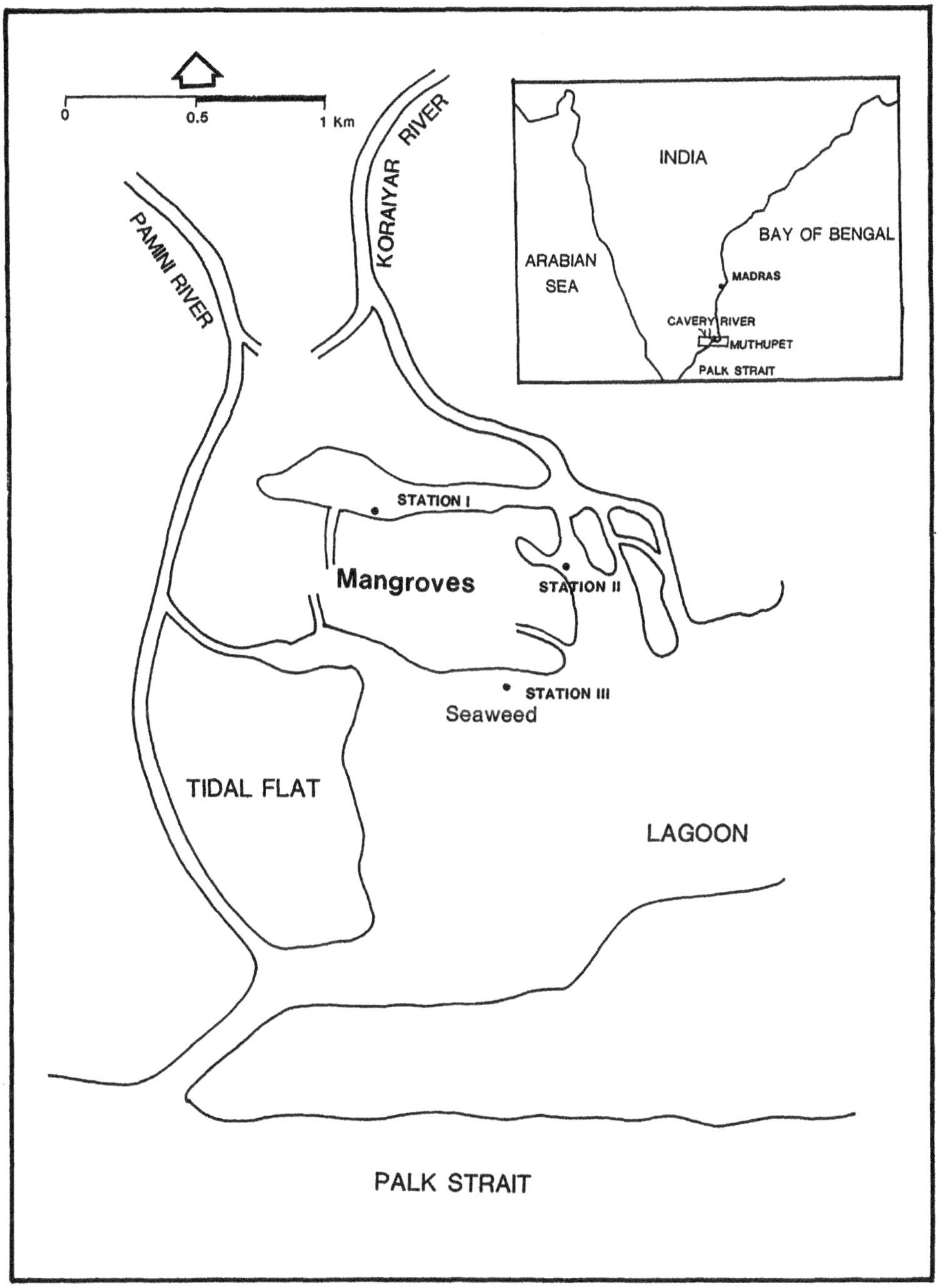

Fig. 1. Location of the shrimp collection sites in the mangrove areas at Muthupet, Tamil Nadu, India.

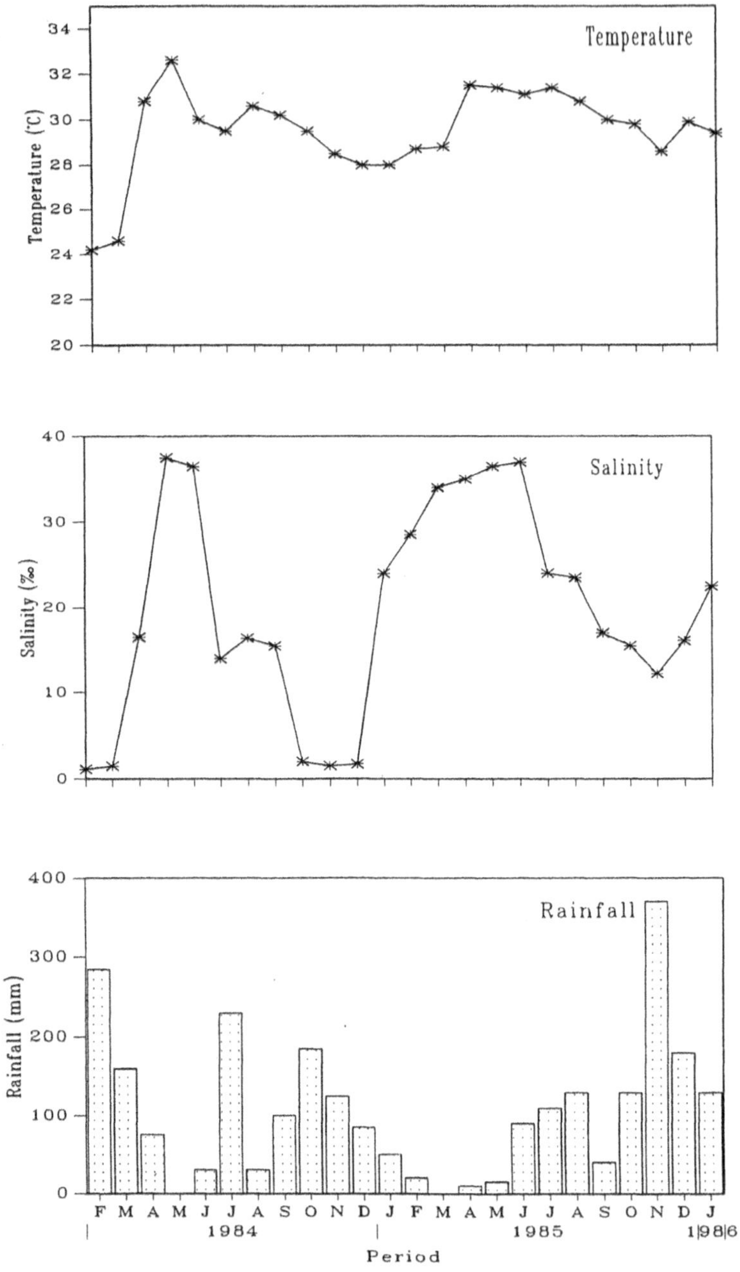

Fig. 2. Monthly water temperatures (°C), salinities (‰) and rainfall (mm) recorded at Muthupet, India during the period from February 1984 to January 1986.

Table 1. Analysis of Variance (ANOVA) of the effects of year, month, station and interactions on temperature and salinity in the mangroves of Muthupet, India from February 1984 through January 1986. (d.f. = degrees of freedom, SS = sums of squares, MS = SS/d.f., F = MSamong/MS within

Source	d.f	SS	F
Temperature			
Year	1	29.26	4221.3**
Month	11	189.74	2488.37**
Station	2	0.01	0.98
Year ⊗ Month	11	47.97	629.16**
Year ⊗ Station	2	0.01	0.54
Month ⊗ Station	22	0.27	1.79
Error	22	0.15	
Salinity			
Year	1	2090.89	31244.98**
Month	11	6706.33	9110.49**
Station	2	0.08	0.62
Year ⊗ Month	11	1771.44	2406.49**
Year ⊗ Station	2	0.19	1.45
Month ⊗ Station	22	1.58	1.08
Error	22	1.47	

** = $P < 0.01$

ly variations in temperature. Station differences were never greater than 0.4 °C and ANOVA test yielded no significant variation.

Salinity ranged from 1.1 to 37.6‰ in 1984–85 and 12.2 to 37.6‰ in 1985–86. In general, annual low salinity occurred in monsoon months from October through December. Following the monsoon season, salinity steadily increased and reached its maximum in May. Then, there was a gradual decline in salinity from June to a minimum in the monsoon months. The decrease in salinity from June was due to river discharge and land drainage following the opening of Mettur dam for irrigation in the Cavery basin during southwest monsoon rains in the nearby western state of Karnataka. However, low salinity during February, March and July 1984 was apparently due to unprecedented heavy rains and floods. ANOVA test indicated highly significant ($P<0.01$) effect of year and month on salinity (Table 1). The variation in salinity among the stations was not significant.

The postlarvae and juveniles of eight species of penaeid shrimps viz. *P. indicus*, *P. merguiensis*, *P. monodon*, *P. semisulcatus* De Haan, *M. dobsoni*, *M. monoceros* (Fabricus), *M. brevicornis* (H. Milne Edwards) and *M. affinis* (H. Milne Edwards) were recorded. The study revealed the predominance of *P. indicus* (39.31%), *M. dobsoni* (27.66%), *P. merguiensis* (17.82%) and *P. monodon* (19.62%), totalling 95.41%. The remaining 4.59% was comprised of *P. semisulcatus* (1.33%), *M. monoceros* (2.27%), *M. affinis* (0.40%) and *M. brevicornis* (0.59%). *P. indicus* and *M. dobsoni* dominated the samples and were recorded continuously in all months of the entire study period. The size of the postlarvae recruited in the mangroves varied with species. Most of the *P. indicus* and *P. merguiensis* recruited at 9–11 mm size, *P. monodon* at 12–14 mm size and *M. dobsoni* at 4–6 mm size. The species *P. indicus*, *P. merguiensis* and *M. dobsoni* were more abundant in Stations I and II than Station III. But *P. monodon* catches did not show any variation in distribution among the stations.

P. indicus

The postlarvae and juveniles of this species were collected throughout the study period (Fig. 3). During 1984–85, the mean monthly density was maximum (17.3 m^{-2}) during August and minimum during November (1.5 m^{-2}). In 1985–86, mean monthly density was higher in October (12.3 m^{-2}) and lower in February (3.9 m^{-2}). ANOVA results indicated that the density significantly varied with year, month, station and year-month interaction (Table 2). Large numbers were caught in Stations I (8.8 m^{-2}) and II (7.5 m^{-2}) than Station III (2.4 m^{-2}) (Fig. 4). Tukeys' test (Table 3) showed the collections made at Stations I and II were similar and significantly different from the catches from Station III.

P. merguiensis

This species showed maximum mean monthly density (6.7 m^{-2}) during August and minimum (0.2 m^{-2}) in December during 1984–85. In 1985–86, density maximum was noticed in October (7.1 m^{-2}), decreasing to minimum (3 m^{-2}) in April (Fig. 3). Differences in shrimp abundance were statistically significant ($P<0.05$) with year and highly significant ($P<0.01$) with month and station (Table 2). High concentrations at Stations I (4.2 m^{-2}) and II (3.3 m^{-2}) were evident (Fig. 4) and catches were significantly different from Station III (Table 3).

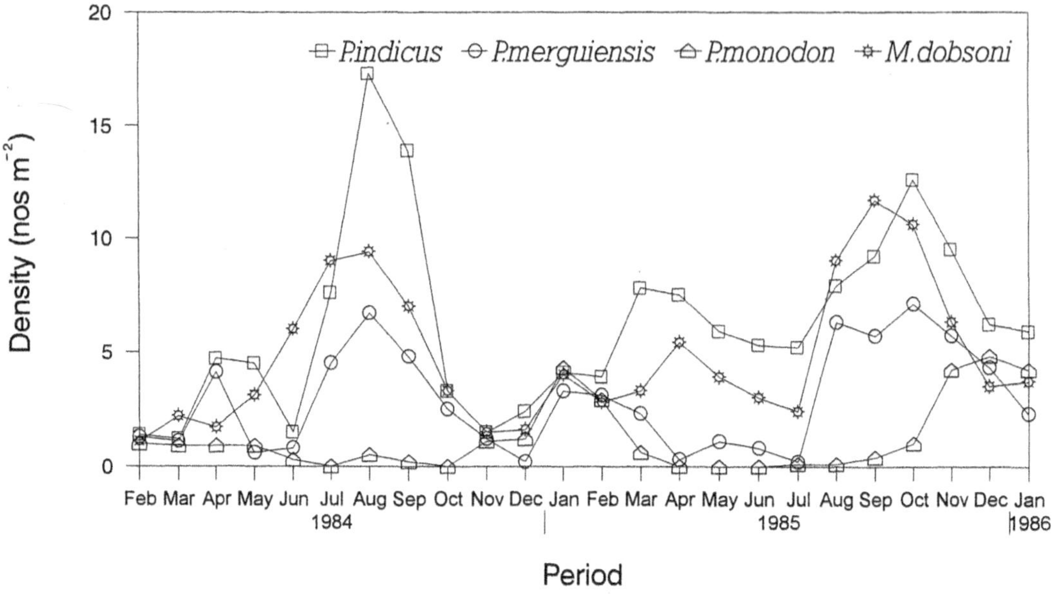

Fig. 3. Monthly mean density (nos m^{-2}) of penaeid shrimp postlarvae and early juveniles in mangroves at Muthupet, India from February 1984 through January 1986.

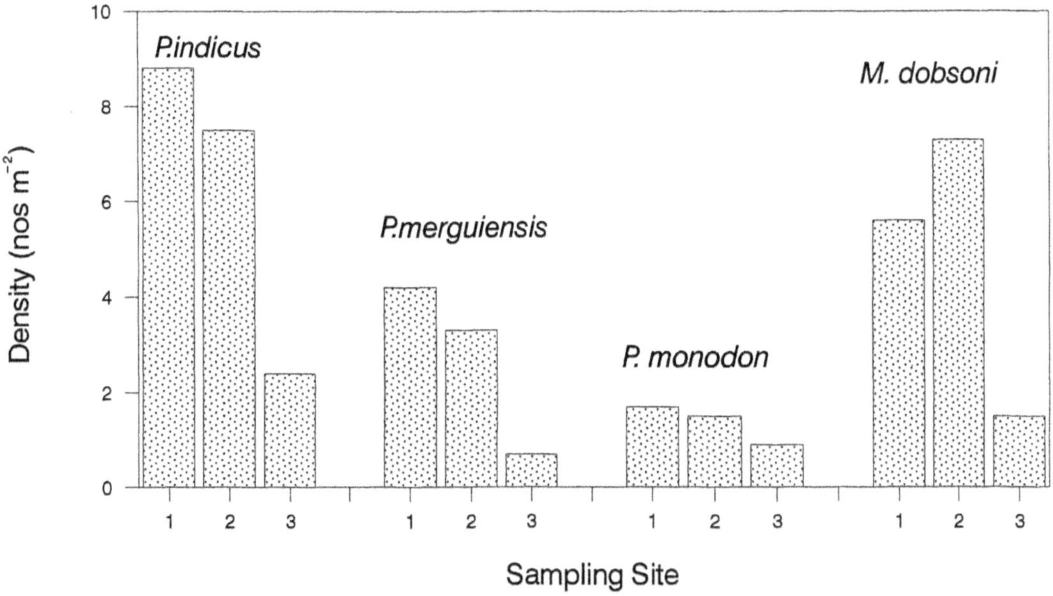

Fig. 4. Mean density (nos m^{-2}) of penaeid shrimp postlarvae and early juveniles collected at three sampling stations in mangroves of Muthupet, India from February 1984 through January 1986.

Table 2. Analysis of Variance (ANOVA) of the effects of year, month, station and interactions on numerical abundance of post larval and early juveniles in the mangroves of Muthupet, India from February 1984 through January 1986. (d.f. = degrees of freedom, SS = sums of squares, MS = SS/d.f., F = MSamong/MS within)

Source	d.f	SS	F
P. indicus			
Year	1	6981.68	8.83**
Month	11	61076.82	7.02**
Station	2	54827. 11	34.65**
Year ⊗ Month	11	46890.15	5.39**
Year ⊗ Station	2	1830.11	1.16
Month ⊗ Station	22	16897.56	0.97
Error	22	17404.56	
P. mergueinsis			
Year	1	1720.89	7.47*
Month	11	24202.11	9.55**
Station	2	15093.36	32.76**
Year ⊗ Month	11	10577.44	4.17**
Year ⊗ Station	2	1000.53	2.17
Month ⊗ Station	22	9445.97	1.86
Error	22	5068.14	
P. monodon			
Year	1	112.50	2.63
Month	11	19512.33	41.44**
Station	2	768.25	8.97**
Year ⊗ Month	11	2362.50	5.02**
Year ⊗ Station	2	102.25	1.19
Month ⊗ Station	22	2300.42	2.44
Error	22	941.75	
M. dobsoni			
Year	1	3120.50	7.52*
Month	11	40951.33	8.97**
Station	2	42672.00	51.42**
Year ⊗ Month	11	22772.17	4.99**
Year ⊗ Station	2	481.00	0.58
Month ⊗ Station	22	8261.67	0.90
Error	22	9129.33	

** $= P < 0.01$, * $= P < 0.05$

P. monodon

In both years, maximum mean monthly catches of 5.2 m^{-2} (1984–85) and 4.8 m^{-2} (1985–86) occurred in December (Fig. 3). Very few *P. monodon* were caught from April through July in both years. Highly significant ($P<0.01$) variation with month, station and year-month interaction was found in ANOVA test (Table 2).

Table 3. Comparison of shrimp postlarval and early juvenile mean densities (nos m^{-2}) at three sampling stations. Values in parenthesis are number of samples on which each mean is based. Means that are associated with the same letter (A or B) do not differ significantly as determined by Tukey's Studentized Range (HSD) test for mean comparisons (α = 0.05).

Species	Station I	Station II	Station III
P. indicus	8.8(24) A	7.5(24) A	2.4(24) B
P. mergueinsis	4.2(24) A	3.3(24) A	1.9(24) B
P. monodon	1.7(24) A	1.6(24) A	0.9(24) A
M. dobsoni	5.6(24) A	7.3(24) A	1.5(24) B

The mean density (Fig. 4) for Station I, II and III were 0.9 m^{-2}, 1.1 m^{-2} and 1.5 m^{-2}, respectively. This spatial distribution was not significantly varied as indicated by ANOVA (Table 2) and Tukeys' test (Table 3).

M. dobsoni

The mean monthly density of postlarvae and juveniles of *M. dobsoni* showed marked temporal and spatial variation. The monthly abundance (Fig. 3) was higher during August (9.4 m^{-2}) in 1984–85 and September (11.7 m^{-2}) in 1985–86. Minimum monthly density was observed in February in both years. ANOVA test (Table 2) indicated that the variation in shrimp abundance was significant ($P<0.05$) with year and highly significant ($P<0.01$) with month and station. Mean catch rates (Fig. 4) in Stations I (5.6 m^{-2}) and II (7.3 m^{-2}) were significantly higher and different from Station III (1.5 m^{-2}) (Table 3).

Discussion

In the present study the monthly variation in density of *P. indicus*, *P. merguiensis* and *M. dobsoni* was broadly similar. Major peak occurred from July through September in 1984–85 and from August to October in 1985–86. Minor peak appeared around March and April in both years. It is obvious that timing of the peak occurrence was more associated with moderate salinity (14–24‰) and that the poor recruitments were during the periods of both low and high salinity. Proper environmental conditions and some minimum standards in the health of the environment are required for the successful settlement of the penaeid seeds (Easo &

Mathew, 1989). Salinity differences have been found to influence postlarval recruitment of many penaeid shrimps (Garcia & Le Reste, 1981). Curtailment of *P. merguiensis* postlarval recruitment due to the onset of freshwater condition was reported in the Gulf of Carpentaria (Staples & Vance, 1987). A similar influence of salinity on the postlarve and juveniles was also observed in Cochin backwaters (Easo & Mathew, 1989) and Parangipettai coastal waters (Natarajan et a/, 1986). It has been reported that salinity more than temperature influenced the settlement of postlarvae in the Cochin backwaters (Kuttyamma, 1980). It is concluded that moderate rainfall extends the nursery area with preferred salinity regimes for successful settlement and growth (Barrett & Ralph, 1976; Garcia & Le Reste, 1981).

The density of *P. monodon* was found to be seasonal, occurring from December through February, which indicates seasonal spawning. The abundance of *P. monodon* was higher in 1985–86 than 1984–85. This may be due to existence of near freshwater conditions during the monsoon period of the year 1984–85 and supports the view that moderate salinity supports higher recruitment and settlement of postlarvae and juvenile penaeid shrimps.

It is clear from the data for the three sampling stations, the abundance of *P. indicus*, *P. merguiensis*, and *M. dobsoni* were more concentrated in Station I and II than in Station III. The substrate characteristics constitute the greatest physical variation in the sampling stations. Station I and II were characterised with detritus rich muddy substratum whereas in Station III, the substratum was sandy covered with dense seaweed. The detritus rich muddy substratum appears to be the preferred nursery habitat for postlarval and early juvenile shrimp of the above three species. The species *P. monodon* did not show any preference for types of substrate; its distribution being the same over the three stations.

Acknowledgments

The authors gratefully acknowledge the financial assistance of the Indian Council of Agricultural Research, New Delhi, India.

References

Babu, K. S. & K. S. Babu, 1986. Recruitment patterns of penaeid prawn post larvae into the Upputeru Estuary, India. In M. F. Thomson, R. Sarojini, R. Nagabhushanam, (eds.), Indian Ocean – Biology of Benthic Marine Organisms. Oxford & IBH Publishing Co., New Delhi: 345–350.

Barrett, B. B. & E. J. Ralph, 1976. Environmental conditions relative to shrimp production in coastal Louisiana. La. State Dep. Wildl. Fish. Tech Bull. 21: 1–20.

Blasco, F., 1976. Outlines of ecology, botany and forestry of the mangals of the Indian sub-continent. In V. J. Chapman (ed.) Ecosystems of the World, I. Wet Coastal Ecosystems. Elsevier, New York: 241–260.

Chakroborty, R. K., D. K. De & M. Subrahmanyam, 1982. Observations on the availability of Bagda *Penaeus monodon* (Fabricius) and Chapra, *Penaeus indicus* (H. Milne Edwards) seed in the Hoogly Estuary around Narpur and Uluberia, West Bengal. J. Inland. Fish. Soc. India 9: 181–183.

Chong, V. C., A. Sasekumar, M. U. C. Leh & R. D'Cruz, 1990. The Fish and Prawn Communities of a Malaysian Coastal Mangrove System, with Comparisons to Adjacent Mud Flats and Inshore Waters. Estuar Coast. Shelf Sci. 31: 703–722.

Daugherty, H. E., 1975. Human impact on the mangrove forests of El Salvador. In G. E. Walsh, S. C. Snedaker & H. J. Teas (eds.), Proc. Int. Symp. on Biology and Management of Mangroves, 1974. University of Florida, Gainesville: 816–824.

de Freitas, A. J., 1986. Selection of nursery areas by six southeast African Penaeideae. Est. Coast. Shelf Sci. 23: 901–908.

Easo, S. & K. J. Mathew, 1989. Seasonal variations in the abundance of penaeid prawn seed in relation to environmental parameters in the seed grounds of Cochin backwaters. Indian J. Fish. 36: 125–132.

Garcia, S., 1988. Tropical Penaeid Prawns. In J. A. Gulland (ed.) Fish Population Dynamics 2nd edition, John Wiley, Chichester: 219–249.

Garcia, S. & L. Le Reste, 1981. Life Cycles, dynamics, exploitation and management of coastal penaeid shrimp stocks. FAO Fish Tech. Pap. No. 302, 215 pp.

ICAR, 1978. Collected Reports from the Third Workshop of the All India Co-ordinated Research Project, Brackish Water Prawn and Fish Farming. Indian Council of Agricultural Research Publication.

Kuttyamma, V. J., 1980. Studies on the prawns and prawn larvae of the Kayamkulam Lake and Cochin backwaters. Bull. Dept. Mar. Sci., Univ. Cochin 11: 1–38.

Longhurst, A. R. & D. Pauly, 1987. Ecology of Tropical Oceans. Academic Press, London, 407 pp.

Macintosh, D. J., 1982. Fisheries and Aquaculture Significance of Mangrove Swamps with Special Reference to the Indo-West Pacific Region. In J. F. Muir & R. J. Roberts (eds.) Recent Advances in Aquaculture, Croom Helm, London: 3–85.

McNae, W., 1974. Mangrove forests and fisheries. Indian Ocean Progress Publications. (IOFC/DEV/74/34) Indian Ocean Fishery Commission, Rome, 35 pp.

Martosubroto, P. & N. Naamin, 1977. Relationship between tidal forest (mangrove) and commercial shrimp production in Indonesia. Mar. Res. Indonesia 18: 81–86.

Motoh, H & P. Buri, 1980. Identification of the post larval *Penaeus* (Crustacea, Decapoda, Penaeidae) appearing along shore waters. Q. Res. Rep. Aquacult. Dep. Southeast Asian Fish. Dev. Cent., 4: 15–19.

Muthu, M S , 1978 Specific identity of penaeid post larvae found in brackish areas CMFRI Bull 28 86–90

Natarajan, R , V S Chandrasekaran & S Shanmugam, 1986 Observations on the Ingress and Abundance of post-larvae of the penaeid prawn, *Penaeus indicus*, in Parangipettai, India coastal waters In M F Thompson, R Sarojini & R Nagabhushanam (eds), Indian Ocean – Biology of Benthic Marine Organisms Oxford & IBH Publishing Co , New Delhi 365–372

Rao, G S , 1990 An assessment of the penaeid prawn seed resource of the Godavari estuary and the adjacent backwaters Indian J Fish 37 99–108

Sambandam, K P , L Lawrence, J Noble, 1982 Some observations on penaeid prawn seed resources in the Vellar estuarine system (Porto Novo) Proc Symp Coast Aquacult 1 308–313

SAS, 1985 SAS User's Guide Statistics, Version 5 Edition, SAS Institute Inc Cary, NC 956 pp

Silas, E G , 1986 Significance of the Mangrove Ecosystem in the Recruitment of Fry and Larvae of Finfishes and Crustaceans along the East Coast of India, particularly the Sunderbans In Report of UNDP/UNESCO Workshop on the conversion of mangrove areas to aquaculture UNDP/UNESCO Regional Project RAS/79/002 19–34

Staples, D J & D J Vance, 1987 Comparative recruitment of the banana prawn, *Penaeus merguiensis* in five estuaries of the south-eastern Gulf of Carpentaria, Australia Aust J mar Fresh wat Res 38 29–45

Staples, D J , D J Vance & D S Heales, 1985 Habitat requirements of juvenile penaeid prawns and their relationship to offshore fisheries In P C Rothlisberg, B J Hill & D J Staples (eds), Second Australian national prawn seminar, Cleveland 47–54

Stoner, A W & R J Zimmerman, 1988 Food pathways associated with penaeid shrimp in a mangrove-fringed estuary Fish Bull USA 86 543–551

Subramanian, P , 1987 Spawner-recruit distribution of *Penaeus indicus* in Parangipettai coastal ecosystem J mar biol Ass India 29 23–36

Suseelan, C & M Kathirvel, 1980 Prawn seed calendars of Cochin backwaters In Proc Symp coast Aquacult 1 173–182

Turner, R E , 1977 Intertidal vegetation and commercial yields of penaeid shrimp Trans am Fish Soc 106 411–416

U S Department of Commerce, 1988 Fisheries of the United States, 1987 Current Fishery Statistics No 8700, Natl Mar Fish Serv , NOM, Wash , DC 115 pp

Vance, D J , M D E Haywood & D J Staples, 1990 Use of a Mangrove Estuary as a Nursery Area by Postlarval and Juvenile Banana Prawns, *Penaeus merguiensis* de Man, in Northern Australia Estuar coast Shelf Sci 31 689–701

Hydrobiologia **295**: 193–201, 1995.
Y. S. Wong & N. F. Y. Tam (eds), Asia-Pacific Symposium on Mangrove Ecosystems.

Community structure and standing crop biomass of a mangrove forest in Futian Nature Reserve, Shenzhen, China

N. F. Y. Tam[1], Y. S. Wong[2*], C. Y. Lan[3] & G. Z. Chen[3]
[1]*Department of Biology and Chemistry, City Polytechnic of Hong Kong, Tat Chee Avenue, Kowloon, Hong Kong*
[2]*Research Centre/Biology Department, Hong Kong University of Science and Technology, Clearwater Bay, Kowloon, Hong Kong*
[3]*Research Institute of Environmental Sciences, Zhongshan University, PRC*
(author for correspondence)*

Key words: mangrove forest, standing crop biomass, community structure, allometric regression

Abstract

The community structure and standing crop biomass of a mangrove forest in Futian Nature Reserve, Shenzhen, the People's Republic of China was studied. This mangrove, located towards the northern latitudinal limit of the mangrove development (22°N), had relatively simple structure and low diversity (the Shannon-Wiener index was 0.78). The three dominant species, namely *Aegiceras corniculatum*, *Kandelia candel* and *Avicennia marina*, possess importance values of 72, 19 and 9%, respectively. The average height of the mangrove community was 4.5 m with no vertical stratification. The values of tree density of *A. corniculatum* and *A. marina* were found to be 5290 and 260 ha^{-1}, respectively. The biomass of both *A. corniculatum* and *K. candel* was best estimated from regression equations using a combination of height and diameter at breast height as the independent variables. For *Avicennia marina*, there was no simple correlation between biomass and height or diameter. The regression models suggested by previous workers did not give satisfactory estimation of biomass of *A. marina* in this mangrove forest. The total biomass of this mangrove forest was 12.1 kg m^{-2}, with 73% of such production contributed by *A. corniculatum* and 8% by *A. marina*. Because of its small percentage, the inaccuracy in estimating biomass of *A. marina* did not affect the overall determination of biomass of the whole community. Average above-ground biomass was 8.7 kg m^{-2} (72% of the total biomass) and the major components were aerial woody tissues, stems and branches. The root:shoot ratio of this plant community was 0.4:1.

Introduction

Mangrove swamps are typical wetland ecosystems found in coastal deposits of mud and silt throughout the tropics and some distances into the subtropical latitudes. The total world-wide mangrove area, which is estimated at about 170 000 km^2, dominates approximately 75% of the world's coastline between latitudes 25 °N and 25 °S (McGill, 1958). Mangrove communities are normally characterized by high productivity, high biomass and litter production when compared to other terrestrial plant communities (Lugo & Snedaker, 1974; Mann, 1982). High rates of primary production and rapid decomposition of the mangrove litter result in an important food and energy source for a variety of aquatic organisms and have positive impacts on coastal productivity, including cultures of commercial shellfish such as oysters, clams, crabs, shrimps, and fishes (Chapman, 1983; Steinke & Ward, 1988).

The structure and productivity of mangrove ecosystems are significantly affected by climatic conditions resulting in geographical variations in biomass and primary production. Productivity appears to decrease with an increase in latitude but is enhanced with an increase in tidal amplitude (Woodroffe *et al.*, 1988). It has also been well-documented that mangrove swamps reach their greatest structural and floristic diversity in the tropics and decline with increasing latitude. Most of the previous studies on mangrove ecology focused on tropical mangroves with relatively few references on sub-tropical regions. Mangroves in Shenzhen Special Economic Zone, the People's Republic of China, at 22 °32′N, is located near the northern limits of the mangrove development. Its flora and fauna, the productivity and environmental factors are different from other mangrove stands located in the tropics. Despite their importance in the conservation and ecology of estuarine ecosystems, very little information is available on the community structure and standing crop biomass of the mangrove forests in this region. The present study therefore aims to (1) provide an acceptable method for the estimation of the above and below ground biomass of the mangrove community in Shenzhen, the PRC; and (2) determine the community structure and standing crop biomass of this mangrove forest.

Materials and methods

Study area

The study was carried out at the Futian Nature Reserve of Shenzhen Special Economic Zone, the PRC (22°31′N, 114°05′), which is contiguous with the Mai Po Nature Reserve of Hong Kong. Futian Reserve stretches 11 km from east to west, and covers an area of 304 hectares. Along the muddy shore of this reserve, 112 ha are mangrove forests dominated by *Aegiceras corniculatum* (L.) and *Kandelia candel* (L.) Blanco. The average height of these two dominant species are 4–5 m with no vertical stratification. Very few seedlings and almost no saplings are found on the forest ground. The forest is 20 to 22 years old. In addition to these two species, some *Avicennia marina* (Forsk.) Vierh and fewer individuals of *Excoecaria agallocha* are also found. At the gap or the open area of the mangrove forest, where more light reaches the forest ground, *Acanthus ilicifolius* Linn. are commonly found. The mangrove ecosystem is inundated by incoming tide twice a day and the largest tidal range, at high spring tides, is about 2.8 m. The climate of this area is humid, with mean annual relatively humidity of 80%. The mean annual air temperature is 22 °C, with the minimum and maximum temperatures at 0.2 and 38.7 °C, respectively. The lowest mean monthly temperature is 14.1 °C (in January) and the highest value is 28.1 °C (in July). Annual precipitation is 1926.7 mm and the rainy months are from May to September (78.4% falls in this wet season).

Harvesting for biomass estimation

Fifteen trees of *Aegiceras corniculatum* (2 to 12 cm DBH, diameter at breast height), nineteen trees of *Kandelia candel* (2 to 15 cm DBH) and 6 trees of *Avicennia marina* (8–15 cm DBH) were randomly chosen. The individual tree was felled and harvested, the height (HT) and diameter at breast height (DBH) were measured. The above-ground part was divided into stem (trunk), branch and twig, leaf, inflorescence and fruiting bodies. The fresh weight of these components was measured. Subsamples were then dried at 80 °C to determine their dry weights. Roots of three trees from each species were excavated manually to a depth of approximately 500 mm to give an estimate of below-ground biomass. All root materials were washed carefully under a jet of water to remove adhering mud. The fresh and oven-dried weight of the roots collected were determined. Various allometric regression equations using either tree diameter or height, or a combination of diameter and height as the independent variables, and the weight of the specific components as the dependent variable were developed (Table 1; Snedaker & Snedaker, 1984). The significance of the regression equation was assessed by the correlation coefficient value (r) and the t-tests on the slope or regression coefficient values (b).

Community structure and standing crop biomass of Futian mangrove forest

Two belt transects 100 m apart within the Futian mangrove forest were laid perpendicular to the shoreline to determine the structural feature and standing crop biomass of the mangrove plant communities. Transects A and B were 90 m × 10 m and 60 m × 10 m, respectively, and the total area covered was 1500 m^2. Fifteen quadrats, each of 10 m × 10 m, were sampled. Within each quadrat, all trees were identified and the number of trees of each species was recorded. The height (HT), DBH, maximum canopy width and length of each tree were measured. The relative density, frequency, dominance, basal area, average diameter and importance values were calculated for each species. The species diversity of the mangrove stand was determined in terms of Shannon-Wiener index and evenness. The biomass of each species and the plant community were estimated by the allometric regression equations formulated from the harvest data.

Table 1. Regression equations used to estimate the relationship between the dependent variable (weight of specific component, y) and the independent variables (height and/or diameter, x).

Regression equations	Independent variable (x)	Regression equations	Independent variable (x)
$y = ax^b$	$DBH^2 * HT$	$y^{1/3} = a + bx$	DBH
$y = ax^b$	DBH^2	$y^{1/3} = a + bx$	HT
$y = ax^b$	DBH	$y = a + bx_1 + cx_2$	x_1: DBH; x_2: HT
$y = ax^b$	HT	$y = a + bx_1 + cx_2$	x_1: DBH^2; x_2: HT
$y = a + bx$	$DBH^2 * HT$		
$y = a + bx$	DBH		
$y = a + bx$	HT		

DBH: Diameter at breast height (m); HT: Tree height (m); a, b and c are constants of the regression model.

Table 2. Harvest data of 15 individuals of *Aegiceras corniculatum* used to calculate the regression equations for biomass estimation.

DBH (cm)	HT (m)	Biomass (Oven-dry weight, kg) Stem biomass	Branch biomass	Leaf biomass	Total aerial biomass	Root biomass
15.1	4.7	7.18	6.50	1.20	14.88	ND
20.2	4.3	7.38	6.64	1.35	15.82	ND
21.7	4.2	9.25	7.57	1.88	18.70	ND
15.8	4.2	6.13	5.53	1.02	12.68	ND
9.9	4.3	2.18	1.97	0.56	4.71	ND
24.6	4.7	10.38	6.60	1.20	18.18	ND
26.4	4.6	7.63	6.74	1.03	15.40	ND
11.3	4.0	3.18	2.49	0.43	6.10	ND
3.7	4.6	1.68	1.34	0.31	3.33	ND
11.9	4.0	3.50	3.03	0.41	6.94	ND
22.7	4.3	7.13	6.38	0.83	14.34	ND
7.1	4.7	3.38	2.26	0.30	5.94	ND
17.6	4.2	6.38	5.68	0.89	12.95	5.46
36.9	4.7	11.38	8.13	2.30	21.81	8.32
20.9	4.3	8.50	6.72	1.36	16.55	4.99

ND: Not determined.

Results

The dry weights of different components of *Aegiceras corniculatum*, *Kandelia candel* and *Avicennia marina* are recorded in Tables 2 to 4, respectively. The dry weight of the trees and their components were best estimated as a power function of the square of the diameter multiplied by tree height ($y = ax^b$ where y = biomass and $x = DBH^2 * HT$). The r values for *A. corniculatum* and *K. candel* were greater than 0.85 and 0.78, respectively (Table 5). In these two species, all other regression models listed in Table 1 were found to be less significant than this power function and the r values were less than 0.7. This indicates that plant biomass could not be satisfactorily estimated by only stem height or diameter. In the case of *A. marina*, none of the regression models being tested provided significant estimates of biomass, and the r values of the best

Table 3. Harvest data of 19 individuals of *Kandelia candel* used to calculate regression equations for biomass estimation.

DBH (cm)	HT (m)	Biomass (Oven-dry weight, kg) Stem biomass	Branch biomass	Leaf biomass	Other biomass	Total aerial biomass	Root biomass
5.7	4.4	4.84	2.80	1.42	0.043	9.10	4.23
6.1	4.5	5.24	3.64	1.01	0.064	9.95	4.26
4.9	4.5	3.07	2.54	0.95	0.068	6.63	ND
6.9	4.5	7.09	5.96	2.17	0.240	15.46	ND
12.6	5.4	14.76	22.30	3.75	1.777	42.59	ND
4.6	4.5	2.13	1.75	0.45	0.054	4.38	ND
9.4	4.3	5.31	2.40	0.80	0.039	8.55	ND
6.0	5.1	4.96	3.87	0.74	0.204	9.77	ND
7.5	4.1	5.08	6.39	1.72	2.146	15.34	6.66
6.9	3.4	2.24	1.52	0.68	0.136	4.58	ND
6.9	3.4	2.01	1.37	0.68	0.036	4.09	ND
4.4	4.9	2.95	1.76	0.57	0.157	5.44	ND
7.6	4.8	4.96	4.93	1.67	0.261	11.82	ND
6.1	4.8	4.72	4.10	1.24	0.275	10.34	ND
7.4	5.2	8.27	8.45	2.14	ND	18.86	ND
9.4	4.8	15.45	22.32	4.31	ND	42.08	ND
7.6	4.7	8.79	17.01	3.38	ND	29.18	ND
5.4	5.1	4.02	2.00	0.94	ND	6.96	ND
8.1	5.5	7.37	8.16	2.88	ND	18.41	ND

ND: Not determined.

Table 4. Harvest data of 6 individuals of *Avicennia marina* used to calculate regression equations for biomass estimation.

DBH (cm)	HT (m)	Biomass (Oven-dry weight, kg) Stem biomass	Branch biomass	Leaf biomass	Total aerial biomass	Root biomass
10.1	5.6	12.09	24.73	2.85	39.67	ND
12.7	3.1	3.29	6.95	1.81	12.05	ND
8.3	4.9	9.80	16.33	2.16	28.24	ND
12.5	5.0	12.49	18.77	2.14	33.40	4.28
14.3	4.7	12.70	22.15	2.68	37.53	5.89
9.9	4.8	9.86	11.88	1.54	23.28	3.62

ND: Not determined.

regression model (the same power function as the other two species) were less than 0.5 (Table 5). This suggests that the biomass of *A. marina* had no simple relationship with its diameter and tree height. This might be related to the unique structure and shape of this species in Futian. *A. marina* found in this mangrove stand was stunted, irregular in shape and started forking at very low level of the tree trunk.

The Futian mangrove community was rather simple with low species diversity. The Shannon-Wiener index and evenness were 0.78 and 48.98%, respectively, lower than that of the sub-tropical evergreen broad-leaf forest commonly found in the southern part

Table 5. Allometric regression equations, based on the harvest data shown in Tables 1-3, for estimating biomass of specific components of three dominant mangrove plant species.

Plant species	Plant components	Regression Equations	*r* value	F and significance of F value
Aegiceras corniculatum	Stem	$\text{Log AS}_W = 1.198 + 0.464 \text{ Log (DBH}^2 * \text{HT)}$	0.93	84.4, $P<0.0000$
	Branch	$\text{Log AB}_W = 1.110 + 0.463 \text{ Log (DBH}^2 * \text{HT)}$	0.93	87.0, $P<0.0000$
	Leaf	$\text{Log AL}_W = 0.393 + 0.475 \text{ Log (DBH}^2 * \text{HT)}$	0.88	43.2, $P<0.0000$
	Total aerial	$\text{Log AT}_W = 1.496 + 0.465 \text{ Log (DBH}^2 * \text{HT)}$	0.94	95.2, $P<0.0000$
	Root	$\text{Log AR}_W = 0.967 + 0.303 \text{ Log (DBH}^2 * \text{HT)}$	0.93	NC
Kandelia candel	Stem	$\text{Log KS}_W = 2.162 + 0.869 \text{ Log (DBH}^2 * \text{HT)}$	0.84	40.9, $P<0.0000$
	Branch	$\text{Log KB}_W = 2.741 + 1.253 \text{ Log (DBH}^2 * \text{HT)}$	0.80	30.4, $P<0.0000$
	Leaf	$\text{Log KL}_W = 1.706 + 0.943 \text{ Log (DBH}^2 * \text{HT)}$	0.78	26.9, $P = 0.0001$
	Total aerial	$\text{Log KT}_W = 2.814 + 1.053 \text{ Log (DBH}^2 * \text{HT)}$	0.83	38.8, $P<0.0000$
	Root	$\text{Log KR}_W = 2.433 + 0.990 \text{ Log (DBH}^2 * \text{HT)}$	0.95	NC
Avicennia marina	Stem	$\text{Log VS}_W = 1.643 + 0.544 \text{ Log (DBH}^2 * \text{HT)}$	0.39	0.75, $P = 0.44$
	Branch	$\text{Log VB}_W = 1.897 + 0.567 \text{ Log (DBH}^2 * \text{HT)}$	0.46	1.05, $P = 0.36$
	Leaf	$\text{Log VL}_W = 0.690 + 0.287 \text{ Log (DBH}^2 * \text{HT)}$	0.47	1.11, $P = 0.35$
	Total aerial	$\text{Log VT}_W = 2.092 + 0.529 \text{ Log (DBH}^2 * \text{HT)}$	0.45	1.02, $P = 0.37$
	Root	$\text{Log VR}_W = 1.361 + 0.615 \text{ Log (DBH}^2 * \text{HT)}$	0.92	NC

DBH: diameter at breast height (m); HT: Height (m); *r*: correlation coefficient between Log biomass and Log ($DBH^2 * HT$), NC: not calculated.

Table 6. Community structure of the mangrove forest in Futian Nature Reserve.

Plant species	Number of Individuals[a] (alive)	Number of Individuals[a] (dead)	Average DBH (cm)	Average Height (m)	Basal Area[b] (m^2)	Relative Density[c]	Relative Frequency[d]	Relative Dominance[e]	Importance Value[f]
Aegiceras corniculatum	794	99	19.56	3.85	1.219	84	42	89	71.67
Kandelia candel	117	0	7.76	4.20	0.163	12	36	9	19.00
Avicennia marina	39	6	13.63	4.35	0.042	4	22	2	9.33

[a] Number of individuals was measured based on a sampling plot size of 1500 m^2;
[b] Basal Area (m^2) = Σ ($DBH^2 * \pi$)/(4 * 10,000);
[c] Relative Density = (number of individuals of a species / total number of individuals) * 100;
[d] Relative Frequency = (frequency of a species / sum frequency of all species) * 100;
[e] Relative Dominance = (total basal area of a species / basal area of all species) * 100;
[f] Importance Value = (Relative Density + Relative Frequency + Relative Dominance) / 3.

of China (Zhang *et al.*, 1989). The mangrove forest was dominated by *A. corniculatum*, with an importance value of 71.67, which was much higher than *K. candel* and *A. marina* (Table 6). The tree density of *A. corniculatum* (5290 trees ha^{-1}) was also larger than the other two species, being 6 times higher than that of *K. candel* (780 trees ha^{-1}). The average tree density of this mangrove community was around 6000 ha^{-1}, and was within the documented values for mangroves reported elsewhere (1000 to 30 000 trees ha^{-1}, Day *et al.*, 1987; Komiyama *et al.*, 1987). However, in this mangrove stand, about 12.5% of the *A. corniculatum* trees were dead. Even for the living ones, a lot of them had dead lower branches and young leaves were found

Table 7. Total biomass and the partitioning of standing biomass in Futian mangrove forest.

Plant species	Stem		Branch		Leaf		Aerial sub-total		Root		Total	Dead material
	Biomass ($kg\ m^{-2}$)	% of Total	Biomass ($kg\ m^{-2}$)	% of Total	Biomass ($kg\ m^{-2}$)	% of Total	Biomass ($kg\ m^{-2}$)	% of Total	Biomass ($kg\ m^{-2}$)	% of Total	Biomass ($kg\ m^{-2}$)	Biomass ($kg\ m^{-2}$)
Aegiceras corniculatum	3.434	35.95	2.745	28.73	0.527	5.52	6.807	71.26	2.746	28.74	9.553	0.849
Kandelia candel	0.464	28.77	0.429	26.59	0.124	7.69	1.058	65.59	0.555	34.41	1.613	ND
Avicennia marina	0.291	29.82	0.493	50.51	0.062	6.35	0.849	86.99	0.127	13.01	0.976	0.078
Total	4.189	34.50	3.667	30.20	0.713	5.87	8.714	71.77	3.428	28.23	12.142	0.927

ND: Not detected.

Table 8. Total and above-ground biomass ($t\ ha^{-1}$) of mangrove forests in the world.

Location	Latitude	Mangrove species	Structure or average height (m)	Above-ground biomass	Total biomass	Source of reference
Panama	9 °N	*Rhizophora brevistyla*	30-40	279.2	468.9	Golley *et al.*, 1975
Thailand	8 °N	*Rhizophora apiculata*	11	159.0	ND	Christensen, 1978
Malaysia	5 °N	*Rhizophora apiculata*	ND	185.3	202.5	Gong & Ong, 1990
Sri Lanka, Dutch Bay	8 °N	*Rhizophora mucronata* and *Avicennia marina*	4.5	71.0	ND	Amarasinghe & Balasubramaniam, 1992b
Sri Lanka, Puttalam lagoon	8 °N	*Avicennia marina*	7.3	201.7	ND	Amarasinghe & Balasubramaniam, 1992b
Puerto Rico	18 °N	*Rhizophora mangle*	8	62.9	112.9	Golley *et al.*, 1962
Philippines	12 °N	Mixed mangrove species	ND	45.9	ND	de la Cruz & Banaag, 1967
China, Hainan Island	19 °N	*Bruguiera sexaugula*	ND	248.5	420.3	Lin *et al.*, 1990
China, Fujian	24 °N	*Kandelia candel*	5	93.4	162.6	Lin, 1989
Hong Kong	22 °N	*Kandelia candel*	ND	129.6	ND	Lee, 1990
Taiwan	24 °N	*Kandelia candel*	ND	ND	143.9	Chen, 1982
USA, Florida	26 °N	*Rhizophora mangle*	fringe	118.9	ND	Lugo & Snedaker, 1974
		Rhizophora mangle	scrub	7.9	ND	Lugo & Snedaker, 1974
		Mixed species	island	49.0	57.0	Lugo & Snedaker, 1974
Japan	24 °N	*Rhizophora mucronata*	5-6	108.1	ND	Suzuki & Tagawa, 1983
		Bruguiera gymnorhiza	5-6	97.6	ND	Suzuki & Tagawa, 1983
Australia, Sydney	34 °S	*Avicennia marina*	6-8	128.3	ND	Briggs, 1977
Australia, Westernport Bay	38 °S	*Avicennia marina*	2-4	86.0	232.0	Clough & Attiwill, 1975
New Zealand	37 °S	*Avicennia marina*	2.5-4	104.1	ND	Woodroffe, 1985
		Avicennia marina	<1	6.8	ND	Woodroffe, 1985
China, Shenzhen	22 °N	*Aegiceras corniculatum* and *Kandelia candel*	4-5	87.1	121.4	This study

mainly at the top of the canopy. The average height of *A. corniculatum* was 3.9 m, slightly shorter than *K. candel* and *A. marina*.

The mean total biomass and the partitioning of biomass were calculated based on regression equations shown in Table 5. Table 7 reveals that *A. corniculatum* had the highest biomass (9.55 kg m^{-2}), 6 times of that found in *K. candel*. In both species, the aerial woody tissues (stem and branch) represented the greatest percentage of the biomass (64.7% and 55.4% for*A. corniculatum* and *K. candel*, respectively) while leaves made up 5.5 and 7.7%, respectively. A similar pattern was also found for *A. marina*. The mean total above-ground biomass contributed more than 65% of the total biomass. The root:shoot ratio of the community was 0.4:1. The highest root:shoot ratio was found *K. candel*, followed by *A. corniculatum* and *A. marina* had the lowest root:shoot ratio (Table 7).

Discussion

The mangrove stand in Futian Nature Reserve of the Shenzhen Special Economic Zone, the PRC had a relatively simple community structure with low species diversity. It was dominated by *Aegiceras corniculatum*, with few individuals of *Kandelia candel* and *Avicennia marina*. There was no zonation of species within this mangrove stand. It has been stated that mangroves are naturally stressed ecosystems. The survival of individual species and thus zonation within a mangrove stand depend on the following environmental factors: frequency of tidal flushing, soil type, soil salinity, drainage plant and animal interactions (Kènneally, 1982). In the present study, the mangrove had very flat topographic slopes and the frequency of tidal flushing in the upslope was low; the upper part of the shore was only flooded by tidal water during the very high spring tide. These created a condition similar to the mainland fringing habitat which was relatively unfavourable to support a large number of species. Amarasinghe and Balasubramaniam (1992a) reported that the mainland fringing mangrove stands had a simpler community structure and dominated by single species than the estuarine sites, the latter habitats had oscillating flushing conditions with more favourable nutrient balances which supported higher species diversity.

The present mangrove stand had a total biomass of 12.142 kg m^{-2} and about 72% of this was contributed by the above-ground portions. This value is comparable with those found in other mangroves of similar latitude such as Putero Rico (Golley *et al.*, 1962) and Japan (Suzuki & Tagawa, 1983), but much lower than those found in tropical regions (Table 8). In general, the species diversity, tree height and density, and the biomass increase with decrease in latitude. The tropical mangroves usually have higher standing crop biomass and more complicated structure than those in sub-tropical regions. However, the biomass of a mangrove forest is not only affected by the geographic location and the macro-climate. It is also related to the species composition, community structure, growth forms and the age of the plant community (Lugo & Snedaker, 1974; Knox, 1986). The biomass values of two geographically close mangrove forests could be very different. For instance, the biomass obtained in this study was significantly less than that reported in Mai Po, Hong Kong (Lee, 1990) although these two mangroves are contiguous to each other. The difference in biomass estimation of these two sites was mainly due to the fact that the other study took place in the tidal ponds (called '*gei wai*') where the mangrove forest was dominated by *Kandelia candel* (Lee, 1990). The enclosed situation and infrequent inundation in the tidal ponds restricted the dispersal of propagules and enhanced the retention of the majority of the nutrients for *in situ* recycling, therefore supporting larger tree density (51770 ha^{-1} in '*gei wai*' ponds and 6333 ha^{-1} in the present study) and higher biomass. Woodroffe (1985) concluded that the biomass depends very much on tree height and density. For the same *Avicennia* mangrove stands with a low and sparse population, the above-ground biomass was less than 2 t ha^{-1} while areas of tall dense trees had value of 104.13 t ha^{-1}. Moreover, *K candel*, at its mature stage, is taller and have more dense canopy cover than *A. corniculatum*. This explained the importance of factors other than macroclimate on the mangrove forest structure and productivity. Similarly, the biomass value of this mangrove stand was lower than those found in Taiwan (Chen, 1982) and Fujian, the PRC (Lin, 1989) with a location at even higher latitude (about 25 °N). These two mangrove stands were dominated by *Kandelia candel*, which had taller and more dense canopy cover than *A. corniculatum* at its mature stage. Furthermore, as discussed previously, a significant percentage of dead individuals of *A. corniculatum* was found in Futian mangrove stand, and most trees had dead leafless branches below the top canopy.

The mean root:shoot ratio of this community was 0.4:1, ranged from 0.53:1 in *K. candel* and 0.15:1 in *A. marina*. These values were similar to that reported by previous researchers, including Golley *et al* (1962, 0 44 1 and 1975, 0 68 1), Tamai *et al* (1986, 0 44 1) and Silva *et al* (1991, 0 25 1), but much lower than that of Briggs (1977, 1 02–1 28 1) and Clough & Attiwill (1975, 1 69 1) Mangrove communities were usually found to have high root shoot ratios (Knox, 1986), and the community with taller trees tended to have a lower root shoot ratio than the community of smaller trees (Briggs, 1977) In this study, the root biomass might have been underestimated especially for *A marina* because of the following reasons First-ly, the excavation for root biomass estimation was only limited to the top 50 cm and the deep roots might have been lost Secondly, the root systems of the mangrove trees, in particular, *A marina*, were very extensive and irregular so field collection of root material might be incomplete Thirdly, the fine roots might have been missed during clearing of the adhering mud and soil Lastly, only three trees were excavated and harvested for root biomass estimation, thus large error might have been generated Actually, good root biomass data in mangroves studies are rare because of the difficulty in sampling roots quantitatively (Knox, 1986) More in-depth study on the root biomass production and its importance in the functioning, in particular, cycling of organic and inorganic material, of the mangrove ecosystem would be essential

In the present study, the biomass of different components of *A corniculatum* and *K candel* were best estimated by power curves using a combination of tree height and diameter at breast height ($DBH^2 * HT$) as the independent variables It was found that estimate of biomass based on only a single parameter, either height (HT) or diameter (DBH), as suggested by many researchers (Amarasinghe & Balasubramaniam, 1992b, Day *et al*, 1987, Golley *et al*, 1962, Woodroffe, 1985), was less significant This implies that regression models developed by previous workers for estimating biomass could not be applied directly without modification for mangrove stands in Futian, PRC Woodroffe (1985) found that the regression model and the importance of the tree height or diameter as the independent variable to estimate biomass value would vary between mangrove stands Even for the same mangrove stand, the regression models would be different for areas of different tree height and density In this study, there was no close relationship between height (and/or diameter) and biomass of *A marina*, and none of the regression models suggested by previous workers was successful Other parameters such as canopy diameter and width, which provide better description on the growth forms of *A marina*, might need to be explored The method developed by Briggs (1977), which calculated the mean tree biomass of *A marina* from the mean volume and mean wood density, and the photosynthetic (leaf and petiole) biomass from the regression equation using stem diameter and non-photosynthetic biomass as independent variables, might be worthwhile for consideration Nevertheless, *A marina* was the least important species in Futian mangrove and it only contributed a small percentage to the total biomass value of the whole community (about 8% of total biomass) The unsatisfactory results in estimating biomass of this species from the harvest data would not affect the overall standing crop biomass values of this mangrove stand

Acknowledgments

The authors wish to express their appreciation to all who assisted with field work, in particular, Dr M S Li and Mr S H Li We would also like to thank the officers and technicians in the Futian Nature Reserve, the PRC, and the financial support from the Hong Kong University of Science and Technology, the Croucher Foundation and the Hong Kong Research Grant Council

References

Amarasinghe, M D & S Balasubramaniam, 1992a Structural properties of two types of mangrove forest stands on the northwestern coast of Sri Lanka Hydrobiol 247 17–27

Amarasinghe, M D & S Balasubramaniam, 1992b Net primary productivity of two mangrove forest stands on the northwestern coast of Sri Lanka Hydrobiol 247 37–47

Briggs, S V, 1977 Estimates of biomass in a temperate mangrove community Aust J Ecol 2 369–373

Chapman, V J 1983 Mangroves in New Zealand In H J Teas (ed), Biology and Ecology of Mangroves Tasks for Vegetation Sciences Series Vol 8, Dr W Junk Publishers, Lancaster 81–86

Chen, M Y, 1982 Ecological study of the mangroves of Tan shui Estuary Quart J Chinese Forestry 15 17–25

Christensen, B , 1978 Biomass and primary production of *Rhizophora apiculata* Bl in a mangrove in southern Thailand Aquat Bot 4 43–52

Clough, B F & P M Attiwill, 1975 Nutrient cycling in a community of *Avicennia marina* in a temperate region of Australia In G E Walsh, S C Snedaker & H J Teas (eds), Proceedings International Symposium on the Biology and Management of Mangroves, Honolulu Vol 1, Univ Florida, Gainesville 137–146

Day, Jr J W, W H Conner, F L Lou, R H Day & A M Navarro, 1987 The productivity and composition of mangrove forests, Laguna de Terminos, Mexico Aquat Bot 27 267–284

Dela Cruz, A. A. & J. F. Banaag, 1967. The ecology of a small mangrove patch in Matabungkay Beach, Batangas Province. Nat. Appl. Sci. Bullet. 20: 486–494.

Gong, W. K. & J. E. Ong, 1990. Plant biomass and nutrient flux in a managed mangrove forest in Malaysia. Estuar. coast. Shelf Sci. 31: 519–530.

Golley, F. B., H. T. Odum & R. F. Wilson, 1992. The structure and metabolism of a Puerto Rican red mangrove forest in May. Ecology 43: 10–19.

Golley, F. B. J. T. McGinnis, R. G. Clements, G. I. Child & M. J. Duever, 1975. Mineral Cycling in a Tropical Moist Forest Ecosystem. Univ. Georgia Press, Athens Ga, 248 pp.

Kenneally, K. F., 1982. Mangroves of Western Australia. In. B. F. Clough (ed.), Mangrove Ecosystem in Australia: Structure, Function and Management, Australian Institute of Marine Sciences, Australia: 95–110.

Komiyama, A., K. Ogino, S. Adsornkoe & S. Sabhasri, 1987. Root biomass of a mangrove forest in southern Thailand, I. Estimation by the trench method and the zonal structure of root biomass. J. Trop. Ecol. 3: 97–108.

Knox, G. A., 1986. Estuarine Ecosystems: A System Approach, Vol. I. CRC Press Inc., Florida: 59–78.

Lee, S. Y., 1990. Primary productivity and particulate organic matter flow in a estuarine mangrove wetland in Hong Kong. Mar. Biol. 106: 453–463.

Lin, P., 1989. Biomass and element cycle of *Kandelia* forest in China. In P. Lin (ed.), Mangrove Research Papers (1980–1989), Xiamen Univ. Press, Xiamen: 143–149.

Lin, P., C. Y. Lu, G. L. Wang & H. X. Chen, 1990. Biomass and productivity of *Bruguiera sexangula* mangrove forest in Hainan Island, China. J. Xiamen Univ. (Nat. Sci.) 29: 209–213.

Lugo, A. E. & S. C. Snedaker, 1974. The ecology of mangroves. Annu. Rev. Ecol. Syst. 5: 39–64.

Silva, C. A. R., L. D. Lacerda, L. F. F. Silva & C. E. Rezenda, 1991. Forest structure and biomass distribution in a red mangrove stand in Sepetiba Bay, Rio de Janeiro. Revta Brasil. Bot. 14: 21–25.

Snedaker, S. C. & J. G. Snedaker, 1984. The Mangrove Ecosystem: Research Methods. UNESCO published, the Chaucer Press Ltd., U.K.

Steinke, T. D. & C. J. Ward, 1988. Litter production by mangroves, II. St. Lucia and Richards Bay. S. Afr. J. Bot. 54: 445–454.

Suzuki, E. & H. Tagawa, 1983. Biomass of a mangrove forest and a sedge marsh on Ishigaki Island, South Japan. Jpn. J. Ecol. 33: 231–234.

Tamai, S., T. Nakasuga, R. Tabuchi & K. Ogino, 1986. Standing biomass of mangrove forests in Southern Thailand. J. Jpn. For. Soc. 68: 384–388.

Woodroffe, C. D., 1985. Studies of a mangrove basin, Tuff Crater, New Zealand: I. Mangrove biomass and production of detritus. Estuar. coast. Shelf. Sci. 20: 265–280.

Woodroffe, C. D., K. N. Bardsley, P. J. Ward & J. R. Hanley, 1988. Production of mangrove litter in a Macrotidal Embayment, Darwin Harbour, N.T., Australia. Estuar. coast. Shelf Sci. 26: 581–598.

Zhang, H. D., B. S. Wang & Y. J. Hu, 1989. Hong Kong Vegetation. Suppl. ACTA Scientiarium Naturalium Universitatis Sunyatseni 8: 118–140.

Hydrobiologia **295**: 203–212, 1995.
Y. S. Wong & N. F. Y. Tam (eds), Asia-Pacific Symposium on Mangrove Ecosystems.

Mangrove outwelling: a review

S.Y. Lee
Department of Zoology and The Swire Marine Laboratory, the University of Hong Kong, Cape d'Aguilar, Sheko, Hong Kong

Key words: mangrove detritus, mass balance, tidal support

Abstract

The export of detritus and faunal biomass from mangroves has long been considered as an important support for offshore biological production and has been widely used as an argument for mangrove conservation. This functional role of the mangroves, like many other paradigms in mangrove ecology, has seldom been put to rigorous test since the hypothesis was postulated about 25 years ago. Past studies on which the hypothesis was based were mostly carried out in mangrove or other wetland environments, little is known about the fate and effects of outwelled detritus on oceanic, offshore, communities. Mass balance studies carried out in the last 15 years tend to suggest that export is common from tidal mangroves, the direction of flow depends, however, on the identity of the chemical species in question. Pore water and groundwater flow can affect tidal material exchange but are poorly studied. Generally, tracer methods using stable isotope ratios or other signatures have suggested that outwelling may be much less significant than expected. Further, most past studies have focussed on particulate matter while it is increasingly apparent that dissolved organics may play a more important role in matter exchange between offshore and mangrove communities. There is also evidence that benthic biomass and richness may not bear any positive or significant relationship with detritus availability.

Introduction

The export of plant detritus and faunal biomass to support offshore consumers has been considered an important functional role of mangroves as well as a strong argument for their conservation (e.g. Snedaker, 1978; Macintosh, 1981). This "outwelling" characteristic of coastal wetlands was first postulated by E. P. Odum (1968) about 25 years ago (Nixon (1980) attributed the idea to Teal (1962)), originally based upon observations on salt marsh ecosystems. Golley *et al.* (1962) made the first study of tidal exchange of materials between a mangrove and the offshore community. Despite the fact that only a small data set was used to substantiate their argument, their work nevertheless was seminal in the study of mangrove outwelling. Whereas export of plant litter is probably more conceivable in mangrove systems than in salt marshes where many producers retain their senescent leaves (Fallon *et al.*, 1985; Newell *et al.*, 1985), the outwelling behaviour of mangroves has seldom been put to test. Similar to many other paradigms of mangrove ecology, extrapolation or analogies are commonly made from findings from the temperate salt marshes.

There are, however, a few past studies which specifically addressed the outwelling problem, e.g. Nixon (1980) discussed salt marsh outwelling in detail and Ong (1984) focussed mainly on the relationship between mangrove occurrence and offshore fisheries production. Odum (1984) himself also briefly reviewed the status his hypothesis but concluded that outwelling from estuarine wetlands, while generally exists, greatly depends on the physical characteristics of the wetland. Direct evidence of mangrove outwelling is scarce and few studies comparable to the flux measurements performed on salt marshes are available. Recently, Robertson *et al.* (1992) reviewed carbon fluxes and food chain relationships in tropical mangrove ecosystems and it seems that more data are becoming available to allow a better assessment of the role played by mangroves in supporting tropical offshore consumers.

Table 1. Factors which may affect the "outwelling behaviour" of mangroves and salt marshes differently.

	Mangroves	Salt marshes
Fate of senescent plant biomass	abscissed, higher chance of export	retained, decomposed *in situ*
Turnover of component which can be exported	high, higher export possible	lower, lower degree of export
Tidal regime	usually stronger tidal energy, higher export	mostly with weak tidal energy, lower export
Litter quality	high levels of secondary compounds, low utilization by detritivores	lower level of secondary compounds, easier utilization by detritivores

Useful tools (e.g. stable isotope tracers) and novel methods (e.g., flume techniques) have become available for the study of the outwelling problem in the last 25 years. This paper examines the recent developments of the Outwelling Hypothesis with special reference to the mangroves in the light of recent data collected from such techniques and suggests research future areas to test the applicability of the hypothesis to Asia-Pacific mangroves.

Development of the outwelling hypothesis

Nixon (1980) suggested that the concept of offshore consumer communities being supported by exported organic matter from productive coastal ecosystems such as salt marshes was first hinted, albeit lightly, in John Teal's (Teal, 1962) analysis of a Georgia salt marsh. It was, however, E. P. Odum who first put it as a formal scientific proposal a few years later. Odum (1968) tried to relate the apparent high productivity of waters adjacent to highly productive systems such as coral reefs, seagrass beds and salt marshes to the export, i.e. outwelling, of nutrients and organic detritus by these systems. It is worth pointing out, however, that little direct evidence was available to support his idea at that time. His paper also initiated numerous studies on the productivity of the coastal salt marshes, eventually leading to the formulation of many of the paradigms in wetland ecology. The concept was, however, extensively applied to the mangrove ecosystem only much later. As a result, the number of studies of materials exchange between mangroves and their surrounding water bodies is much less than that on salt marshes. A search for reports specifically discussing export of organic matter from mangroves published in the last 15 years yielded only 34 records out of a total of 1659 papers. This is another example of mangrove ecology borrowing ideas from the salt marshes. The two systems are fundamentally different in several important aspects (Thom, 1982), e.g. the fate of senescent litter, geographical influence on tidal regimes and detritus quality, most of which affecting the exchange of materials with offshore waters (Table 1). There is therefore a need to establish an outwelling picture for mangroves based on data collected from themselves.

The mass balance approach

Similar to most hypotheses in ecology, formulation of the Outwelling Hypothesis seems to have been an intuitive proposal based on the close observations of a talented ecologist on a system. Both J. M. Teal and E. P. Odum based their outwelling idea on the speculation that the high productivity of the salt marshes could probably support consumers offshore. At their time, there were few studies to give support to this relationship, but gradients in offshore production have been demonstrated by a few studies, e.g. Thomas (1966) and have been used as arguments for the hypothesis. A rapid rise in productivity studies followed the postulation of the hypothesis, which allowed Nixon (1980) to review the role of salt marshes and mangroves in exporting nutrients and organic matter in support of offshore communities. His review was mainly based on the results of mass balance studies.

Nixon (1980) concluded from 12 mass balance studies carried out in salt marshes that there seemed to be no agreement regarding whether salt marshes are net exporters or importers of organic matter, a conclusion reiterated by the later review by Hopkinson (1988) on the same topic. Rather, the actual relationship will depend strongly upon the inundation regime, the geomorphology and the form of organic matter (dissolved versus particulate) or nutrient (carbon versus nitrogen

Table 2 Reports on outwelling from mangroves published during the period 1971–92.

1 Transport of detritus and nutrients					
System studied	Physical characteristics	Materials	Export/import	Rate	Reference
Australia	tidal forest	detritus	Export	no rate determined conclusion based on hydrodynamics predictions	Wolanski *et al* (1980)
Northeastern Australia	tidal forest	POC	Export	$10 kgC ha^{-1} day^{-1}$	Boto & Bunt (1981)
Everglades, USA	*Avicennia germinans* Basin forest	POC/DOC	Export	$63\ 7 gC m^{-2} y^{-1}$ 75% as DOC	Twilley (1983)
Westernport Bay Australia	*Avicennia marina* tidal forest	POC	Export	40% of all litter crabs can consume 50% of unbagged litter	van der Valk & Attiwill (1984)
India			Export	$261\ tC\ y^{-1}$ to estuarine waters	Subramanian *et al* (1984)
Malaysia		N species	Export/Import	actual direction of flow varied with tides and freshwater influx	Wong (1984)
Andamen Sea Thailand		POC	Export	only estimated for a 2-month period limited data	Chansang & Poovachiranon (1985)
Eastern Australia		OC, ON, IN	Export	inferred export based on concentrations on OC, ON & IN in soil and water	Clark (1985)
Tuff Crater New Zealand	forest with restricted tidal flow	POC, TSS, ISS, OSS	weak export	<2% of detritus produced $< 3 kgC ha^{-1} day^{-1}$ exported	Woodruffe (1985a, b)
Northeastern Australia	tidal *Rhizophora* forest	POC	Export	significant in situ consumption by crabs (30–80%), tidal export much less than expected	Robertson (1986)
Matang mangrove	estuarine influence	suspended solids, POC,PON,POP, IN,IP	mangrove as sink	not calculated	Nixon *et al* (1980)
El Verde Lagoon Mexico	lagoon with ephemeral opening	POC	Export/storage	90% of accumulated litter exported during opening of inlet	Flores-Verdugo *et al* (1987)
Northeastern Australia	tidal forest	OC, NO_x, P	?	No significant net flux recorded for DOC, NO_x & soluble reactive PO_4	Boto & Wellington (1988)

or phosphorus) concerned and may also be highly variable temporally even for the same wetland (e.g., Heinle & Flemer, 1976; Childers *et al.*, 1993). Such factors were also addressed by W. E. Odum's earlier paper on the same topic (Odum *et al.*, 1979) although with less supporting data. Wolanski (1992) examined the hydrodynamics of a mangrove-nearshore environment and concluded that stable coastal boundary layer water can be formed in shorelines fringed by straight mangroves sheltered by headlands. This coastal boundary layer will effectively reduce the extent of outwelling from mangroves to the offshore areas. Under such conditions, outwelling effects will only be present over restricted distances.

Twilley (1988) in his review of mangrove mass balance studies, stated that probably net export of organic matter is a common feature of most mangroves, because stronger tidal exchange, regular rainstorms and floating and non-retained litter (cf. standing dead litter in the salt-marsh) are all conducive to this difference from the saltmarsh wetlands (e.g., Chalmers *et al.*, 1985; Jordan *et al.*, 1986). The huge amounts of nutrients and carbon associated with well developed mangrove forests have also been increasingly appreciated, e.g. Gong & Ong (1990) and Twilley *et al.* (1992), making the mangroves a highly potential exporter for these materials. A survey of the literature for papers published between 1971–92 discussing exchange of materials between mangroves and offshore areas sug-

Table 2 cont

Great Barrier Reef Australia	mangrove & coral reef	OC	Export	mangroves estimated to export 12523tCy^{-1}	Robertson (1988)
Northeastern Australia	tidal forest	POC	Export	in situ crab consumption of litter affects exported	Robertson & Daniel (1989)
Hong Kong	high-zoned *Kandelia candel* in ponds	POC	Storage Import	low export, high accumulation and in situ processing, <1% exported	Lee (1990)
Klong Ngoa Estuary Thailand		N	Export	Expect outwelling based on hydrodynamics model (NO_2, NO_3)	Wattayakorn *et al* (1990)
Crab Cay Bahamas	mangrove-offshore *Rhisophora mangle*	OM	Export	mangrove DOM makes up 10% of all DOM at 1km from forest	Moran *et al* (1991)
II Export of animal biomass from mangroves to offshore areas in the literature					
System	physical characteristics	Group studied	Export/import	Rate	Reference
Mexico	tidal forest	*Aratus pisonii*	Export	assumed export of larvae but no measurement	Beever *et al* (1979)
Northeastern Australia	mangrove estuary	fish		No significant nursery function for most commercially important species	Robertson & Duke (1990)
Malaysia	tidal forest and mudflat	prawns and fish		Mangrove as nursery site for prawns but not fish	Chong *et al* (1990)
Costa Rica	tidal forest	crab zoea		No significant trend for all taxa studied	Dittel & Epifanio (1991)
Costa Rica	tidal forest	crab zoea	net export	net export of zoea of some groups	Dittel *et al* (1991)

gests that export is a feature of most tidally inundated mangroves (Table 2). The total number of such papers is small, only in 34 of the 1659 papers published in the period (2.05%) was export or outwelling from mangrove discussed or measured. A survey of the 34 reports on mangrove materials exchange seems to support Twilley's view of mangrove generally acting as exporters of organic matter and nutrients (Table 2). There are, however, reports documenting mangroves acting as net importers of organic carbon and retain large proportions of the litter production for in situ consumption, mainly as a result of a restricted inundation regime (e.g., Twilley *et al.*, 1986; Flores-Verdugo *et al.*, 1987; Lee, 1990). Again, factors such as geomorphology and the strength and frequency of tidal inundation will affect the behaviour of the mangrove in relation to their potential in exporting the production.

One finding arising from recent mass balance studies is the overwhelming importance of dissolved forms of organic matter in the materials exchanged between the wetland and offshore communities. Twilley (1985) estimated that up to 75% of all carbon exchanged between *Avicennia* forests and a Florida estuary was in the dissolved form. Robertson (1987) also stressed that the role played by dissolved materials represents one aspect overlooked by most past studies in mangrove environments. Recent reports have suggested rapid utilisation of dissolved organic carbon or nitrogen by estuarine macrofauna and microfauna, e.g., Benner *et al.* (1986); Camilleri & Gibi (1986) and such utilisation may dominate carbon metabolism in aquatic systems (Wetzel, 1984). The behaviour of mangroves may differ with different elements important to ecosystem metabolism, e.g. carbon versus nitrogen export. Recent studies incorporating the dissolved organic matter fraction in salt marshes have indicated that DOC may not be always be significantly exported from the wetland but DON is consistently exported (Whiting *et al.*, 1987; Dame *et al.*, 1991). The same variability has also been documented for mangroves (e.g., Wong, 1984).

There are also other "missing links" in most past mass balance studies. In most earlier studies of mangrove export, it was assumed that in situ consumption of litter in tidal mangroves was insignificant. Thus Boto & Bunt (1981) have estimated that about 70% of all litter produced between high tides are exported from an Australian mangrove forest and many other studies also acknowledge a large proportion of the litter is exported offshore. Recent reports have, however, suggested that some macrodetritivores such as grapsid crabs could process significant portions of freshly fallen litter, either by direct consumption or storage, before it is exported by the next incoming tide (40% – van der Valk & Attiwill, 1984; 30–80% – Robertson, 1986; Robertson and Daniel, 1989). Other reports have also demonstrated the importance of fresh mangrove litter to these crabs (e.g. Malley, 1978; Lee, 1989).

Retention of nutrients and organic matter by porewater and transportation through groundwater flow have largely been overlooked in most of the mass balance studies. These storage compartments and transportation pathways can sometimes contribute significantly to the outwelling behaviour of mangrove swamps, e.g. Ovalle *et al.* (1990) reported upon the influence of porewater as a trap for nutrients to result in reduced net tidal outwelling from the mangrove; and Mazda *et al.* (1990) stressed the role played by groundwater movements in overall nutrient export during the tidal cycle. In other cases, in situ processing of some nutrients may confuse the apparent mass balance picture. Boto & Robertson (1990), for example, measured nitrogen fixation rate in a tropical Australian mangrove and found that nitrogen fixation (acetylene reduction) from the sediment, algal mats, decomposing logs and algal-covered prop roots closely matches previously documented nitrogen export rates. The exact role and contribution of these compartments and links in the mangrove mass balance picture require more data to refine.

Another important missing link in mangrove mass balance studies is the movement of animal biomass. Again, mangroves have been widely cited for their ability to provide refuge to adult as well as larval stages of fish and crustaceans, many of which are commercially important (Teas, 1981; Macintosh, 1981). Those species which spend their larval stages in the mangrove environment but migrate offshore as adults, e.g. most penaeid shrimps, may represent one important form of outwelled carbon originated from mangrove primary production. Rozas *et al.* (1988) investigated use of salt marsh microhabitats and found high abundance and biomass of fish associated with tidal creeks and rivulets. The relative importance of this form of outwelling has, however, rarely been acknowledged, addressed or evaluated. Lindall *et al.* (1973) demonstrated a higher abundance of juvenile crustaceans and fish in areas adjacent to mangroves and Robertson & Duke (1987) also concluded from their study in Australia that while many mangroves do provide nursery site to crustacean and fish species, the nursery value may vary considerably between particular sites. Beever *et al.* (1979) discussed the export of larval stages of the tree crab *Aratus pisonii* as one form of outwelling from mangroves but did not measure the actual value. Dittel *et al.* (1991) and Dittel & Epifanio (1991) measured movement of crab zoeae in Costa Rican mangroves and concluded that whereas significant tidal movement of zoeae were found, there was no unequivocal evidence to suggest uni-directional movement of the larvae through the mangrove (Table 2, II). The flux of assimilated mangrove-derived carbon in the form of animal biomass is still largely unknown.

Stable isotope and other techniques

Whereas new data have become available to evaluate the fate of mangrove production within the mangrove forests, little new light is available to explain the value of such organic matter on offshore communities. Challenging the assumption that the excess primary production from wetlands is always exported, Haines (1977, 1979) employed the stable isotope technique to trace the origin of organic matter in a Georgia estuary. There was no direct evidence from the stable isotope studies to support a consumer community based upon organic matter exported from coastal salt-marshes. Moran *et al.* (1991) directly measured the extent of the export of mangrove-derived organic matter and concluded that dissolved organic matter from mangroves may support, despite only for up to ca. 1km, offshore bacterioplankton growth. Similar findings have been reported by Rodelli *et al.* (1984) using stable carbon isotopes to study outwelling effects of mangroves in Malaysia. Drawing evidences from various studies making use of stable isotopes as tracers, Gearing (1988) concluded that in most cases terrestrial organic matter is exported no more than a few kilometers offshore.

Few stable isotope analyses have been performed to understand trophic relationships between mangroves and offshore consumers. Fleming *et al.* (1990) documented some export of mangrove detritus from Florida mangroves to support nearshore consumers but, similar to the findings of Moran *et al.* (1991), the geographical extent of export is small. As pointed out by Fry & Sherr (1984), Twilley (1988) and Gearing (1988, 1991), the use of stable isotopes in trophic analyses is still problematic and cannot be used as unambiguous evidence in support of outwelling, especially when more than two potential sources of similar isotopic ratios are present. This latter situation is likely to be the norm rather than the exception, as one feature often

Table 3. Documented effects of "outwelled" mangrove detritus on consumer communities.

System	physical characteristics	Group studied	Enrichment effects	Reference
Florida, USA	mangrove estuary	all macrofauna	90% of all species had significant amounts of mangrove detritus in gut, suspected trophic dependence	Odum & Heald (1975)
Northeastern Australia	laboratory tests + field sampling	meiofauna	mangrove-derived tannins negatively affected meiofaunal density	Alongi (1987)
Australia	soft mud in estuary	bacteria	bacterial respiration increased with detritus exported but no significant net DOC flux across sediment despite high DOC level	Alongi *et al.* (1989)
Australia	soft mud	bacteria	concentration of OC and TN (but not P) proportional to mangrove litter abundance bacterial production increased with proximity to mangrove	Alongi (1990a)
Australia	soft mud	ciliates	exported detritus increased densities by changing edaphic characteristics, e.g. C/N ratio	Alongi (1990b)
Northeastern Australia	creek-bottoms in tidal mangrove, estuary & open embayments	epibenthos	positive effect on epibenthos but may be due to the provision of refuge	Daniel & Robertson (1990)
Great Barrier Reef Lagoon Australia	shallow sand, silt bottom	infauna (macro & meio)	densities negatively correlated with detritus and tannin levels, so were diversity and evenness	Alongi & Christoffersen (1992)

quoted of mangroves is their diverse array of primary producers (Odum *et al.*, 1982). Fry and Sherr (1989) suggested that ^{13}C fractionation increases across trophic levels steadily in offshore food chains, making it difficult to use this method to trace feeding relationships. In some studies utilising the stable isotope approach, e.g. Rezende *et al.* (1990), mangrove carbon was suggested to contribute to highly variable proportions of the total POC available and oceanic carbon may make up significant portions of the carbon pool.

More glamourous arguments for mangrove outwelling has, however, been cited based on the positive correlation relationship between mangrove areal extent and offshore shrimp catch in southeast Asia (Martosubroto & Naamin, 1977) and the salt marshes in the Gulf of Mexico (Turner, 1977).

Fate of detritus and impact on consumer communities

Another assumption of the Outwelling Hypothesis is the benign and beneficial nature of mangrove detritus to consumer communities. Whereas this may be true from an intuitively appealing, there is few evidence to suggest that mangrove detritus, when exported offshore, will enhance secondary productivity. Alongi (1987) and Alongi *et al.* (1989) investigated the effects of exported detritus on a bacterial community adjacent to a mangrove and concluded that high concentrations of tannins associated with mangrove detritus may deter meiofaunal and macrofaunal colonisation and bacterial production. It is likely that positive associations between detritus and macrofaunal standing crop result from the provision of extra spatial heterogeneity rather than a nutritional source by the mangrove detritus (Daniel & Robertson, 1990). Recent studies of the effect of outwelled mangrove detritus on the benthic community is summarized in Table 3. As a matter of fact, most past reports on decomposition and enrichment effects of mangrove litter to detritivores are based on findings of studies carried out in the mangrove environment, not in the recipient communities, i.e., more oceanic communities.

Odum & Heald (1975) first provided evidence that most estuarine consumers ingest significant amounts of mangrove detritus and suggested that this formed the trophic basis of the community. Subsequent studies on the nutritional value of detritus to consumers have indicated, however, that mangrove or marsh plant detritus usually has poor nitrogen content and may only be assimilated at low efficiencies by estuarine consumers, e.g., Kreeger *et al.*, (1990); Langdon & Newell, (1990). Assimilation efficiency was greatly enhanced by the colonisation of the detritus by bacteria, increasing from 3% to 10% in *Crassostrea virginica* (Langdon & Newell, 1990). Such microbial enrichment action is probably required in order that most macro-detritivores can beneficially utilise mangrove detritus. Such microbial enrichment has been shown to be important for detritivores to derive nutrition from vascular plant detritus (Bärlocher, 1985; Bärlocher *et al.*, 1989; Graça *et al.*, 1993). Most detritivores probably cannot effectively utilise unconditioned mangrove detritus containing high levels of refractory compounds, e.g. lignin. One counter example is the active consumption of fresh mangrove detritus by mangrove grapsids, which are apparently unaffected by the lack of microbial enrichment (Malley, 1978; Robertson, 1986; Lee, 1989). The exact digestive mechanism of these intertidal macro-detritivores still needs to be examined.

What is still unknown ?

Most past studies on mangrove outwelling were conducted in macro-tidal, "typical" mangrove communities. Whereas it seems likely from the literature that most tidal mangroves do serve as net exporters of organic carbon and nutrients, future studies should encompass the whole range of tidal regimes experienced by natural mangroves before a complete picture may be known.

Material exchange studies conducted in mangroves mostly also lack the resolution attained by studies with similar aims conducted in salt marshes. The role of the mangrove is often ascribed based upon only a crude mass balance calculation involving mostly only the particulate forms, and with little information on the extent of spatial and temporal integration and variability in the exchange pattern. This general low resolution of the mangrove studies may have been a result of 1) a more complex hydrodynamic environment in the mangroves, leading to flux calculations using the "flume" type design difficult; 2) a much shorter history of investigations on mangrove outwelling as compared with those on the salt marshes; and 3) a tendency to overstress the importance of the much more apparent particulate fraction and neglecting "invisible" exchanges taking place through water flow. This last point is of particular interest, as field biologist are often inclined to studying subjects which are "apparent". In the salt marsh where the bulk of the senescent materials are retained in the marsh and undergo in situ decomposition, this apparency problem is less important and might have facilitated the transfer of attention to the soluble fraction.

The mechanism of utilisation of mangrove detritus by offshore as well as mangrove consumers represents another gap of our knowledge. It has been shown that whereas most detritivores will prefer detritus enriched by micro-organisms, some consumers are adapted to utilising fresh detritus in particular environments such as the intertidal region, where they compete with the tide for organic matter.This competition between export and in situ consumers for organic matter in the tidal mangrove deserves more attention. Several questions are of interest: 1) how is adequate nutrition achieved by the detritivore if no prior microbial enrichment is present? 2) How important generally is in situ consumption and storage compared with tidal export? Since sesarmid crabs (agents reportedly consume significant amounts of litter) are a common and abundant feature of the Asia-Pacific mangroves (Jones, 1984), the effects on export should be assessed for more mangrove systems in the region.

The fate of outwelled detritus in offshore environments should also be scrutinized. Most past speculative studies on the beneficial effects of outwelling extrapolate from nutrient release data based on works carried out in the mangrove environment itself. By definition, however, outwelled detritus have their transformation and subsequent utilization in the oceanic or estuarine environment with a different suite of physical and chemical characteristics as well as different communities of microbial organisms. This extrapolation is inappropriate and the value of outwelled detritus to non-mangrove consumers can only be inferred from studies carried out in the receiving environments.

Acknowledgements

This work is supported by a grant from the Research Grants Council, Hong Kong. I thank Dr. J. E. Ong for his constructive criticisms of the manuscript.

References

Alongi, D M , 1987 The influence of mangrove-derived tannins on intertidal meiobenthos in tropical estuaries Oecologia 71 537–540

Alongi, D M , K G Boto & F Tirendi, 1989 Effect of exported mangrove litter on bacterial productivity and dissolved organic carbon fluxes in adjacent tropical nearshore sediments Mar Ecol Prog Ser 56 133–144

Alongi, D M , 1990a Effect of mangrove detrital outwelling on nutrient regeneration and oxygen fluxes in coastal sediments of the central Great Barrier Reef Lagoon Estuar Coast Shelf Sci 31 581–588

Alongi, D M , 1990b Abundances of benthic microfauna in relation to outwelling of mangrove detritus in a tropical coastal lagoon Mar Ecol Prog Ser 63 53–63

Bärlocher, F, S Y Newell & T L Arsuffi, 1989 Digestion of *Spartina alterniflora* Loisel material with and without fungal constituents by the periwinkle *Littorina irrorata* Say (Mollusca Gastropoda) J exp mar Biol Ecol 130 45–53

Beever, J W, D Simberloff & L L King, 1979 Herbivory and predation by the mangrove tree crab *Aratus pisonii* Oecologia 43 317–328

Benner, R , E R Peele & R E Hodson, 1986 Microbial utilisation of dissolved organic matter from leaves of the red mangrove, *Rhizophora mangle*, in the Fresh Creek Estuary, Bahamas Estuar coast shelf Sci 23 607–619

Boto, K G & J S Bunt, 1981 Tidal export of particulate organic matter from a Northern Australian mangrove system Estuar coast shelf Sci 13 247–255

Boto, K G & A I Robertson, 1990 The relationship between nitrogen fixation and tidal exports of nitrogen in a tropical mangrove system Estuar Coast Shelf Sci 31 531–540

Boto, K G & J T Wellington , 1988 Seasonal variations in concentrations and fluxes of dissolved organic and inorganic materials in a tropical, tidally-dominated, mangrove waterway Mar Ecol Prog Ser 50 151–160

Camilleri, J C & G Ribi, 1986 Leaching of dissolved organic carbon (DOC) from dead leaves, formation of flakes from DOC, and feeding on flakes by crustaceans in mangroves Mar Biol 91 337–344

Chalmers, A G , R G Wiegert & P L Wolf, 1985 Carbon balance in a salt marsh interactions of diffusive export, tidal deposition and rainfall-caused erosion Estuar coast shelf Sci 21 757–771

Chansang, H & S Poovachiranon, 1985 Fate of mangrove litter in a mangrove forest at Ko Yao Yai, southern Thailand Estuaries 8 106A

Childers, D L , S Cofer-Shabica & L Nakashima, 1993 Spatial and temporal variability in marsh-water column interactions in a southeastern USA salt marsh estuary Mar Ecol Prog Ser 95 25–38

Chong, V C , A Sasekumar, M U C Leh & R D'Cruz, 1990 The fish and prawn communities of a Malaysian coastal mangrove system, with comparisons to adjacent mudflats and inshore waters Estuar Coast Shelf Sci 31 703–722

Clark, P J , 1985 Nitrogen pools and soil characteristics of a temperate estuarine wetland in eastern Australia Aquat Bot 23 275–290

Dames, R F, J D Spurrier, T M Williams, B Kjerfve, R G Zingmark, T G Wolaver, T H Chrzanowski, H N McKellar & F J Vernberg, 1991 Annual material processing by a salt marsh-estuarine basin in South Carolina, USA Mar Ecol Prog Ser 72 153–166

Daniel, P A & A I Robertson, 1990 Epibenthos of mangrove waterways and open embayments community structure and the relationship between exported mangrove detritus and epifaunal standing stocks Estuar coast shelf Sci 31 599–619

Dittel, A I & C E Epifanio, 1990 Seasonal and tidal abundance of crab larvae in a tropical mangrove system, Gulf of Nicoya, Costa Rica Mar Ecol Prog Ser 65 25–34

Dittel, A I , C D Epifanio & O Lizano , 1991 Flux of crab larvae in a mangrove creek in the Gulf of Nicoya, Costa Rica Estuar Coast shelf Sci 32 129–140

Fallon, R D , S Y Newell & L C Groene, 1985 Phylloplane algae of standing dead *Spartina alterniflora* Mar Biol 90 121–127

Fleming, M , G Lin & L da Silveira Lobo Steinberg, 1990 Influence of mangrove detritus in an estuarine ecosystem Bull mar Sci 47 663–669

Flores-Verdugo, F J , J W Jr Day, R Briseno-Duenas, 1987 Structure, litterfall, decomposition, and detritus dynamics of mangroves in a Mexican coastal lagoon with an ephemeral inlet Mar Ecol Prog Ser 35 83–90

Fry, B & E B Sherr, 1984 $\delta^{13}C$ measurements as indicators of carbon flow in marine and freshwater ecosystems Cotnrib mar Sci 27 13–47

Fry, B & E B Sherr, 1989 $\delta^{13}C$ measurements as indicators of carbon flow in marine and freshwater ecosystems In P W Rundal, J R Ehleringer & R A Nagy (eds) Stable Isotopes in Ecological Research Springer-Verlag, New York 196–229

Gearing, J N , 1988 The use of stable isotope ratios for tracing the nearshore–offshore exchange of organic matter In B O Jansson (ed) Coastal-offshore Ecosystem Interactions Springer-Verlag, Berlin 69–100

Gearing, J N , 1991 The study of diet and trophic relationships through natural abundance ^{13}C In Coleman, D C & Fry, B (eds) Carbon Isotope Techniques Academic Press, New York, 201–218

Golley, F, H T Odum & R F Wilson, 1962 The structure and metabolism of a Puerto Rico mangrove forest in May Ecology 43 9–19

Gong, W K & J E Ong, 1990 Plant biomass and nutrient flux in a managed mangrove forest in Malaysia Estuar Coast Shelf Sci 31 519–530

Graça, M A S , L Maltby & P Calow, 1993 Importance of fungi in the diet of *Gammarus pulex* and *Asellus aquaticus* I feeding strategies Oecologia 93 139–144

Haines, E B , 1977 The origins of detritus in Georgia salt marsh estuaries Oikos 29 254–260

Haines, E B , 1979 Interactions between Georgia salt marshes and coastal waters a changing paradigm In R J Livington (ed) Ecological Processes in Coastal and Marine Systems Plenum Press, New York 35–46

Heinle, D R & D A Flemer, 1976 Flows of materials between poorly flooded tidal marshes and an estuary Mar Biol 35 359–373

Hopkinson, C S, 1988 Patterns of organic carbon exchange between coastal ecosystems the mass balance approach in salt marsh ecosystems In B -O Jansson (ed) Coastal-offshore Ecosystem Interactions Springer Verlag, Berlin 122–154

Jordan, T E , J W Pierce & D L Correll, 1986 Flux of particulate matter in the tidal marshes and subtidal shallows of the Rhode River Estuary Estuaries 9 310–319

Kreeger, D A , R I E Newell & C J Langdon, 1990 Effect of tidal exposure on utilization of dietary lignocellulose by the ribbed mussel *Geukensia demissa* (Dillwyn) (Mollusca Bivalvia) J exp mar Biol Ecol 144 85–100

Langdon, C J & R I E Newell, 1990 Utilization of detritus and bacteria as food sources by two bivalve suspension-feeders, the oyster *Crassostrea virginica* and the mussel *Geukensia demissa* Mar Ecol Prog Ser 58 299–310

Lee, S Y, 1989 The importance of Sesarminae crabs *Chiromanthes* spp and inundation frequency on the decomposition of mangrove (*Kandelia candel* (L) Druce) leaf litter in a Hong Kong tidal shrimp pond J exp mar Biol Ecol 131 23–43

Lee, S Y, 1990 Primary productivity and particulate organic matter flow in an estuarine mangrove-wetland in Hong Kong Mar Biol 106 453–463

Macintosh, D J, 1981 The importance of mangrove swamps to coastal fisheries and aquaculture In Proceedings of the Seminar on Some Aspects of Inland Aquaculture Mangalore, Karnataka, 14 and 15 July, 1980 27–33

Malley, D F, 1978 Degradation of mangrove leaf litter by the tropical sesarmid crab *Chiromanthes onycohophorum* Mar Biol 49 377–386

Martsubroto, P & N Naamin, 1977 Relationship between tidal forests (mangroves) and commercial shrimp production in Indonesia Mar Res Indonesia 18 81–86

Mazda, Y, H Yokochi & Y Sato, 1990 Groundwater flow in the Bashita-Minato mangrove area, and its influence on water and bottom mud properties Estuar Coastal Shelf Sci 31 621–638

Moran, M A, R J Wicks & R E Hodson, 1991 Export of dissolved organic matter from a mangrove swamp ecosystem evidence from natural fluorescence, dissolved lignin phenols, and bacterial secondary production Mar Ecol Prog Ser 76 175–184

Murray, F 1985 Cycling of fluoride in a mangrove community near a fluoride emission source J Appl Ecol 22 277–285

Newell, S Y, R D Fallon, R M Cal Rodriguez & L C Groene, 1985 Influence of rain, tidal wetting and relative humidity on release of carbon dioxide by standing-dead salt marsh plants Oecologia 68 73–79

Nixon, S W, 1980 Between coastal marshes and coastal waters – a review of twenty years of speculation and research on the role of salt marshes in estuarine productivity In P Hamilton & K B MacDonald (eds), Estuarine wetland Processes Plenum Publishing Corporation, New York 437–520

Nixon, S W, B N Furnas, V Lee, N Marshall, J E Ong, C H Wong, W K Gong & A Sasekumar, 1980 The role of mangroves in the carbon and nutrient dynamics of Malaysia estuaries In Soepadmo, E, A N Rao & D J Macintosh (eds) Mangrove environment research and development University of Malaya and UNESCO 534–544

Odum, E P, 1968 A research challenge evaluating the productivity of coastal and estuarine water In Proceedings of the Second Sea Grant Conference, University of Rhode Island 63–64

Odum, E P, 1984 The status of three ecosystem-level hypotheses regarding salt marsh estuaries tidal subsidy, outwelling and detritus-based food chains In V S Kennedy (ed) Estuarine Perspectives Academic Press, New York 485–495

Odum, W E & E J Heald, 1975 The detritus-based food web of an estuarine mangrove community In Cronin, L E (ed) Estuarine Research, Volume I Academic Press, New York 265–286

Odum, W E, J S Fisher & J C Pickral, 1979 Factors controlling the flux of particulate organic carbon from estuarine wetlands In Livingston, R J (ed) Ecological Processes in Coastal and Marine Systems Plenum Press, New York 69–80

Ong, J E, 1984 Mangrove outwelling? In J E Ong & W K Gong (eds) Productivity of the Mangrove Ecosystem Management Implications Universiti Sains Malaysia, Penang 30–36

Ovalle, A R C, C E Rezende, L D Lacerda & C A R Silva, 1990 Factors affecting the hydrochemistry of a mangrove tidal creek, Sepetiba Bay, Brazil Estuar Coast Shelf Sci 31 639–650

Rezende, C E, L D Lacerda, A R C Ovalle, C A R Silva & L A Martinelli, 1990 Nature of POC transport in a mangrove ecosystem a carbon stable isotope study Estuar Coast Shelf Sci 30 641–645

Robertson, A I, 1986 Leaf-burying crabs their influence on energy flow and export from mixed mangrove forests (*Rhizophora* spp) in northeastern Australia J exp mar Biol Ecol 102 237–248

Robertson, A I, 1987 The determination of trophic relationships in mangrove dominated systems areas of darkness In C D Field & A J Dartnall (eds) Mangrove Ecosystems in Asia and the Pacific status, utilisation and management Australian Institute of Marine Science, Townsville 292–304

Robertson, A I & P A Daniel, 1989a The influence of crabs on litter processing in high intertidal mangrove forests of tropical Australia Oecologia 78 191–198

Robertson, A I & P A Daniel, 1989b Decomposition and the annual flux of detritus from fallen timber in tropical mangrove forests Limnol Oceanogr 34 640–646

Robertson, A I & N C Duke, 1987 Mangroves as nursery sites of fish and crustaceans in mangrove and other nearshore habitats in tropical Australia Mar Biol 96 193–205

Robertson, A I & N C Duke, 1990 Recruitment, growth and residence time of fishes in a tropical Australian mangrove system Estuar Coast Shelf Sci 31 723–743

Robertson, A I & N C Duke, 1990 Mangrove fish-communities in tropical Queensland, Australia Spatial and temporal patterns in densities, biomass and community structure Mar Biol 104 369–379

Robertson, A I, D M Alongi, P A Daniel, K G Boto, 1990 How much mangrove detritus enters the Great Barrier Reef lagoon? In Choat, J H, D Barnes, M A Borowitzka, J C Coll, P J Davies et al (eds) Proceedings of the Sixth International Coral Reef Symposium, Townsville, Australia, 8th–12th August 1988 volume 2 Contributed Papers 1988 601–606

Robertson, A I, D M Alongi & K G Boto, 1992 Food chains and carbon fluxes In Robertson, A I & D M Alongi (eds) Tropical mangrove ecosystems American Geophysical Union, Washington, D C 293–326

Rodelli, M R, J N Gearing, P J Gearing, N Marshall & A Sasekumar, 1984 Stable isotope ratio as a tracer of mangrove carbon in Malaysian ecosystems Oecologia 61 326–333

Rozas, L P, C C McIvor, & W E Odum, 1988 Intertidal rivulets and creekbanks corridors between tidal creeks and marshes Mar ecol Prog Ser 47 303–307

Snedaker, S C 1978 Mangroves their value and perpetuation Nat Resour 14 6–13

Subramanian, P, M J Prince Jeyaseelan, K Krishnamurthy, 1984 The nature of biodegradation of vegetation in mangrove ecosystem Chem Ecol 2 47–68

Teal, J M, 1962 Energy flow in the salt marsh ecosystem of Georgia Ecology 43 614–624

Thom, B G 1982 Mangrove ecology – a geomorphological perspective In Clough, B F (ed) Mangrove ecosystems in Australia Australian National University Press, Canberra 3–17

Thomas, J P, 1966 The influence of the Altamaha River on primary production beyond the mouth of the river M S Thesis, University of Georgia, Athens 1–88

Turner, R E, 1977 Intertidal vegetation and commercial yields of Penaeid shrimp Trans am Fish Soc 106 411–416

Twilley, R R, 1983 Litter dynamics and organic carbon exchange in black mangrove (*Avicennia germinans*) basin forests in a southwest Florida estuary Ph D dissertation, University of Florida, Gainesville, 274 pp

Twilley, R R, 1985 The exchange of organic carbon in basin mangrove forests in a southwest Florida estuary Estuar coast shelf Sci 20 543–557

Twilley, R R, 1988 Coupling of mangroves to the productivity of estuarine and coastal waters In B -O Jansson (ed) Coastal-offshore Ecosystem Interactions Springer-Verlag, Berlin 155–180

Twilley, R R, R H Chen & T Hargis, 1992 Carbon sinks in mangroves and their implications to carbon budget of tropical coastal ecosystems Water, Air and Soil Pollut 64 265–288

Twilley, R R, A E Lugo & C Patterson-Zucca, 1986 Litter production and turnover in basin mangrove forests in southwest Florida Ecology 67 670–683

Valk, A G Van der & P M D Attiwill, 1984 Decomposition of leaf and root litter of *Avicennia marina* at Westernport Bay, Victoria, Australia Aquat Bot 18 205–221

Wattayakorn, G, E Wolanski & B Kjerve, 1990 Mixing, trapping and outwelling in the Klong Ngao mangrove swamp, Thailand Estuar Coast Shelf Sci 31 667–688

Wetzel, R G, 1984 Detrital dissolved and particulate organic carbon functions in aquatic ecosystems Bull mar Sci 35 503–509

Whiting, G J, H N McKellar, B Kjerfve & J D Spurrier, 1987 Nitrogen exchange between a southeastern USA salt marsh ecosystem and the coastal ocean Mar Biol 95 173–182

Wolanski, E, 1992 Hydrodynamics of mangrove swamps and their coastal waters In Jaccarini, V & E Martens (eds) The Ecology of Mangroves and Related Systems Kluwer Academic Publishers, Belgium 141–161

Wolanski, E, M Jones & J S Bunt, 1980 Hydrodynamics of a tidal creek-mangrove swamp system Aust J mar Freshwat Res 31 431–450

Wong, C -H 1984 Mangrove aquatic nutrients In Ong, J -E & W -K Gong (eds) Proceedings of the Workshop on Productivity of the Mangrove Ecosystem Management Implications 60–67

Woodruffe, C D 1985a Studies on a mangrove basin, Tuff Crater, New Zealand I Mangrove biomass and production of detritus Estuar Coast Shelf Sci 20 265–280

Woodruffe, C D 1985a Studies on a mangrove basin, Tuff Crater, New Zealand II The flux of organic and inorganic particulate matter Estuar Coast Shelf Sci 20 447–462

Hydrobiologia **295**: 213–220, 1995.
Y. S. Wong & N. F. Y. Tam (eds), Asia-Pacific Symposium on Mangrove Ecosystems.

An ecological study on the Mollusca in mangrove areas in the estuary of the Jiulong River

J. X. Jiang & R. G. Li
Third Institute of Oceanography, SOA, 361 005, Xiamen, People's Republic of China

Key words: Mollusca, mangrove, Jiulong River, estuary, ecology

Abstract

Based on the surveying data obtained in February, May, August and November, 1987, the ecology of Mollusca in mangrove areas in the estuary of the Jiulong River in Fujian Province was studied. The results are as follows: There are 52 species of Mollusca in the areas, and species and abundance distribution increase with salinity. Biomass and density average 7.99 g m^2 and 25 ind. m^2 respectively, with higher values in autumn and winter, and lower in spring and summer. Vertical distribution of biomass is highest in high tide zones and lowest in low tide zones, and density is highest in mid-tide zones and lowest in low tide zones. The relationship between the distribution of Mollusca and the environmental factors are discussed.

Introduction

As a type of ecosystem in the world of many subtropic and tropical estuaries and intertidal of bays, mangrove ecosystem owns high primary production. The deposit in the sediments from mangal leaves, steams and flowers, etc. is gradually decomposed due to the interactions of physical and microbiologic factors, which makes up the bases of detritus food web. Marine invertebrates form an important linkeges between the primary detritus at the bases of food web and the consumers in the higher tropic levels. However, little is known about the invertebrates in the mangrove ecosystem (Well, 1990). It is known that molluscs and crustaceans are important animals in the mangrove area.

Gao & Li (1985) reported that among macrofaunas in the mangrove in the Jiulong River Estuary there were 22 species of molluscs accounting for 33.3% of the total species, among which 20 species were gastropods and 2 species, bivalves. Yipp (1982) analysed the distributions of 5 gastropods and their food compositions in the mangrove of Hong Kong where the relations of food to the distributions of animals was discussed. Cha (1992) studied the factors, mainly desiccation tolerance, governing the vertical distributions of 4 gastropods. He ranged out the distributing zones for those species. Well (1984) compared the distribution of macrofauna with Molluscs and crustaceans in mangrove in the NW of Australia, where 21 molluscs and 16 crustaceans were found. Well (1985) pointed out that six of nine species of Potamidiae were living in the mangrove area in Hong Kong. Well (1990) emphasized the importance of molluscs in the mangrove after he studied the distributions of invertebrates in Hong kong. In this area, 24 species of molluscs accounts for 75% the total marine invertebrates with density and biomass at 93.3% and 94.8% of the whole, respectively.

In this paper, the species distribution, seasonal variations and vertical distributions of molluscs in the mangrove of the Jiulong River Estuary and their relations to the environment are discussed.

Material and methods

This paper reports data obtained in February, May, August and November, 1987. The surveying sites were located in the mangrove areas of Fugang, Haimen and Zhen'an in the estuary of the Jiulong River (Fig. 1).

For quantitative benthic sampling, 5 sampling stations were established, eight samples at each station, at interval of 5 m, taken using a quantitative frame of $25 \times 25 \times 35$ cm^3, and were set from high tide shore-

Table 1. Chraacteristics of sediments of surveying sites

Sites			Sediments	Grain size coefficient		pH	OM (%)
				(MdΦ)	(QDΦ)		
Fugong	high-tide zone		clay-silt	7.80	1.67	6.65	3.24
	mid-tide zone	upper part	clay-silt	7.85	1.61	6.84	2.53
		middle part	clay-silt	7.63	1.55	7.01	2.05
		lower part	clay-silt	7.10	1.84	7.37	1.91
	low tide zone		clay-silt	7.38	1.61	7.40	2.13
Haimen	high-tide zone		gravel-sand	1.47	1.18	–	0.19
	mid-tide zone	upper part	clay-silty-sand	5.16	2.90	7.13	1.20
		middle part	sand-clay-silt	7.00	2.22	7.30	2.11
		lower part	clay-sandy-silt	5.24	2.33	7.35	2.00
	low tide zone		clay-sandy-silt	5.98	2.52	7.44	1.87
Zhen'an	high-tide zone		clay-sandy-silt	0.74	1.72	–	0.51
	mid-tide zone	upper part	middle-fine-sand	1.96	0.68	7.36	0.33
		middle part	clay-silty-sand	2.91	3.09	7.31	0.90
		lower part	clay-silt	7.14	1.77	7.27	1.28
	low tide zone		silt-clay-sand	6.01	3.16	7.30	1.20

MdΦ = medium grain size
QDΦ = 1/2 (Φ 75– Φ 25)
OM = organic matter

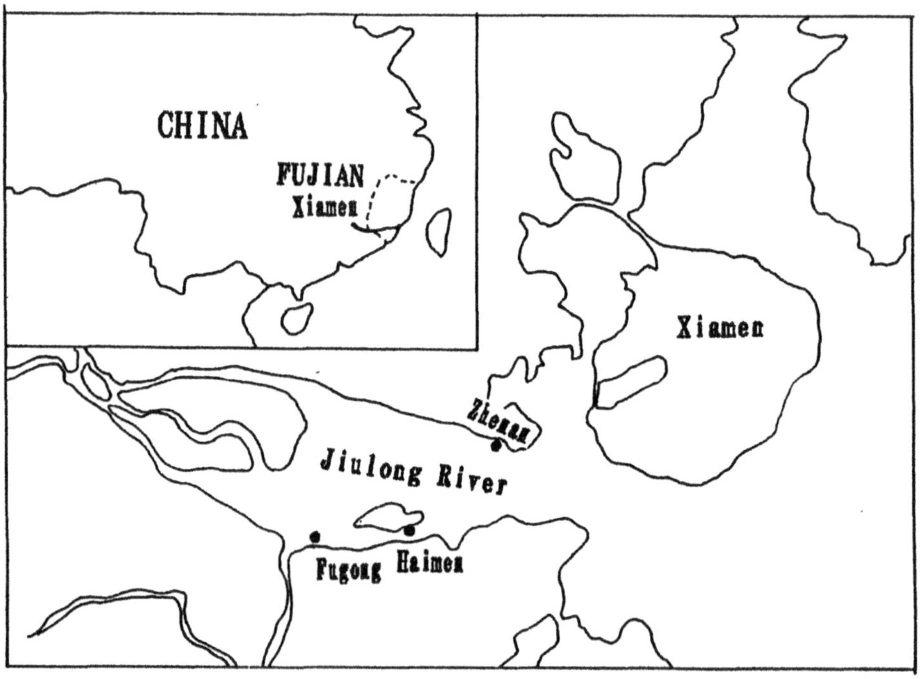

Fig. 1. Sites for mollusc surveying in the mangrove areas in the estuary of Jiulong River of Fujian.

Table 2 Species list and distribution of Molluscs in the mangrove areas in the Jiulong River estuary, Fujian

Species	Fugong	Haimen	Zhen'an
Bivalvia			
Mytilidae			
Modolius sp	+		
Musculista senhousia (Benson)		+	+
Tellinidae			
Moerella iridescens (Benson)	+	+	
Moerella sp	+	+	+
Nitidotellina minuta (Lischke)		+	+
Nitidotellina nitidula (Lischke)	+	+	+
Angulus vestalioides (Yokoyama)	+	+	+
Angulus restalis (Hanley)			+
Pulvinus micans (Hanley)			+
Tellina sp		+	+
Merisca sp		+	+
Macoma sp			+
Semelidae			
Theora sp		+	
Psammobiidae			
Hiatula sp	+		
Pharellidae			
Sinonovacula constricta (Lamarck)	+	+	+
Pharalla acutidens (Broderip et Sowerby)	+		
Corbiculidae			
Corbicula fluminea (Muller)	+	+	
Geloina coaxans (Gmelin)		+	
Veneridae			
Cyclina sinensis (Gmelin)			+
Clausinella isabellina (Philippi)			+
Glaucomyidae			
Clauconme chinensis Gray		+	+
Corbulidae			
Potamocorbula ustulata (Reeve)	+	+	
Laternulidae			
Laternula (Exolaternula) marilina (Reeve)			+
Thraciidae			
Trigonothracia jinxingae Xu,			+

line to low tide shoreline on each side For the benthic fauna classification, sediment samples were put into a Spiral Benthos Apparatus and then sifted on a 1 mm sieve mesh

Physical and chemical factors measured include temperature, salinity, dissolved oxygen, oxygen consumption, pH and alkalinity, grain size, pH and organic matter content in sediments They were determined simultaneously during benthic sampling

Results

Natural environment

The Jiulong River estuary of Xiamen is in the subtropic area in SN Asia and is characterized by a warm and humid climate, ample sunshine, strong monsoon season and frequent typhoons It is not hot in summer and is warm in winter Precipitation comes with

Table 2 cont

Species	Fugong	Haimen	Zhen'an
Gastropoda			
Prosobranchia			
Turbinidae			
Lunella coronata granulata (Gmelin)			+
Neritidae			
Nerita sp	+	+	+
Nerita (Ritena) yoldii (Recluz)			+
N (Ampinerita) polita Linnaeus	+		+
Littorinidae			
Littoraria (Palustorina) melanostoma Gray	+	+	+
Stenothyridae			
Stenothyra glabar A Adams	+		
Assimineidae			
Assiminea brevicula Pferffer	+	+	+
Potamodidae			
Cerithidea sp			+
Cerithidea ornata (A Adams)	+	+	+
C cingulata (Gmelin)		+	+
Batillaria sp			+
Batillaria cumingi (Crosse)			+
B zonalis (Bruguere)			+
Epitoniidae			
Spiniscala sp			+
Malanellidae			
Eulima maria (A Adams)			+
Naticidae			
Lunatica gilva (Philippi)			+
Muricudae			
Thais gradata Jonas			+
Pyrenidae			
Mitrella bella Reeve		+	+
Nassariidae			
Nassarius (Zeuxis) dorsatus (Roeding)		+	
N (Reticunassa) festivus (Powys)			+
Nassarius sp		+	+

the monsoon. There are distinct dry and wet seasonal alternations usually with a rainy season from April to August that accounts for 67% of annual average rainfall.

Annual average air temperature is 20.9 °C with the highest 24.8 °C and the lowest 18.1 °C. The highest annual air temperature is 38.5 °C and the lowest, 2.0 °C. Annual average relative humidity is 77% and annual mean rainfall is 1183 mm. The highest one-year recorded rainfall is 1771.8 mm.

It is estimated that 14.8 billion cubic meters of runoff come from the Jiulong River per year and that the amount of sand transported by the river is 3.07 million tons with average 0.207 kg m^3. A large amount of mud and sand is deposited in the estuarine area to the west of Jiyu Islet.

According to the hydrological data obtained by a local hygrological station from 1957 to 1971, the annual mean water temperature at 0.5 m depth was 21.1 °C with the highest one 31.4 °C and the lowest, 10.2 °C.

Table 2 cont..

Species	Fugong	Haimen	Zhen'an
Opisthobranchia			
Pyramidellidae			
Pyramidella sp.			+
Ringiculidae			
Pseudoringicula sinensis Lin			+
Atyidae			
Bullacta exarate (Philippi)		+	+
Retusidae			
Retusa (Coelophysis) boenensis (A. Adams)			+
Triclidae			
Ecoylichna braunsi (Yokoyama)			+
Acteocinidae			
Acteocina sp.			+
Pulmonata			
Ellobiidae			
Pythia cecillei (Philippi)			+
Onchidiidae			
Oncidium verruculatum Cuvier			+
Scaphopoda			
Dentaliidae			
Graptacme buccinulum (Gould)			+

Physical and chemical characteristics in the surveying areas

The tide in the estuary of the Jiulong River is a regular semidiurnal tide with different tide heads each day.

1. Water temperature. It is similar at Fugong and Hiamen, ranging from 16.50 to 32.20 °C in a year. It ranges from 16.20 to 30.40 °C at Zhen'an.
2. Salinity. It is lowest at Fugong with wide variation (2≈18.2). It is 6.4~25.8 at Haimen and high at Zhen'an with a narrow variation (22.4~26.8).
3. Dissolved oxygen (mg l^{-1}). All values at the three sites are over 5.0 except for one, which is 4.5 at Zhen'an in August.
4. Oxygen consumption (mg l^{-1}). All values at the three sites are below 3.0 except for one, which is 4.0 at during low tide at Fugong.
5. pH. All pH values are (7.5~8.5).
6. Alkalinity (mg mol l^{-1}). It is (0.710~2.113 mg mol l^{-1}) in the surveying areas.

Species composition and distribution

54 Mollusc species were collected in the mangrove areas at Fugong, Haimen and Zhen'an in the survey. Among them, there are 24 species of Bivalvia which accounted for 44.4% of all species, 21 species of Prosobranchia accounting for 39%, 6 species of Opisthobranchia accounting for 11%, 2 species of Pulmonata, and 1 species of Scaphopoda (Table 2).

Mollusc in the surveying area are mainly composed of near-shore euryhalinous species. The main representatives include *Musculista senhousia (Benson), Nitidotellina nitidula (Lischke), Angulus vestalioides (Yokoyama, Sinonovacula constricta (Lamarck), Trigonothracia jinxingae Xu, Cerithidea cingulata (Crosse), C. ornata (A. Adams), Batillaria zonalis (Bruguere)*, and *Nassarius (Reticunassa) festivus (Powys)*.

Estuarine low-salinity species are mostly found at Fugong and Haimen where it is low in salinity. The main representative species are *Moerella iridescens (Benson), Pharalla acutidens (Broderip et Sowerby), Potamocorbula ustulata (Reeve)* and *Stenothyra glabra A. Adams*.

Table 3. Seasonal variation in species number of molluscs in the mangrove areas in the Jiulong River estuary

Site	Seasons				
	Spring	Summer	Autumn	Winter	Annual
Fugong	7	8	6	7	16
Haimen	8	10	8	10	23
Zhen'an	16	19	24	22	44

Table 4. Seasonal variation of biomass of molluscs in the mangrove areas in the Jiulong River estuary

Number	Seasons				
	Spring	Summer	Autumn	Winter	Annual
Biomass (g m^2)	5.45	5.51	10.25	10.73	7.99
Density (ind. m^2)	17	16	22	45	25

The species numbers increase remarkably with increasing salinity. There are 16 species at Fugong with salinity 10.87, 23 at Haimen with 15.78, and 44 at Zhen'an with 23.50 (Fig. 2), which shows the distribution of Mollusc in this subtropical estuary.

The seasonal variation for the mollusc species are small from site to site. It differs insignificantly between Fugong and Haimen. However, a large difference is found in autumn and winter compared with that in spring and summer at Zhen'an (Table 3).

The results show that there are 34 mollusc species in the mangrove area which 18 species appear only inside the mangrove area and 35 in the non- mangrove area. In the latter area, 19 species are exclusively non-mangrove species. Besides, 16 are common species both in mangrove and non-mangrove areas.

The distribution of dominant species varies from site to site. At Fugong, they are *Moerella iridescens, Pharalla acutidens* and *Potamocorbula ustulata*, at Haimen, *Glauconme chinensis, Potamocorbula ustulata, Moerella iridescens, Angulaus vestalioides* and *Cerithidae cingulata*, and at Zhen'an, *Glauconme chinensis, Angulaus vestalioides, Cerithidea cingulata, C. ornata, Batillaria cumingi* and *Retusa (Coelophysis) boenensis*.

Table 5. Vertical distribution of mollusc numbens in mangrove areas in the Jiulong River estuary

Locations	Biomass (g m^2)	Density (ind. m^2)
High tide zones	11.39	24
Mid-tide zones		
upper	15.62	34
middle	4.52	40
lower	2.76	18
mean	7.63	31
Low tide zones	2.09	9

Abundance distribution and seasonal variation

Biomass and density of mollusc in the surveying area are low with the annual mean 7.99 g m^2 and 25 ind. m^2, respectively. Both biomass or density are higher in autumn and winter than in spring and summer (Table 4).

The mollusc abundance in different locations in the estuarine mangrove area differs significantly. Both biomass and abundance of the animals are higher in the mangrove of Zhen'an than that of Fugong and Haimen. It is considered that the differences are the results of salinity gradients and sediment types (Table 1). The sediment at Fugong and Haimen is dominated by the fine grains while that at Zhen'an is mainly coarse grains with some fine grains. The highest biomass is found at Zhen'an with value of 19.29 g m^2, and at Haimen and Fugong, 2.89 g m^2 and 1.79 g m^2, respectively. The densities are 59, 11 and 5 ind. m^2 at Zhen'an, Haimen and Fugong in sequence.

The vertical distribution of bivalve biomass in the mangrove of the Jiulong River estuary decrease from the high tide zone to the low tide zone. The vertical distribution of density decreases from the mid-tide zone to the high tide zone with the lowest density found in low zone (Table 5).

Discussion

Relationship between species distribution and salinity

The surveyed areas located in the estuary of the Jiulong River where environmental factors change greatly

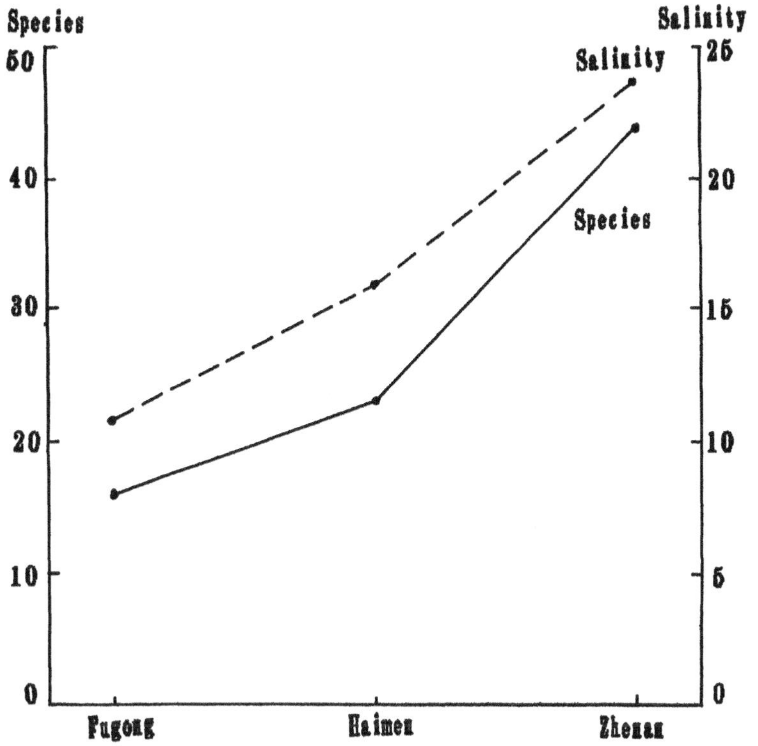

Fig. 2. Relationship between species distribution of molluscs and sanility in the mangrove areas in the estuary of Jiulong River.

in intertidal zone. The salinity is the most important one compared with other factors such as temperature, tide, sediments, etc. The numbers of mollusc species in the areas decrease with the dropping of salinity. As shown in Fig. 2, the salinity is the highest (23.80) with 44 species at Zhen'an, the second (15.78) with 23 species at Haimen and the lowest (10.87) with 16 species at Fugong. As the surveyed areas are influenced by varing runoff (148 × 108 m-3 yearly in average) of the Jiulong River, the salinity varies seasonally. Remarkable differences in salinity limit the distribution of most species. For example, some species such as *Batillaria zonalis, Nassarius (Reticunassa) festivus* and *Retusa (Coelophysis) boenensis* living at Zhen'an do not appear at Haimen and Fugong. In contrast, some low salinity species like *Potamocorbula ustulata, Corbicula fluninea* and *Stenothyra glabar* appear only at Haimen and Fugong.

Relationship between vertical distribution of species and the tide

The vertical distribution of molluscs, characterizing by their desiccation resistance, is closely related to tide which submerged and exposed the animals regularly. The results of the lab and field experiments (Cha, 1992) showed that animals living in high zone had strong resistance to dissociation and heat than those in middle and low tide zones. Yipp (1982) believed that gastropods in mangroves overlaped in their distribution ranges but each species has its own. For instance, *Terebralia sulcata* and *Cerithidea rhizophorarum* are found in the high zones which *C. cingulata, C. djad jariensis* and *Batillaria zonalis* in the middle tide zone. The results of this study shows that the ranges of vertical distribution for molluscs in the surveyed areas vary with different species, which is the result of long time adaptation by evolution, showing the adaptive capability to the environment as well as the rigidity in distribution. For example, species occurring in high tide zone may include bivalve *Pharalla acutidens* and *Glauconme chinensis* and gastropod *Cerithidea ornata*; in middle tide zone, bivalve *Moerella iridescens, Angulus vestalioides* and *Laternula marilina*, gastropod *Assiminea brevicula, Cerithidea cingulata, Batillaria cumingi, B. zonalis, Nassarius festivus* and *Retusa boenensis*; and in the areas from lower middle tide zone to low tide zone, *Potamocorbula ustulata, Trigonoth- racia jinxingae, Eulima maria, Eocylichna braunsi* and *Graptacme buccinulum.*

Relationship between species distribution and sediments

The size distribution of the settled particles in the mudflat of interidal zone determines the composition, feeding behavior, numerical development and distribution of the animals in these habitats. The intertidal sediments at Fugong and Haimen are composed of fine particles, mainly clay-silt (Table 1), and high in organic content with value of 1.01–3.24% for the former and 1.87–2.11% for the latter.

They are dominated by infauna such as *Moerella iridescens, Angulus vestalioides, Nitidotellina nitidula, Pharalla acutidens, Potamocorbula ustulata* and *Glauconme chinensis.* The intertidal sediments at Zhen'an are more complex. They are composed of clay-sand-silt in high tide zone and middle fine-sand in upper middle tide zone with low organic content (0.33–0.51%), and dominated by epifauna such as *Cerithidea ornata, C. cingulata, Batillaria cumingi, B. zonalis, Nassarius festivus* and *Nerita yoldi.*

The study by Wells (1990) in the mangrove area at Sai Keng, New Territories, Hong Kong showed that among 23 species of molluscs the main species were epifauna (18), and less for infauna (2). In 6 feeding types he found there were mainly deposit feeders (9 species), surface raspers (7) and filter feeders (5), accounting for 91.3% of total species. He found that in fact there was little infaunal mollusc in mangrove areas, without exception in his studies. However, the result of our study is different. We found that there were 34 species of molluscs in mangrove areas, among them 23 were epifauna and 11, infauna. For the reason of this difference, we suggest that, apart from surveyed areas, it is mainly due to the selection of sampling section, sampling representativeness as well as the sample handling *in situ.* For the selection in estuarine areas, salinity gradient, various sediment types and niches must be considered, and so to the sampling representativeness and the sample handling *in situ.* Samples should not be taken only in dense mangrove areas or in acid sediments in order to present the whole biodiversity of invertebrate in the mangroves.

It is worth to point out that Gao & Li (1985) did similar work on the fauna in mangrove areas in the estuary of the Jiulong River. However, their results were quite different from ours. They reported 22 species of molluscs, 20 for epifauna and 2 for infauna, which was less than what we found. Although they selected more sections (5) with large areas they found less molluscs. The main reason may due to their improper selection of sections and sampling representativeness.

References

Cha, M. W., 1992. Factors affecting the vertical distribution of four mangrove gastropods in Hong Kong. Marine Flora and Fauna of Hong Kong and South China III. (Edited by Brian Morton). Hong Kong University Press: 373–382.

Gao, S. H. & F. X. Li, 1985. Community ecology of grund-dwelling macrofauna of mangrove in the Jiulong River Estuary, Fujian. Taiwan Strait. 4: 179–191.

Wells, F. E., 1984. Comparative distributions of macromolluscs and macrocrustaceans in a northwestern Australian mangrove system. Aust. J. mar. Freshwat. Res. 35: 591–596.

Wells, F. E., 1985. The Potamidiae (Mullusca: Gastropoda) of Hong Kong, with mangrove system. In Proceedings of the Second International Workshop on the Malacofauna of Hong Kong and Southern China, Hong Kong, 1983. (Eds. B. Morton & D. Dudgeon). Hong Kong University Press: 135–154.

Wells, F. E., 1990. Distribution of Marine Invertebrates in a Hong Kong mangrove with emphasis on Molluscs. Proceedings of the Second International Marine Biological Workshop: The Marine Flora and Fauna of Hong Kong and Southern China, Hong Kong, 1986. (Ed. B. Morton) Hong Kong: Hong Kong Press: 783–793.

Yipp, M. W., 1982. The distribution of ground-dwelling gastropods in a small mangrove stand in Hong Kong. In Proceedings of the First International Marine Biological Workshop: The Marine Flora and Fauna of Hong Kong and Southern China, Hong Kong, 1980. (Eds. B. Morton & C. K. Tseng). Hong Kong University Press: 705- 720.

Hydrobiologia **295**: 221–230, 1995.
Y. S. Wong & N. F. Y. Tam (eds), Asia-Pacific Symposium on Mangrove Ecosystems.

The temporal changes in benthic abundances and sediment nutrients in a mudflat of the Chuwei Mangrove Forest, Taiwan

Cheng, I-Jiunn
Institute of Marine Biology, College of Fishery Sciences, National Taiwan Ocean University, Keelung, Taiwan, 20224, R.O.C.

Abstract

The temporal variation of the surface deposit feeders in a mudflat of Chuwei Mangrove Forest showed a significant input of populations from the recruitments in the spring. The chemical analysis of the sediment in the top 8 cm showed that the sedimentary organic content had a subsurface maximum value in the deeper layer in most cases. Most of the temporal variations in the organic carbon and nitrogen contents occurred in the top few centimeters. The high C:N ratio and low organic nitrogen content in the sediment suggests the supply of carbon (or energy) to the benthos was low. The increase of labile protein content in the deep layer might be due to the downcore transport of sediment nutrients by the bioturbation. Laboratory growth experiment suggests that some the chemical characters examined in this study may represent the sediment nutrients to the animals.

Introduction

Mangrove forest is one of the most productive fringe ecosystems in coastal regions. It serves as a hatching and nursery grounds for large numbers of commercial species. Despite the fact that most deposited litter is refractory in nature, it serves as a major food source for the local benthos (Alongi, 1990; Alongi & Christofferson, 1992; Fleming *et al.*, 1990; Kristensen *et al.*, 1991; Zieman *et al.*, 1984). Mangrove-derived tannins, though exerts a negative impact on benthic animals, are leached and lost from the mangrove litter within a month or two (Alongi & Christofferson, 1992; González & Mee, 1988). In addition, mangrove stands also receive nutrients from other sources, such as terrestrial runoff and estuary outflows, depending on the geomorphology of the basin and tidal amplitude (Alongi, 1990; Boto & Bunt, 1981; Fleming *et al.*, 1990; Twilley, 1985). The mangrove forest of the Chuwei swamp is a semi-closed brackish marshland. It receives inorganic nutrients such as N, K, Ca, from the Tamshui Estuary during the flood tide and exports dissolved organic matter such as $NO^{-}{}_{3}$ during the ebb tide (Chow & Bi, 1990).

Most freshly deposited organic matter are either utilized by benthic organisms or resuspended and exported during the tidal phase. Only a small fraction is buried and lost permanently from the detrital food web. Bioturbation plays an important role in vertical distribution of sediment nutrients (Rhoads & Boyer, 1982). Local sedimentation rate can also influence the residence time of the deposited organic matter at specific depth. It is probably true that the abundance of surface deposit feeders reflect the nutritional value of the organic matter, even though the bioturbation can limit the accessibility of the sediment nutrients to these animals (Cheng *et al.*, 1993).

Detritus, despite of its low utilization, may not be a limiting factor for deposit feeders because of its large storehouse in sediments (Cheng & Lopez, 1991; Levinton & Stewart, 1988; Lopez & Levinton, 1987). Microbial foods, such as diatoms and sedimentary bacteria, are highly utilizable, also unlikely to be a limiting factor (e.g. Cheng & Lopez, 1991; Lopez & Levinton, 1987; Phillips, 1984). Bioavailable nitrogen, on the other hand, can limit the growth of the animals (Rice *et al.*, 1986). Cheng *et al.* (1993) found that in a temperate saltmarsh mudflat, the sediment nutrients were exhausted in the summer, due to the heat and high decomposition, and recharged in the late fall and winter, due to the low temperature and animal activities.

In this study, by employing a series of chemical analysis and a bioassay method, we tested the hypoth-

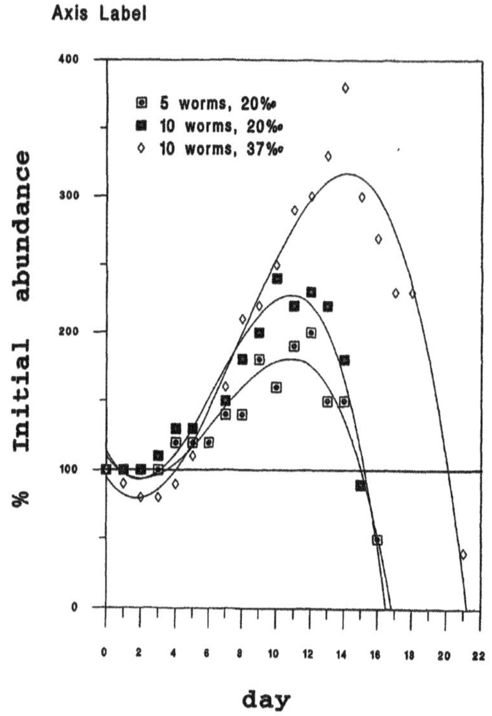

Fig. 1. Preliminary growth experiment of *Paranais litoralis*. The experiment was designed for three conditions, 5 worms in 20% salinity seawater, 10 worms in 20% salinity seawater, and 10 worms in 37% salinity seawater.

esis that in a subtropical mudflat of Chuwei Mangrove Forest, the temporal change of the surface deposit feeders reflect the changes of sediment nutrients. Bioturbation (e.g. crab bioturbation) may transport nutrients to the deeper layers (Kristensen *et al.*, 1992). This implies that there will be a subsurface maximum in the vertical distribution of the sediment nutrients.

Study site

The study site is a mudflat located in Chuwei Mangrove Forest on the eastern shore of the relatively well-developed Tamshui Estuary, Taipei County (25 °9′N to 25 °10′N, 121 °26′E to 27′E). According to the degree of tidal action and vegetation pattern, the forest is subdivided into six habitats, tidal creek mangrove, riverside mangrove, mixed stand, grassland, dwarf mangrove, and sandy riverside (Chou & Bi, 1990). The mudflat stands in the middle of the tidal creek mangrove and the sediment is composed mainly of the poorly to very poorly sorted, fine to very fine sand, with mud content ranges from 26 to 39% total sediment weight. The mean grain size ranges from 70 to 137 μm in diameter (Cheng, unpublished data).

Materials and methods

Sediment collection and storage

From March 1992 to April 1993, animal and sediment samples were collected from a mid-intertidal field site on the mudflat, usually every 4 mo. for analysis of sediment properties, macrofauna abundance, and biogeochemical properties. Cores of 3.5 cm in diameter and few centimeters in depth were collected ($n = 20$). In order to determine the abundance of surface deposit feeders, the sediment in the top 2 cm were sieved through a 0.5 mm mesh. We counted only the most common invertebrates in theses depth intervals, the oligochaetes, *Paranais litoralis* (Müller), and *Monopylephours parvus* (Ditlevsen), and the polychaete *Capitella* sp. For all samples, sediment cores were subdivided into 1 cm depth intervals to establish the depth occurrences of these invertebrates.

Chemical analysis

In order to establish the change of sediment nutrients with depth, three sediment cores were subdivided into 0–1, 1–2, 2–3, 4–5, and 7–8 cm depth intervals. In each depth, sediment was analyzed for the total organic content, the organic carbon and nitrogen contents, and the labile protein content. Total organic content was determined by the weight loss of sediment upon ignition. Organic carbon and nitrogen contents were analyzed by a CHN-elemental analyzer (Perkin-Elmer Model 2400 Series II). Samples were acidified with the concentrated HCl vapor overnight to remove inorganic carbonate before analysis. Labile protein was measured using the method of Mayer *et al.* (1986), which assays those polypeptides amenable to enzymatic hydrolysis.

Bioassay method: laboratory assays of population growth

Bioassay approach such as population growth is one of the meaningful methods to characterize the sediment nutrients (Cheng, *et al.*, 1993; Levinton & Stewart, 1988). The population responses and grow well

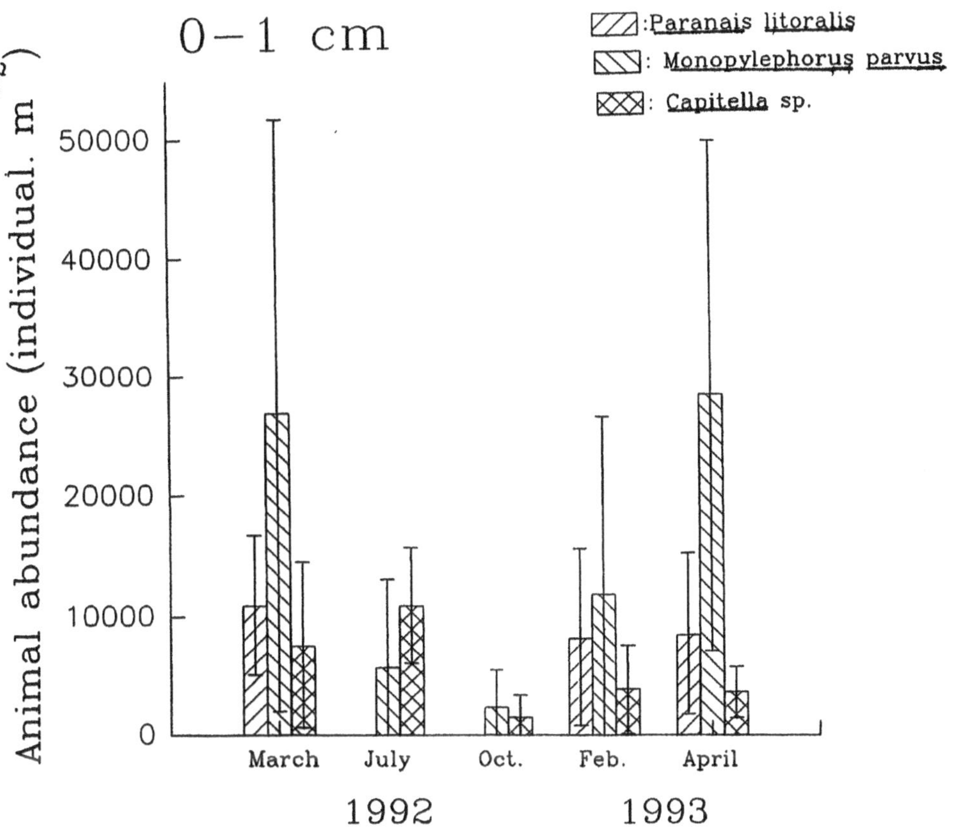

Fig. 2. Temporal variations of three common surface deposit feeders in the top (0–1 cm) layer of the Chuwei Mangrove Forest mudflat.

in the nutrient-rich sediment. The population, however, tends to shrink and decrease in nutrient-poor sediment. In this study, naid oligochaete *Paranais litoralis* was chosen to assess the population responses to sediments collected in different seasons and at different levels below the sediment-water interface. *P. litoralis* grows by naid paratomy, an asexual form of reproduction involving addition of segments, and budding off of a daughter individual. This form of growth makes this species ideal for laboratory growth studies (Cheng *et al.*, 1993). Because the population may responds to different salinity and food abundance differently, in the preliminary studies, animals were grown at random combinations of salinities and animal densities (temperature was not considered, because *P. litoralis* flourishes only within a narrow temperature range in the field; Cheng, *et al.*, 1993). The results showed that, (Fig. 1) despite the difference in the salinity and animal density, the population reached the maximum growth rate between day 4 to day 10. However, day 7 was selected as a optimal date, because in all cases, the population can sustains its maximum growth rate at this date. Therefore, in the study, replicates ($n=3$) with 10 worms in each dish were allowed to grow in the sediment for 7 days before the final abundance was measured. The growth rate is expressed as: (final abundance/initial abundance) × 100%.

Results and discussions

All three principal surface deposit feeders increased in abundance in late winter and early spring, and declined to the lowest level by the fall (Fig. 2). *Paranais litoralis* was absent in the July and October samples. This result could be related to the characters of its life history. In

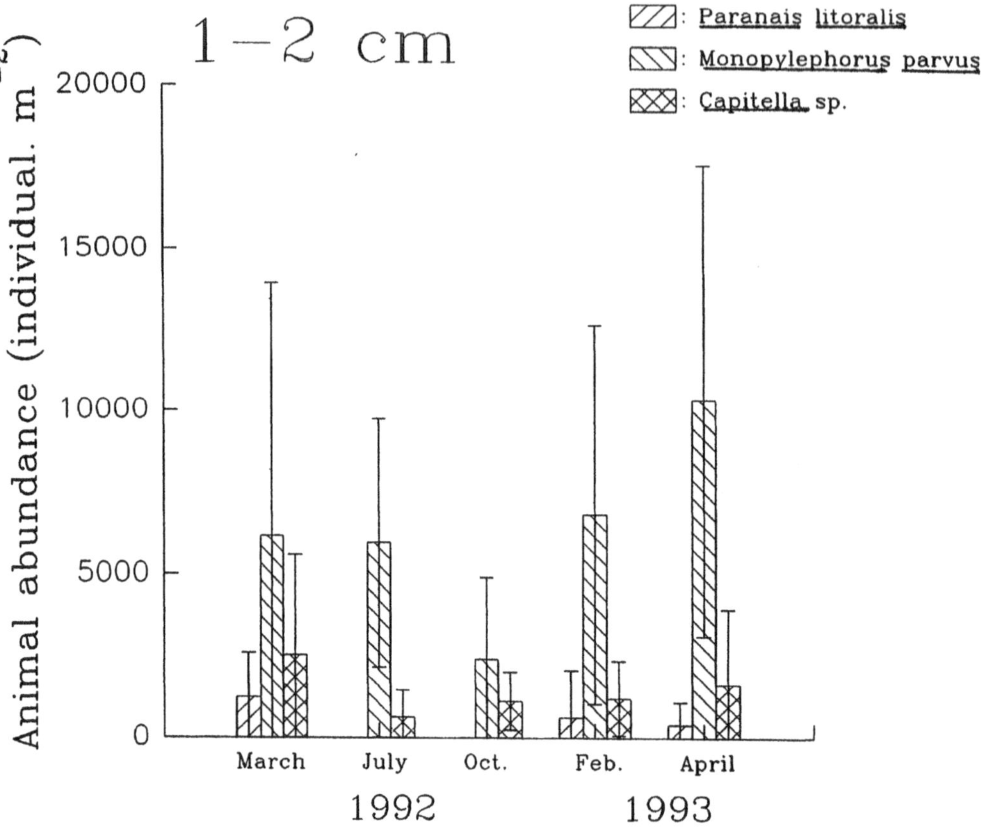

Fig. 3. Temporal variations of three common surface deposit feeders in the 1–2 cm layer of the Chuwei Mangrove Forest mudflat.

a New England mudflat, this animal crashed and disappeared from the field after the summer each year. The population, however, reappeared in the field early next March (Cheng *et al.*, 1993). Despite the absent of *P. litoralis* in these two months, it still be considered as a dominant species because no other animals except *Monopylephours parvus* and *Capitella* sp. can be as abundant as this animal when it appeared in the field. The spring increment of *M. parvus* and *C.* sp. represent the recruitment from juveniles emerging from eggs deposited by adults (Personal observation). Peak recruitment of *M. parvus* occurred in March 1992 and in April 1993. *P. litoralis* seems to reproduce only asexually in the study site as reported in the Flax Pond, New York (Cheng *et al.*, 1993), so population growth can most likely be explained only by the expansion of the local population *vie* asexual reproduction. Peak recruitment of this animal occurred in March 1992 and in February 1993. In spite of the little change in sediment properties in the top 2 cm (Cheng, unpubl. data), abundance of all three species decreased significantly ($p<0.001$, student t-test) from the top to 1–2 cm layer (Figs 2 & 3). Although there were no temporal variations in animal abundances at 1–2 cm, 2-way ANOVA showed strong season and depth interactions ($p<0.001$) for all 3 species. It is thus suspected that factors other than the sedimentary characters, such as oxygen and sediment nutrients availability, must contributed to the depth correspondence of these animals.

Sedimentary organic content (Fig. 4) decreased to the lowest value at all depth in July 1992. Temporal variation of this parameter in the surface layer cannot explain the corresponding change of animal abundances. The organic content showed the subsurface maximum in most cases. Sedimentary organic carbon

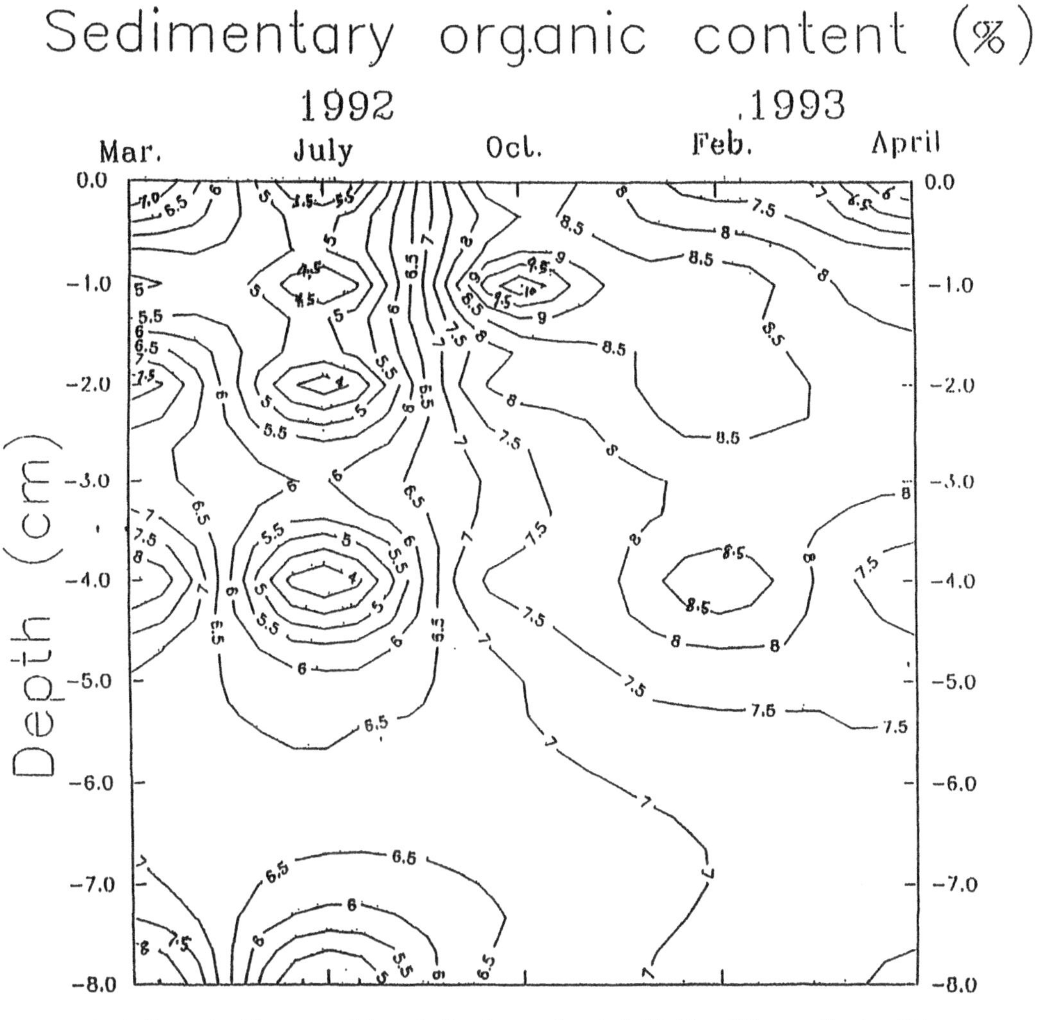

Fig. 4. Temporal and downcore variations of sedimentary organic content of the Chuwei Mangrove Forest mudflat.

and nitrogen, however, decreased with depth in all cases, with the exception of the organic nitrogen in July 1992 (Figs 5 & 6). This discrepancy suggests that both the organic carbon and nitrogen may not be the major components of the organic matter in the sediment. Other substances, such as sulfate may contribute to the subsurface maximum of the sedimentary organic matter.

Sedimentary organic carbon (TOC) (Fig. 5) showed significant temporal variations in the top 2–3 cm sediment, with the value high in March 1992 and February 1993, and low in April 1993 ($p<0.001$, one-way ANOVA). Similar results were found for the sedimentary organic nitrogen (TN) (Fig. 6)($p < 0.01$). TOC and TN decreased with depth in all cases, with the exception of TN in July 1992. There were strong season and depth interactions on these two parameters ($p<0.001$, 2-way ANOVA). These results suggest that most of the temporal variations in the TOC and TN occurred mainly in the top few centimeters. Very little change of these

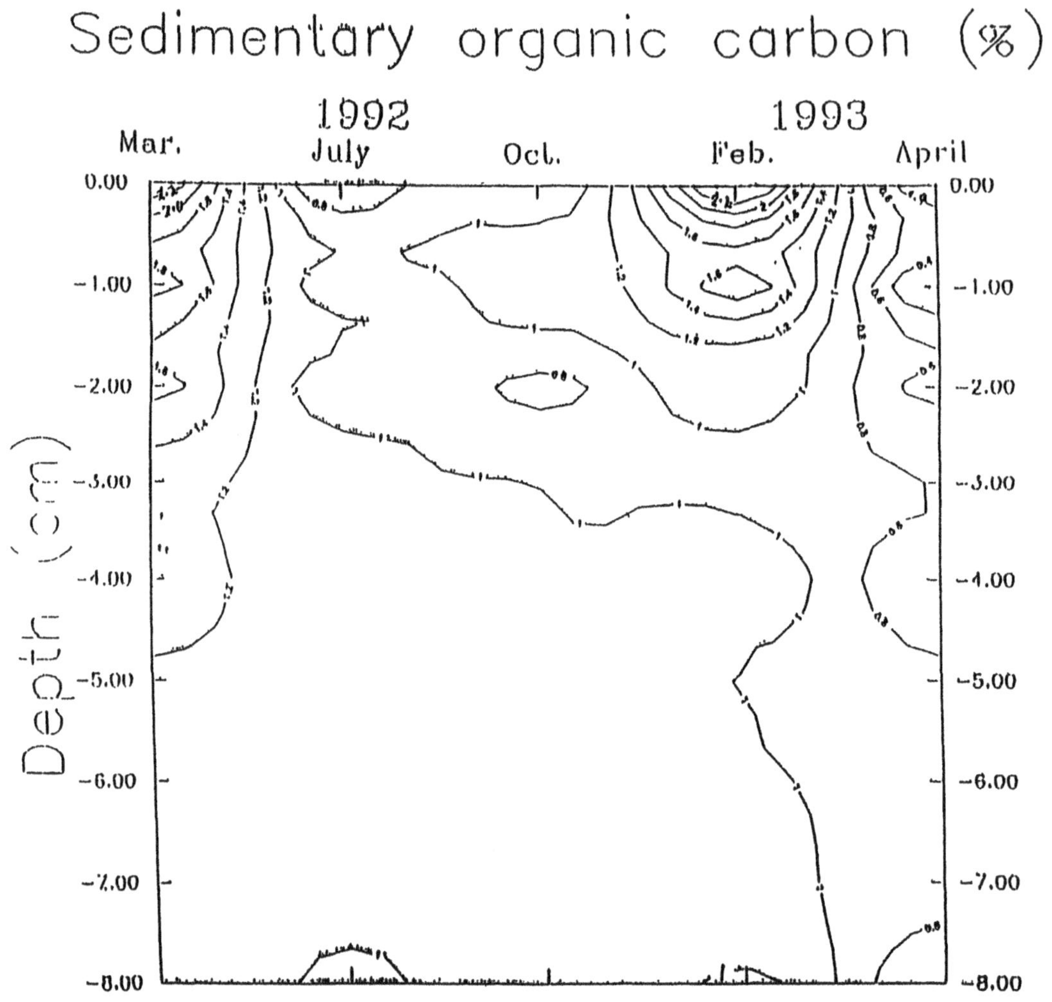

Fig. 5 Temporal and downcore variations of sedimentary organic carbon (TOC) of the Chuwei Mangrove Forest mudflat

parameters in the deeper layer. TN was high at 4–5 cm in July 1992. There is no good explanation for this result.

The C:N ratio ranged from 5 to 24, with most values fall between 12 and 21. High C:N ratio suggests that most of the sedimentary organic matter in top 8 cm may comprised mainly of detritus derived from the mangrove forest (Fleming *et al.*, 1990; Zieman *et al.*, 1984). Except for the high TN resulted in the low ratio at 4–5 cm in July 1992, the low value of C:N ratio in the other seasons were due mainly to the drop of the TOC relative to the TN value. Thus, the decrease of C:N ratio in this study might mean that the supply of the carbon (or energy) to the benthos was low.

There was little temporal variation in the labile protein content in the top layer (Fig. 7). The vertical profiles of this parameter showed subsurface high values in the deeper layers. This parameter measures the labile polypeptides, and subsurface high value suggests downward transport of sedimentary proteinous mate-

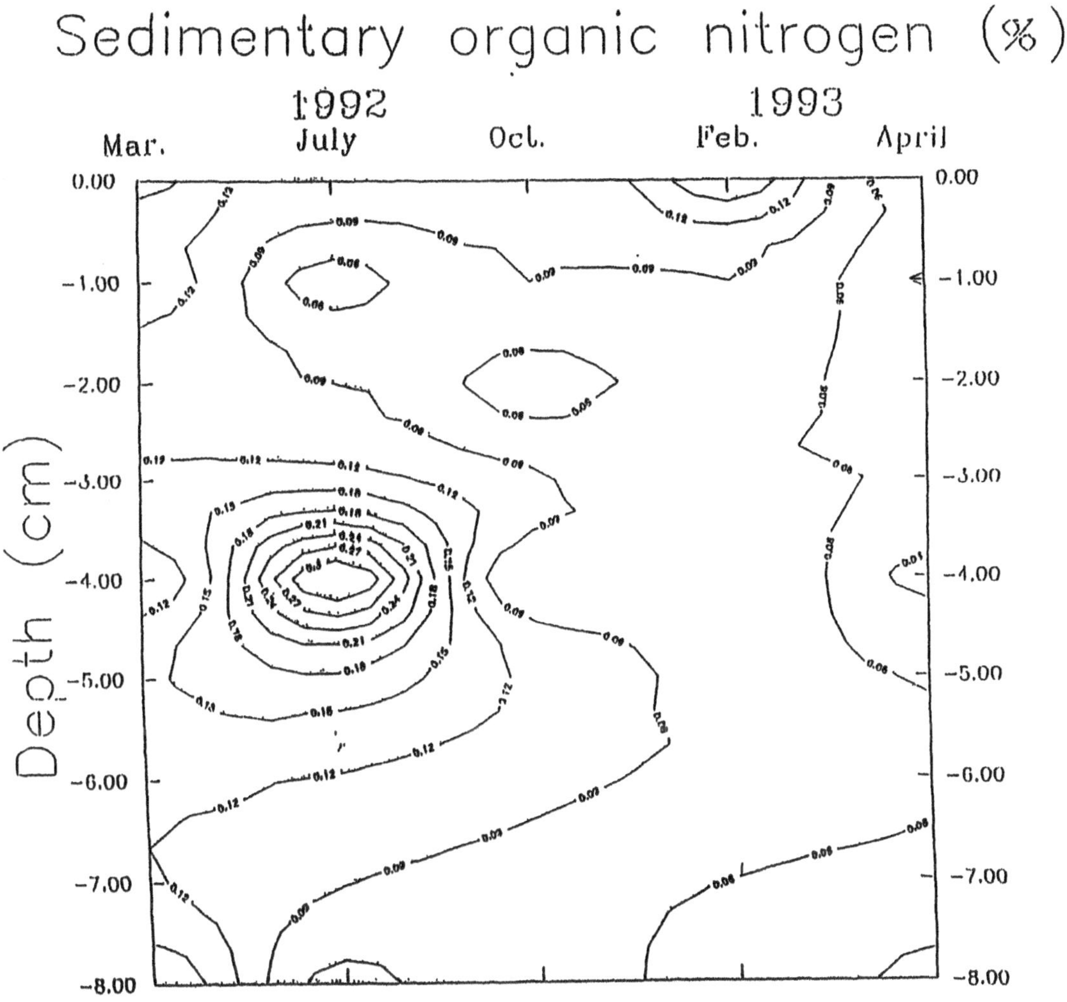

Fig. 6 Temporal and downcore variations of sedimentary organic nitrogen (TN) of the Chuwei Mangrove Forest mudflat.

rial to the deeper layer by the bioturbation (e.g. labile protein content at 7–8 cm reached high value comparable to the top layer in most cases). The crabs in the study site were found inhabit at least a few centimeters below the surface. Large macrofauna, such as Nereidae polychaetes were also encountered frequently in the deeper sediment. This parameter decreased from the top to the 1–2 cm layer. It could be one of the reasons explaining why animal abundance decreased significantly from 0–1 to 1–2 cm layers. Labile protein content attained a high value at all depths in February 1993. Both TOC and TN were also found high in the surface sediment of this month. Because all three common species dropped to the lowest values in October 1992 (Figs 1 & 2), it is suspected that the decrease of animal abundances in the fall may promote the accumulation of sediment nutrients in the winter. Both *Paranais litoralis* and *Monopylephours parvus* increased their abundances in March 1992 and February 1993.

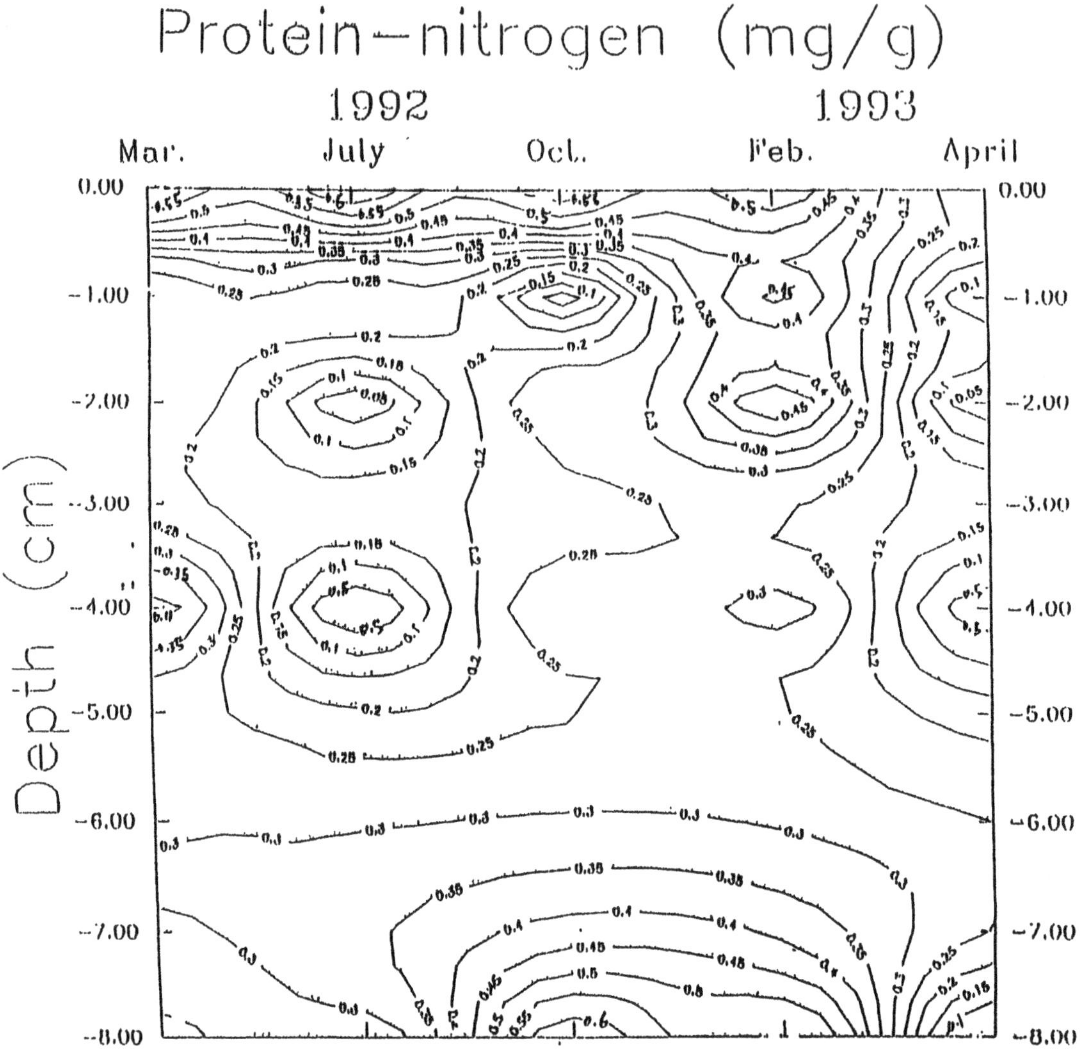

Fig. 7. Temporal and downcore variations of labile protein content of the Chuwei Mangrove Forest mudflat.

The labile protein content, TOC, and TN dropped to the lowest value at all depths in April 1993. The result of the growth experiment (Fig. 8) also suggest that the sediment nutrients was higher in February than in April 1993 at all depths. *Monopylephours parvus*, however, reached its peak abundance in this month. Most animals collected in the field appeared unhealthy (personal observation). Thus, it is possible that the sediment chemical characters examined in this study may represent part of the nutritive values of the sediment. The spring peak of *M. parvus* in 1993 could then imply the overshooting of the population from the recruitment. The growth experiments were conducted only in these two months because the worm culture was not established until the beginning of this year.

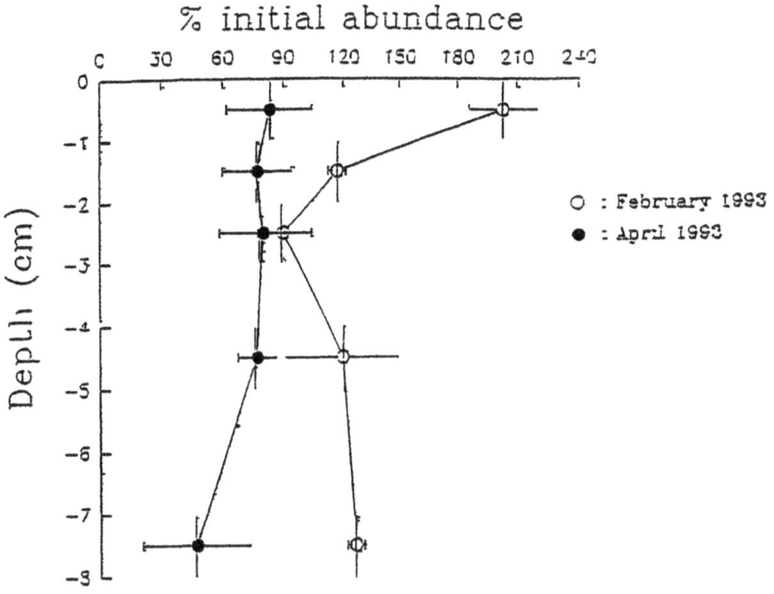

Fig 8 The comparisons of the growth response of *Paranais litoralis* on the sediment collected at different depth between February and April 1993

Conclusions

The results of this study support partially both hypotheses The subsurface maximum of the labile protein content might due to the downcore transport of labile material by the bioturbation The change in the vertical profile of the sediment chemical characters over the time failed to reflect the temporal variation of surface deposit feeders These chemical materials in the surface layer, however, may represent part of the sediment nutrients to the animals This study suggests that, other factors such as seasonal recruitment may also contribute to the spring peak of the animal abundances

Acknowledgments

The author thanks Drs W-L Shai and Mr M R Milligan for the identification of the macrofauna The author also thank Ms Wha-Jen Shu, Ms Pin-Chin Chang, and Mr Tan-shi Chen for the field and laboratory assistance This study was supported by a grant from the National Science Council, ROC (Grant No NSC 82-0209-B-019-027) to I-J Cheng

References

Alongi, D M, 1990 Effect of mangrove detrital outwelling on nutrient regeneration and oxygen fluxes in coastal sediments of the Central Great Barrier Reef Lagoon Estuar coast mar Sci 31 581–598

Alongi, D M & P Christofferson, 1992 Benthic infauna and organism-sediment relations in a shallow tropical coastal area influence of outwelled mangrove detritus and physical disturbance Mar Ecol Prog Ser 81 229–245

Boto, K G & J S Bunt, 1981 Tidal export of particulate organic matter from a Northern Australian mangrove system Estuar coast Shelf Sci 13 247–255

Cheng, I-J & G R Lopez, 1991 Contribution of bacteria and sedimentary organic matter to the diet of *Nucula proxima*, a deposit-feeding protobranchiate bivalve Ophelia 34 157–170

Cheng, I-J, J S Levinton, M McCartney, D Martinez & M J Weissburg, 1993 A bioassay approach to seasonal variation in the nutritional value of sediment Mar Ecol Prog Ser 94 275–285

Chou, C H & C C Bi, 1990 Dynamic distribution of nutrients and variation of environmental factors in Tamshui Estuary ecosystem Proc Natl Sci Counc B ROC 14 131–141

Fleming, M, G Lin & L de S L Steinberg, 1990 Influence of mangrove detritus in an estuarine ecosystem Bull mar Sci 47 663–669

Gonzalez, F & L D Mee, 1988 Effect of mangrove humic-like substances on biodegradation rate of detritus J exp mar Biol Ecol 119 1–13

Kristensen, E, M Holmer & N Bussarawit, 1991 Benthic metabolism and sulfate reduction in a Southeast Asia mangrove swamp Mar Ecol Prog Ser 73 93–103

Kristensen, E., A. H. Devol, S. I. Ahmed & M. Saleem, 1992. Preliminary study of benthic metabolism and sulfate reduction in a mangrove swamp of the Indus Delta, Pakistan. Mar. Ecol. Prog. Ser. 90: 289–297.

Mayer, L. M., L. L. Schick & F. W. Setchell, 1986. Measurement of protein in nearshore marine sediment. Mar. Ecol. Prog. Ser. 30: 159–165.

Levinton, J. S. & S. Stewart, 1988. Effects of sediment organic, detrital input, and temperature on demography, production, and body size of a deposit feeder. Mar. Ecol. Prog. Ser. 49: 259–266.

Phillips, N. W., 1984. Role of different microbes and substrate as potential supplies of specific essential nutrients to marine detritivores. Bull. mar. Sci. 35: 283–298.

Rice, D. L., T. S. Bianchi & E. H. Roper, 1986. Experimental studies of sediment reworking and growth of *Scoloplos* spp. (Orbiniidae: Polychaeta). Mar. Ecol. Prog. Ser. 30: 9–19.

Twilley, R. R., 1985. The exchange of organic carbon in basin mangrove forests in a Southwest Florida Estuary. Estuar. coast Shelf Sci. 20: 543–557.

Zieman, J. C., S. A. Macko & A. L. Mills, 1984. Role of seagrass and mangrove in estuarine food webs: temporal and spatial changes in stable isotope composition and amino acid content during decomposition. Bull. mar. Sci. 35: 380–392.

Hydrobiologia **295**: 231–241, 1995.
Y. S. Wong & N. F. Y. Tam (eds), Asia-Pacific Symposium on Mangrove Ecosystems.

Mangrove soils as sinks for wastewater-borne pollutants

N. F. Y. Tam[1] & Y. S. Wong[2]
[1]*Department of Biology and Chemistry, City Polytechnic of Hong Kong, Tat Chee Avenue, Kowloon, Hong Kong;* [2]*Research Centre/Biology Department, Hong Kong University of Science and Technology, Clear Water Bay, Kowloon, Hong Kong*

Key words: mangrove, soil/sediment, wastewater, sink, nutrients, heavy metal

Abstract

Soil column leaching experiments were conducted to assess the retention of nutrients and heavy metals in two types of mangrove soils receiving strong wastewater throughout a period of 5 months. NH_4^+-N was the dominant form of nitrogen, nitrite and nitrate were in relatively low concentrations in all leachate collected. The concentrations of NH_4^+-N in leachate collected from columns packed with Sai Keng of Hong Kong mangrove soil were higher than those packed with soils collected from Shenzhen of China. The leachate NH_4^+-N contents of Shenzhen columns were significantly lower than that of the synthetic wastewater even at the end of the experimental period, indicating Shenzhen soils had very high capacity to bind nitrogen, and the amount of ammonium added from wastewater did not exceed the binding capacity of mangrove soil. The data also suggest that soils collected from Shenzhen mangrove swamp had higher capacity in retaining wastewater nitrogen than the Sai Keng soils. In contrast, leachate from Sai Keng columns had significantly lower ortho-P contents than those from Shenzhen columns. Actually, the leachate P concentrations of the Sai Keng columns treated with wastewater were similar to those receiving seawater (< 0.1 mg l^{-1}). This finding implies Sai Keng soils were very effective in retaining wastewater P. Throughout the experiment, most heavy metals, including Cu, Zn, Cd, Ni and Cr were not detected in all leachate samples by flame atomic absorption spectrophotometry, indicating that both types of mangrove soils were capable of trapping wastewater-borne heavy metals. The study demonstrates that mangrove soils were good traps to immobilize wastewater-borne phosphorus and heavy metals but they were less efficient in retaining nitrogen from wastewater.

Introduction

Wetland (both natural and constructed) have been used as biological systems for effluent purification and are attractive to the industry as they provide an alternative low-cost, low-maintenance and simple method for domestic and industrial sewage treatment (Clough *et al.*, 1983; Harbison, 1986; Conley *et al.*, 1991). The entire wetland soil-plant-water system, which provides both aerobic and anaerobic conditions, is important in the reduction of nutrients, heavy metals and even organic pollutants from wastewater (Ambus & Lowrance, 1991; Dunbabin & Bowmer, 1992). The relative significance of plants and soils in the removal of pollutants in wetland is a subject of debate. Rogers and co-workers (1991) found that in a constructed wetland, plant uptake accounted for 85% of the total nitrogen reduction and thus the importance of plant component in the whole system was emphasized. However, due to senescence, vegetative uptake could only be considered as a temporary storage mechanism and required removal through harvesting. Therefore the soil and its indigenous microbial populations would ultimately immobilize the added nutrients (Burgoon *et al.*, 1990). Johnston (1991) stated that the total amounts of nutrients stored in the soil fraction were far higher than other storage compartments of a wetland system, and once the nutrients entered the soil component, they were retained there for an extended period. Soils of coastal marshes can also be acted as sinks for heavy metals such as Zn, Pb and Cd (Banus *et al.*, 1975).

The mechanisms involved in the immobilization of pollutants in the soil component of a wetland system include adsorption on ion exchange sites, binding

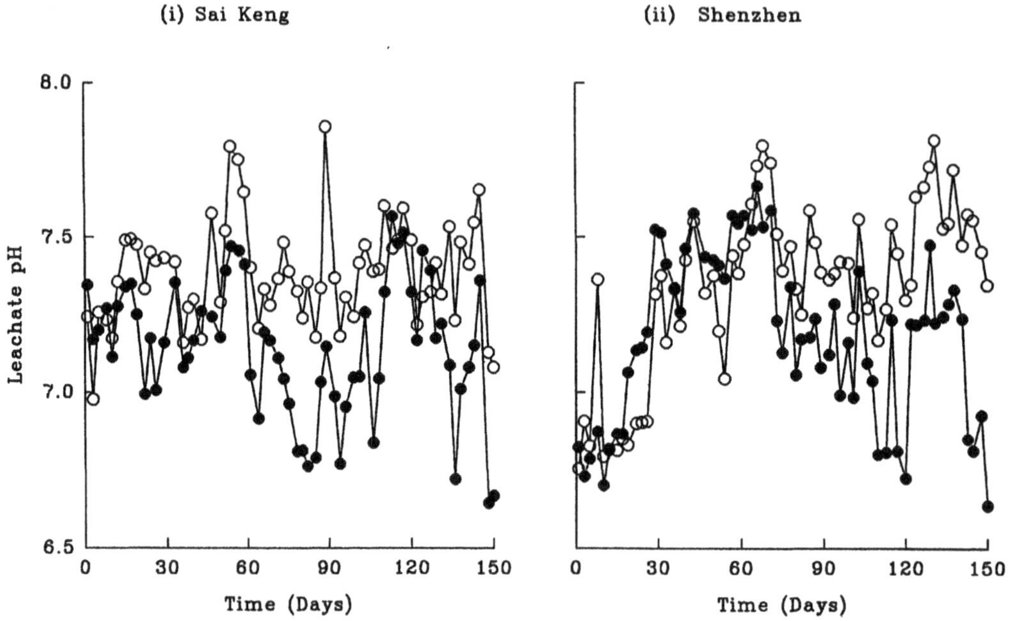

Fig. 1. pH values (mean of 3 replicates) of leachate collected from the control and treated columns of Sai Keng and Shenzhen soils. (o: seawater control; •: wastewater treatment; ANOVA results showed that there was no significant difference between control and treated columns, and also no difference between the two soil types)

to organic matter, incorporation into lattice structures, and precipitation into insoluble compounds (Chan *et al.*, 1982; Dunbabin & Bowmer, 1992). Microbial-mediated reactions including oxidation and reduction, nitrification and denitrification also determined the fate of pollutants in wetland system (Reed *et al.*, 1988; Dunbabin & Bowmer, 1992). All these processes are affected by the biological, chemical and physical properties of wetland soils, in particular pH values, texture, porosity, cation exchange capacity, redox potential, salinity, contents of organic matter, oxides and hydroxides of iron, manganese and aluminium, carbonates, phosphates and sulphide, microbial population sizes and activities (DeBustamante, 1990). Different wetland systems of varied soil characteristics will have different capacity to trap nutrients and heavy metals from wastewater. Mangrove ecosystems, one of the major types of natural wetland in tropical regions, appear to possess high capacity to retain wastewater-borne pollutants. This might be attributed to their anaerobic and reduced conditions, periodic flooding by incoming and outgoing tides, and high clay and organic matter content. Over the past few decades, mangrove swamps have often been deliberately used as convenient dumping sites for waste and sewage in many developing countries (Clough *et al.*, 1983; Dwivedi & Padmakumar, 1983). Although it seems possible to employ mangrove ecosystems as wastewater treatment facilities, the significance of mangrove soil in retaining nutrients and heavy metals from wastewater is poorly understood, especially in sub-tropical mangrove swamps. The present soil column study is therefore aimed (1) to examine the ability of mangrove soils in purifying nutrients and heavy metals from wastewater of high strength; and (2) to compare the effectiveness of two mangrove soils in retaining these wastewater-borne pollutants.

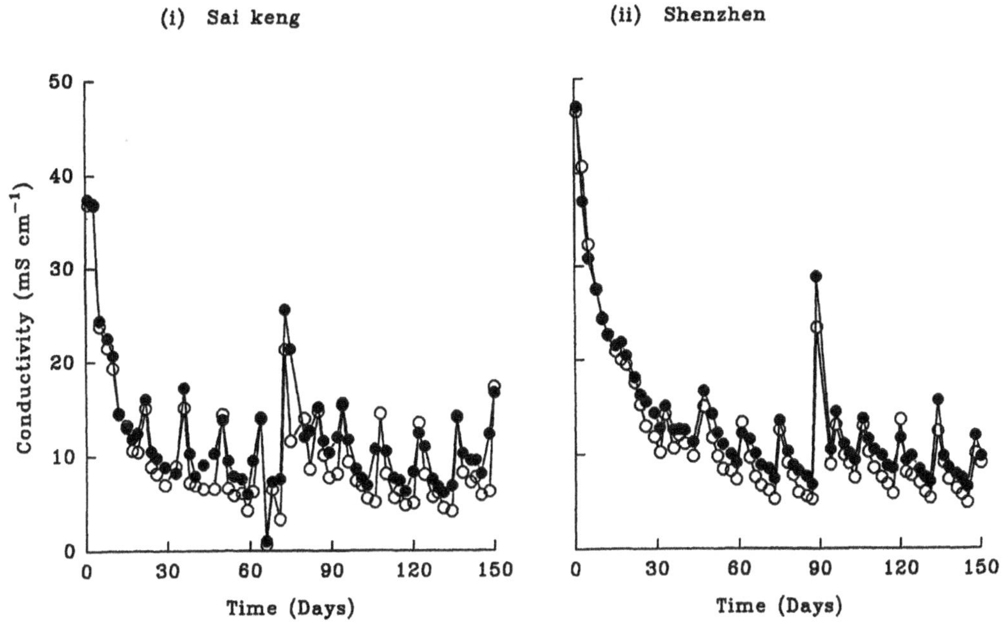

Fig. 2. Conductivity values (mean of 3 replicates) of leachate collected from the control and treated columns of Sai Keng and Shenzhen soils. (o: seawater control; •: wastewater treatment; ANOVA results showed that there was no significant difference between control and treated columns, and also no difference between the two soil types)

Table 1. Composition of the stock solution of artificial wastewater.

Components	Quantity (g)	Components	Quantity (g)
KH_2PO_4	4.39	$CuSO_4 \cdot 5H_2O$	0.39
KCl	7.13	$MnSO_4 \cdot 4H_2O$	2.00
$NaNO_3$	0.61	$ZnSO_4 \cdot 7H_2O$	2.20
NH_4Cl	15.28	$3CdSO_4 \cdot 8H_2O$	0.02
Bactopeptone	5.88	Deionized water	100 ml
Glucose	20.01		

Materials and methods

Mangrove soils

Bulk samples of surface soil (0–10 cm) were collected from two mangrove swamps, one in Sai Keng, Hong Kong, and the other in Futian Nature Reserve, Shenzhen Special Economic Zone, the People's Republic of China. These two sites were selected for comparison because the former one was a small, disturbed and immature mangrove with stony substrate while the latter one was a protected and mature mangrove with muddy soil. The soil was air-dried, ground to pass a 2-mm sieve, thoroughly mixed and stored at room temperature. Subsamples of the soils were analyzed for pH and conductivity (1:10 water extract), texture (wet sieving), organic matter (loss on ignition), ammonium, nitrite and nitrate (1:4 2M KCl extraction followed by Kjeldahl steam distillation), Kjeldahl nitrogen (micro-Kjeldahl digestion and distillation) and ortho-phosphorus (Olsen *et al.*, 1954). Total P content was determined by analyzing the Kjeldahl digest using ascorbic acid-molybdenum blue colorimeter method (Murphy & Riley, 1962). Total K and heavy metals (Cu, Zn, Mn, Cd and Ni) concentrations were measured by dissolving the ash (after ignition at 550 °C for 8 hours) in 6M HCl overnight, followed by flame atomic absorption spectrophotometry. The available heavy metal content (1:10 1M ammonium acetate pH 4 extraction) was also determined by flame atomic absorption spectrophotometry.

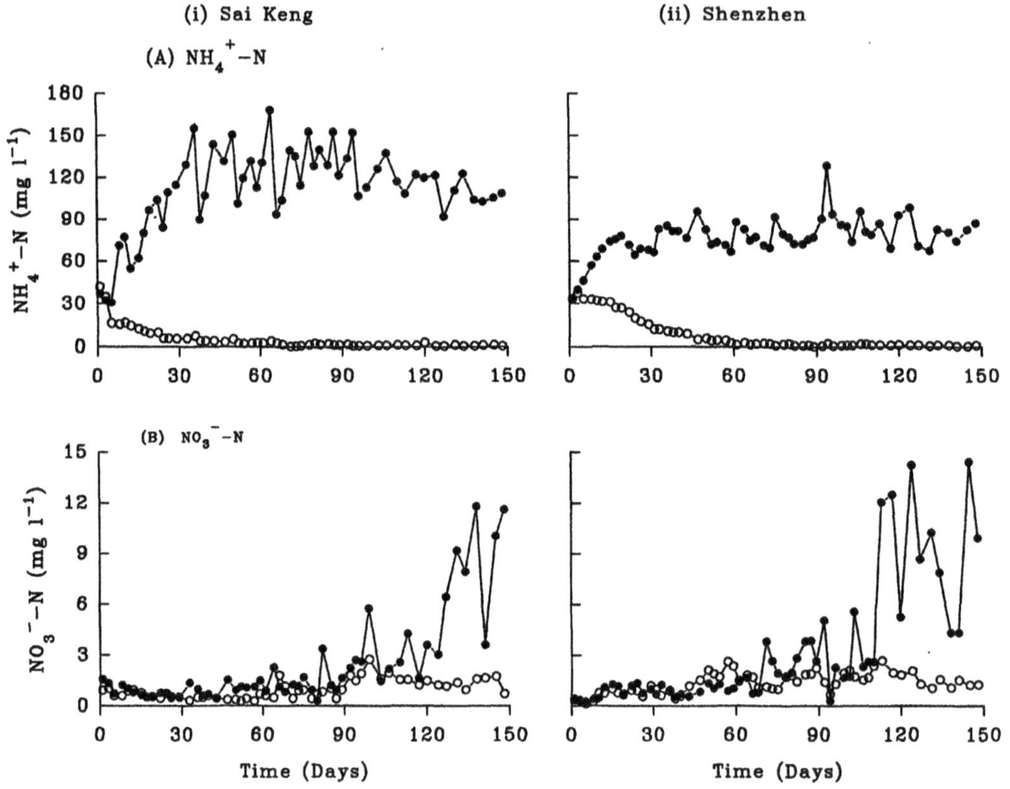

Fig. 3. NH_4^+- and NO_3^--N concentrations (mean of 3 replicates) of leachate collected from the control and treated columns of Sai Keng and Shenzhen soils. (o: seawater control; •: wastewater treatment; For NH_4^+-N: ANOVA results showed that the treated columns had significantly higher ammonium level than the control columns, and significant difference was found between the two soil types; For NO_3^--N: no significant difference was found between the two soil types, and the treated columns had significantly higher nitrate level than the control from day 90 onwards)

Set-up of soil column leaching experiment

Polyvinyl chloride columns (70 cm in length by 10 cm in diameter) were packed uniformly with air-dried soil to a 20- cm depth with a bulk density of 1.02 g cm^{-3}. The columns were initially saturated with artificial seawater for few days. The artificial seawater was obtained by dissolving commercially available salts in deionized water. A complete randomized block design of triplicates was used, and a total of 12 columns were set up for two types of mangrove soil from two sites, namely Sai Keng and Shenzhen, each leached with either artificial seawater (control) or wastewater (treated). The artificial wastewater was prepared by diluting the stock solution (Table 1) 250 times with deionized water and its pH was adjusted to 6.5 ± 0.5. The wastewater consisted of 160 mg l^{-1} NH_4^+-N, 4 mg l^{-1} NO_3^--N, 48 mg l^{-1} organic N, 40 mg l^{-1} PO_4^{3-}-P, 320 mg l^{-1} TOC (total organic carbon), 1094 mg l^{-1} COD (chemical oxygen demand), 200 mg l^{-1} K, 4 mg l^{-1} Cu, 20 mg l^{-1} Zn, 20 mg l^{-1} Mn and 0.4 mg l^{-1} Cd. Its strength was four times of that found in local municipal sewage (Wu, personal communication). A quantity of 200 ml freshly prepared wastewater was applied to each treated column thrice a week. The water was allowed to percolate through and discharge from the

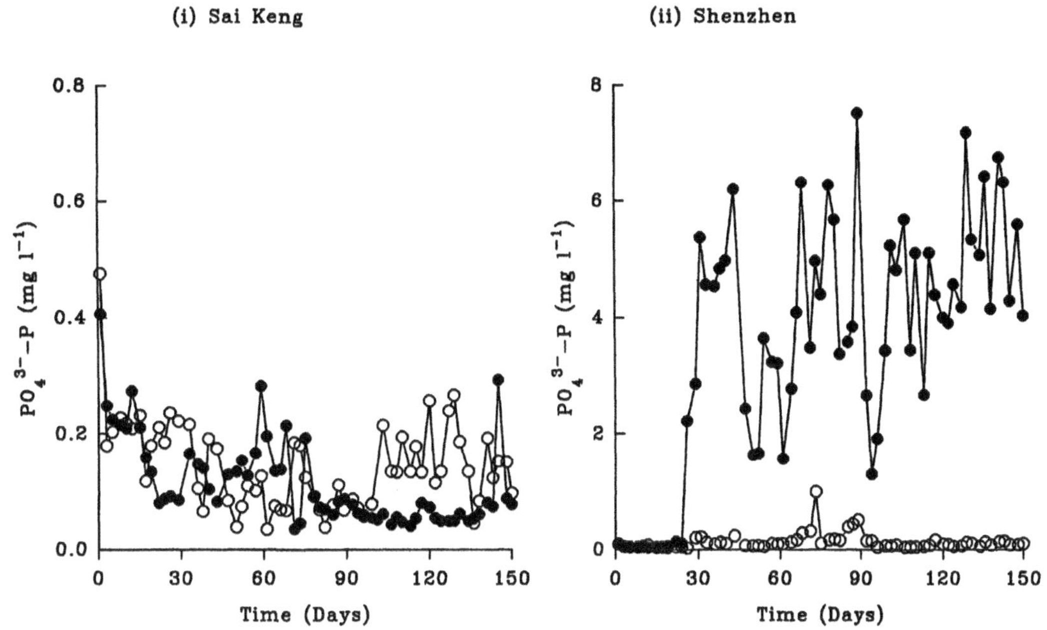

Fig. 4. PO_4^{3-}-P concentrations (mean of 3 replicates) of leachate collected from the control and treated columns of Sai Keng and Shenzhen soils. (o: seawater control; •: wastewater treatment; ANOVA results showed that there was no significant difference between the control and treated columns in Sai Keng soils which had significantly lower phosphorus level than Shenzhen (note the difference in PO_4^{3-}-P scale between the two graphs), the treated columns of Shenzhen had significantly higher phosphorus level than the control)

bottom of the column by gravitational force. For the control columns, artificial seawater, prepared by dissolving the commercially available salts in deionized water with pH adjusted to 6.5 ± 0.5, instead of wastewater was added. After the water had totally drained from the column, the soil was left to dry for 8–10 hours. The column was then re-filled with seawater and left to soak overnight. The seawater was then allowed to drain through the columns, collected and reused. This operation scheme simulated the natural periodic wetting (flooded by incoming tide) and drying (exposed due to outgoing tide) of a mangrove ecosystem. The seawater imitated the flooding tidal water was changed at every two weeks intervals. The whole experiment lasted for 150 days.

Analysis of leachate samples

The leachate collected from each column were analyzed for pH, electrical conductivity, total organic carbon (TOC), NH_4^+-, NO_2^- and NO_3^--N, PO_4^{3-}-P, K, Cu, Zn, Mn, Ni, Cr and Cd, according to the standard methods for water and wastewater analyses (APHA *et al.*, 1989). The mean and standard deviations of the triplicates were calculated. All data on the leachate properties were analyzed by 3-way analysis of variance (ANOVA): the difference between the treated and the control columns, between soil types, and among sampling time were determined. The percentage of the wastewater-borne pollutants retained in mangrove soil was computed based on the difference between quantities added to the column and the quantities of solute in the leachate.

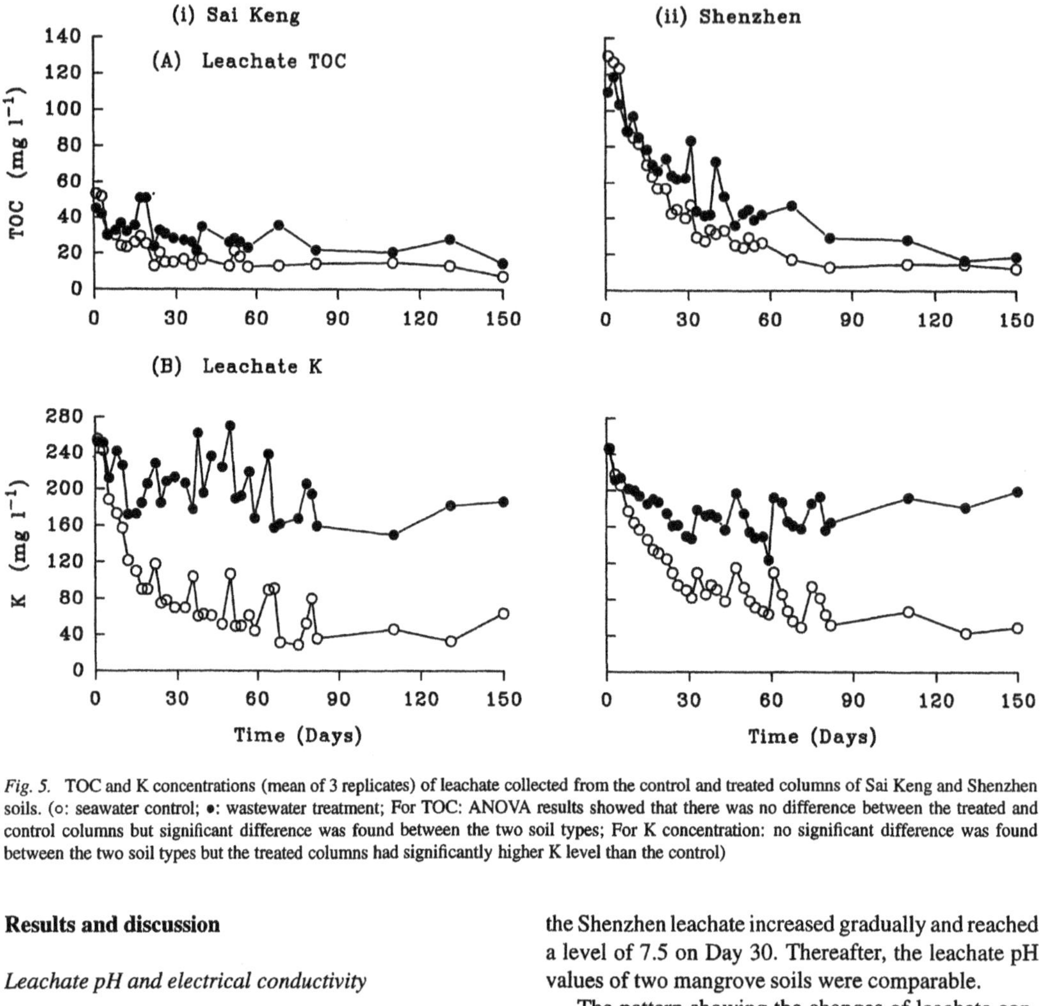

Fig. 5. TOC and K concentrations (mean of 3 replicates) of leachate collected from the control and treated columns of Sai Keng and Shenzhen soils. (o: seawater control; •: wastewater treatment; For TOC: ANOVA results showed that there was no difference between the treated and control columns but significant difference was found between the two soil types; For K concentration: no significant difference was found between the two soil types but the treated columns had significantly higher K level than the control)

Results and discussion

Leachate pH and electrical conductivity

In all columns, the leachate pH values remained at 6.5–7.8 throughout the experimental period (Fig. 1). In both Sai Keng and Shenzhen soil types, the columns treated with wastewater had pH values similar to the control (leached with seawater) in the first few weeks. But after the initial phase, pH of the wastewater treated columns was lower than that of the controls. The high ammonium content in the wastewater might cause nitrification which released H^+ and resulted a slight decrease in leachate pH. Leachate of Shenzhen columns had lower pH (around 6.8) than those of Sai Keng (around 7.2) in the first 3 weeks of study. From Week 3 onwards, pH of the Shenzhen leachate increased gradually and reached a level of 7.5 on Day 30. Thereafter, the leachate pH values of two mangrove soils were comparable.

The pattern showing the changes of leachate conductivity with experimental time was the same in all columns except that the Shenzhen ones had significantly higher initial conductivity values than those of Sai Keng (Fig. 2). The conductivity declined dramatically in the first 30 days, indicating the soluble salts of the mangrove soils were rapidly washed out. After this initial drop, the conductivity of all columns was maintained at around 10 mS cm^{-1}, with repeated cycles of slight increase and decrease throughout the rest of the study period. This suggests that the steady and equilibrium state of the soil column was achieved after 30 days leaching. In general, application of wastewater to

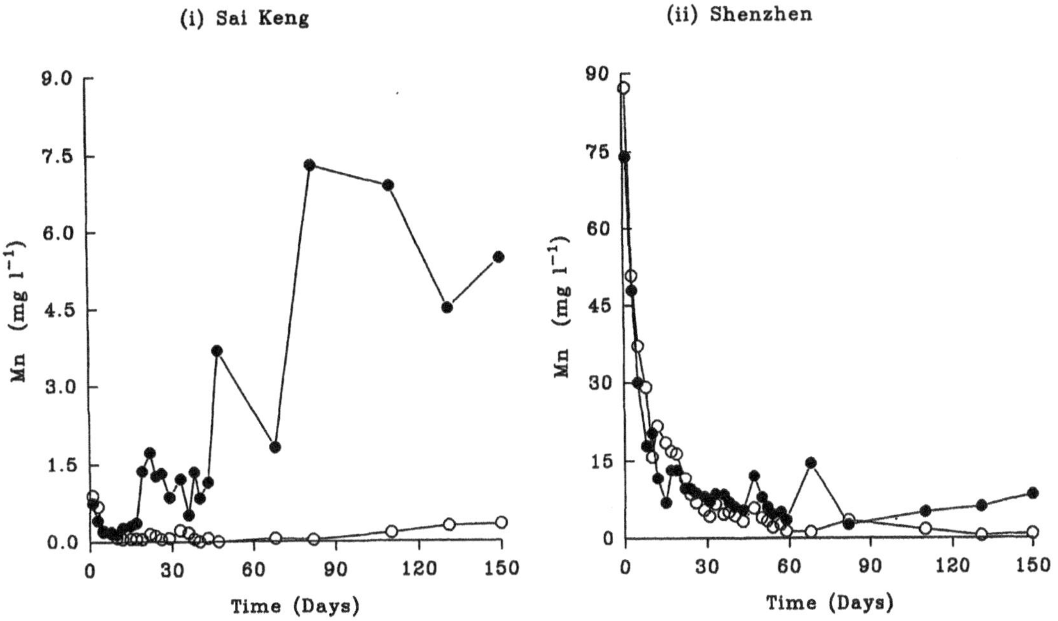

Fig. 6. Mn concentrations (mean of 3 replicates) of leachate collected from the control and treated columns of Sai Keng and Shenzhen soils. (o: seawater control; •: wastewater treatment; ANOVA results showed that the treated columns of Sai Keng had significantly higher Mn level than the control, but the Sai Keng soils had significantly lower Mn level than Shenzhen (note the difference in Mn scale between the two graphs), there was no difference between the treated and control columns of Shenzhen soil)

mangrove soil did not cause any increase in leachate salt content and there was no significant difference in leachate conductivity between the control and the wastewater treated columns.

Leachate inorganic nitrogen content

The NH_4^+-N concentrations of leachate collected from the control columns of both Sai Keng and Shenzhen soils decreased gradually with time and the NH_4^+-N content was less than 2 mg l^{-1} from Day 60 onwards (Fig. 3). This indicates that NH_4^+ originally bound to the soil especially the Sai Keng Mangrove was washed out rapidly by seawater. On the contrary, NH_4^+-N content of leachate discharged from columns treated with wastewater showed a gradual increase during the initial phase of the study. The initial ammonium rise period of Sai Keng wastewater treated columns lasted for about 35 days, reached a peak value of about 150 mg l^{-1}, then fluctuated at a level of around 130 mg l^{-1} throughout the rest of the experimental period (Fig. 3A (i)). The level was lower than that of the applied wastewater, indicating the mangrove soil had some capacity in immobilizing ammonium nitrogen. A similar pattern was found in columns of Shenzhen soil with two exceptions: (a) the initial phase of increase only lasted for 20 days and the NH_4^+-N concentration recorded on Day 19 was 78 mg l^{-1} (about half of that found in Sai Keng leachate); and (b) the NH_4^+ was maintained at around 75 mg l^{-1} after the initial rise, much lower than that of the Sai Keng columns (Fig. 3A(ii)). These results indicated that the Shenzhen columns reached the steady and equilibrium state more quickly than those of Sai Keng soil. Shenzhen soil also seemed to be more effective in retaining wastewater-borne ammonium as the leachate NH_4^+-N concentration was less than half of that found in wastewater (160 mg l^{-1} NH_4^+-N). Table 2 shows that about 40% of the added ammonium was immobilized in Sai Keng soil column while 58% was recorded in Shenzhen soils. The higher capacity of Shenzhen soil in trapping ammonium nitrogen may be because Shenzhen soil was more clayey than Sai Keng soil (Table 3). The former soil contained 55.5% clay while the latter only 11.6%. Shenzhen soil also had

Table 2. Nutrient budget of wastewater-borne N and P in column leaching study.

Treatment	Amounts of Nutrient Load		Amounts of Nutrient Released		Retention in Soil
	Wastewater (mg)	Soil(mg)	Leachate (mg)	Leached Seawater (mg)	(% of total nutrient added)
NH_4^+-N					
Sai Keng					
Control	ND	49.71	47.53	5.54	NC
Treatment	2048	49.71	1016.45	237.75	40.21
Shenzhen					
Control	ND	109.36	91.47	10.04	NC
Treatment	2048	109.36	766.79	143.51	57.81
NO_3^--N					
Sai Keng					
Control	ND	3.42	8.64	1.94	NC
Treatment	51.20	3.42	22.21	7.18	46.19
Shenzhen					
Control	ND	ND	13.80	4.01	NC
Treatment	51.20	ND	30.82	13.16	14.10
PO_4^{3-}-P					
Sai Keng					
Control	ND	26.40	1.66	0.21	NC
Treatment	512	26.40	1.32	0.21	99.72
Shenzhen					
Control	ND	133.34	1.61	0.56	NC
Treatment	512	133.34	42.39	7.17	92.32

(The nutrient budget was calculated based on the mean values of triplicate samples; ND: below the detection limits; NC: not calculated)

relatively higher cation exchange capacity (31.5 meq 100 g^{-1} soil). All these properties were in favour of the adsorption of NH_4^+ to the negatively charged colloids. The high organic matter content of Shenzhen soil (9.2%) might also explain its high NH_4^+ binding ability. Loss of wastewater ammonium through ammonia volatilization was unlikely in the present study because pH of the soil and leachate was not alkaline enough to stimulate volatilization.

Compared with NH_4^+-N, all leachate samples contained very low NO_3^--N content and nitrite was not even detected. During the first three months of study, the leachate NO_3^--N concentrations in all columns were maintained at relatively low values (< 2 mg l^{-1}) and the columns treated with wastewater had similar nitrate content to the control (Fig. 3B). This suggested that either nitrification was not significant or denitrification exceeded nitrification, or both. The mangrove soils, due to periodic flooding (wetting) and drying, were highly reduced and anaerobic which favoured denitrification. The high content of soluble organic carbon (glucose and peptone) in artificial wastewater further enhanced the efficiency of denitrification (Yamaguchi *et al.*, 1990). Moreover, the added glucose and peptone provided carbon and energy sources for heterotrophic bacteria and encouraged nitrogen mineralization and ammonification. The total heterotrophs were likely to have more competitive for limiting amounts of ammonium than autotrophic nitrifiers, and nitrifiers would probably have survived as viable inactive cells (Verhagen *et al.*, 1992). Figure 3B depicts that nitrate accumulated in leachate collected from wastewater-treated columns in the last two months and NO_3^--N concentrations of treated columns were significantly higher than that of the control, indicating that a certain amount of nitrification had taken place towards the end of the experiment. The nitrification was not steady, and large fluctuations in leachate

Table 3. General properties of mangrove soils.

Properties	Sai Keng	Shenzhen
pH	5.57 ± 0.09	5.39 ± 0.07
Conductivity (mS cm^{-1})	2.54 ± 0.06	4.33 ± 0.17
Texture: Sand (%)	73.65 ± 0.40	12.91 ± 0.21
Silt (%)	14.79 ± 0.43	32.41 ± 0.30
Clay (%)	11.56 ± 0.23	54.68 ± 0.33
CEC (meq 100 g^{-1} soil)	3.27 ± 0.16	31.50 ± 2.66
Nutrients		
Organic matter (%)	1.77 ± 0.07	9.16 ± 0.47
NH_4^+-N (mg Kg^{-1})	31.07 ± 0.80	68.35 ± 3.83
NO_3^--N (mg kg^{-1})	2.135 ± 0.004	ND
Total N (%)	0.053 ± 0.002	0.173 ± 0.011
PO_4^{3-}-P (mg kg^{-1})	16.50 ± 4.57	83.41 ± 0.18
Total P (%)	0.019 ± 0.003	0.215 ± 0.012
Ext. K (mg kg^{-1})	11.35 ± 0.56	222.91 ± 14.36
Total K (%)	0.166 ± 0.02	0.79 ± 0.23
Heavy metals (mg kg^{-1})		
Ext. Cu	1.27 ± 0.11	1.56 ± 0.005
Total Cu	3.56 ± 0.36	8.28 ± 0.53
Ext. Zn	2.28 ± 0.09	15.92 ± 1.29
Total Zn	8.13 ± 1.02	63.15 ±11.41
Ext. Mn	2.81 ± 0.34	157.89 ± 7.59
Total Mn	10.70 ± 0.06	198.10 ±12.63
Ext Cd	0.08 ± 0.03	0.08 ± 0.04
Total Cd	0.25 ± 0.01	0.25 ± 0.06
Ext Ni	0.66 ± 0.05	1.30 ± 0.16
Total Ni	1.66 ± 0.16	10.86 ± 0.64

(mean and standard deviation values of 3 replicates were shown; ND: below the detection limit)

Table 4. Strength of artificial wastewater used in this study.

Composition	Conc. (mg l^{-1})	Composition	Conc. (mg l^{-1})
NH_4^+-N	160	K	200
NO_3^--N	4	Cu	4
organic N	48	Zn	20
PO_4^{3-}-P	40	Mn	20
TOC	320	Cd	0.4

NO_3^--N content were also recorded. These findings suggest that nitrifiers might be able to establish their activities and populations within some oxidized microenvironment after a certain period of time. No major difference in nitrate level was found between Shenzhen and Sai Keng soil columns except that nitrification seemed to start earlier in the former soil type. When the nitrate budget was calculated, 29 mg nitrate was recovered in leachate from a total nitrate load of 55 mg in Sai Keng columns treated with wastewater. This indicates 46% of the nitrate was either retained in the soil or denitrified as N_2 or N_2O gases and released back to the atmospheric environment (Table 2).

Leachate phosphorus content

Figure 4(i) shows that in both control and wastewater treated Sai Keng columns, the PO_4^{3-}-P content was very low (less than 0.3 mg l^{-1}). Addition of wastewater containing 40 mg l^{-1} PO_4^{3-}-P to Sai Keng soil columns did not cause any increase in leachate ortho-P concentrations, indicating that the soil of Sai Keng was very effective in retaining wastewater P. More than 99.5% phosphorus in the wastewater was removed after percolating through Sai Keng soil, probably due to complexation with humic substances and physical/chemical adsorption on sites such as hydroxides and oxides of Al and Fe, carbonates of Ca and layer silicate minerals (Mansell *et al.*, 1985). Periodic flooding and drying of the soil might also increase a soil's P-binding capacity. The results also suggest that Sai Keng had very large capacity in trapping and immobilizing through-flowing P. The sorption sites of this soil type were not saturated with phosphorus even after 150 days of wastewater application.

A very different picture was recorded in Shenzhen soil columns. The P content of leachate collected from both seawater and wastewater columns remained very low (< 0.15 mg l^{-1}) in the first month but the P concentrations of wastewater treated columns increased rapidly from Day 24 onwards, reached a high value of 5.4 mg l^{-1} on Day 29, and thereafter fluctuated at a level of 4.5 ± 2 mg l^{-1} (Fig. 4(ii)). This level was much lower than the added wastewater P concentration (40 mg l^{-1}), indicating that although Shenzhen soil was less effective in immobilizing P than Sai Keng soil, over 90% wastewater P was bound in Shenzhen soils (Table 2). Pell and Nyberg (1989) employed sandfilters as wastewater purification systems and reported that the P adsorption process consisted of two phases: an initial phase during which all of the wastewater P was rapidly adsorbed to the soil particles, followed by a second phase characterized by less effective sorption. The lower P retention capacity of Shenzhen soil might be related to its high level of native P (Table 3). Differences in redox potential between Shenzhen and Sai Keng soil might also explain their difference in P-binding capacity: the former soil had higher redox potential. Patrick and Khalid (1974) showed that soils with low redox potentials could bind more P from soil solutions high in soluble PO_4^{3-} than could soils with high redox potentials.

Leachate TOC and K content

Leachate collected from Sai Keng columns had significantly lower initial TOC content than those of the Shenzhen leachate. A rapid drop of TOC was found in the first 30 days in all Shenzhen columns and the differences in TOC concentrations between the two soil types became less obvious from Day 30 onwards (Fig. 5A). The trend of TOC changes with experimental time was similar to that of conductivity values (Fig. 2). In both soil types, the TOC trend of the wastewater treated columns was the same as the control except that the treated ones had slightly higher TOC values from Week 4 onwards (Fig. 5A). The TOC content of the wastewater treated columns reached a low and steady state at around 20 mg l^{-1} towards the latter part of the study. More than 90% wastewater-borne TOC was removed by the soil mass. The added glucose and peptone of the wastewater might have been quickly degraded and utilized by soil microorganisms or adsorbed on the soil particles, as observed by previous studies (Gambrell *et al.*, 1987; Pell & Nyberg, 1989). A comparison of Shenzhen and Sai Keng leachate TOC concentrations indicates that the effectiveness of the two mangrove soil types in the purification of wastewater organic matter was similar.

Potassium, a mobile cation, was consistently leached out by seawater in the control columns of both Sai Keng and Shenzhen soils, declined from initial 260 mg l^{-1} to 40 mg l^{-1} in the first 3 months, and maintained at this low level thereafter (Fig. 5B). In both soil types, the columns treated with wastewater had significantly higher K concentrations in the leachate. After the early drop of potassium, the leachate K of the treated columns reached a steady level of about 180 mg l^{-1}, very similar to the wastewater K concentration (200 mg l^{-1}). This suggests that both types of mangrove soils were not effective in binding wastewater K. It has been reported that the K retained in the clay structure might have been expelled in the presence of NH_4^+ ions. Decomposition of organic matter and destruction of silicates through the action of organic acids might also release K from soil (DeBustamante, 1990).

Leachate heavy metal content

Heavy metals including Cu, Zn and Cd were not detected in any leachate samples, indicating the wastewater heavy metals were completely removed and immobilized in the soil mass. Previous studies have shown that soils of salt marshes were reduced and acted as sinks for heavy metals discharged from sewage effluents (Boto & Patrick, 1978; Harbison, 1986). In addition to physical adsorption on clay particles and chemical complexation on hydrous iron and manganese oxide, oxide formation is also important in precipitating metals from solution (Dunbabin & Bowmer, 1992). Cr, Ni and Fe could not be detected in any of the leachate either, indicating that the mangrove soils themselves were not contaminated. The only heavy metal detected in leachate samples was manganese. Figure 6 exhibits that the leachate of Shenzhen columns had very high initial Mn content (about 90 mg l^{-1}), which decreased rapidly in the first 2 weeks. The wastewater treated columns followed the same pattern and had fairly similar Mn values as the control, suggesting that Shenzhen soil was capable of retaining wastewater Mn. On the contrary, leachate of Sai Keng soil column started with very low Mn level but Mn concentrations increased gradually in wastewater treated samples (Fig. 6). Nevertheless, the Mn content of Sai Keng treated columns was comparable to that of the Shenzhen leachate, both having Mn concentrations lower than that of the artificial wastewater.

Conclusion

Mangrove soils are effective in removing TOC, P and heavy metals from wastewater of high strength. The columns reached steady state after some time of equilibration. Leachate samples collected from wastewater treated columns contained very low concentrations of PO_4^{3-}-P, TOC and heavy metals during the steady state. The reduction of wastewater NH_4^+-N was less satisfactory, and more than half of the wastewater NH_4^+-N were leached out from the columns. The degree of wastewater pollutant removal depends on the type of mangrove soil as the mechanisms involved in binding nutrients and heavy metals are significantly affected by soil properties, especially soil texture (the clay content), cation exchange capacity, contents of organic matter, carbonates, hydrous and oxides of iron, manganese and aluminium, and soil microbial activities. In the present study, Shenzhen mangrove soil was found to be more clayey It had higher cation exchange capacity and contained more organic matter, phosphorus and manganese, and thus was more effective in retaining NH_4^+ than that of Sai Keng On the contrary, Sai Keng soil had higher capacity in trapping P In general, there is apparent difference between the two soil types in terms of wastewater TOC, K and Mn removal Both soils had very high capacity in immobilizing wastewater-borne heavy metals Metals such as Cu, Zn, Ni, Fe, Cr and Cd were not detected in leachate samples during the 150 days of experimental period The only heavy metal found in leachate was Mn Nevertheless, Mn concentrations of leachate collected from wastewater treated columns were very low although higher than those of the controls These results indicate that discharge of wastewater containing high concentrations of heavy metals over a 150-day period did not cause any contamination problem in the effluent if the wastewater was percolated through the mangrove soil However, the maximum capacity of the mangrove soil to retain wastewater-borne heavy metals and its long-term effects require further investigation

Acknowledgements

The authors would like to thank Miss P T Chan and L C Ling for their assistance in laboratory work We would also like to express our gratitude to the Research Centre, Hong Kong University of Science and Technology, for their financial support

References

Ambus, P & R Lowrance, 1991 Comparison of denitrification in two Riparian soils Soil Sci Soc Am J 55 994–918

APHA, AWWA & WPCF (American Public Health Association, American Water Works Association & Water Pollution Control Federation), 1989 Standard Methods for the Examination of Water and Wastewater (17th ed) American Public Health Association, Washington, D C

Banus, M D , I Valiela & J M Teal, 1975 Lead, zinc and cadmium budgets in experimentally enriched salt marsh ecosystems Est Coast Mar Sci 3 421–430

Boto, K G & W H Jr Patrick, 1978 Role of wetland in the removal of suspended sediments In P E Greeson & J R Clarks (eds), Wetland Functions and Values The State of Our Understanding American Water Resources Association 479–489

Burgoon, P S , T A DeBusk, K R Reddy & B Koopman, 1990 Vegetated submerged beds with artificial substrate I BOD removal J Environ Eng 117(4) 394–407

Chan E , T A Bursztynsky, N Hantzsche & Y J Litwin, 1982 The Use of Wetland for Water Pollution Control Municipal Environmental Research Laboratory, U S EPA–600/2–82–086

Clough, B F , K G Boto & P M Attiwill, 1983 Mangroves and sewage a re evaluation In H J Teas (ed), Biology and Ecology of Mangroves Tasks for Vegetation Science Series Vol 8, Dr W Junk Publishers, Lancaster 151–162

Conley, L M , R I Dick & L W Lion, 1991 An assessment of the root zone method of wastewater treatment Res J Wat Pollut Cont Fed 63(3) 239–247

DeBustamante, I , 1990 Land application its effectiveness in purification of urban and industrial wastewaters in La Mancha, Spain Environ Geol Wat Sci 16(3) 179–185

Dunbabin, J S & K H Bowmer, 1992 Potential use of constructed wetland for treatment of industrial wastewaters containing metals The Sci Tot Environ 111 151–168

Dwivedi, S N & K G Padmakumar, 1983 Ecology of a mangrove swamp near Juhu Beach, Bombay with reference to sewage pollution In H J Teas (ed), Biology and Ecology of Mangroves Tasks for Vegetation Science Series Vol 8, Dr W Junk Publishers, Lancaster 163–170

Gambrell, R P , Khalid, R A & W H Jr Patrick, 1987 Capacity of a swamp forest to assimilate the TOC loading from a sugar refinery wastewater stream J Wat Pollut Cont Fed 59(10) 897–904

Harbison, P , 1986 Mangrove muds – a sink and a source for trace metals Mar Pollut Bullet 17(6) 246–250

Johnston, C A 1991 Sediment and nutrient retention by freshwater wetland effects on surface water quality Crit Rev Environ Cont 21(5&6) 491–565

Mansell, R S , P J McKenna, E Flaig & M Hall, 1985 Phosphate movement in columns of sandy soil from a wastewater irrigated site Soil Sci 140(1) 59–68

Murphy, J & J P Riley, 1962 A modified single solution method for the determination of phosphate in natural wastes Anal Chem Acta 27 31–36

Olsen, S R , C V Cole, F S Watanabe & L A Dean, 1954 Estimation of available phosphorus in soils by extraction with sodium bicarbonate U S Dept Agric Circ 939 1–19

Patrick W H Jr & R A Khalid, 1974 Phosphate release and sorption by soils and sediments Effects of aerobic and anaerobic conditions Science 186 53–55

Pell, N & F Nyberg, 1989 Infiltration of wastewater in a newly started pilot sand-filter system I reduction of organic matter and phosphorus J Environ Qual 18(4) 451–457

Reed S C , E J Middlebrooks & R W Crites, 1988 Natural Systems for Waste Management and Treatment McGraw-Hill Book Co , New York, N Y

Rogers K H , P F Breen & A J Chick, 1991 Nitrogen removal in experimental wetland treatment systems evidence for the role of aquatic plants J Wat Pollut Cont Fed 63(7) 934–941

Verhagen, F J M , H Duyts & H J Laanbroek, 1992 Competition for ammonium between nitrifying and heterotrophic bacteria in continuously percolated soil columns Appl Environ Microbiol 58(10) 3303–3311

Yamaguchi, T , P Moldrup, S Teranishi & D E Rolston, 1990 Denitrification in porous media during rapid, continuous leaching of synthetic wastewater at saturated water flow J Environ Qual 19(4) 676–683

Hydrobiologia **295**: 243–254, 1995.
Y. S. Wong & N. F. Y. Tam (eds), Asia-Pacific Symposium on Mangrove Ecosystems.

Effect of wastewater discharge on nutrient contamination of mangrove soils and plants

Y. S. Wong[1], C. Y. Lan[2], G. Z. Chen[2], S. H. Li[2], X. R. Chen[3], Z. P. Liu[3] & N. F. Y. Tam[4]
[1]*Research Centre/Biology Department, Hong Kong University of Science and Technology, Clear Water Bay, Kowloon, Hong Kong*
[2]*Research Institute of Environmental Science, Zhongshan University, People's Republic of China*
[3]*Futian National Nature Reserve, Shenzhen Special Economic Zone, People's Republic of China*
[4]*Department of Biology and Chemistry, City Polytechnic of Hong Kong, Tat Chee Avenue, Kowloon, Hong Kong*

Key words: mangrove, soil/sediment, plant, wastewater, sink, nutrients.

Abstract

The ecological impact of sewage discharges to a mangrove wetland in Futian National Nature Reserve, the People's Republic of China was assessed by comparing the plant community, plant growth and nutrient status of soils and vegetation of a site treated with settled municipal wastewater (Site A) with those of a control adjacent site (Site B) which did not receive sewage. During the one year study, the total and available N and P, and organic carbon concentrations of mangrove soils in Site A did not significantly differ from those of Site B. In both sites, the soil organic C, total N, NH_4^+-N and total P content exhibited a descending trend from landward to seaward regions, with the lowest measurements obtained from the most foreshore location. Seasonal variation in N content of soil samples was more obvious than any difference between wastewater treated and the control sites. The soil N content was lower in spring and summer. This was attributed to the higher temperature in these seasons which facilitated degradation of organic matter and absorption of nitrogen by the plants for growth. No significant difference in plant community structure, plant growth (in terms of tree height and diameter) and biomass was found between Sites A and B. Leaf samples of the two dominant plant species, *Kandelia candel* and *Aegiceras corniculatum* collected from Site A had comparable content of organic carbon, N, P and K to those of Site B. These preliminary results indicated that the discharge of a total volume of 2600 m^3 municipal wastewater to an area of 1800 m^2 mangrove plants over the period of a year did not produce any apparent impact on growth of the plants. The soils and plant leaves of Site A were not contaminated, in terms of nutrient content, by the discharged sewage.

Introduction

Wastewater treatment and recycling have long been serious environmental problems encountered in both developed and developing countries. Conventional wastewater treatment in the activated sludge process or trickling filter system relies solely on bacteria and the associated protozoa. These processes are not only costly in construction and operation, but also far from satisfactory in terms of nutrient and heavy metal removal. The concept of wetland (both constructed and natural) as an alternative biological treatment system for effluent purification has developed rapidly over the last 25 years (Gersberg *et al.*, 1986; Kadlec, 1987; Wathugala *et al.*, 1985). It has been reported that the P removal rates in pots planted with *Phragmites*, *Zizania* and *Typha* were very high, almost up to 100% reduction (Wathugala *et al.*, 1985). In 1986, Gersberg and his co-workers recorded that 94% nitrogen was removed from wastewater in constructed bulrush wetland. Similar findings were obtained by Gambrell *et al.* (1987) and their constructed wetland resulted in nearly 100% removal of COD from wastewater.

Wetland systems are well suited for receiving large volumes of nutrient-laden water for three reasons: (1) the wetland system disperses waste from a point source over a large area; (2) the vegetation itself is adapted to filter nutrients from the water (Boyt *et al.*, 1977), actually the nutrient uptake for most wetland plants is via the roots and pore water; (3) the absorption and adsorption by wetland soils. Pollutants are removed through a complex variety of biological, physical and chemical processes, in particular, the combination of saturated soil, plants, and microorganisms which provide both aerobic and anaerobic conditions for the reduction of pollutants from the overlying wastewater (Dunbabin & Bowmer, 1992; Gale *et al.*, 1993). Different wetland systems of varied plant species and productivity, distinct soil properties and different interactions between soil-plant-water would have different capacity in reducing nutrients from discharged sewage. Mangrove ecosystems, the most common natural marine wetland found in the tropics and some distance into the sub-tropical latitudes, consist of perennial plants of potentially high productivity and biomass, which have extensive root systems that may be very effective in trapping nutrients from wastewater. The plants are also specially adapted to the stressed environment characterized by high temperatures, fluctuating salinities, shifting anaerobic/aerobic substrate. The anaerobic nature, periodic drying and wetting, and the high content of organic matter and sulphide in mangrove soil also contribute to its capacity in retaining pollutants from wastewater. Historically, mangrove swamps have often been used as disposal sites for domestic and industrial wastes (Clough *et al.*, 1983). It has been suggested that mangrove ecosystems not only resist pollution, but also decrease the concentrations of wastewater-borne suspended solids, nutrients and other contaminants (Por, 1984). Nutrient enrichment from discharged sewage may even have a beneficial effect on the productivity of mangrove ecosystems, particularly in areas where soil nutrient status is low (Boto, 1992; Boto & Wellington, 1983; Clough *et al.*, 1983). Despite the fact that mangrove swamps have been used as dumping ground, very little scientific research has been conducted, and there are no published data on the ecological impacts of sewage discharges to mangrove wetland. The present project is therefore aimed to investigate (1) the effects of sewage addition on the nutrient status of mangrove soils and plants, and (2) the impact of sewage discharge on the plant growth and primary productivity of the mangrove ecosystem.

Materials and methods

Site description

The study area is an extensive mangrove wetland system, located at the Futian National Nature Reserve (300 hectares, 11 km long and 300 m wide), Shenzhen Special Economic Zone, the People's Republic of China. The Nature Reserve is geographically contiguous to the Mai Po Marshes in Hong Kong (114°5′E and 22°32′N). The mangrove swamps are dominated by *Aegiceras corniculatum*, *Kandelia candel* and *Avicennia marina*, the average height of the mangrove trees is 4–5 m. Within the mangrove swamp, two elongated sites 150 m apart were chosen for experimental work; one for the wastewater treatment experiment (Site A) and the other one as the control (Site B). Each site was about 180 m × 10 m. The experimental site was separated from the rest of the mangrove swamp by several hundreds of continuous concrete plates (5 cm thick and 50 cm above ground level) fixed on two longitudinal sides of the site. The plates were firmly sealed to form a continuous embankment. Since September 1991, municipal wastewater collected from the surrounding premises was firstly settled overnight in a sedimentation tank (4 m (×> 2 m (×> 8 m) followed by discharging to the landward edge (the back of the shore) of Site A during the low tide period. The sewage was discharged three times a week for one year with a hydraulic loading for each discharge of about 20 m^3. A total volume of 2600 m^3 municipal wastewater was discharged to Site A from September 1991 to October 1992. The characteristics of the settled municipal sewage were analyzed according to the Standard Methods for Water and Wastewater (APHA *et al.*, 1989).

Soil sampling and analyses

Surface soil samples (0–20 cm) were collected at regular intervals from landward to seaward regions of both Sites A and B in April and September 1991 (prior to the discharge of wastewater and used as baseline data), and at 3-month intervals after the commencement of wastewater irrigation, i.e. January, April, July and October 1992. Triplicate samples were collected, air-dried, passed through a 2-mm sieve and analyzed for pH and electrical conductivity (1:5 soil:water extract), organic carbon (Walkley & Black, 1934), NH_4^+-, NO_2^-- and NO_3^--N (extracted with 2N KCl at 1:4 ratio, followed by steam distillation), total Kjeldahl nitrogen (Micro-Kjeldahl digestion followed by steam distillation; Keeney & Nelson, 1982), available-P (Olsen *et al.*, 1954) and total P (Murphy & Riley, 1962).

Vegetation survey and plant tissue analyses

In parallel to soil sampling, samples (in triplicate) of mature leaves were collected from the two most dominant plant species, *Aegiceras corniculatum* and *Kandelia candel* at 2, 10, 20, 40, 60, 100, 140 and 180 m away from the landward edges of both sites. The leaf samples were oven-dried at 60 °C, then ground and analyzed for total carbon, nitrogen and phosphorus content according to the standard methods described in Allen *et al.* (1974). In April 1991 (before wastewater discharge) and May 1993 (one and a half year after the wastewater treatment), an ecological survey was carried out in both Sites A and B. The variety of plant species, the number of individuals of each plant species, the height, breast height diameter and the crown size (width and length) of each tree were measured (Snedaker & Snedaker, 1984).

Statistical analyses

The mean and standard deviation values of the triplicate samples were calculated. The data collected were treated with a parametric three-way analysis of variance (ANOVA) to test the significant differences between Sites A and B, between locations from landward to seaward regions, and between sampling times. The least significant difference (lsd) values at 5% probability level were calculated if the results of the ANOVA indicated significant differences. All statistical analyses were done with an IBM-compatible computing package 'SPSS'.

Results and discussion

Characteristics of municipal sewage

The characteristics of the municipal sewage collected from local premises and after sedimentation for 12–18 hours are shown in Table 1. When compared with the sewage collected from Shatin (half a million population), one of the largest satellite towns of Hong Kong, it is quite clear that the municipal wastewater discharged to the mangrove site was a relatively weak sewage. Its BOD_5 (5-day biochemical oxygen demand) concentration was only one third of that found in Hong Kong. Although still beyond the discharging standard, the nutrient and heavy metal contents of the sewage collected from the Shenzhen area were also significantly lower than those reported in the literature (Irvine *et al.*, 1987). The low strength of the wastewater collected and used in this study reflected the fact that Shenzhen has only been developed in recent years and relatively little industrial activity has been carried out in the premises surrounding the Futian Nature Reserve. Therefore, the total pollutant loadings, including BOD_5, N, P and heavy metal, to the mangrove site over the one-year study were quite low.

Effects of wastewater treatment on mangrove soil properties

Mangrove soil pH and conductivity. Figure 1 shows the soil pH values of Site A were comparable to those of Site B during the one year wastewater discharge experiment. It seemed that addition of wastewater to Site A did not cause any observable change in soil pH even at the very back of the shore (2 m from the sewage discharge points). Although it has been reported that addition of sewage material would cause a reduction in soil pH, probably ascribed to the nitrification of the applied ammonia nitrogen and organic nitrogen, from oxidation of sulphide and acid production from organic carbonaceous matter (Subbiah & Ramulu, 1979), the weak strength of the sewage and the shifting anaerobic-aerobic condition of the mangrove swamps may inhibit the nitrification activities. In most of the samples collected, soils were slightly to highly acidic. Seasonal variation was obvious in both sites: soil samples collected in January and April 1992 had significantly lower pH than those obtained from July and October 1992. The lowest pH values were recorded in the middle part of the experimental site (20–60 m from the land) in January and April. In the other two sampling times, however, pH fluctuation from landward to seaward regions was not significant. Similar to pH values, addition of wastewater did not cause any change in soil electrical conductivity (Fig. 2). This may be explained by the fact that the mangrove soils, being inundated by incoming tides, had relatively high salt contents, therefore addition of weak sewage with low anions and cations would not cause any significant increase in electrical conductivity. The variations in conductivity from landward to seaward were small, although the most foreshore location had lower conductivity values. The soil samples collected in July 1992 had

Table 1. Properties of settled municipal sewage prior to discharge into the experimental site (Site B) of Futian Mangrove Swamp.

Parameters	Settled sewage collected in Futian, the PRC			
	Mean (s.d.)	Range	Settled seawage in Shatin, H. K.	Pollutants quantity (kg)*
pH	7.52 (0.28)	7.28–7.83	7.09 (0.19)	NC
Conductivity (μS cm^{-1})	102.10 (22.07)	82.0–128.0	ND	NC
TKN (mg l^{-1})	24.58 (9.02)	13.4–39.6	38.4 (0.5)	63.91
TP (mg l^{-1})	1.23 (0.77)	0.41–2.67	5.01 (0.10)	3.20
COD (mg l^{-1})	122.5 (11.94)	111.1–139.0	351.3 (43.7)	318.4
BOD_5 (mg l^{-1})	55.9 (21.4)	37.3–89.9	158.1 (19.1)	145.6
Pb (μg l^{-1})	3.85 (1.66)	1.07–6.26	22.6 (19.6)	0.010
Zn (μg l^{-1})	161.30 (139.3)	13.4–143.0	226.2 (50.9)	0.419
Cu (μg l^{-1})	10.36 (8.73)	2.47–30.30	532.9 (279.5)	0.027
Cr (μg l^{-1})	34.55 (23.94)	7.02–71.70	79.6 (48.5)	0.090
Cd (μg l^{-1})	0.09 (0.04)	0.06–0.16	3.48 (0.89)	0.00002
Mn (μg l^{-1})	293 (223)	126–606	ND	0.762

*Pollutant Quantity (kg) mean the total amount of pollutant discharged to Site A of Futian mangrove swamp over the experimental period (one year).
Mean and s.d. (standard deviation) of 18 samples were shown; ND: not determined; NC: not calculated.
TKN: Total Kjeldahl Nitrogen; TP: Total Phosphorus; COD: Chemical Oxygen Demand; BOD_5: 5-day Biochemical Oxygen Demand.

significantly lower conductivity than at other sampling times, probably due to dilution effect by rain: July is the wettest month in Shenzhen and the total amount of rainfall in this month was around 330 mm, 18% of the total annual rainfall.

Soil organic carbon content. In the most landward locations, near the wastewater discharge points, carbon concentrations of Site A were not significantly different from those of Site B. Although at the regions of 10–100 m away from the land, Site A seemed to have higher levels of organic carbon than Site B at four sampling times (Fig. 3), the baseline data collected in April and September 1991 (before the discharge of wastewater) also showed that Site A had more organic carbon than Site B at these locations (Tam *et al.*, 1993). This suggested that discharging municipal wastewater to Site A for one year did not increase the soil organic carbon concentrations. Seasonal variations of organic carbon content were not obvious. In all seasons, the carbon concentrations decreased gradually towards the seashore, with very low C level at the foreshore location.

Soil nitrogen and phosphorus content. Wastewater discharge had no effect on soil nitrogen, total N, NH_4^+- and NO_3^--N concentrations (Fig. 4). Similarly, addition of sewage did not cause any change in soil total and available P content (Fig. 5). Boyt and his co-workers (1977) also reported that there was no evidence of a greater build up of nutrients in the sediments of a wastewater treated experimental swamp than in a control swamp. However, Cooke *et al.* (1990) found that after a decade of receiving sewage effluent, a sewage-wetland soil had much higher nutrient content (especially the plant available forms of nitrogen and phosphorus) and the sewage-wetland soil contained two orders of magnitude more ammonium and inorganic phosphorus than that of the control wetland. Boto and Wellington (1983) also reported that addition of 400 Kg P per hectare (as superphosphate) to an Australian mangrove forest, over a 12-month period, increased the amount of weak acid extractable P level in the soil. The discrepancy between the present results and the previous researchers (Boto & Wellington, 1983; Cooke *et al.*, 1990) could be attributed to a number of reasons. Firstly, in our experiment, the hydraulic loading of

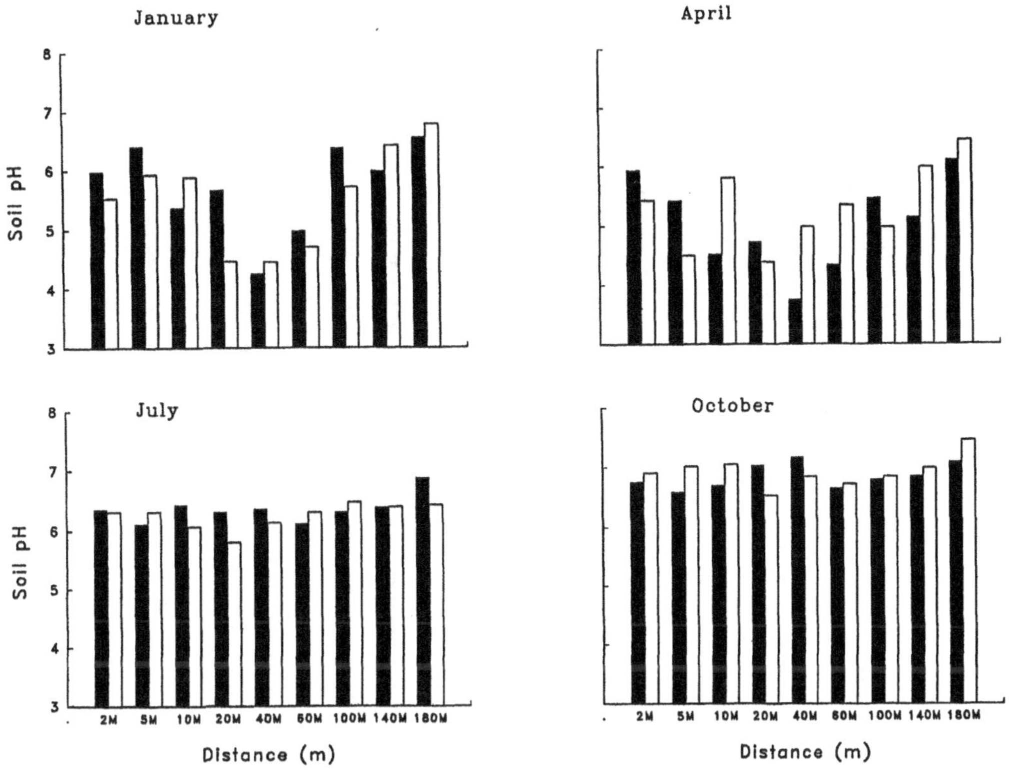

Fig. 1. Effects of municipal wastewater discharge on the profile of soil pH at four different sampling dates in 1992. (■: Site A; □: Site B).

Table 2. The average concentrations of different forms of nitrogen in mangrove soils of Sites A and B.

Sampling Time	Total N (%)		NH_4^+-N (mg kg^{-1})		NO_3^--N (mg kg^{-1})	
	Site A	Site B	Site A	Site B	Site A	Site B
April 1991	0.189^a	0.169^a	6.64^a	9.63^a	1.77^a	1.35^a
Sept. 1991	0.177^a	0.166^a	6.78^a	4.34^a	2.55^b	2.13^a
Jan. 1992	0.247^b	0.219^b	30.90^b	28.01^b	2.43^b	2.54^{ab}
April 1992	0.175^a	0.128^c	21.98^c	21.88^c	2.65^b	2.74^{ab}
July 1992	0.167^a	0.184^a	31.37^b	27.65^b	3.21^c	2.08^a
Oct. 1992	0.221^b	0.223^b	18.81^c	18.24^c	2.57^b	3.22^b

There was no difference between Sites A and B at a probability level of 0.05.
Different superscripts within columns indicate which means were significantly different at $P<0.05$.

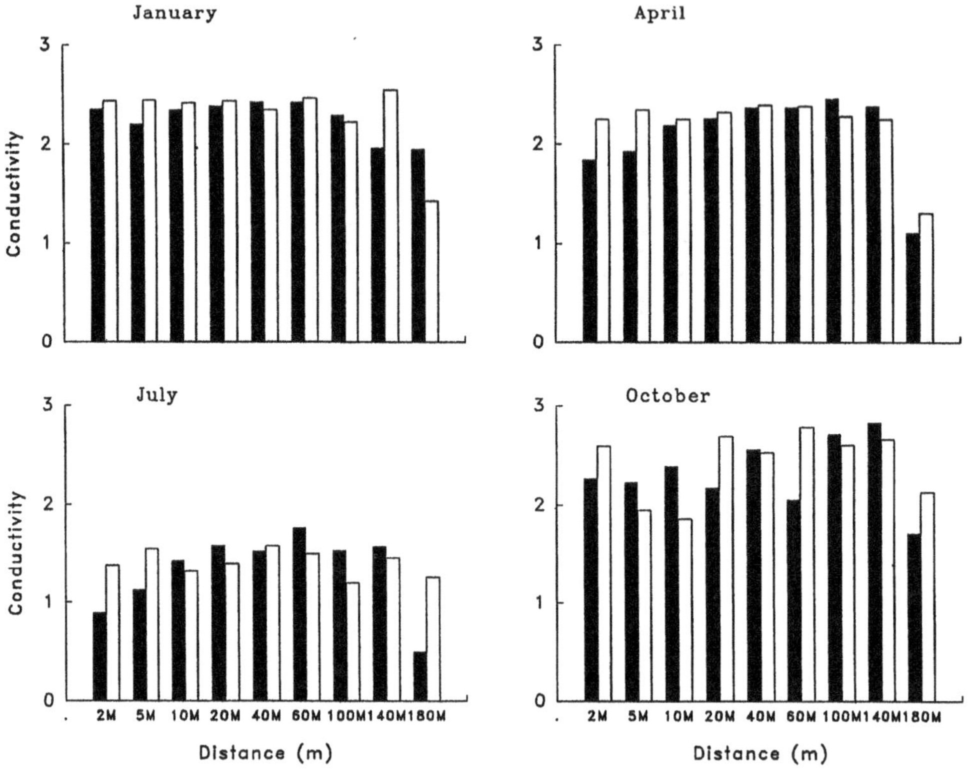

Fig. 2. Effects of municipal wastewater discharge on the profile of soil electrical conductivity (mS cm^{-1}) at four different sampling dates in 1992. (■: Site A; □: Site B).

Table 3. The average concentrations of total and extractable P in mangrove soils of Sites A and B.

Sampling Time	Total P (%)		Extractable P (mg kg^{-1})	
	Site A	Site B	Site A	Site B
April 1991	0.157^a	0.143^a	16.55^a	13.46^a
Sept. 1991	0.126^a	0.127^a	15.12^a	15.29^a
Jan. 1992	0.134^a	0.129^a	19.23^b	18.83^a
April 1992	0.126^a	0.106^{ac}	17.06^a	20.13^b
July 1992	0.197^b	0.231^b	15.37^a	15.96^a
Oct. 1992	0.088^c	0.083^c	19.36^b	23.08^a

There was no difference between Sites A and B at a probability level of 0.05.

Different superscripts within columns indicate which means were significantly different at $P < 0.05$.

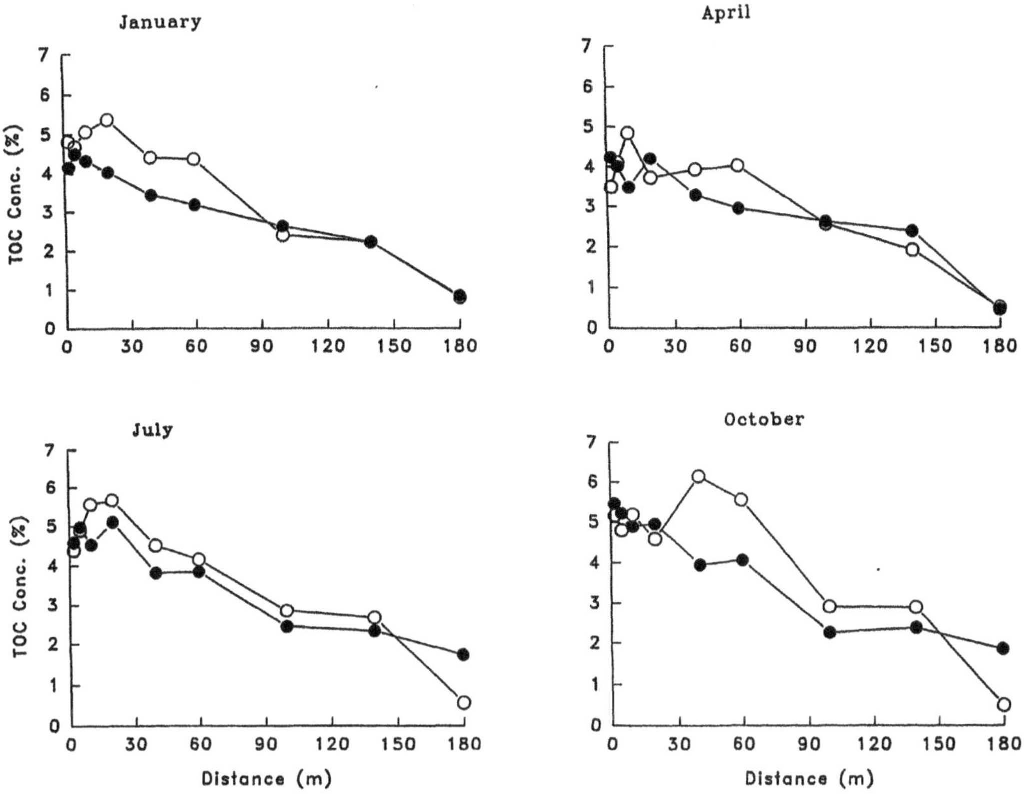

Fig. 3. Effects of municipal wastewater discharge on the changes of soil total organic carbon (TOC) content at four different sampling dates in 1992. (o: Site A; •: Site B).

municipal sewage was relatively low, about 0.0039 m^3 m^{-2} d^{-1}, or less than 4 kg wastewater in total per unit area of Site A. Secondly, the municipal sewage in our experiment had relatively low concentrations of organic matter, N and P. The total loadings of nitrogen and phosphorus to Site A over one-year study were 63.9 and 3.2 kg of N and P, respectively (Table 1), and the daily loadings were 97.3 mg N m^{-2} and 4.87 mg P m^{-2}. These values were far lower than the inherent total N and P concentrations (as indicated in Tables 2 and 3 where the background N and P content was 0.2 and 0.1%, respectively), therefore addition of this kind of sewage would be unlikely to contribute much to the N and P level of the soil. Thirdly, mangrove soils have a very large capacity to assimilate nutrients from wastewater. Tam and Wong (1993) found that addition of strong municipal wastewater containing 50 mg l^{-1} N and 10 mg l^{-1} P to mangrove soil columns did not increase its total N and P content. Fourthly, mangrove swamps are very productive ecosystems, with high biomass turnover and litter production (Lugo, 1980). The net primary productivity of this mangrove swamp was 27.41 t ha^{-2} (Lee, 1993) which suggested the mangrove plants would have very high capacity to absorb and store nutrients from the soil and the discharged wastewater. Clough *et al.* (1983) have estimated that nutrient uptake by mangroves can lead to the immobilization of significant amounts of N and P by incorporation into the plant tissue, around 150 to 250 kg N per hectare and 15 to 20 kg P per hectare annually (Boto, 1992). The significance of plant component in a wetland system used for wastewater purification has been emphasized by previous researchers. It has been reported that plant uptake accounted for 85% of

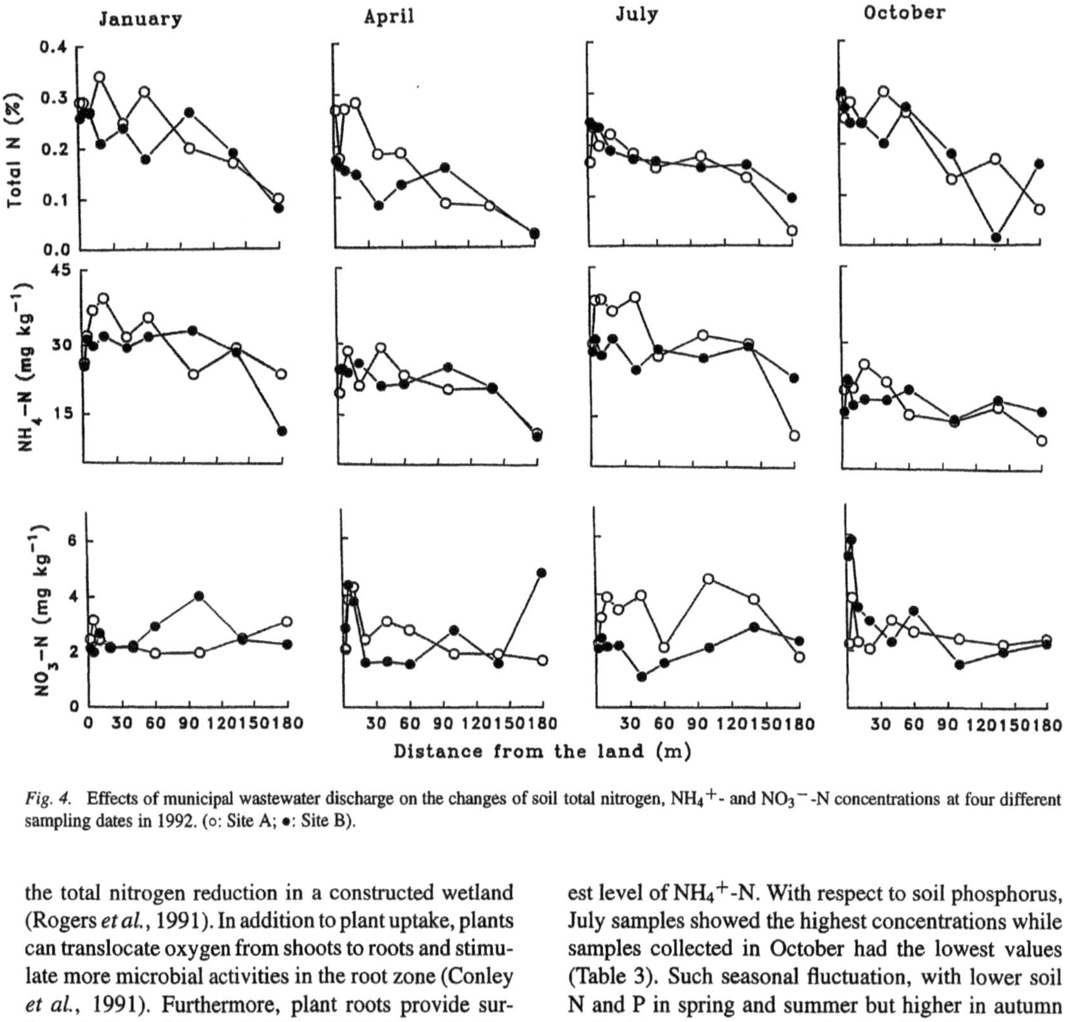

Fig. 4. Effects of municipal wastewater discharge on the changes of soil total nitrogen, NH_4^+- and NO_3^--N concentrations at four different sampling dates in 1992. (o: Site A; •: Site B).

the total nitrogen reduction in a constructed wetland (Rogers *et al.*, 1991). In addition to plant uptake, plants can translocate oxygen from shoots to roots and stimulate more microbial activities in the root zone (Conley *et al.*, 1991). Furthermore, plant roots provide surfaces for bacterial growth. More detailed investigation should be performed to evaluate the total amounts of nutrients taken up by the plants.

Most of the nitrogen presented in the soil was organically bound and the proportion of inorganic N was low. Nitrite was not detected in this study and ammonium-N was more dominant than nitrate-N in all soil samples (Table 2). Like organic C, total N and NH_4^+-N concentrations decreased with distance from the land (Fig. 4). Seasonal variations of soil N and P concentrations were obvious. Samples collected in April and July were found to have lower total N concentrations while October samples had the lowest level of NH_4^+-N. With respect to soil phosphorus, July samples showed the highest concentrations while samples collected in October had the lowest values (Table 3). Such seasonal fluctuation, with lower soil N and P in spring and summer but higher in autumn and winter seasons, seemed to be directly related to the growth and reproduction periods of the mangrove plants. Vegetative growth, requiring large quantities of N, occurred mainly in spring and summer months, therefore plants took up more N from the soil during this period. Boto and Wellington (1983) found that the soil ammonium and nitrate concentrations of a mangrove forest in northern Australia dominated by *Rhizophora*, *Bruguiera* and *Ceriops* species was significantly lower during periods of rapid plant growth. Towards the end of the summer and at the beginning of autumn, plants entered the reproductive phase which required more P for flowering and fruit formation.

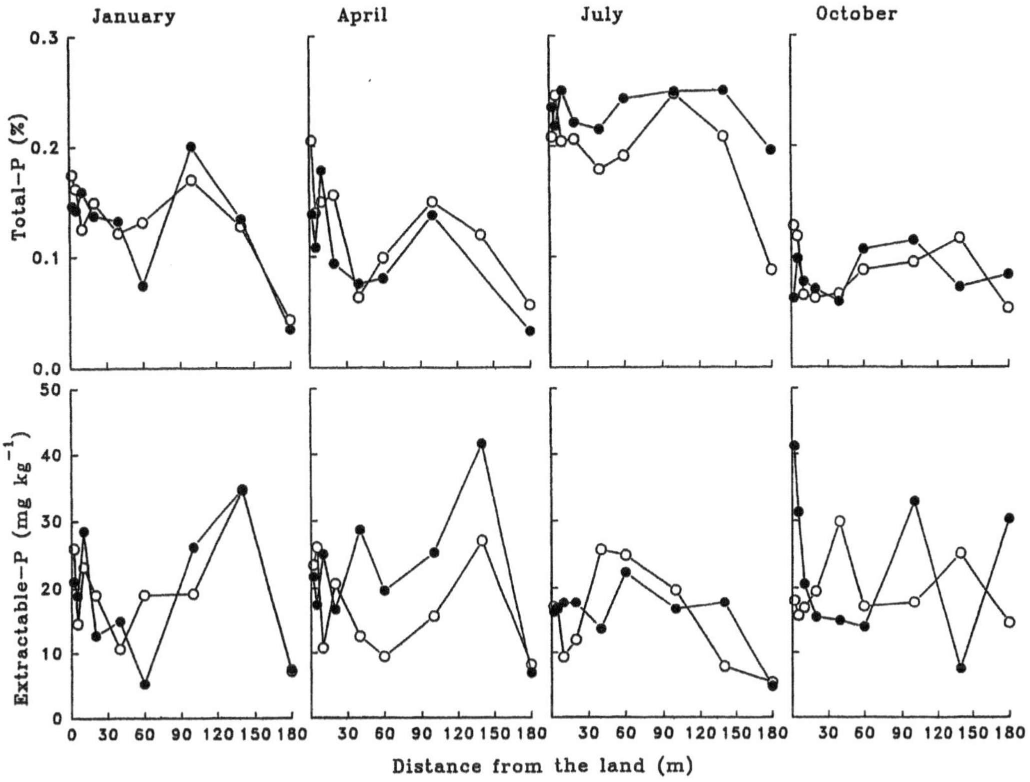

Fig. 5. Effects of municipal wastewater discharge on the changes of soil total phosphorus and available P concentrations at four different sampling dates in 1992. (o: Site A; •: Site B).

Table 4. The height and crown diameter of mangrove plants before and after one year of wastewater discharge to Site B of Futian mangrove swamp.

Plant species	Average tree height (m)			Average crown diameter (length × width) (m)		
	April 1991	May 1993	Increase	April 1991	May 1993	Increase
Ac	3.85	4.20	0.35	1.80 × 1.85	2.30 × 2.50	0.50 × 0.65
Kc	4.20	4.63	0.43	2.65 × 2.75	3.50 × 3.60	0.85 × 0.85
Am	4.35	4.60	0.25	3.00 × 3.20	3.50 × 3.40	0.50 × 0.20

Ac: *Aegiceras corniculatum*; Kc: *Kandelia candel*; Am: *Avicennia marina*.

Effects of sewage discharge on mangrove plant community and plant nutrient content The plant community structure, the species diversity, frequency and density of Site A recorded in May 1993 (after one and a half year discharge of wastewater) were similar to the baseline data obtained in April 1991. The growth of the plant communities, measured in terms of height, breast height diameter, and the crown size of the mangrove trees, in Sites A and B between April 1991 to May 1993 was comparable. Table 4 reveals

Table 5. Effect of wastewater discharge on leaf nutrient concentrations (% dry weight) of *K. candel* and *A. corniculatum*.

Species & Time	Nitrogen		Phosphorus		Potassium		Carbon	
	Site A	Site B	Site A	Site B	Site A	Site B	Site A	Site B
Kandelia candel								
April 1991	1.33	1.45	0.130	0.131	0.63	0.55	44.7	40.8
Jan. 1992	1.01	1.00	0.133	0.148	ND	ND	44.9	44.5
April 1992	1.30	1.28	0.162	0.151	0.48	0.55	47.8	47.5
July 1992	1.07	1.13	0.119	0.110	0.61	0.58	42.4	42.3
Aegiceras corniculatum								
April 1991	1.33	1.31	0.120	0.159	0.49	0.56	42.5	44.1
Jan. 1992	0.94	0.97	0.137	0.156	ND	ND	49.1	47.6
April 1992	1.02	1.20	0.117	0.133	0.60	0.66	50.7	50.2
July 1992	1.09	1.08	0.127	0.119	0.64	0.76	45.5	45.6

ND: not determined.

that both the height and the crown size of *Aegiceras*, *Kandelia* and *Avicennia* increased after sewage irrigation, suggesting that the growth of the mangrove plants was not affected by the discharged wastewater. Mangrove plants have been known to be very tolerant to stressed environment, including high and fluctuated salinities, high temperatures, unstable substrate, alternating wetting (anaerobic) and drying (aerobic) conditions (Por, 1984). They can also withstand high concentrations of pollutants such as Zn, Cd, Pb and Hg (Lin & Chen, 1990; Zheng *et al.*, 1992). Clough *et al.* (1983) pointed out that sewage discharge produced a beneficial effect on plant growth and productivity of the mangrove ecosystems due to the nutrient supply. In addition to increased mangrove growth, fertilization would also lead to elevated tissue nutrient content. Boto and Wellington (1983) observed significant increases in foliar N and P for new leaves of *Rhizophora* in Australian mangrove at the fertilized site (after ammonium and phosphate enrichment for one year) compared with adjacent untreated site. The increased growth coupled with the increased tissue nutrient levels would lead to very substantial increases in plant uptake and immobilization of added nutrients.

Like the baseline study by Tam *et al.* (1993), the present experiment showed that the nutrient concentrations of mangrove plant leaves were relatively constant, with very little variations from the landward to seaward regions or with season. The average concentrations of C, N and P of leaves collected in Site A were similar to those of Site B, indicating that the sewage discharge did not cause any change in leaf nutrient content in either *Kandelia* or *Aegiceras* (Table 5). Whether sewage addition would cause changes in the soil and vegetation of a wetland would largely depend on the initial trophic status of the wetland. Clough *et al.* (1983) pointed out that the adsorption maxima of the undisturbed mangrove sediment were in the range of 250 to 700 μg P per gram dry weight but a lower maximum was reported for a mangrove sediment which had received treated sewage effluent for 20 years, however the latter sediment had much higher total P (1720 μg P per gram dry weight). In most mangrove forests especially those with nutrient limitation, nutrient additions from wastewater should lead to greater nutrient availability which will result in greater cycling, decomposition, and ultimately, to a change in community composition to species more tolerant to high nutrient loads (Cooke *et al.*, 1990). However, such changes are likely to be subtle rather than dramatic. The added nutrients would gradually accumulate in the soil and plant tissues to a level that would affect plant growth and the community structure of the mangrove ecosystem. It has been suggested that more profound ecological changes in vegetation were difficult to detect on a short time scale and could take 20 years or more to become apparent (Odum, 1987). In the present study, only one and a half year observation was made and the load of sewage added was low. Thus subtle differences in plant community, nutrient status of soils and plant tissue would not be detected. Nevertheless, these results demonstrate that the mangrove forest being studied had a potential to retain nutrients and pollutants as the soil and plants seemed to have high capacity to adsorb and absorb nutrient and heavy metal content from the discharged sewage

Conclusion

The present study indicates that the municipal wastewater collected from the premises close to the Futian mangrove swamp was relatively weak The low hydraulic and nutrient loadings did not cause any significant effect on the plant community structure of the wastewater-treated mangrove sites Nutrient concentrations, including organic C, total N, NH_4^+- and NO_3^--N, total and available P, of the leaf samples of *Kandelia* and *Aegiceras* collected in the treated site were comparable to those of the control site Leaf nutrient content was relatively constant and no significant difference was found between different sampling times and different locations from landward to seaward regions However, seasonal variations of the soil N and P concentrations were obvious, and statistically significant differences between sampling time were detectable Only subtle differences in soil nutrients and heavy metal concentrations were found between Sites A and B The soil C, N and P levels declined gradually with the distance away from the land In spite of the fact that discharge of locally collected municipal sewage to the mangrove ecosystem did not cause any significant change in the soil and plant nutrient status within a short time scale (less than one and a half year) and the mangrove ecosystem seemed to be able to purify the nutrients and organic matter from the wastewater, this did not imply that addition of sewage especially that with high strength would have no adverse effect on the mangrove ecosystem on a long term basis The experimental set-up in Futian should be monitored for at least 3–4 years to verify this issue The fate of the nutrients after discharging from wastewater to various components of the mangrove ecosystems must also be understood The amounts of nutrient absorbed and stored in plant tissues, retained in the soils, immobilized by microorganisms associated with the root zone area and soil, and resuspended and released to the tidal water must also be investigated

Acknowledgements

The authors would like to thank the technicians in Futian Nature Reserve, Shenzhen Special Economic Zone and Zhongshan University, Guangzhou, the PRC for their assistance in field sampling and laboratory analyses We would also like to express our gratitude to the Research Centre, Hong Kong University of Science and Technology, for their financial support

References

Allen, S E, H M Grimshaw, J A Parkinson & C Quarmby, 1974 Chemical Analysis of Ecological Materials Blackwell Sci Publ, Oxford

APHA, AWWA & WPCF (American Public Health Association, American Water Works Association & Water Pollution Control Federation), 1989 Standard Methods for the Examination of Water and Wastewater American Public Health Association, Washington, D C

Boto, K G & J T Wellington, 1983 Nitrogen and phosphorus nutritional status of a northern Australian mangrove forest Mar Ecol Press Ser 11 63–69

Boto, K G, 1992 Nutrients and mangroves In Connel D W & D W Hawker (eds), Pollution in Tropical Aquatic Systems CRC Press Inc, Ann Arbor, London 129–145

Boyt, F L, S E Bayley & J Jr Zoltek, 1977 Removal of nutrients from treated municipal wastewater by wetland vegetation J Wat Pollut Cont Fed 49 789–799

Clough, B F, K G Boto & P M Attiwill, 1983 Mangroves and sewage a re-evaluation In H J Teas (ed), Biology and Ecology of Mangroves Tasks for Vegetation Science Series Vol 8, Dr W Junk Publishers, Lancaster 151–162

Conley, L M, R I Dick & L W Lion, 1991 An assessment of the root zone method of wastewater treatment Res J Wat Pollut Cont Fed 63(3) 239–247

Cooke, J G, A B Cooper & N M U Clunie, 1990 Changes in the water, soil, and vegetation of a wetland after a decade of receiving a sewage effluent New Zealand J Ecology, 14 37–47

Dunbabin, J S & K H Bowmer, 1992 Potential use of constructed wetland for treatment of industrial wastewaters containing metals The Sci Tot Envir 11 151–168

Gale, P M, K R Reddy & D A Graetz, 1993 Nitrogen removal from reclaimed water applied to constructed and natural wetland microcosms Wat Envir Res 65(2) 162–168

Gersberg, R M, B V Elkins, S R Lyon & C R Goldman, 1986 Role of aquatic plants in wastewater treatment by artificial wetland Wat Res 20(3) 363–368

Gambrell, R P, R A Khalid & W H Jr Patrick, 1987 Capacity of a swamp forest to assimilate the TOC loading from a sugar refinery wastewater stream J Wat Pollut Cont Fed 59(10) 897–904

Irvine, R L, D V S Murthy, M L Arora, J L Copeman & J A Heidman, 1987 Analysis of full-scale SBR operation at Grundy Centre, Iowa J Wat Pollut Cont Fed 59 132–138

Kadlec, J A, 1987 Nutrient dynamics in wetlands In Reddy K R & W H Smith (eds), Aquatic Plants for Water Treatment and Resource Recovery Magnolia Publishing Inc, Orlando, Florida 393–413

Keeney, D R & D W Nelson, 1982 Nitrogen – inorganic forms In A L Page, R H Miller & D R Keeney (eds), Methods of Soil

Analysis. Part 2: Chemical and Microbiological Properties No. 9 (Part 2) in the Series of Agronomy, Wisconsin, USA: 643–698.

Lee, M. S., 1993. Research on Mangrove Ecosystem in Futian, Shenzhen. Ph. D. Thesis, Zhongshan University, Guangzhou, the PRC.

Lin P. & R. Chen, 1990. Role of mangrove in mercury cycling and removal in the Jiulong estuary. Acta Oceanologica Sinica 9(4): 622–624.

Lugo, A. E., 1980. Mangrove ecosystems: successional or steady state? Biotropica 12: 65–72.

Murphy, J. & J. P. Riley, 1962. A modified single solution method for the determination of phosphate in natural wastes. Anal. Chem. Acta 27: 31–36.

Odum, W. E., 1987. Ecological considerations. In: Reddy K. R. & W. H. Smith (eds.), Aquatic Plants for Water Treatment and Resource Recovery, Magnolia Publishing Inc., Orlando, Florida: 1016–1017.

Olsen, S. R., C. V. Cole, F. S. Watanabe & L. A. Dean, 1954. Estimation of available phosphorus in soils by extraction with sodium bicarbonate. U.S. Dept. Agric. Circ. 939: 1–19.

Por, F. D., 1984. The ecosystem of the mangrove forests. In: F. D. Por & I. Dor (eds), Hydrobiology of the Mangal. Dr W. Junk Publishers, The Hague: 1–14.

Richardson, C. J. & D. S. Nichols, 1985. Ecological analysis of wastewater management criteria in wetland systems. In: P. J. Godfrey (ed.), Ecological considerations in wetland treatment of municipal wastewaters, Van Nostrand Reinhold, New York: 351–391.

Rogers, K. H., P. F. Breen & A. J. Chick, 1991. Nitrogen removal in experimental wetland treatment systems: evidence for the role of aquatic plants. J. Wat. Pollut. Cont. Fed. 63(7): 934–941.

Snedaker, S. C. & J. G. Snedaker, 1984. The Mangrove Ecosystem: Research Methods. United Nations Educational, Scientific and Cultural Organization Published, Richard Clay (The Chaucer Press) Ltd., U.K.

Subbiah, S. & U. S. S. Ramulu, 1979. Influence of sewage wastes addition on the soil characteristics: I. Effect on physico-chemical properties of soils of Tamil Nadu (India). Mysore J. Agric. Sci. 13: 408–414.

Tam, N. F. Y. & Y. S. Wong, 1993. Retention of nutrients and heavy metals in mangrove sediment receiving wastewater of different strengths. Envir. Technol. 14: 719–729.

Tam, N. F. Y., S. H. Li, C. Y. Lan, G. Z. Chen, M. S. Lee & Y. S. Wong, 1993. Nutrients and heavy metal contamination of plants and sediments in Futian mangrove swamps. Proc. Asian-Pacific Symp. on Mangrove Ecosystems, Hong Kong, Sept. 1993 (in press).

Walkley, Y. A. & I. A. Black, 1934. An examination of the Detjareff method for determining soil organic matter and a proposed modification of the chronic acid titration method. Soil Sci. 37: 29–38.

Wathugala, A. G., T. Suzuki & Y. Kurihara, 1985. Studies on the removal of phosphorus from the wastewater using sand filtration pot systems with *Phragmites australis*, *Zizania latifolia* and *Typha latifolia*. In: Laster, J. N. & P. W. W. Kirk (eds), Proc. Int. Conf. on Management Strategies for Phosphorus in the Environment, Publisher Selper, London: 332–335.

Zheng, F. Z., P. Lin, W. J. Zheng & Z. X. Zhuang, 1992. Study on the absorption and removal of *Kandelia candel* for pollutant cadmium. Acta Phytoecologica et Geobotanica Sinica 16(3): 220–226.

Hydrobiologia **295**: 255–261, 1995.
Y. S. Wong & N. F. Y. Tam (eds), Asia-Pacific Symposium on Mangrove Ecosystems.

The use of demographic studies in mangrove silviculture

Gong Wooi Khoon[1] & Ong Jin Eong[2]
[1]*School of Biological Sciences and* [2]*Centre for Marine & Coastal Studies, Universiti Sains Malaysia, 11800 Penang, Malaysia*

Key words: mangrove, demography, silviculture, management

Abstract

The Matang Mangrove Forest Reserve in Malaysia has been managed for timber production since the beginning of the century and is reputedly the best managed mangrove forest in the world. The present management plan is a 30-year rotation period with two thinnings, at 15 and 20 years. However, there has been a decline in yield from 299 t ha^{-1} of green-wood from virgin stands to the second generation yields of 158 t ha^{-1} in 1967–69 to an even lower 136 t ha^{-1} in 1970–77.

This study on the demography of the forest was conducted to try to determine ways to improve the silviculture and management system. The species of the tree, whether it was living or dead, and the girth at breast height were recorded for all trees in selected representative plots covering a range of ages (5, 8, 13, 18, 23 and 28 years). The standing biomass of these plots was calculated using previously obtained allometric regressions.

The high density of 15 030 *Rhizophora apiculata* trees per hectare in the 5 year-old stand and the sharp decrease to 9810 in the 8 year-old stand indicate that the initial stocking was too high. We suggest that artificial regeneration should be carried out at 1.2 m intervals only if the natural regeneration is less than 50% (rather than 90% as is the present practice). Extremely high mortality occurred in the 13 year-old and the 18 year-old stands where 43% and 29% respectively of the *Rhizophora* trees were dead. We therefore suggest that the thinnings be carried out earlier – at 12/13 and 17/18 years (instead of 15 and 20 years) to reduce this wastage due to natural thinning. An additional silvicultural thinning could be carried out at 8–9 years to remove non-*Rhizophora* trees and to reduce stand density to around 8000 ha^{-1} to allow better growth. The standing biomass of the trees did not increase from 23 years (155 t ha^{-1}) to 28 years (153 t ha^{-1}). Based on biomass, we suggest that a rotation of 25 years be used instead of the present 30. This is also supported by size distribution of the stems which showed slow increase in the girth after 18 years.

Introduction

The Matang mangrove Forest Reserve, reputedly the best managed mangrove forest in the world, has been managed primarily for timber production (mainly *Rhizophora apiculata* Bl.) since the beginning of this century. Management is carried out by the State Forestry Department, Perak, Malaysia, through the implementation of constantly updated management plans ranging from one of 40-year rotation period with three thinnings carried out at approximately 15, 25 and 30 years to the present plan of a 30-year rotation period with two thinnings at 15 and 20 years (Ong, 1982). After final felling, sites which are considered inadequately stocked are replanted. Despite the intensive management, there has been a decline in yield from 299 t ha^{-1} of green-wood from virgin stands to the second generation yields of 158 t ha^{-1} in the period 1967–69 and 136 t ha^{-1} in the period 1970–77 (Tang *et al.*, 1984).

The decline in yield from virgin stands to second generation stands is understandable as the trees in the virgin stands were much older (perhaps 50–70 years) and therefore bigger than the second generation harvest at 30 years. However, there appears to be a decline between yields in the seventies (136 t ha^{-1}) compared to yields in the sixties (158 t ha^{-1}). If the decline is real, what are the reasons for this decline? Is it caused by a decrease in soil fertility with number of crops? Is it the result of competition due to successful invasion by other species, like the fern *Acrostichum*? Is it the result of the length of the rotation or the thinning regime? All these need to be considered to fully understand the reasons for the decline; but a quick synoptic look at the demography at different stages of the rotation should provide indications as to how the various silvicultural practices and management systems may have contributed to this decline.

This study on various aspects of the demography of *R. apiculata* at Matang was therefore carried out to obtain more information on what actually takes place in different aged stands of a 30-year rotation, which will help establish what is really happening through the rotation and suggest possible reasons for the decline in yield. These information can then be used in the silviculture so that the forest can be managed better to ensure sustained yield.

The study site

The study site is located near Kuala Sepetang (Port Weld), in the Matang Mangrove Forest Reserve (4 ° 50′N, 100 ° 36′E) which covers an area of 40 000 ha. This forest has been managed by the Perak State Forestry Department since the early part of this century, for timber production (mainly for charcoal and firewood), and the preferred species is *R. apiculata*.

At present, the forest is logged on a 30-year cycle (Haron, 1981), and about 1000 ha are clear-felled every year in patches of a few hectares. The slash is left to decompose and after about two years, the area is checked for the stocking of seedlings. In some 50% of these clear-felled areas, manual planting (at 1.2 m intervals) is carried out. The seedlings are left to grow till around 15 years when the first thinning is carried out using a 1.2 m stick so that any tree within a 1.2 m radius of a selected central tree is removed (and the bigger trees sold as poles). Thus theoretically, with a 1.2 m interval planting and a 1.2 m stick thinning, there will be no poles to harvest. In practice, seedling densities are increased considerably by natural recruitment even after manual planting which results in a harvest of about half of the trees during the first thinning. A second thinning takes place at about 20 years when a 1.8 m stick is used. This thinning also removes about half of the trees.

As the Forestry Department keeps quite good records, it is possible to estimate the age of any particular stand to within three years of the actual age.

Methods of study

In November 1980, 5, 8,13, 18, 23 and 28 year-old stands were selected for study so that a whole range of different aged stands within the 30-year rotation period was covered. The stands selected were close together and of the same soil-type and tidal inundation class (Watson 1928's Inundation Class 3) so as to minimise differences due to physical characteristics and history. Varying number and size of plots were set up in the different aged stands depending on the density of trees. Four 10 m × 10 m plots were established in the 5 year-old stand; six 10 m × 10 m plots in the 8, and 13 year-old stands; three 20 m × 20 m plots in the 18 year-old stands; and four 20 m × 20 m plots in the 23 and 28 year-old stands. The species of every tree (defined as plants above 2 m in height) in every plot and whether the tree was dead or alive were noted, and the GBH (girth at breast height *i.e.* 1.3 m) of every tree measured.

The total above-ground weight and trunk weight of every tree was calculated using the regressions obtained by Ong *et al.* (1985) for *Rhizophora apiculata* in the Matang Mangrove Forest Reserve. These are:

$$\underset{(\text{kg})}{W_{\text{ag(total above-ground weight)}}} = 0.0135\,\underset{(\text{cm})}{\text{GBH}}^{2.4243}$$

and

$$\underset{(\text{kg})}{W_{\text{trunk (total trunk weight)}}} = 0.0067\,\underset{(\text{cm})}{\text{GBH}}^{2.5414}$$

where GBH = girth at breast height (1.3m)

Results and discussion

Table 1 summarises the density of *R. apiculata* and *Bruguiera parviflora* Wight and Arnold ex Griffith trees, both living and dead, in the six different aged stands. Using these data, we can look at the initial stocking, as well as the changes in stand density, size distribution, mortality rates, and increment in biomass with age of stand.

Table 1. Density (no. ha^{-1}) of living and dead *Rhizophora apiculata* (Ra) and *Bruguiera parviflora* (Bp) trees in different aged stands of the Matang Mangrove Forest. Values are means ± s.e.

Stand (Age in yrs)	5	8	13	18	23	28
Trees						
Living Ra	15030	9810	9250	2740	2190	2550
	±2910	±2730	±1170	±770	±480	±290
Dead Ra	2380	700	6970	1120	380	110
	± 660	±200	±990	±180	±60	±30
Living Bp	1800	17	517	592	19	0
	±945	±17	±119	±211	±12	
Dead Bp	125	0	50	150	6	0
	±95		±34	±95	±6	

Initial stocking

In the 5 year-old stand (Table 1), the density of live *R. apiculata* trees was 15 030 trees ha^{-1}. There were 2380 dead trees (14% of the total standing stems). We estimate that at this age, dead trees do not remain standing for more that about a year so most of the standing dead trees would have died within the last year. It is reasonable to assume the presence of some mortality in the first few years as well, perhaps in the region of 3000 to 5000 trees. At the same time, there will also be natural recruitment over the first three years or so. From these, we estimate an initial stocking of some 20 000 seedlings ha^{-1}.

We would like to point out here that variability is high. With the four adjacent 10 m × 10 m plots we obtained a *R. apiculata* density of 15 030±s.e. 2910. With non-adjacent plots the standard error would no doubt be higher. A more intensive sampling regime would be needed for a more definitive study but the present data gives a reasonable indication of trends.

The estimated initial stocking of 20 000 is much higher than the initial stocking of 12 500 estimated by Srivastava & Daud (1978) in their study in Matang, as sufficient for 'a fully stocked stand at the end of a rotation if there is no large scale mortality'. According to Haron (1981), the mortality rate of *Rhizophora* seedlings may be more than 50% in Watson's (1928) Inundation Class 1, 4 and 5 (Haron, 1981). These areas are not optimum for *R. apiculata* (too wet in Class 1 which is inundated by all high tides, and too dry in Class 4 and 5 which are inundated by spring tides and exceptionally high tides respectively). In Inundation Class 3 (inundated by all high tides) where our stands were located, the mortality rate would be much lower. Thus, in these stands, it would appear that the problem of initial stocking did not occur. That the initial stocking and recruitment in the early years might in fact be too high is also shown by the fact that the high density of *R. apiculata* at 5 years dropped to 9810 ha^{-1} (about 35% mortality) at 8 years.

Under the present management scheme in Matang, artificial regeneration is carried out if stocking of natural regeneration two years after final felling is less than 90% (Haron, 1981). The recommended spacing for artificial regeneration is 1.2 m × 1.2 m, which allows a planting density of 6726 seedlings in a hectare (Haron, 1981). From the high density in the 5 year-old stand (Table 1), it would appear that there is quite a lot of natural recruitment. To minimise wastage of propagules, we suggest that artificial regeneration at 1.2 m intervals should only be carried out if natural regeneration is less than 50% (*i.e.* less than 3400 seedlings ha^{-1}) at two years after final felling. When the natural regeneration is between 50 to 90%, we suggest that enrichment planting to bring the density to 6726 seedlings ha^{-1} be carried out. This reduction in the number of propagules needed for artificial regeneration is important especially now that insufficient seed is a problem in certain areas (Haron & Cheah, 1979).

Stand density

From Table 1, it can be seen that the density of *B. parviflora* is extremely variable being 11, 2, 5, 18, 1 and 0% of the total living trees in the 5, 8, 13, 18, 23 and 28

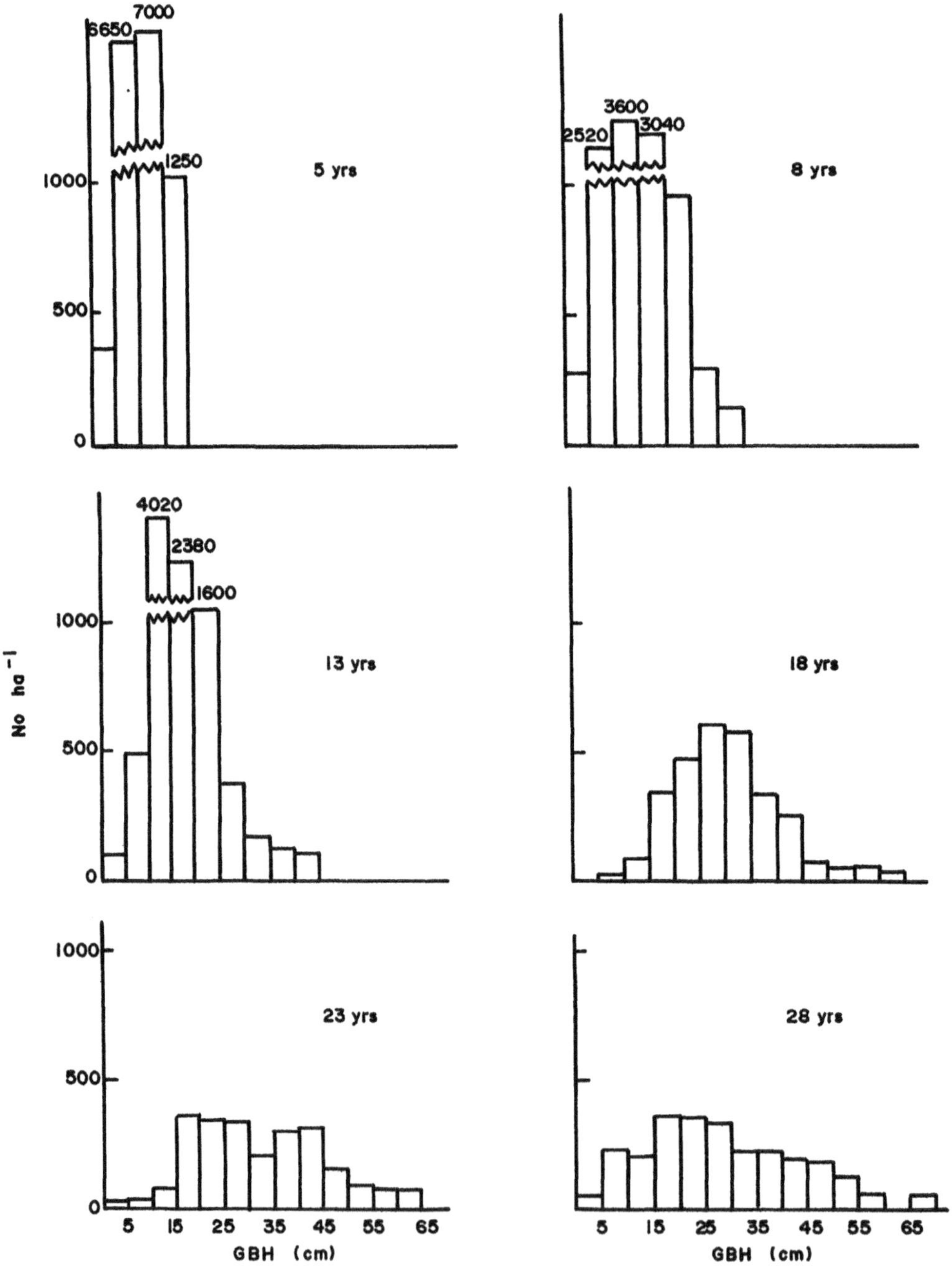

Fig. 1. Size distribution of *Rhizophora apiculata* trees in different aged stands of the Matang Mangrove Forest.

year-old stands respectively. Since *B. parviflora* constituted a significant proportion of the stand density (at least in the 5 and 18 year-old plots), we have included this species in our discussion on stand density.

The high density of *R. apiculata* of 15 030 trees ha^{-1} in the 5 year-old stand dropped to 9810 ha^{-1}, or 65%, in the 8 year-old stand (Table 1), but dropped only slightly to 9250 ha^{-1} in the 13 year-old stand. However, care must be taken in interpreting these data from different stands as they may have started off with different densities. That this was indeed the case here is borne out by the fact that the total density (of live and dead *R. apiculata* trees) in the 8 and 13 year-old stands were 10 510 and 16 220 trees ha^{-1} respectively, indicating a much higher initial (before death) density in the 13 year-old stand. The number of dead *R. apiculata* stems as a percentage of the total living and dead stems in the 5, 8 and 13 year-old stands was 14, 7 and 43% respectively. Taking *Bruguiera* trees into consideration as well, the density of living trees was 16 830; 9827 and 9767 ha^{-1} in the 5, 8, and 13 year-old stands respectively and the corresponding percentage of dead stems were 13, 7 and 42%.

The high percentage of dead stems (over 40%) in the 13 year-old stand is wasteful. We therefore suggest that the first thinning should be carried out earlier than 15 years (the present practice) to decrease wastage due to natural thinning. Admittedly, the size of poles at this stage would be less than what is presently considered as desirable, but it is likely that trees of 12/13 years that had been planted at a lower density would attain a bigger girth than trees under the present silvicultural practice. In addition, we suggest that the Forestry Department look into the possibility of a purely silvicultural thinning at between 8 and 9 years, during which the density of the stand should be reduced to around 8000 ha^{-1} to enable remaining stems to grow better so that trees of 12/13 years would attain commercially acceptable pole size. In this silvicultural thinning, trees of other species (other than *R. apiculata* and *Rhizophora mucronata* Lamk.) should be removed; in the case of a pure stand of *Rhizophora*, the smaller trees should be removed. Although the data in Table 1 suggest that a higher mortality of *R. apiculata* occurred where living *B. parviflora* density is high, the high variability encountered reduces the confidence of such an interpretation. The question of whether *B. parviflora* is a significant competitor merits study.

The density of *R. apiculata* trees in the 18 year-old stand dropped to 2740 ha^{-1} or 30% of that in the 13 year-old stand (Table 1). A large part of this decrease is the result of first thinning at 15 years, when a 1.2 m stick is used and usually, about half of the stems is removed. There were 1120 dead trees, or 29% of the total *R. apiculata* trees. This high percentage of dead trees is wasteful and we suggest that the second thinning should be carried out at 17/18 years (by which time the trees would definitely have reached desirable pole size even under the present management plan) rather than the present practice of second thinning at 20 years. We should also point out here that there were 592 live and 150 dead *B. parviflora* in this 18 year-old stand. Thus, this species constituted a fair proportion of the standing density (18%) and would result in lower density for the preferred species, *R. apiculata*. Haron (1981) had in fact suggested that areas heavily infested with *B. parviflora* after final felling should be treated to convert them to *R. apiculata* areas. As further recruitment of *B. parviflora* and other species is possible after the initial treatment, we have earlier suggested that a further silvicultural thinning which includes removal of non-*Rhizophora* species be carried out at 8–9 years.

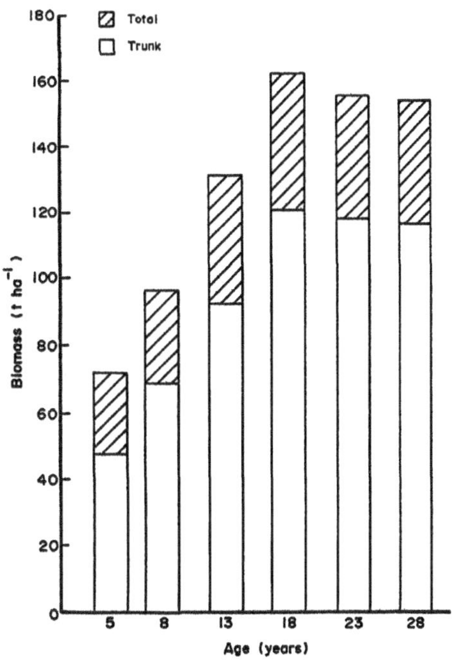

Fig. 2. The total above-ground and trunk biomass of *Rhizophora apiculata* trees in different aged stands of the Matang Mangrove Forest.

The density of *R. apiculata* in the 23 year-old stand was 2190 trees ha^{-1} or 80% of that in the 18 year-old. Taking *B. parviflora* into consideration as well, the density of the 23 year-old stand was 66% that of the 18 year-old stand. Under the present management plan, a second thinning would have been carried out at 20 years, with a 1.8 m stick which would have removed about half of the stems.

In the 28 year-old stand, the density of *R. apiculata* trees increased to 2550 ha^{-1}. The size distribution (see next section) of the trees suggest that the increase in density was due to new recruitment.

Size distribution

Figure 1 shows the size distribution (5 cm GBH intervals) of *R. apiculata* trees in the different aged stands. In the 5 year-old stand, trees were found only in the first four size classes (*i.e.* GBH up to 20 cm), with most of the trees in the 5–15 cm range. In the 8 year-old stand, the trees had become bigger with some trees in the 30–35 cm girth class, but with most in the 5–20 cm range. The biggest trees had reached the 40–45 cm girth class in the 13 year-old stand, with the majority in the 10–25 cm range. In the 18 year-old stand, the biggest trees had attained girths between 60–65 cm, but most of the trees were in the 20–35 cm range. At this stage, there were no trees in the smallest size class and indeed very few trees less than 15 cm girth because thinning at 15 years had removed most of the small trees. A noteworthy point about the 23 year-old stand was that there appeared to be no increase in girth; the biggest trees were still between 60–65 cm in girth and most of the trees were in the 15–30 cm range (rather than in the expected larger size classes of, say, 25–40 cm). One possible explanation for this is that at the second thinning at 20 years, many of the bigger trees might in fact have been removed (instead of being left behind for final harvest). This is not a problem with the silvicultural practice but with the management. We agree with Haron (1981) that there should be stricter control to ensure that the stands are not degraded because of the type of thinning ('for purely commercial purposes') practised. In the 28 year-old stand, the biggest trees had reached the 65–70 cm size class but the mode was still between 15–30 cm. There were also quite a number of trees less than 15 cm, suggesting that some recruitment had occurred after 18 years, and especially between 23 and 28 years. However, a lot of this recruitment is wasted for as Haron (1981) pointed out, during final felling, a lot of the small trees are destroyed despite the fact that manual (with a chain-saw) felling is carried out.

Biomass increment

Figure 2 shows the total above-ground and trunk biomass in the different aged stands. The biomass increased rapidly from 72 t ha^{-1} at 5 years to 131 t ha^{-1} at 13 years. The biomass increase between the 13 and the 18 year-old stands was masked as thinning was carried out at 15 years. The standing biomass decreased from 161 t ha^{-1} to 155 t ha^{-1} at 23 years. Again, this could be attributed largely to the thinning carried out at 20 years. The biomass did not increase at 28 years (153 t ha^{-1}). From these biomass figures, it appears that a 30-year rotation may be too long as the same standing biomass may be obtained at 23 years. This is supported by Ong *et al.* (1984) who showed that the mean annual increment of *R. apiculata* trees at Matang peaked at 10 years at 18 t ha^{-1} and decreased at 15 years to 13 t ha^{-1} and to 12 t ha^{-1} at 25 years.

The trunk biomass (which is essentially the biomass of importance to the fuel wood industry) followed the same trend as the total above-ground biomass, increasing from 50 t ha^{-1} in the 5 year-old stand to 70 t ha^{-1} in the 8 year-old stand and 95 t ha^{-1} in the 13 year-old stand before levelling off at 120 t ha^{-1} in the 18 year old stand. The slight drop in biomass at 23 years may be the result of thinning at 20 years, but, once again, there was no increase in biomass between 23 and 28 years. So, once again, based on the trunk biomass data, it would appear that the rotation cycle should be shortened, perhaps to 25 years. There is however the point that there is a demand for larger-sized charcoal. However, as we had earlier discussed, there was no marked increase in GBH after 18 years (Fig. 1). The Forestry Department has to weigh the advantages of the larger-sized charcoal against the total volume produced and the shorter rotation period.

Conclusion

This study has illustrated how a quick demographic study, using plots of different ages at one instant of time rather than following the fate of permanent plots, does give a good idea of possible improvements in the silviculture, especially with respect to the density of initial stocking, the spacing for artificial regeneration, the age of thinning and the rotation cycle. Further study to look into details (e.g. causes of mortality, factors affecting the rate of growth at different ages) is needed to further improve the silviculture. In the more detailed study, it is necessary to have more plots because of the high variability and it is also necessary to have plots under different conditions (e.g. inundation classes) to cover the range existing in Matang. It should be noted that not all of these problems are ecological, some involve management decisions that may be commercially-based. The Forestry Department has to look into all these now to ensure the sustained yield management of this important resource.

Acknowledgments

We would like to thank the Forestry Department, Perak, Malaysia for assistance rendered throughout the duration of this study. The study was made possible with a grant from the UNESCO/MAB Programme on Man and the Biosphere in Malaysia and infrastructural support from the Universiti Sains Malaysia. We are grateful to Dr Wong Chee Hoong who participated in the earlier part of this study and to several laboratory and student assistants for their cheerful participation in the field work.

References

Haron, H. A. H., 1981. A Working Plan for the Second 30 years Rotation of the Matang Mangrove Forest Reserve, Perak. The First 10-year Period, 1980–1989. State Forestry Department, Perak, 115 pp.

Haron, H. A. H. & L. C. Cheah, 1979. Sustained yield management of the mangrove forest of Peninsular Malaysia with special reference to the Matang Mangrove. Paper presented at the 7th Forestry Seminar, Forestry Department, Malaysia, Kuala Lumpur, Malaysia.

Ong, J. E., 1982. Mangroves and aquaculture in Malaysia. Ambio 11: 252–257.

Ong J. E., W. K. Gong & C. H. Wong, 1985. Seven years of productivity studies in a managed Malaysian mangrove forest, then what? In: K. N. Bardsley, J. D. S. Davie & C. D. Woodroffe (eds), Coasts and Tidal Wetlands of the Australian Monsoon Region. Australian National University North Australia Research Unit, Darwin: 213–223.

Ong J. E., W. K. Gong, C. H. Wong & G. Dhanarajan, 1984. Contribution of aquatic productivity in a managed mangrove ecosystem in Malaysia. In: E. Soepadmo, A. N. Rao & D. J. Macintosh (eds), Proceedings of the UNESCO Asian Symposium on Mangrove Environment: Research and Management. University of Malaya, Kuala Lumpur: 209–215.

Srivastava, P. B. L. & K. Daud, 1978. Progress of natural regeneration after final felling under the current silvicultural practices in Matang Mangrove. Pertanika: 126–135.

Tang, H. T., H. A. H. Haron & L. C. Cheah, 1984. Mangrove forests of Peninsular Malaysia – a review of management and research objectives and priorities. In: E. Soepadmo, A. N. Rao & D. J. Macintosh (eds), Proceedings of the UNESCO Asian Symposium on Mangrove Environment: Research and Management. University of Malaya, Kuala Lumpur: 796–808.

Watson, J. G., 1928. Mangrove Forests of the Malay Peninsula. Malayan Forest Records No. 6, 275 pp.

Hydrobiologia **295**: 263–273, 1995.
Y. S. Wong & N. F. Y. Tam (eds), Asia-Pacific Symposium on Mangrove Ecosystems.

Effect of synthetic wastewater on young *Kandelia candel* plants growing under greenhouse conditions

G. Z. Chen[1], S. Y. Miao[1], N. F. Y. Tam[2], Y. S. Wong[3*], S. H. Li[1] & C. Y. Lan[1]
[1]*Research Institute of Environmental Science, Zhongshan University, People's Republic of China*
[2]*Department of Biology and Chemistry, City Polytechnic of Hong Kong, Tat Chee Avenue, Kowloon, Hong Kong*
[3]*Research Centre/Biology Department, Hong Kong University of Science and Technology, Clear Water Bay, Kowloon, Hong Kong*
*(*author for correspondence)*

Key words: Mangrove, *Kandelia candel*, seedling, growth, wastewater, pollution

Abstract

A greenhouse experiment was performed to evaluate the effects of synthetic wastewater in three different strengths, NW, MW and CW, on the growth of the one-year old *Kandelia candel* (L.) Druce plants. NW had the characteristics and strength similar to natural municipal wastewater while MW and CW contained five and ten times of the nutrients and heavy metals in NW, respectively. Artificial seawater was used as the control. During one year wastewater treatment experiment, *Kandelia* were found to withstand wastewater of high strength and toxic symptoms were not detected in all plants. Synthetic wastewater with strength similar to the natural municipal sewage (NW) stimulated plant growth. The plants treated with NW had significantly higher aerial and root biomass, taller stem than those found in the CW, MW and the control. Maximum growth, in terms of both stem height and total biomass, of all plants occurred in summer months, from June to September. With respect to the physiological and biochemical activities, CW and MW treated plants had significantly lower levels of chlorophyll *a*, total chlorophyll and catalase activity than those found in NW and control groups. In contrast, the proline content of plants treated with wastewater was similar to that of the control. These results suggest that normal wastewater (NW), attributed to its nutrients and trace elements, enhanced plant growth. The medium (MW) and concentrated wastewater (CW) supported similar amount of plant growth as the control but the physiological and biochemical parameters indicate that these treated plants might have been exposed to some kind of stress, probably due to the excess heavy metals present in MW and CW.

Introduction

Many investigations have shown that wetland (both natural and constructed) has a potential for treating wastewater (De Jong, 1976; Gambrell *et al.*, 1987; Gersberg *et al.*, 1986; Wathugala *et al.*, 1987). In a wetland system, higher aquatic plant is the most obvious biological component and is of great significance in removing pollutants from sewage. The aquatic macrophytes such as *Phragmites australis*, *Typha capensis* and *T. latifolia* have been successfully employed to purify wastewater, through systems of reed ponds (De Jong, 1976; Rogers *et al.*, 1991), sand or gravel filtration (Conley *et al.*, 1991; Gersberg *et al.*, 1986; Wathugala *et al.*, 1987), and natural marsh (Boyt *et al.*, 1977; Reed *et al.*, 1988). However, the ability of plants to remove nutrients and pollutants from wastewater has not yet been fully examined. The major mechanism for wastewater purification includes the plant uptake and the microbial transformation by bacteria associated with the plant rhizosphere (Gersberg *et al.*, 1986; Reed *et al.*, 1988; Rogers *et al.*, 1991). Different plant species with varied growth rates and primary productivity levels would have different abilities to remove nutrients and pollutants from discharged sewage. Mangrove plants, which are intertidal and which dominate the estuarine shores within tropical and subtropical regions, are well known for their high standing biomass and productivity (Lugo, 1980). The plants are perennial, consist of extensive root system and specially adapted to the harsh environment of shifting aerobic and anaerobic, as well as alternating wetting and drying conditions (Lugo & Snedaker, 1974). These features enable the mangrove communities to withstand and retain wastewater-borne nutrients and pollutants (Nedwell, 1974). Clough *et al.* (1983) estimated that mangrove plants, by incorporation into the plant tissues, can annually immobilize very significant amounts of N and P with values around 150 to 250 kg N ha^{-1} and 10 to 20 kg P ha^{-1}, respectively.

On the other hand, sewage effluent will affect the plant communities of wetlands such as *Eleocharis sphacelala* (sedge) and *Typha orientalis* (bulrush), primarily through changing the hydrological characteristics, the nutrient status and the pollutant levels (Cooke *et al.*, 1990; Kadlec, 1987). Mangrove plants, well-known to be resistant to environmental stresses such as high salinity and waterlogging, are also able to tolerate high concentrations of nutrients (Clough *et al.*, 1983; Nedwell, 1974) and heavy metals (Lin & Chen, 1990). It has been suggested that the input of inorganic nutrients and trace elements from sewage discharges are not likely to be deleterious to mangroves, and indeed, may well be beneficial to their growth, especially in mangrove communities which are limiting in nutrients (Boto, 1992; Clough & Attiwill, 1982; Onuf *et al.*, 1977). Boto and Wellington (1983) reported that addition of N and P to the mangrove communities resulted in 30% increase in mangrove growth coupled with a higher nutrient content in plant tissues. However, it has also been reported that sewage is an important biotic factor contributing to the gradual disappearance of mangrove vegetation near Bombay (Navalkar, 1951). The effects of sewage addition to mangrove plants may differ significantly among different mangals as they vary widely in their physical, chemical and biological properties (Clough *et al.*, 1983). Moreover, the addition of sewage may cause sub-lethal damages in plants such as changes in photosynthetic rates, chlorophyll concentrations and enzymatic activities (Culic, 1984; Peng, 1990), which can be used as an early warning to wastewater pollution problems. The present study is therefore aimed (1) to examine the possible effects of sewage discharge on the growth, and the physiological and biochemical responses of young *Kandelia candel* plants growing under greenhouse conditions; and (2) to compare the impact of different strengths of wastewater on mangrove plants.

Materials and methods

Experimental set-ups

Twelve polyethylene tanks ($0.7 \times 0.5 \times 0.4$ m^3) were filled to a depth of 30 cm with 100 kg of soil collected from a mangrove swamp in Futian National Nature Reserve, Shenzhen Special Economic Zone, the People's Republic of China. The mangrove soil was previously air-dried, ground and passed through a 2-mm mesh sieve before placed in the tide tank. One-year old *Kandelia candel* plants were transplanted from Futian mangrove swamp to the tanks. Each tank was flooded with artificial seawater (prepared by dissolving commercially available salts in deionized water and the salinity was 1.2–1.6%) twice a day (at 10 am to 2 pm and 10 pm to 2 am) to simulate high tide. This gave about 16 hours exposure every day which simulated the natural tidal regime of Shenzhen area. During high tide period (about 8 hours per day), the base of the plants (about 2 cm from the soil surface) were submerged in the artificial seawater. Fifteen plants were grown in each tank in a greenhouse for about 9 months to acclimatize to the greenhouse environment prior to treatment with wastewater. At the beginning of the wastewater treatment, the young plants reached an average height of 17.7± 1.6 cm, stem diameter of 6.1± 0.5 mm, leaf number of 10± 2, and the average leaf, stem, hypocotyl, root and total biomass were 1.74, 1.50, 6.06, 1.92 and 11.22 g dry weight, respectively. The twelve tide tanks planted with *Kandelia* were divided into four groups (each group was triplicated). Three groups were irrigated with synthetic wastewater of three different strengths, namely natural wastewater (NW), medium wastewater (MW) and concentrated wastewater (CW), twice a week with the hydraulic loading of 1.75 l per tank per irrigation during the exposure period (*i.e.* at 5 pm of the day). The reason for adding wastewater during the exposure period was to increase the amount of wastewater infiltrated into the soil and to enhance the treatment efficiency. The remaining group (SW) was irrigated with artificial seawater and was used as a control. NW represented the synthetic wastewater having its composition similar to the natural municipal sewage (Table 1), MW and CW had 5 and 10 times the amount of nutrients and heavy metals of the NW, respectively. In between two wastewater irrigation, all the tanks were flooded with seawater according to the tidal regime. This experiment lasted for one year under the greenhouse conditions.

Table 1. Characteristics of artificial wastewater with composition similar to the natural municipal wastewater (NW).

Nutrient	Compound used	Concentration ($mg\ l^{-1}$)	Nutrient	Compound used	Concentration ($mg\ l^{-1}$)
COD	glucose	500.0	Cu^{2+}	$CuCl_2.2H_2O$	2.0
NH_4^+-N	NH_4Cl	40.0	Zn^{2+}	$ZnSO_4.7H_2O$	5.0
NO_3^--N	$NaNO_3$	1.0	Cd^{2+}	$CdCl_2.2H_2O$	0.1
Organic N	urea	10.0	Ni^{2+}	$NiCl_2.6H_2O$	1.0
PO_4^{3-}-P	KH_2PO_4	10.0	Pb^{2+}	$Pb(NO_3)_2$	1.0
K^+	KCl	50.0	Cr^{6+}	$K_2Cr_2O_7$	0.5
Fe^{3+}	$FeCl_3.6H_2O$	30.0	Mn^{2+}	$MnCl_2.4H_2O$	5.0

Determination of physiological/biochemical responses and growth of Kandelia plants

Plant growth
Plant growth, in terms of stem height (H) and diameter at the base of the stem (D), was measured every month. The biomass values of different plant parts were obtained by a non-destructive allometric technique (Snedaker & Snedaker, 1984) and was calculated every month. Twenty-one young *Kandelia candel* plants (about two-years old) were collected from Futian mangrove swamp. Stem height and diameter at the base of the stems were determined. The young plants were divided into stem, leaves, hypocotyl, and root components, oven-dried (105 °C) and their weights were recorded (Table 2). The least-square regression equations using diameter and height as independent variables were calculated according to: log biomass $= a + b$ log (D^2H) (Snedaker & Snedaker, 1984). The allometric relations were highly significant (Table 3) and were used for estimating biomass of various plant parts during the experiment.

Physiological/biochemical measurements
At the end of one-year treatment with wastewater, five to ten mature leaves (the third pair of the leaves from the apical part of the stem) of every plant from each tank were collected, mixed and analyzed for various parameters. The content of chlorophyll *a* and *b* (extracted by acetone and measured at wavelength of 663 and 645 nm; Zhu *et al.*, 1990), concentration of free proline (extracted by glacial acetic acid, triketohydrindene and benzene, followed by measuring at wavelength of 515 nm; Zhu, 1983), and catalase activity (with H_2O_2 as the substrate and titrated with $KMnO_4$; Huang & Chen, 1990) were determined. The photosynthetic rate of five mature leaves from every plant of each treatment were measured by a portable photosynthetic system (LI-6200, LI-COR Inc.) and the data were calculated based on the light flux density of 175.0 μ mol $m^{-2}\ s^{-1}$, CO_2 concentration of 380 mg l^{-1} and air temperature at 293.15 °K.

Statistical analyses

The mean and standard deviation of the triplicates of all measured parameters in each group were calculated. The results collected except the plant growth data were treated with a parametric one-way analysis of variance (ANOVA) to test the significant difference between four treatments (three types of wastewater and the control). The plant growth, in terms of monthly increment in stem height and biomass, was also evaluated by two-way ANOVA, with treatment and time as the main effects. The least significant difference (lsd) values at 5% probability level was calculated if the results of the ANOVA indicated significant differences. All statistical analyses were performed by means of an IBM-compatible computing package 'SPSS'.

Results and discussion

Plant growth during one-year treatment with wastewater

Stem height and diameter
The monthly accumulation of stem height indicates that irrigation with wastewater of strength similar to the natural municipal sewage (NW) led to best plant

Table 2. Stem height (H), diameter at the base of the stem (D) and the biomass (oven dried weight, W) of 21 young *Kandelia candel* plants for regression analyses.

			Biomass (g)				
Plant No.	D (cm)	H (cm)	Stem	Leaf	Hypocotyl	Root	Total
1	0.678	6.8	0.7	1.1	5.3	1.3	8.4
2	0.538	15.0	0.8	0.5	6.2	0.4	7.9
3	0.478	13.0	0.8	1.7	5.4	1.2	9.1
4	0.656	19.4	1.6	1.4	5.9	1.5	10.4
5	0.548	12.3	0.9	1.6	4.4	0.9	7.8
6	0.872	36.3	5.4	5.5	8.6	5.7	25.2
7	0.564	20.3	1.7	0.9	6.6	2.3	11.5
8	0.720	26.0	2.4	3.1	5.5	3.7	14.7
9	0.724	22.3	1.9	1.8	6.3	2.2	12.2
10	0.623	16.8	1.2	1.0	5.7	2.6	10.5
11	0.432	10.2	0.5	1.0	5.8	0.5	7.8
12	0.825	21.4	2.7	2.9	8.4	2.9	16.9
13	0.401	3.8	0.3	0.5	5.5	1.0	7.3
14	0.729	22.2	2.6	1.5	6.8	2.8	13.7
15	0.353	7.0	0.2	0.5	4.8	0.6	6.1
16	0.708	17.8	1.6	2.1	6.8	3.5	14.0
17	0.822	30.4	3.8	1.4	8.9	3.9	18.0
18	0.551	8.2	2.1	0.8	3.8	1.2	7.9
19	0.583	16.9	1.2	1.2	3.6	1.7	7.7
20	0.669	14.5	1.3	1.4	5.3	2.1	10.1
21	0.788	23.2	4.4	0.6	6.7	4.2	15.9

Table 3. The regression equations showing the relationships between biomass (W), stem diameter (D) and stem height (H) of *Kandelia plants.*

Biomass (dry weight, g)	Regression equation	Correlation Coefficient (r)
Stem	$\log W_S = -0.47 + 0.79 \log (D^2H)$	0.93**
Leaf	$\log W_L = -0.26 + 0.96 \log (D^2H)$	0.70
Aerial (stem + leaf)	$\log W_A = -0.05 + 0.63 \log (D^2H)$	0.93**
Hypocotyl	$\log W_H = 0.66 + 0.15 \log (D^2H)$	0.62
Root	$\log W_R = -0.22 + 0.61 \log (D^2H)$	0.82*
Total biomass	$\log W_T = 0.78 + 0.33 \log (D^2H)$	0.89**

The regression analyses were calculated based on data from 21 plants shown in Table 2; * and ** indicated the r values were significant at probability levels of 0.05 and 0.01, respectively, the r values without any asterisk meant statistically not significant (probability greater than 0.05).

growth, followed by CW, and MW treatment shared similar height increment as the control (Fig. 1). The growth of individual plants within the same treatment group tended to differ profoundly. Variations among 15 individuals in the same tide tank and among the triplicated treatment group were very obvious. Only plants from tanks treated with NW had significantly higher amount of stem growth than the other treated groups. At the end of the one year wastewater treatment experiment, the stem height of the plants from

NW tanks increased from an initial 17.74 cm to the final 28.14 cm, with an average of 10 cm total increase in stem height, whereas the total stem increment was around 7 cm in MW and the control (Table 4). The plants treated with concentrated wastewater (CW) had intermediate level of growth. Similarly, the average stem diameter of *Kandelia* followed the descending order of NW>CW>MW, control (Table 5). These results suggest that addition of wastewater stimulated stem growth as the mangrove soil is often deficient in nutrients and mangrove plants always response positively to sewage discharge (Boto & Wellington, 1983; Clough *et al.*, 1983). Tam *et al.* (1993) reported that the amount of bioavailable P and N in the native soil of the Futian mangrove fell within the range recorded in other mangrove ecosystems and could be considered as nutrient deficiency. There was no inhibition in stem growth when the plants were irrigated with wastewater containing very high content of nutrients and heavy metals such as MW and CW, implying that the mangrove plants, even at a relatively young growing stage, possessed resistance to wastewater pollution.

The monthly increment in stem height shows that, as expected, more active growth occurred during the summer months, from July to October (some even extended to November), with maximum growth in August in both treated and control plants (Fig. 2). Very little growth occurred during the winter season. This growth pattern reflects that the ambient temperature is one of the most important factors affecting the growth of mangrove plants. Indeed, the distribution of mangals is closely related to temperature as they can only be found in tropical and subtropical regions, at latitudes between 25 °N and 25 °S (Lugo & Snedaker, 1974). Wastewater addition did not alter the growth pattern measured in terms of the monthly increase in stem height.

Plant biomass

Figure 3 shows that the pattern of the cumulative increase in total biomass of plants during the one year study which was similar to that found in the cumulative stem height. The plants treated with NW had the highest total biomass, followed by CW, and MW and the control were similar. The differences between the wastewater treated groups and the control were very small during the initial 5 months (from March to July 1992), but the plants irrigated with NW had much more growth than the other three groups from August onwards (Fig. 3). At the end of the study, total biomass production in NW group was 66% more than that of the control. Clough *et al.* (1983) concluded that addition of wastewater significantly increased the concentrations of available nutrient such as NH_4^+-N and PO_4^{3-}-P of the mangrove soil, and enhanced both the plant productivity and the nutrient content of the plant tissues. Therefore it is not surprised to find that discharge of NW resulted increases in both stem height and biomass production. Similar research work examined the effects of nutrient enrichment of *Rhizophora mangle* in Florida also indicated that the plants in fertilized sites had higher growth rates, greater additions of leaves, reproductive parts and new lateral branches, and larger increases to existing stems (Boto & Wellington, 1983; Onuf *et al.*, 1977). On the other hand, wastewater containing very high concentrations of N, P and various heavy metals such as CW did not cause any reduction in biomass. This could be related to the fact that the mangrove soils were capable to immobilize excess N, P and heavy metals and made them less available to plant uptake. Tam & Wong (1992 & 1993) found that most of the P and heavy metals added to mangrove soil from discharged sewage were bound to the mangrove soils either by adsorption on ion-exchange sites, incorporation into lattice structures or by precipitation as insoluble sulphide. Only a very small portion of these elements were bio-available. Similar findings were also reported by Boyt *et al.* (1977) and Cooke *et al.* (1990). This explained why the growth of the plants was not inhibited by irrigation of CW. Furthermore, the mangrove environment is highly variable owing to a combination of periodic fluctuations and extremes in physico-chemical parameters. This variability also enables the mangrove flora and fauna to become highly adaptable to adverse conditions. It has been reported that the growth of the seedlings of *Avicennia alba*, *Rhizophora muoronata* and *R. mangle* were not inhibited by Zn, Pb, Cd and Hg (Chen & Lin, 1988).

With respect to monthly increment, Fig. 4 reveals that the maximum biomass production occurred from June to August, a bit earlier than the maximum increases in stem height (Fig. 2). As the biomass was calculated according to the regression equation based on stem height and diameter, this meant that the plants started to have rapid growth by first expanding their diameter followed by an increase in stem height. The average monthly increases in biomass and stem height (over 12 months of the experimental period) were shown in Table 4. The root biomass contributed to quite a significant portion of the total biomass, and the aerial

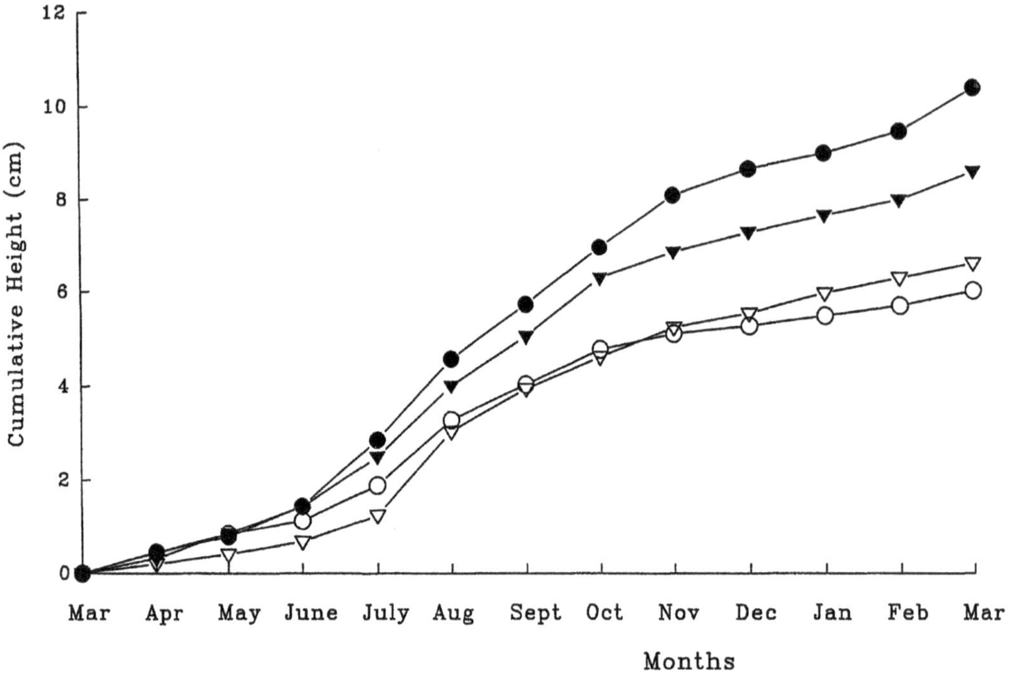

Fig. 1. Effects of wastewater in different strengths on cumulative increases in stem height during one year study. (○: control; ●: NW; ▽: MW; ▼: CW).

Table 4. Monthly increments in stem height, aerial (stem + leaves), root and total biomass of *Kandelia candel* plants treated with various kinds of wastewater.

Treatment	Height (cm) Monthly Increment			Biomass (g dry wt. per plant) Monthly Increment		
	Mean	Minimum	Maximum	Aerial	Root	Total
Control	0.488^b (0.063)	0.208	1.142	0.158^a (0.037)	0.119^a (0.020)	0.249^a (0.049)
NW	0.867^b (0.368)	0.183	2.350	0.208^b (0.038)	0.180^b (0.013)	0.385^b (0.027)
MW	0.552^a (0.135)	0.117	1.625	0.152^a (0.016)	0.121^a (0.010)	0.246^a (0.038)
CW	0.717^a (0.293)	0.150	2.183	0.180^a (0.053)	0.159^a (0.053)	0.305^a (0.088)

The mean and standard deviation values (in brackets) of 12 months data on 45 individual plants of each treatment were calculated. The means followed by different superscripts within each column indicated that they were significantly different at a probability level of 0.05 according to ANOVA test.

to root biomass ratio was around 1.55:1. This reflects that mangrove plants had very extensive root system in order to adapt to the stress environment.

Physiological/biochemical responses

Chlorophyll content

In all wastewater treated plants, the chlorophyll *a* and *b* content was significantly lower than that of the control,

Table 5. Changes of stem diameter (cm) under different wastewater treatments during one year study.

Months	Control	NW	MW	CW
March'92	0.610	0.610	0.610	0.610
April	0. 603	0.621	0.616	0.626
May	0.610	0.628	0.614	0.620
June	0.673	0.663	0.700	0.684
July	0.712	0.717	0.733	0.712
August	0.730	0.736	0.725	0.707
September	0.733	0.736	0.711	0.729
October	0.732	0.748	0.716	0.724
November	0.741	0.781	0.726	0.754
December	0.749	0.788	0.727	0.763
January 93	0.751	0.793	0.725	0.761
Feb	0.752	0.801	0.742	0.772
March	0.753	0.822	0.747	0.774

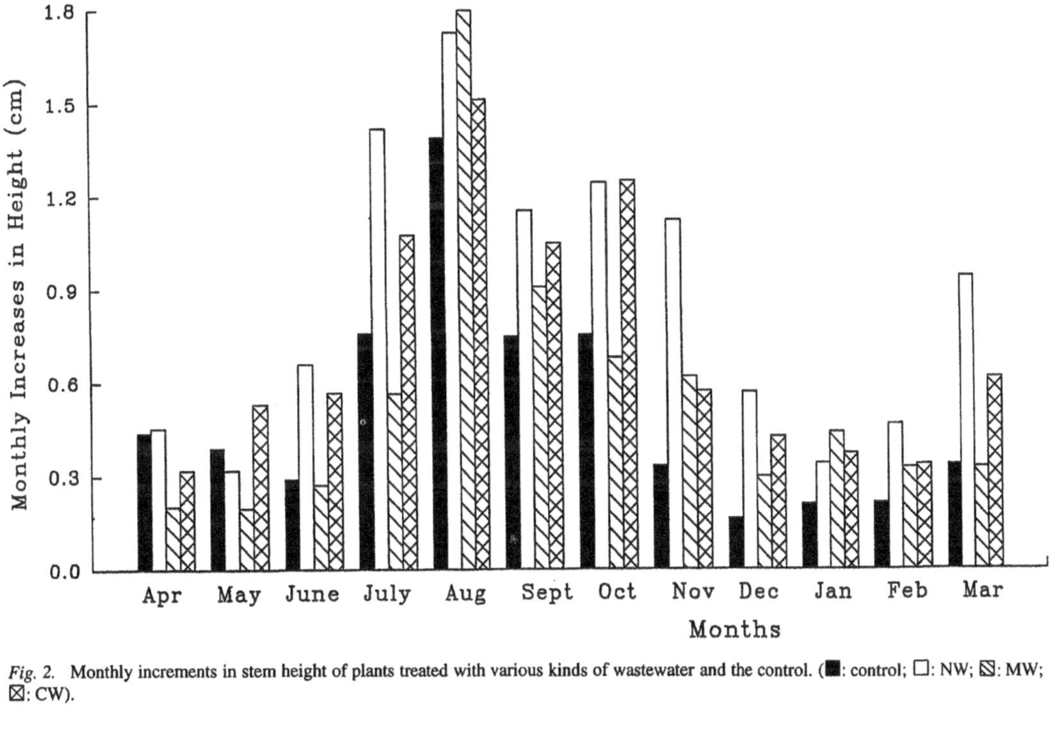

Fig. 2. Monthly increments in stem height of plants treated with various kinds of wastewater and the control. (■: control; □: NW; ▧: MW; ⊠: CW).

with a descending order of: control>NW>CW>MW (Table 6). The Chl *a* and *b* contents of plants treated with CW were 65% and 50% of the control, respectively. Wastewater treatment caused a bigger decline in Chl *b* than Chl *a*. The ratio of Chl *a* to Chl *b* was the highest in plants treated with CW (Table 6), and the ratio increased with the strengths of wastewater. The effect of wastewater discharge on chlorophyll content

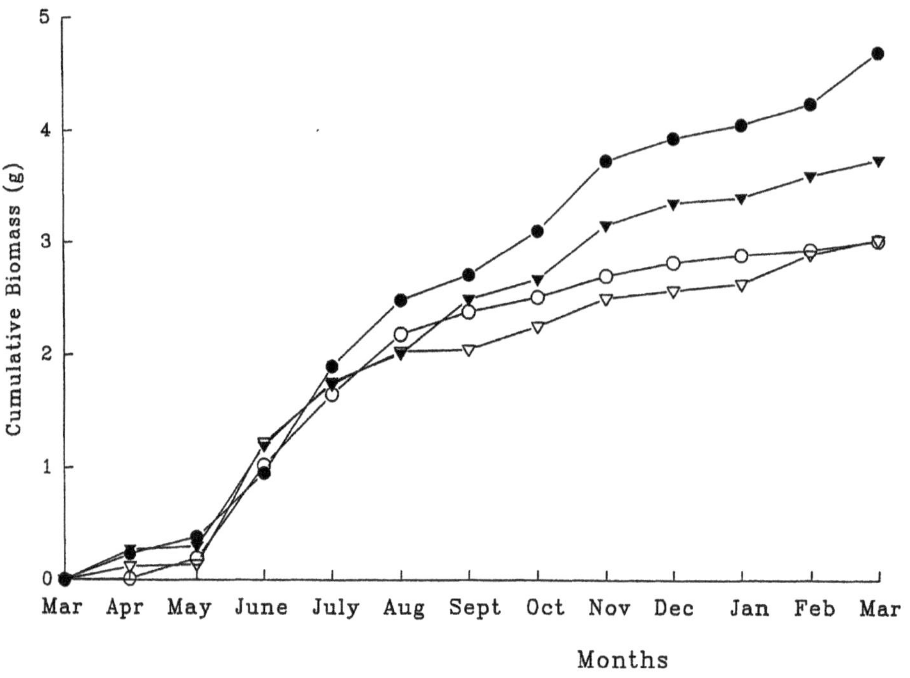

Fig. 3. Effects of wastewater in different strengths on cumulative increases in total biomass during one year study. (○: control; ●: NW; ▽: MW; ▼: CW).

was similar to other studies on stress physiology of mangrove plants. It has been found that the environmental stress not only caused a drop in chlorophyll content (Culic, 1984; Lin *et al.*, 1984), it also changed the ratio of chlorophyll *a* and *b* (Wang, 1990). Lin *et al.* (1984) found that under high soil salinity (1% salinity), the synthesis and accumulation of chlorophyll in *Kandelia candel* were reduced and the ratio of Chl *a* to Chl *b* raised. The reduced chlorophyll level of plants treated with wastewater suggest that the plant vigour and their physiological activity might have been affected although the growth of these plants was not reduced by wastewater treatment.

Photosynthetic rates

Despite the lower chlorophyll content in plants treated with wastewater, the photosynthetic rates measured in the treated groups, NW and CW, were significantly higher than that of the control (Table 7). This suggests the nutrient input from wastewater treatment enhanced the photosynthesis, in particular, the photosynthetic rate per unit chlorophyll molecule. Lugo *et al.* (1976) also concluded that gross photosynthesis appears to be sensitive to terrestrial nutrient input and that the development of mangrove biomass is dependent on the quantity of nutrients.

Proline content and catalase activity

All treated and control plants had similar level of free proline but those treated with MW and CW had significantly lower catalase activity than NW and the control (Table 7). Catalase is an enzyme commonly found in plant tissue and its activity is closely related to the degree of pollution. It has been found the higher the pollutant concentration, the more the leaves were stressed, the lower the activity of catalase (Peng, 1990). This suggests that the young plants treated with MW and CW might have been subjected to some kind of stress. The high heavy metal concentrations of these wastewater treatment might have created the stressed environment which required further investigation. The present results also suggest that catalase activity might be a more sensitive parameter in indicating cellular damages caused by wastewater treatment than free proline level, another parameter commonly used to relate the stress resistance of many plants (Peng, 1990).

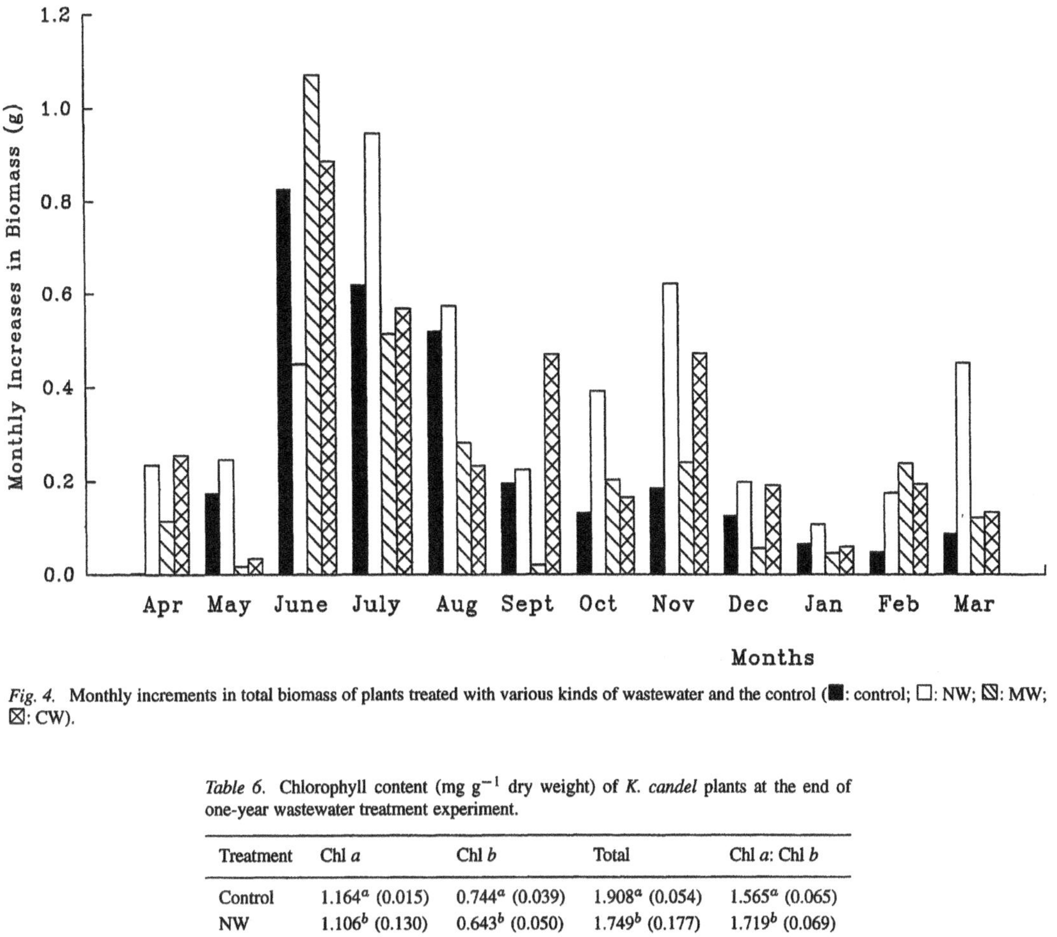

Fig. 4. Monthly increments in total biomass of plants treated with various kinds of wastewater and the control (■: control; □: NW; ▨: MW; ⊠: CW).

Table 6. Chlorophyll content (mg g^{-1} dry weight) of *K. candel* plants at the end of one-year wastewater treatment experiment.

Treatment	Chl *a*	Chl *b*	Total	Chl *a*: Chl *b*
Control	1.164^{a} (0.015)	0.744^{a} (0.039)	1.908^{a} (0.054)	1.565^{a} (0.065)
NW	1.106^{b} (0.130)	0.643^{b} (0.050)	1.749^{b} (0.177)	1.719^{b} (0.069)
MW	0.543^{c} (0.046)	0.286^{c} (0.079)	0.829^{c} (0.112)	1.899^{b} (0.445)
CW	0.756^{d} (0.090)	0.370^{d} (0.027)	1.126^{d} (0.113)	2.045^{c} (0.107)

The mean and standard deviation values (in brackets) of three parallel groups (with 15 plants in each group) were calculated. The mean values followed by different superscripts within each column indicated that they were significantly different at a probability level of 0.05 according to ANOVA test.

Conclusion

The present greenhouse experiment demonstrates that the young plants of *Kandelia candel*, the most dominant mangrove species in Hong Kong and Southeast China, could tolerate the pollution impact from sewage effluents. Treatment with wastewater of various strengths did not reduce any plant growth, in terms of stem height, diameter, biomass and photosynthetic rates. Addition of wastewater with nutrient concentration similar to the natural municipal sewage stimulated plant growth. This could be attributed to the extra nutrients and trace elements available in NW. Although the *Kandelia* plants appeared to be tolerant to wastewater of high strength (even up to 10 times of the concentrations of the natural municipal sewage) and maintained their growth under severe water pollution situation, irrigation with concentrated wastewater (MW or CW)

Table 7 Physiological and biochemical properties of *K candel* plants after receiving wastewater of different strengths for one year

Treatment	Proline content ($\mu g\ g^{-1}$)	Catalase activity ($mg\ g^{-1}\ min^{-1}$)	Photosynthetic rate ($\mu mol\ CO_2\ m^{-2}\ s^{-1}$)
Control	144 65^a (12 65)	0 334^a (0 029)	3 458^a (0 209)
NW	134 58^a (11 84)	0 348^a (0 008)	4 724^b (0 876)
MW	149 81^a (20 82)	0 301^b (0 016)	3 591^a (0 157)
CW	154 71^a (21 22)	0 289^b (0 007)	4 384^b (0 782)

All data were expressed on fresh weight basis, the mean and standard deviation values (in brackets) of three parallel groups (with 15 plants in each group) were calculated The mean values followed by different superscripts within each column indicated that they were significantly different at a probability level of 0 05 according to ANOVA test

did create a slightly stressed environment The physiological and biochemical indicators such as content of chlorophyll *a* and *b* and catalase activity decreased significantly when the plants were exposed to MW and CW As the present study only lasted for one year, further investigation on the longterm effect of wastewater treatment on growth and physiological activities of mangrove plants should be carried out to further understand the resistance mechanisms of mangroves to high levels of nutrients and heavy metals

Acknowledgments

The authors would like to thank the technicians of Futian Nature Reserve, Shenzhen Special Economic Zone and Zhongshan University, Guangzhou, the PRC for their assistance in field sampling and laboratory analyses We would also like to express our gratitude to the Biotechnology Research Institute and the Hong Kong Research Grant Council for their financial support

References

Boto, K G, 1992 Nutrients and mangroves In Connel D W & D W Hawker (eds) Pollution in Tropical Aquatic Systems CRC Press, Inc Ann Arbor, London 129–145

Boto, K G & J T Wellington, 1983 Nitrogen and phosphorus nutritional status of a northern Australian mangrove forest Mar Ecol Press Ser 11 63–69

Boyt F L, S E Baylay & J Zoltek, 1977 Removal of nutrients from treated municipal wastewater by wetland vegetation J Wat Pollut Cont Fed 49 789–799

Chen, R H & P Lin, 1988 Influence of mercury and salinity on the growth of seedlings of three mangrove species Universitatis Amoiensis Acta Scientiarum Naturalium 27 110–115

Clough, B F & P M Attiwill, 1982 Primary productivity of mangroves In B F Clough (ed) Mangrove Ecosystems in Australia Miami, FL 213–222

Clough, B F, K G Boto & P M Attiwill, 1983 Mangrove and sewage a re-evaluation In H J Teas (ed), Biology and Ecology of Mangroves, Vol 8 Dr W Junk Publishers, Lancaster 151–161

Conley, L M, R I Dick & L W Lion, 1991 An assessment of the root zone method of wastewater treatment Res J Wat Pollut Cont Fed 63 239–247

Cooke, J G, A B Cooper & N M U Clunie, 1990 Changes in the water, soil, and vegetation of a wetland after receiving a sewage effluent New Zealand J Ecol 14 37–47

Culic, P, 1984 The effects of 2,4-D on the growth of *Rhizophora stylosa* Griff seedlings In H J Teas (ed), Physiology and Management of Mangroves, Vol 9 Dr W Junk Publishers, The Hague 57–63

De Jong, J, 1976 The purification of wastewater with the aid of rush or reed ponds In Jr Towbier & R W Pierson (eds), Biological Control of Water Pollution University of Pennsylvania Press, Philadelphia 133–139

Gambrell, R P, R A Khalid & W H Jr Patrick, 1987 Capacity of a swamp forest to assimilate the TOC loading from a sugar refinery wastewater stream J Wat Pollut Cont Fed 59 897–904

Gersberg, R M, B V Elkins, S R Lyon & C R Goldman, 1986 Role of aquatic plants in wastewater treatment by artificial wetlands Wat Res 20 363–368

Huang, X L & R Z Chen, 1990 An Experimental Handbook of Seed Physiology Agriculture Press, Beijing 122–124

Kadlec, J A, 1987 Nutrient dynamics in wetlands In Reddy K R & W H Smith (eds), Aquatic Plants for Water Treatment and Resource Recovery, Magnolia, Publishing Inc, Orlando, Florida 393–419

Lin, P & R Chen, 1990 Role of mangrove in mercury cycling and removal in the Jiulong estuary Acta Oceanologica Sinica 9 622–624

Lin, P, D H Chen & W J Li, 1984 A preliminary study of the interrelationship between the physiological characteristics of certain enzymes in the leaves of mangrove plants and soil salinity in tidal swamps Universitatis Amoiensis Acta Scientiarum 27 110–115

Lugo, A E , 1980 Mangrove ecosystems successıonal or steady state? Bıotropıca 12 65–72
Lugo, A E & S C Snedaker, 1974 The ecology of mangroves Annu Rev Ecol Syst 5 39–64
Lugo, A E , M Sell & S C Snedaker, 1976 Mangrove ecosystem analysıs In B C Patten (ed), Systems Analysıs and Sımulatıon ın Ecology Academıc Press, New York 113–145
Navalkar, B S , 1951 Successıon of the mangrove vegetatıon of Bombay and Salsette Islands J Bombay Nat Hıst Soc 50 157–161
Nedwell, D B , 1974 Sewage treatment and dıscharge ınto tropıcal coastal waters Ecology 5 187–190
Onuf, C P , J M Teal & I Valıela, 1977 Interactıons of nutrıents, plant growth and herbıvory ın a mangrove ecosystem Ecology 58 514–526
Peng, S Q , 1990 The Stress Physıologıcal Fundamentals of Plants Unıversıty of Northeast Forestry Press, Haerbın, 106 pp
Reed, S C , E J Mıddlebrooks & R W Crıtes, 1988 Natural Systems for Waste Management and Treatment McGraw-Hıll Book Co , New York, N Y
Rogers, K H , P F Breen & A J Chıck, 1991 Nıtrogen removal ın experımental wetland treatment systems evıdence for the role of aquatıc plants J Wat Pollut Cont Fed 63 934–941
Snedaker, S C & J G Snedaker, 1984 The Mangrove Ecosystem Research Methods Unıted Natıons Educatıonal, Scıentıfic and Cultural Organızatıon Publıshed, Rıchard Clay (The Chaucer Press) Ltd , U K
Tam, N F Y & Y S Wong, 1992 Nutrıent and heavy metal retentıon ın mangrove sedıment receıvıng wastewater Proc Int Specıalıst Conf on Wetland Systems ın Water Pollutıon Control 30 Nov – 3 Dec 1992, Sydney, Australıa 45 1–45 9
Tam, N F Y & Y S Wong, 1993 Retentıon of nutrıents and heavy metals ın mangrove sedıment receıvıng wastewater of dıfferent strengths Envır Technol 14 719–729
Tam, N F Y , S H Lı, C Y Lan, G Z Chen, M S Lı & Y S Wong, 1993 Nutrıents and heavy metal contamınatıon of plants and sedıments ın Futıan mangrove forest Presented at the Asıa-Pacıfic Symposıum ın Mangrove Ecology, January 1993, Hong Kong
Wang, H X , 1990 Fundamentals of Pollutıon Ecology Yunnan Unıversıty Press, Kunmıng 91–94
Wathugala, A G , T Suzukı & Y Kurıhara, 1987 Removal of nıtrogen, phosphorus and COD from wastewater usıng sand filtratıon system wıth *Phragmıtes australıs* Wat Res 21 1217–1224
Zhu, G L , 1983 Determınatıon of free prolıne ın plants Plant Physıol Commun 1 35–37
Zhu, G L , H W Zhong & A Q Zhang, 1990 Plant Physıologıcal Experıments Beıjıng Unıversıty Press, Beıjıng 122–124

Hydrobiologia **295**: 275–284, 1995.
Y. S. Wong & N. F. Y. Tam (eds), Asia-Pacific Symposium on Mangrove Ecosystems.

Spatial and temporal variations of mangrove fish assemblages in Martinique (French West Indies)

Max Louis, Claude Bouchon & Yolande Bouchon-Navaro
Université des Antilles et de la Guyane, Laboratoire de Biologie Animale, BP 592, 97159 Pointe-à-Pitre cedex (Guadaloupe FWI)

Key words: mangrove, Martinique FWI, Western Atlantic, fish community structure, abundance, spatial and temporal distributions

Abstract

A study of the mangrove fish fauna in a bay of Martinique Island (French West Indies) was carried out at different seasons during two consecutive years. Fishes were sampled with specific hoop-nets in the coastal areas at 8 stations.

A total of 87 species was collected in the bay. Most individuals were represented by small-size specimens and juveniles. The overall species richness varied according to the stations and the sampling periods. The biomass and number of individuals were variable according to the location but remained stable in time. A factor correspondence analysis and a hierarchical clustering with median links were used to follow the evolution of the stations in space and time. Two types of stations were differentiated: the stations characterized by the mangrove and those under the influence of seagrass beds. A seasonal cycle, opposing the dry periods to the others, was observed.

Thus, it seems that the use of the mangrove habitat by the fishes is optimized through a complete reorganization of communities in terms of species composition whereas the overall number and biomass remain stable. This model remains valid even for the most constraining biota of the mangrove ecosystem inhabited by a small number of well adapted species.

Introduction

The present work was part of a multi-disciplinary study on the mangroves of the bay of Fort-de-France, Martinique Island (French West Indies). The mangrove areas, bordering the eastern part of the bay, are presently subjected to human pressures as a result of industrial and urban development. The aims of the study were to describe the fish communities living in the mangrove areas from a spatial and temporal point of view. The results will contribute to a better understanding on the functioning of these communities.

Several authors have shown that mangroves constitute a zone for the development of juveniles (Austin, 1971; Lindall *et al.*, 1973; Phatia, 1976; Krishnamurthy & Prince Jeyaseelan, 1981; Robertson & Duke, 1987) and an important trophic zone for the adults of numerous species (Heald & Odum, 1970; Christensen, 1983; Louis & Guyard, 1982; Louis *et al.*, 1985). The fish fauna find among the mangrove roots optimum conditions for shelter andfood (Bell *et al.*, 1984), thanks to the high turbidity of waters and to the important development of sessile flora and fauna on the roots.

Materials and methods

The study was carried out with 7 successive field-trips (P1 to P7), from October 1989 to June 1991: P1 in October 1989 ; P2 in March 1990 ; P3 in June 1990, P4 in September 1990, P5 in December 1990, P6 in March 1991 and P7 in June 1991. The time interval between each field trip was 3 months, except for the first two field trips which were made at 6 months interval. The samplings were made during the dry season, the wet season as well as intermediate periods.

The samplings were made at 8 stations situated in the mangrove areas in the eastern part of the bay (Fig. 1). Most of the stations were located in zones where the

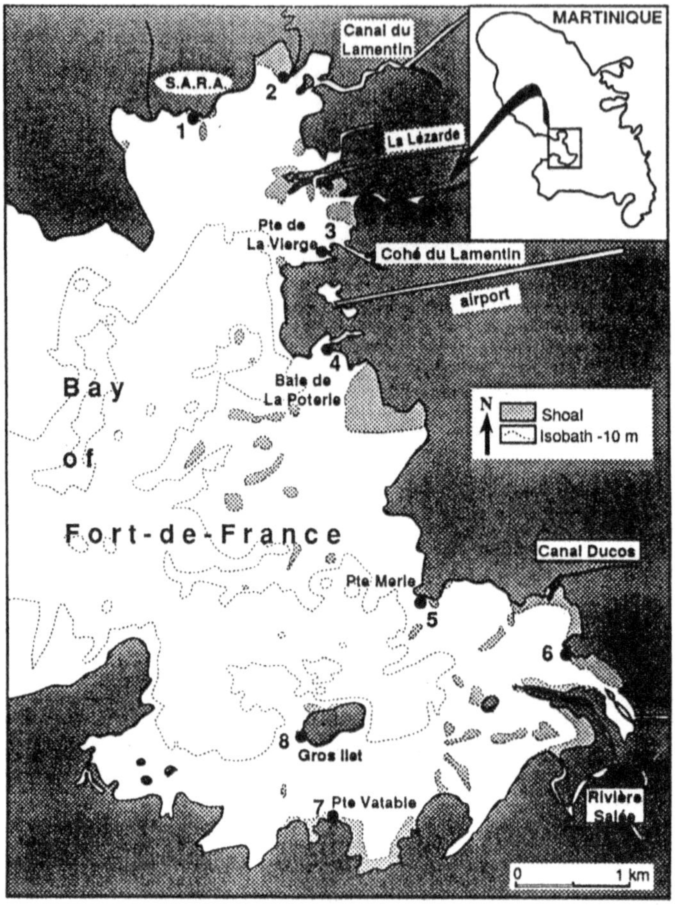

Fig 1 Location of the 8 studied stations in the Bay of Fort-de-France, Martinique

bottom was a thick layer of more or less fluid bare mud At 3 stations (5, 7 and 8) the bottom was colonized by seagrass Two stations (2 and 6) were situated near channel or river mouths

The fishes were collected with a specific fishing gear, called a "capéchade" (Fig 2) It is a stationary device designed to operate in a depth range of 0 5 to 2 m and is composed of a fence net (50 m long, 2 m high, mesh size 13 8 mm) and 3 hoop-nets From the mouth to the extremity of the hoop-nets, the mesh size decreased from 13 8 to 6 mm

This net has already been used in Guadeloupe and is well adapted to the mangrove environment (Lasserre & Toffart, 1977, Louis, 1983, Baelde, 1990, Bouchon *et al*, 1991) The nets were set early in the morning and lifted 24 hours later The fishes were then brought back to the laboratory where they were sorted, counted and weighed

During fish samplings, several parameters, such as temperature, salinity, pH and dissolved oxygen were measured At each station, the sedimentation rate was also estimated using sediment traps Water movement was also estimated using the erosion of an hemisphere of plaster set at 0 3 m above the bottom, following the technique of Doty (1971)

The qualitative results (presence/absence of the species) were analysed with the community index of Jaccard (CJ) (1900, 1908) to measure the similarity between two sampling periods or between two different stations Several ecological indices, including the

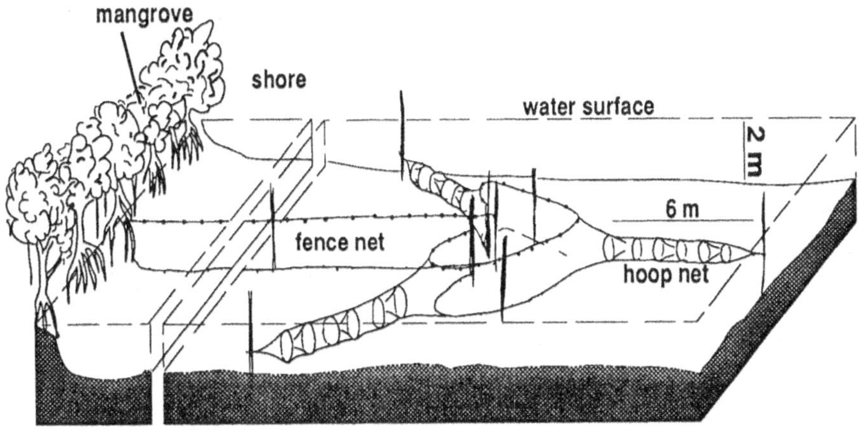

Fig. 2. Diagram of the fishing net ("capéchade") used for sampling in the mangrove.

species richness, the species diversity (H) of Shannon & Weaver (1948) and the evenness (E) of Pielou (1969), were calculated using the quantitative biomass data.

To consider separately the spatial and temporal component of the mangrove fish communities, the different samplings were pooled per station for a spatial study and per sampling period for a temporal study. Non parametric statistical tests were used to compare the samplings because of the absence of normality in the distribution of the data and the low number of series to analyse (8 stations and 7 periods). The Kruskal-Wallis one-way analysis of variance by ranks was used in a first step to compare several series of data. Whenever a difference was found, the test of Mann and Whitney was calculated to search the variables responsible for the observed heterogeneity. In the same way, the Friedman twoway analysis of variance by ranks was used to compare the distribution of biomass in the stations or during the different sampling periods.

Factor correspondence analyses were made, using the data on biomass, for the spatial and the temporal studies. The data concerning the environmental parameters measured were introduced as supplementary factors. A Spearman rank-order correlation coefficient were calculated between these parameters and the coordinates of the samplings on the factor axis. Finally, a hierarchical ascending classification was made to complete the factor analysis.

Results

Composition of the fish community

A total of 87 fish species belonging to 42 families were collected in the different stations (Table 1). The most important families were the Holocentridae, the Carangidae, the Pomadasyidae, the Scaridae (5 species), the Lutjanidae and the Gerreidae (4 species). The cumulative number of species for the catches at each station was highest (49 and 50 at station 8 and station 4) and was lowest (28 at station 6). In the other stations, the number varied between 33 and 44 with respectively 33 species at station 1, 36 species at stations 2 and 3, 42 in station 5 and 44 in station 7. However, the total species richness per station at a given period varied only between 12 and 29, with an average of about 19 species (Table 2). The highest values of the coefficient of variation of the means were recorded in stations 7 and 8 whereas the lowest values were found for stations 2, 3 and 6, showing a greater stability of the species richness in the last stations. Nearly one fourth of the number of species (21) only appeared once in the samplings and can be considered as "rare species". Moreover, no ubiquitous species (present in all the samplings and at any period) was found.

The Jaccard indices calculated between the different stations (all periods pooled) remained in general relatively high. They varied between 68.3% and 26.3%, with an average value of 46.9%. The highest similarities were found between stations 1 and 3 (68.3%) and between stations 2 and 6 (64.1%) where-

Table 1 Fish species sampled in different stations in the Bay of Fort-de-France

Family	Species	Stations								Family	Species	Stations							
		1	2	3	4	5	6	7	8			1	2	3	4	5	6	7	8
Elopidae	*Elops saurus*		+		+		+				*Lutjanus griseus*							+	
Megalopidae	*Megalops atlantica*				+						*Lutjanus synagris*				+				+
Muraenidae	*Gymnothorax moringua*							+			*Ocyurus chrysurus*					+		+	+
	Gymnothorax ocellatus				+		+	+		Gerreidae	*Diapterus rhombeus*	+	+	+	+	+	+	+	+
Congridae	*Conger triponceps*	+	+								*Eucinostomus argenteus*	+	+	+	+	+	+	+	+
	Echiophis cf punctatus		+			+			+		*Eucinostomus gula*	+	+	+	+	+	+	+	+
Clupeidae	*Harengula clupeola*	+	+	+	+	+	+	+	+		*Gerres cinereus*	+	+	+	+	+	+	+	+
	Harengula humeralis	+		+	+	+	+	+	+	Pomadasyidae	*Haemulon aurolineatum*		+		+	+	+		+
	Jenkinsia lamprotaenia					+					*Haemulon bonariense*	+	+	+	+	+		+	
Engraulidae	*Anchoa lyolepis*	+	+	+	+	+	+	+	+		*Haemulon chrysargyreum*			+	+	+		+	+
	Anchovia cf surinamensis	+	+	+	+	+	+	+	+		*Haemulon flavolineatum*			+					+
Synodontidae	*Synodus foetens*	+	+	+	+		+				*Haemulon plumieri*				+				
Antennariidae	*Phrynelox scaber*								+	Sparidae	*Archosargus rhomboidalis*	+		+	+	+		+	+
Hemiramphidae	*Hyporamphus unifasciatus*	+	+	+	+	+	+	+	+	Sciaenidae	*Bairdiella ronchus*	+	+	+	+	+	+	+	
Belonidae	*Strongylura notata*		+	+				+			*Odontascion dentex*			+					
	Tylosurus crocodilus		+	+		+	+	+	+	Mullidae	*Pseudupeneus maculaus*								+
Atherinidae	*Amerinomorus stipes*				+	+		+	+	Ephippidae	*Chaetodipterus faber*	+			+			+	
	Menidia beryllina		+							Chaetodontidae	*Chaetodon capistratus*	+			+				+
Holocentridae	*Adioryx coruscus*				+			+	+	Pomacentridae	*Stegastes diencaeus*								+
	Holocentrus ascencionis				+	+			+		*Stegastes leucostictus*								+
	Holocentrus rufus				+	+		+	+		*Stegastes variabilis*	+							
	Adioryx sp				+			+	+	Mugilidae	*Mugil curema*		+		+	+		+	
	Myripristis jacobus				+					Sphyraenidae	*Sphyraena barracuda*	+	+	+	+	+	+	+	+
Syngnathidae	*Syngnathus cf floridae*							+		Polynemidae	*Polydactylus virginicus*	+	+	+	+		+		
	Hippocampus reidi					+				Scaridae	*Scarus iserti*				+				+
Scorpaenidae	*Scorpaena grandicornis*							+	+		*Sparisoma chrysoptemm*							+	+
Triglidae	*Prionotus rubio*				+		+				*Sparisoma radians*					+		+	+
Dactylopteridae	*Dactylopterus volitans*								+		*Sparisoma rubripinne*				+				
Centropomidae	*Centropomus ensiferus*	+	+	+	+	+	+	+	+		*Sparisoma viride*					+			
	Centropomus pedimaculatus				+					Gobidae	*Gobionellus oceanicus*	+	+	+			+		
	Centropomus undecimalis							+		Acanthuridae	*Acanthurus bahianus*	+	+	+	+	+		+	+
Serranidae	*Hypoplectrus puella*								+		*Acanthurus chirurgus*				+	+			+
	Serranus Naviventus	+	+	+				+	+		*Acanthurus coeruleus*					+			
	Serranus trigrinus								+	Soleidae	*Achirus cf lineatus*	+	+	+	+		+		
Grammistidae	*Rypticus randalli*		+	+		+	+	+			*Citharichthys spilopterus*	+	+	+	+	+	+		
Apogonidae	*Phaeoptyx conklini*			+	+	+			+	Cynoglossidae	*Symphurus plagusia*		+						
	Phaeoptyx pigmentaria	+		+	+					Monacanthidae	*Monacanthus ciliatus*							+	+
	Astrapogon alutus					+			+		*Stephanolepis hispidus*							+	
Carangidae	*Caranx hippos*		+		+	+	+				*Stephanolepis setiler*					+		+	+
	Caranx latus	+	+	+	+	+	+	+	+	Tetraodontidae	*Sphaeroides greeleyi*	+		+	+	+		+	+
	Chloroscombrus chrysurus	+	+								*Sphaeroides spengleri*								+
	Oligoplites saurus	+	+	+	+	+	+	+			*Sphaeroides testudineus*	+	+	+	+	+	+	+	+
	Selene vomer	+	+	+	+		+	+		Diodontidae	*Diodon holacanthus*			+	+	+			+
Lutjanidae	*Lutjanus apodus*	+	+	+	+	+		+	+										

as the lowest values were recorded between stations 6 and 8 (26.3%) and between stations 2 and 8 (27.0%). Moreover, when the total number of species is cumulated for the different periods, the similarities tend to increase and varied between 69.1% (P2 and P7) and 53.9% (P4 and P7), the average value being 61.7%. Thus the species richness tend to become more homogeneous in time than in space.

The results concerning the 25 most abundant species in numbers (99.6% of total number of individuals) and in biomass (98.2% of total biomass) are presented in Table 3. The most abundant species, *Anchovia surinamensis*, which represents alone 48.5% of the numbers and 59.8% of the total biomass was not reported from Guadeloupe, another island of the French West Indies. *A. surinamensis* is a small pelagic

Table 2 Temporal variation of species richness in the studied stations CV = coefficient of variation

Stations	1	2	3	4	5	6	7	8	Mean	ρ	CV	Min	Max
Oct 89	21	17	16	17	20	16	17	23	18 3	2 6	14 2	16	23
Mar 90	17	17	16	21	14	15	13	18	16 4	2 5	15 3	13	21
Jun 90	14	20	18	22	24	18	23	29	21	4 6	21 7	14	29
Oct 90	25	23	20	25	22	15	25	20	21 9	3 5	15 9	15	25
Jan 91	21	21	17	20	24	16	14	19	19	3 2	16 9	14	24
Mar 91	18	18	14	18	15	14	15	13	15 6	2 1	13 2	13	18
Jun 91	19	18	17	20	19	12	19	23	18 4	3 1	16 9	12	23
Mean	19 3	19 1	16 9	20 4	19 7	15 1	18 0	20 7	18 7				
ρ	3 5	2 3	1 9	2 6	4 0	1 9	4 6	5 0		1 9			
CV	18 1	11 8	11 1	12 9	20 4	12 3	25 5	24 1			10 2		
Min value	14	17	14	17	14	12	13	13				12	
Max value 25	23	20	25	24	18	25	29					29	

Table 3 Number (N) and biomass of the main 25 fish species collected

Species	N	%	Species	Weight (kg)	%
Anchovia cf surinamensis	48615	48 5	*Anchovia cf surinamensis*	542 80	59 8
Diapterus rhombeus	19516	19 5	*Bairdiella ronchus*	69 65	7 7
Anchoa lyolepis	9124	9 1	*Diapterus rhombeus*	63 11	7 0
Bairdiella ronchus	5632	5 6	*Sphaeroides testudineus*	50 03	5 5
Harengula clupeola	4533	4 5	*Eucinostomus gula*	26 68	2 9
Eucinostomus gula	2620	2 6	*Gerres cinereus*	21 10	2 3
Gerres cinereus	2037	2 0	*Harengula clupeola*	16 10	1 8
Caranx latus	1903	1 9	*Caranx latus*	16 02	1 8
Sphaeroides testudineus	1860	1 9	*Anchoa lyolepis*	15 06	1 7
Atherinomorus stipes	697	0 7	*Centropomus ensiferus*	14 48	1 6
Centropomus ensiferus	541	0 5	*Archosargus rhomboidalis*	12 46	1 4
Eucinostomus argenteus	435	0 4	*Sphyraena barracuda*	8 00	0 9
Hyporamphus unifasciatus	432	0 4	*Sphaeroides greeleyi*	5 05	0 6
Sphaeroides greeleyi	404	0 4	*Hyporamphus unifasciatus*	3 52	0 4
Oligoplites saurus	309	0 3	*Gobionellus oceanicus*	3 07	0 3
Harengula humeralis	268	0 3	*Echiophis cf punctatus*	3 01	0 3
Haemulon chrysargyreum	203	0 2	*Mugil curema*	2 94	0 3
Gobionellus oceanicus	142	0 1	*Holocentrus ascencionis*	2 90	0 3
Archosargus rhomboidalis	124	0 1	*Eucinostomus argenteus*	2 75	0 3
Holocentrus ascencionis	107	0 1	*Haemulon chrysargyreum*	2 68	0 3
Sphyraena barracuda	90	0 1	*Caranx hippos*	2 23	0 2
Phaeoptyx conklini	76	0 1	*Conger triporiceps*	2 09	0 2
Haemulon bonariense	63	0 1	*Atherinomorus stipes*	1 93	0 2
Haemulon aurolineatum	55	0 1	*Synodus foetens*	1 64	0 2
Citharichthys spilopterus	53	0 1	*Centropomus undecimalis*	1 63	0 2

Table 4 Pielou index calculated from the biomass of collected fishes in the mangrove stations CV = coefficient of variation

Stations	1	2	3	4	5	6	7	8	Mean	ρ	CV	Min	Max
Oct 89	0 56	0 47	0 58	0 65	0 69	0 35	0 71	0 65	0 58	0 1	21 0	0 35	0 71
Mar 90	0 65	0 21	0 58	0 61	0 49	0 24	0 48	0 41	0 46	0 2	35 8	0 21	0 65
Jun 90	0 48	0 44	0 82	0 66	0 68	0 15	0 57	0 64	0 56	0 2	36 2	0 15	0 82
Oct 90	0 38	0 66	0.39	0 69	0 70	0 39	0 75	0 65	0 58	0 2	27 9	0 38	0 75
Jan 91	0 56	0 74	0 68	0 51	0 70	0 12	0 75	0 65	0 59	0 2	35 4	0 12	0 75
Mar 91	0 50	0 77	0 82	0 55	0 59	0 07	0 85	0 74	0 61	0 3	41 9	0 07	0 85
Jun 91	0 16	0 52	0 69	0 74	0 62	0 56	0 75	0.74	0 60	0 2	33 2	0 16	0 75
Mean	0 47	0 54	0 65	0 63	0 64	0 27	0 69	0 64	0 57				
ρ	0 2	0 2	0 2	0 1	0 1	0 2	0 1	0 1		0 1			
CV	34 3	36 0	23 2	12 6	12 3	65 3	18 4	17 5			24 7		
Min value	0 16	0 21	0 39	0 51	0 49	0 07	0 48	0 41				0 07	
Max value	0 65	0 77	0 82	0 74	0 70	0 56	0 85	0 74					0 85

fish of the Engraulididae family, particularly abundant in the muddy stations (1, 2 and 6). The other common species, also abundant in other neighbouring islands, were: *Diapterus rhombeus*, *Bairdiella ronchus*, *Harengula clupeola*, *Eucinostomus gula*, *Sphaeroides testudineus*, *Gerres cinereus*, *Caranx latus* et *Archosargus rhomboidalis*. Thus, the fish community is dominated by a small number of species whose individuals are very abundant. In general, most of the collected fishes were small-size juveniles. In the present study, the average weight of each fish was about 9 g.

STATION	S1	S2	S3	S4	S5	S6	S7	S8
S1		+	+	+	+	–	+	+
S2	-0.174		+	+	+	–	+	+
S3	-1.611	-1.951		+	+	+	+	+
S4	-0.645	-0.907	-2.455		+	–	+	+
S5	-0.321	-0.513	-1.349	-0.257		+	+	+
S6	-2.248	-2.769	-1.622	-2.952	-1.932		+	–
S7	-0.707	-0.772	-0.129	-1.218	-0.770	-1.029		+
S8	-0.578	-0.904	-1.863	-0.129	-0.064	-2.310	-0.965	

Fig 3 Results of the Mann and Whitney tests calculated on the species richness at each station, all periods pooled + no significant difference at 95% level (Ho hypothesis)

Spatial distribution of fishes

The distribution of species varied according to the stations. In stations 1, 2 and 6, the most abundant species belonged to the Engraulidae (*Anchovia surinamensis*), the Sciaenidae (*Bairdiella ronchus*), the Tetraodontidae (*Sphaeroides testudineus*) and the Gerreidae (*Diapterus rhombeus*). Among the other dominant fishes of the community we can note Eucinostomus gula (*Gerreidae*) and *Caranx latus* (Carangidae) in station 3, *Centropomus ensiferus* (Centropomidae) and *Eucinostomus gula* (Gerreidae) in station 4, *Harengula clupeola* (Clupeidae) and *Archosargus rhomboidalis* (Sparidae) in station 5, *Anchoalyolepis* (Engraulidae) and *Eucinostomus gula* (Gerreidae) in station 7, *Sphaeroides greeleyi* (Tetraodontidae) and *Haemulon chrysargyreum* (Pomadasyidae) in station 8. *Bairdiella ronchus* is present in all the stations whereas *Sphaeroides testudineus*, well represented in the first seven stations is totally absent from station 8 where it is replaced by *Sphaeroides greeleyi*.

A statistical significant difference was found between the values of species richness for the 8 stations (test of kruskal-Wallis: H= 14.598; *df*= 7). A comparison of pair values by the Mann and Whitney test showed that station 6, where the lowest values were observed, was responsible for that heterogeneity (Fig. 3).

Concerning biomass, the average catch in a sampling was 16.2 kg (about 1900 fishes), which represented an important abundance. Nevertheless, these values fluctuated according to the stations. The maximum average catch was 48.9 kg (4903 fishes) in station 6 and the minimum was 4.7 kg (740 individuals) in station 7. The highest values recorded at stations 2 and 6 were essentially due to the catch of the pelagic species *Anchovia surinamensis*. A statistical significant difference was found between the 8 stations (test of kruskal-Wallis H= 31.86; *df*= 7).

The average values of the Shannon diversity index varied between 1 and 2.9 (with a maximum per catch of 3.5 at station 7 and a minimun of 0.25 at station 6) The evenness of Pielou presented similar variations (Table 4). The average values fluctuated between 0.27 to 0.69, with a maximum at station 7 (E = 0.85) and a minimun at station 6 (E = 0.07). A significant difference was found between the 8 stations for the different evenness values, using the kruskal-Wallis test (H=23.189; *df*= 7). Station 7, with the highest evenness (0.69) and a coefficient of variation of the means of 18.4, appears as the station having the most stable community, followed by station 3. On the contrary, station 6 with an average evenness of 0.27 and a coefficient of variation of 65.3 possessed the most unstable fish community.

The result of the factor correspondence analysis made with the data on biomass is presented in Fig. 4. The first axis, which gathered 48.3% of the information, opposed stations 5, 7 and 8 to stations 1, 2 and 6. A significant correlation was found between the first axis and dissolved oxygen (Rs = 0.683 ; *df*= 8) and between the second axis and salinity (*Rs* = 0.743; *df*= 8). The species characterizing the first axis were on one side: *Atherinomorus stipes*, *Holocentrus rufus*, *H. ascencionis*, *Adioryx* sp., *Astrapogon alutus*, *Haemulon aurolineatum*, *H. chrysargyreum* and *Sparisoma radians*. These species are generally frequent in the seagrass beds. On the other side of the axis, the contributing species were: *Anchovia surinamensis*, *Gobionellus oceanicus*, *Chloroscomibrus chrysurus*, which are more characteristic of mangrove areas. The first axis then separated the stations with seagrass beds, well oxygenated, to the muddy stations more characteristic of mangrove habitat. The second axis opposed the stations situated near the mouth of drainage channels (3, 4) to that located in the more open sea (8). Examination of the cluster analysis shows that three groups of stations are separated: stations 3 and 4; stations 1, 2 and 6 and stations 5 and 7. Station 8 remains isolated (Fig. 4).

Temporal distribution of fishes

The variations of species richness in time are given in Table 2. A significant difference was found for the values of species richness collected at different periods (Kruskal-Wallis test; H = 18.025 ; df=6). Results of the comparisons between pair values, using the Mann and Whitney test are shown in Fig. 5. In general, a

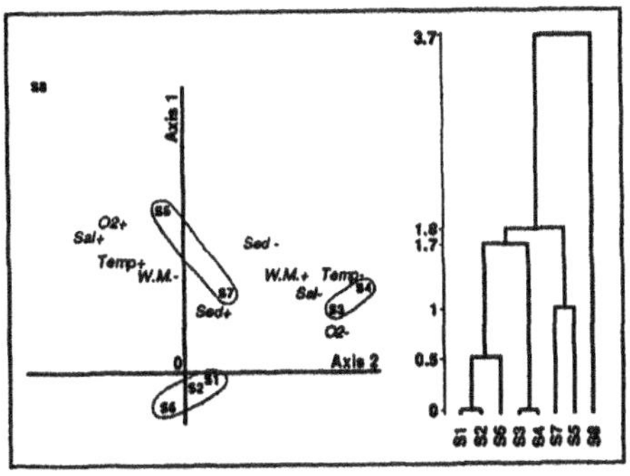

Fig. 4. Results of the factor correspondence analysis and hierarchical classification of the stations calculated on the cumulated biomass, all periods pooled.

significant difference was found between March (*i.e.* during the dry season) and the other periods. During this period, species richness tended to decrease significantly, with values of 16.4 in March 1990 and 15.6 in March 1991, whereas the overall average value was 18.7.

A comparison of the number of fishes collected at different periods by the Kruskal-Wallis test revealed that there was no significant difference between the 7 periods (H= 4.414; *df*= 6). The same result was found concerning biomass(H= 4.414; df=6). Thus, the average overall biomass remained stable in time (Fig. 6), with an average of 16 kg per period, although species richness varied in time. Thus, the variations of overall biomass are less important in time than in space. However, considering the data species by species, an analysis of variance of ranks calculated on the species abundances, with a test of Friedman (*Xr2* = 16.011, *N*=84, *df*=6) showedsignificant differences in their distribution through time.

The variations in time of diversity and evenness indices presented similar characteristics with those observed in space. The comparison of the indices with the Kruskal-Wallis test showed that there was no significant difference (*H*=5.829, *df*=6). However, the great uniformity of coefficients of variation of the mean was found, revealing a greater stability of the distributions of the fish community in time than in space.

A factor correspondance analysis was made with the data on biomass pooled per period (Fig. 7). No statistical correlations were found between the coordinates of the sampling periods on the factorial axes and the environmental parameters. However, the first axis

DATE	Oct 89	Mar 90	Jun 90	Oct 90	Jan 90	Mar 91	Jun 91
Oct. 89		+	+	+	+	+	+
Mar. 90	-1.338		-	-	+	+	+
Jun. 90	-1.479	-2.268		+	+	-	+
Oct. 90	-1.857	-2.586	-0.741		+	-	-
Jan. 91	-0.126	-1.587	-0.896	-1.692		-	+
Mar. 91	-1.800	-0.426	-2.461	-2.971	-2.116		-
Jun. 91	-0.479	-1.639	-1.164	-2.174	-0.477	-2.072	

Fig. 5. Results of the Mann and Whitney test calculated on the species richness at each period, all stations pooled. +: no significant difference at 95% level (Ho hypothesis).

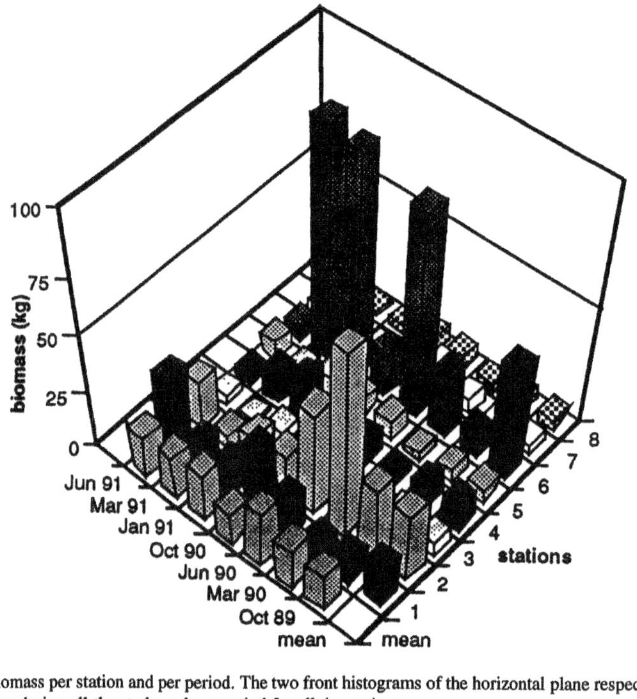

Fig. 6. Variations of the biomass per station and per period. The two front histograms of the horizontal plane respectively represent the average values of biomass per station during all the study and per period for all the stations.

and the water temperature were correlated at the level of 90% (*Rs* = 0.746, *df*= 7).

The first 3 axes explain 80% of the total variability of the data. A Guttman effect was found on the first two axes and authorized only the examination of the factorial plan formed by axis 1 and 3. On the first axis, 2 sampling periods (P1 and P4) were opposed to the other periods. The species characterizing the first axis are, on one side: *Adioryx coruscus*, *Adioryx* sp., *Stegastes variabilis*, *Mugil curema*, and on the opposite side: *Synodus foetens*, *Rypticus randalli*, *Gerres cinereus* and *Chaetodipterus faber*. The first group of species is either present or more abundant during the periods P1 and P4, that is to say during the rainy season, whereas the second group is essentially found during the dry season (P2, P3 and P6). The hierarchical classification confirm these results. The periods P2, P3, P7 and P6 were clustered and the other periods remained isolated. Furthermore, the different points of the factor analysis linked in a chronological order show a clockwise pattern (Fig. 7). In fact, this pattern emphasizes a cyclic phenomenon of the structure of the community in time, which corresponds to the recruitment of fish species in the mangroves.

Discussion and conclusions

From a qualitative point of view, high similarity coefficients were seen between the stations. So, for certain stations considered by pairs (stations 1, 2, 3 and 6), the values of the similarity coefficient was always greater than 50%. These are the stations where the substrate is a rather fluid mud. Stations 5, 7 and 8, colonized by seagrasses also had high similarities. The lowest similarities were found between station 8 and stations 2 and 6. Therefore, the presence or the absence of seagrass beds in the habitat is the most important factor structuring the fish community.

A statistically significant difference in the number of species was noted between the stations. The number of species was lower in the stations devoid of segrasses. In these stations, a small number of species showed a great dominance in abundance and in biomass. Also, their fluctuations through time was of little importance. They are species well adapted to harsh ecological conditions (high sedimentation rate and dissolved oxygen below the average values). This fact has already been reported by Louis *et al.* (1985) for the mangrove lagoon of Guadeloupe island. In the stations situated near seagrass beds, species richness is higher and the renewal of species is more important. Many juveniles are present there, taking advantage of the shelter and the food provided by the seagrasses. They are quantitatively of little importance but contribute to increase the species richness of the community.

The temporal study revealed a tendency to a decrease of species richness during the dry season.The month of March corresponds (in the West Indies) to the beginning of the reproductive period for numerous species (Baelde, 1990) and so the recruitment in the mangroves has not yet occurred. However, although species richness is low during that period, the overall numbers and biomass of fishes remain stable, due to the modifications of the species abundance. The level of dissolved oxygen was an important parameter for the distribution of fishes in space whereas temperature is more important for the temporal variations.

In the Caribbean zone, the mangrove is known to be a trophic and a nursery zone for numerous animals and more particularly for fishes (Louis & Guyard, 1982). According to Collignon (1991), this very productive ecosystem represents an important food resource for

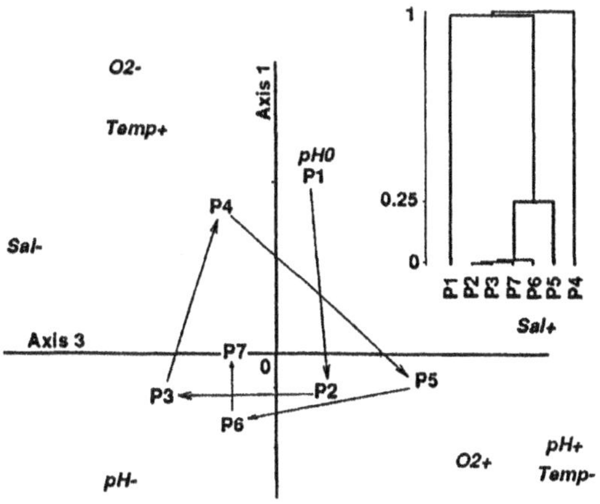

Fig. 7. Factor correspondence analysis and hierarchical classification of the sampling periods. The biomass in the different stations are cumulated for each sampling period.

the mobile carnivorous fauna. Therefore, there are permanently migratory movements of fishes between the mangrove and the neighbouring environments (coral reefs, seagrass beds). These migrations induce permanent variations in the species composition and abundances of the mangrove fish community.

The ichtyofauna of the mangrove in Martinique is characterized by a great number of individuals and a high biomass. However, contrary to the observations of Morton (1990) in Australia and Thollot (1992) in New Caledonia, these mangroves contain but a few species of commercial value. As observed in Guadeloupe (Louis, 1983), most of the fishes were either juveniles, either small size species with a rapid turnover, *i.e.* with "*r*" strategies. In the mangroves of the bay of Fort-de-France, the fishing activity is not much developed and mainly concern species visiting this habitat (small pelagic fishes, used as bait). However, due to its high production, the mangrove constitutes an extremely important network from a trophic point of view. It is a zone of shelter and food for numerous fish species which utilize this habitat during their juvenile stages. The utilisation of the resources of the mangrove seems to be optimized by the fishes, which by adjustments of the number of species and individuals, can maintain a stable biomass all year round. The sedentary species, "*r* strategists" and able of rapid fluctuations, occupy the vacant space let by the juveniles of migrant species which succeed in time.

Acknowledgements

The present study was financed by the Regional Council of Martinique. It was placed under the auspices of U N E.P, from UNESCO. Thanks are expressed to the laboratories IFREMER and ORSTOM (Pôle Océanologique Caraïbe) for their logistic support during the program

References

Austin, H M, 1971 A survey of the ichthyofauna of the mangroves of western Puerto Rico during December, 1967 - August, 1968 Carib J Sci 11 27–39

Baelde, P, 1990 Differences in the structures of fish assemblages in *Thalassia testudinum* beds in Guadeloupe, French West Indies, and their ecological significance Mar Biol 105 163–173

Bell, J D, D A Pollard, J J Burchmore, B C Pease & M J Middleton, 1984 Structure of a fish community in a temperate tidal mangrove creek in Botany Bay, New South Wabs Aust J mar Freshwat Res 35 33–46

Bouchon, C, Y Bouchon-Navaro, D Imbert & M Louis, 1991 Effets de l'ouragan Hugo sur les communautés côtières de Guadeloupe (Antilles françaises) Ann Inst océanogr 67(1) 5–33

Christensen B, 1983 Mangroves What are they worth ? Unasylva 35 2–15

Collignon J, 1991 Ecologie et biologie marine Introduction a l'halieutique Masson, Paris, 312 pp

Doty, M S, 1971 Measurement of water movement in reference to benthic algae growth Botanica marina 14 32–35

Heald E & W E Odum, 1970 The contribution of mangrove swamps to Florida fisheries Proc 22nd Gulf Carib Fish Inst 130–135

Jaccard P, 1900 Contribution au problème de l'immigration post-glacière de la flore alpine Bull Soc vaudoise Sci nat 36 87–130

Jaccard P, 1908 Nouvelles recherches sur la distribution florale Bull Soc vaudoise Sci nat 44 223–270

Krishnamurthy K & M J Prince Jeyaseelan, 1981 The early life history of fishes from Pichavaram mangrove ecosystem of India Rapp P-V Réun Cons int Explor Mer 178 416–423

Lasserre, G & J-L Toffart, 1977 Echantillonnage et structure des populations ichtyologiques des mangroves de Guadeloupe en septembre 1975 Cybium 3 115–127

Lindall W N, P J Reay, W A Fable & L A Collins, 1973 A survey fishes and commercial invertebrates of the shore and estuarine zone between Cape Romano and Cape Sable, Florida Nat mar Fish Serv, US Dep Comm PB-235 215 NTIS, 62 pp

Louis M, 1983 Biologie, écologie et dynamique des populations de poissons dans les mangroves de Guadeloupe (Antilles françaises) Thèse de Doctorat d'état, USTL, Montpellier (France), 275 pp + 29 annexes

Louis M & A Guyard, 1982 Contribution à l'étude des peuplements ichtyologiques dans les mangroves de Guadeloupe (Antilles françaises) Bull Ecol 13(1) 9–21

Louis M, T Lam Hoai & G Lasserre, 1985 Résultats préliminaires sur le recrutement en poissons dans deux lagunes des mangroves de Guadebupe Belle-Plaine et Manche-à-Eau Rev Hydrobiol trop 18 (4) 249–265

Morton R M, 1990 Community structure, density and standing crop of fishes in a subtropical australian mangrove area Mar Biol 105 385–394

Phatia V, 1976 An ecological survey of the seabass fry *Lates calcarifer* (Bloch) in the mangrove areas along the Indian Oceancoasts of Thailand Proc 1 st Thai Nat Sem Mangr Ecol, Phuket 626–644

Pielou E G, 1969 An introduction to mathematical ecology Ed Wiley, Interscience, New-York, 286 pp

Robertson A L & N C Duke, 1987 Mangroves as nursery sites comparisons of the abundance and species of fish and crustaceans in mangroves and other nearshore habitats in tropical Australia Mar Biol 96 193–205

Shannon C E & W Weaver, 1948 The mathematical theory of communication Urbana Univ Press, Illinois, 117 pp

Thollot P, 1992 - Les poissons de mangrove du lagon sud-ouest de Nouvelle-Calédonie Ecologie des peuplements Relations avec les communautés ichtyologiques côtières Thèse Doctorat, Univ Aix-Marseille II, 406 pp

Hydrobiologia **295**: 285–294, 1995.
Y.S. Wong & N.F.Y. Tam (eds), Asia-Pacific Symposium on Mangrove Ecosystems.

An integrated comparative approach to mangrove vegetation mapping using advanced remote sensing and GIS technologies: preliminary results

Josef Aschbacher, Rey Ofren, Jean Pierre Delsol, Tri Binarko Suselo, Suvit Vibulsresth, Thongchai Charrupat
Asian Institute of Technology Bangkok, Thailand; National Research Council of Thailand Bangkok, Thailand; Royal Forestry Department Bangkok, Thailand

Key words: mangroves, remote sensing, GIS

Abstract

This study aims to develop an integrated methodology applying Remote Sensing and Geographic Information Systems (GIS) techniques for the assessment of the ecological status of mangrove forests. The study area is located at Phangnga Bay, Thailand. Various commonly available remote sensing data are evaluated for mangrove vegetation mapping. The satellite sensors used are covering the visible and infra-red (VIR) spectrum up to the microwave region of the electromagnetic spectrum. This study provides recommendations regarding the selection of a single sensor approach or sensor combination that fulfills a minimum requirement for practical mangrove mapping and inventory purposes (e.g. mangrove and non-mangrove areas, varying stocking density, dominant species composition and impact of human activities). Both their technical capabilities and their potentials are presented in correlation with the existing ground conditions.

Introduction

The foremost objective of this study is to develop a digital image analysis methodology to map mangrove forests with Remote Sensing and Geographic Information Systems (GIS) technology. Specifically, the study aims to explore the potentials of space-borne remote sensing radar data in combination with visible and infra-red images for the mangrove ecosystem.

This study focuses on mangrove forests that occur on tidal flats along the sea coast and sometimes extend along brackish water streams. Mangrove forests are a vital resource that serve the inhabitants of coastal areas. They serve as a source of wood products for house construction, firewood, tannin and other domestic products such as medicines. There is also strong correlation between the presence of mangrove ecosystems and the abundance of fish and marine life in coastal areas.

Mangrove forest management is commonly characterized by rapid changes, i.e. conversion into other land uses such as shrimp farming. The availability of information is therefore crucial for the development of policies and their implementation programs. The need for intensive monitoring and appropriate management of this resource is pressing. Thus, remote sensing allows quantitative and qualitative assessments of ground conditions over large and inaccessible areas. Another powerful tool for collecting, storing, retrieving, transforming and displaying spatial data is Geographic Information Systems (GIS).

Results that can be generated from this study could provide an additional opportunity for a better understanding of mangrove ecosystems geared towards its sustainable resource use. On the technical side, this could serve as a guideline in choosing the appropriate remote sensing technology to address its real time problems and generate rational decisions for natural resource management.

Study area

The study area is located in the Phangnga Bay, Province of Phangnga, Southern Thailand, 913 km south of Bangkok. It harbors a large, still intact mangrove forest in the country. The tidal range at the bay is considerably large and very gentle slopes or flat landforms are found. However, the bay has numerous scattered towering limestone islets, with cliffs that rise straight up from the water.

Two distinct management regimes can be found in the study area. On the western side is a declared Ao Phangnga National Park, where resource utilization is prohibited. On the eastern side, the area is utilized as fishing ground, aquaculture area and concessional area for charcoal and pole production. Tourism is an important socio-economic component for the surrounding communities.

Data basis

This study employs a wide variety of satellite data which are used to evaluate their performance for mangrove vegetation mapping. The investigations will determine whether a single sensor approach is sufficient for practical purposes and which sensor or sensor combination fulfills the minimum requirement for a specific type of monitoring. Topographic and soil maps as a form of ancillary data were collected and digitized and include elevation, river and road network. Table 1 shows the available satellite data for the current study.

Table 1. Satellite data used in the current study.

Satellite data	Date of acquisition	Wavelength (micrometer)	Ground Resolution (meters)
SPOT XS	22 March 1992	0.51–0.89	20
LANDSAT TM	20 April 1993	0.45–12.5	30
MOS-1 MESSR	22 February 1989	0.5–1.1	50
JERS-1 SAR	08 September 1992	L-band (23.5 cm)	18.5
ERS-1 SAR	16 July 1992	C-band (5.7 cm)	25
ERS-1 SAR	20 August 1992	C-band (5.7 cm)	25
ERS-1 SAR	29 October 1992	C-band (5.7 cm)	25

Characteristics of optical and radar remote sensing

An increasing amount of valuable environmental and resource information is being acquired by sensors that operate in various portions of the electromagnetic spectrum (EMS). Various spaceborne satellites scan the Earth on a regular basis at various portions of the spectrum. Among them are satellites operating at the visible and infra-red region (0.4–3 micrometer) which rely on reflected sunlight as the source of radiation. The received signal is sensitive to the photosynthetic activity of vegetation, soil moisture, water turbidity, etc. The satellite data used are Landsat Thematic Mapper (TM), SPOT (Systeme Probatoire d'Observation dela Terre) Multi-spectral sensor (XS) and Marine Observing Satellite Multispectral Electronic Self-scanning Radiometer (MOS-MESSR). In terms of spatial ground resolution, these are 30 m, 20 m, and 50 m, respectively.

Radar (radio detection and ranging) remote sensing on the other hand, uses the microwave region of the EMS. The microwave region extends from 0.3 GHz to 300 GHz, or in terms of wavelengths, from 1 mm to 1 m. Its backscattered radar signal is sensitive to the dielectric and geometric properties of the observed object. Here, dielectric property mainly corresponds to the amount of water, such as the soil volume or plant canopy, and geometric property relates to the size, shape, orientation or volume of the backscattering medium.

Specifically, the ERS-1 (First European Remote Sensing Satellite) sensor uses advanced microwave or radar techniques, which enable global measurements independent of cloud and sunlight conditions. Thus, its images can be acquired under any weather condition and at any time of day. The pixel size is 12.5 m, the frequency is 5.3 GHz (C-band). The JERS-1 (Japanese Earth Resources Satellite) Synthetic Aperture Radar (SAR) operates following the same principle as ERS-1 but using a frequency of about 1.275 GHz (i.e. L-band) allowing a deeper penetration into the medium. The information thus results not only from the top but also from below the canopy of trees (i.e., forest floor).

Methodology

The first step in the analysis of satellite data was geo-referencing in order to conform with a common map projection at 20 m resampling size. In the case of radar data, speckles (noise) were first eliminated using the 'Modified Refined Gamma Map Filter' developed at the Asian Institute of Technology (AIT). The algorithm detects groups of homogenous pixels and directional patterns while retaining single pixels that represent a distinctive feature on the ground. Rectification was also applied for the purpose of overlaying these images on top of each other.

A detailed knowledge of the field conditions and ancillary data in the form of maps such as elevation and soil type was needed to carry out a digital classification of satellite data. A supervised classification scheme was applied to extract different mangrove land cover categories while applying a mask to exclude the water areas.

Multi-temporal space-borne radar data were subsequently analyzed. This approach exemplified one important characteristic of radar satellite data in monitoring the land cover dynamics in any season or time of the year.

Taking advantage of the strengths of both sensor types, visible/infra-red and radar remote sensing sensors, a second series of classifications was performed using an overlay of ERS-1 and SPOT XS data for the purpose of assessing the heterogeneity of mangrove vegetation.

Related spatial data such as elevation were used for the refinement of the digital image classification results through GIS analysis, while taking into consideration other spatial data (e.g., soil, tidal inundation) that affect the ecosystem. Field work was done for field familiarization, data collection, ground truthing and verification of results. A detailed flow diagram is shown in Diagram 1.

Results and discussion

SPOT XS digital image analysis

Based on the visual analysis of SPOT (Systeme Probatoire d'Observation dela Terre) imagery, mangrove and non- mangrove areas can easily be discriminated (see Figure 1). A detailed digital analysis using three channels (Band 1 – Visible Green, Band 2 – Visible Red and Band 3 – Near Infra-red) made it possible to discriminate mangrove stands of different stocking densities. Digital classification made it possible to distinguish a pure and mature *Rhizophora* stand that comprises 7.11% of the whole study area and 3.40% of the area covered by Nipa palm. Its changing infra-red reflection reveals a varying heterogenous species composition that ranges from a Medium Density Mixed Mangrove (21.07%), Low Density Mixed Mangrove (10.25%) to a Very Low Density Mixed Mangrove type (2.61%). Inland land-cover (non-mangrove) has been generally categorized as rubber plantation, dryland forest, mixed inland vegetation and bare lands (Figure 2).

The following is a brief description of the major mangrove categories separable in the SPOT classified image:

a. **Homogenous Dense Rhizophora** – characterized by dense, mature stand of *Rhizophora* of at least 4 m in height;
b. **Nipa Palm** – groups of nipa palms of uniform height;
c. **Medium Density Mixed Mangrove** – a heterogenous stand of mature *Avicennia-Sonneratia* species in association with medium height

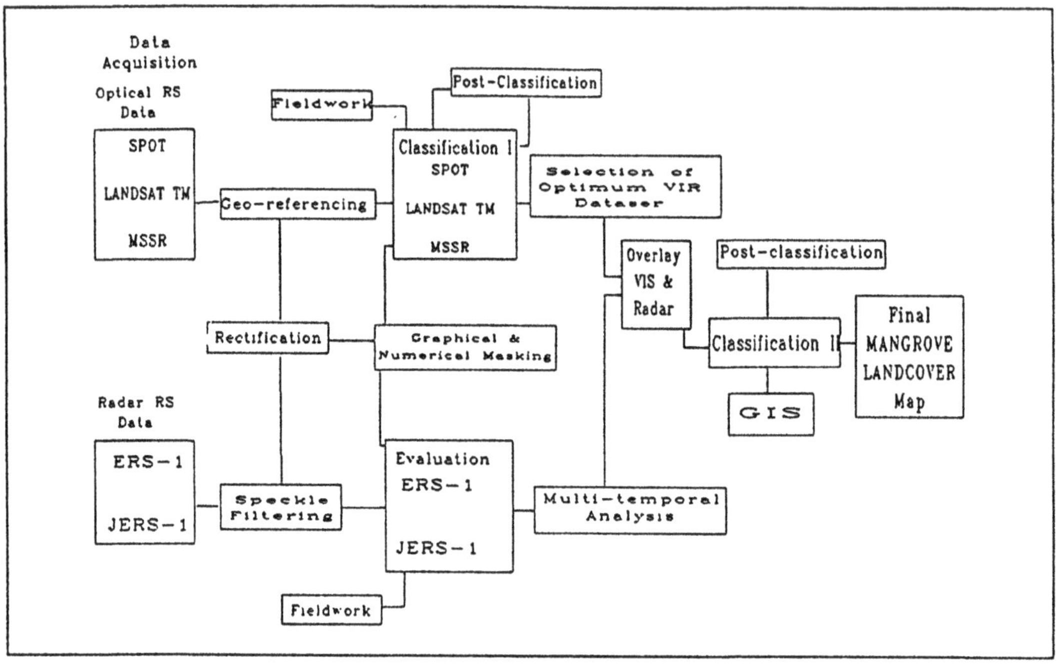

Diagram 1. Methodological framework of the study.

Rhizophora (3–4 m) with 50–58% tree crown coverage;

d. **Low Density Mixed Mangrove** – composed of low *Rhizophora* (1–3 m) with a medium to low density distribution of *Avicennia* and other inland species; nipa is also abundant; crown coverage is 30–40%;
e. **Very low Density Mixed Mangrove** – sporadic distribution of mature *Avicennia* and inland species (e.g. *Heretiera*, *Xylocarpus*, *Instia*) left during the cutting for charcoal where the understory is composed of regeneration of young *Rhizophora* (< 1 m) with shrubs and creepers; crown coverage is less than 30%.

In the process of digital analysis, misclassification was found for *Rhizophora sp.* represented pixels that were located in the mountainous region and represented as natural forest and vice versa. Heterogeneous mangroves were confused with mixed inland vegetation as well as Rubber Plantation, while some Nipa were incorrectly classified as young rubber plantation and other Mixed Inland vegetation like oil and coconut palms. Table 2 shows the irregularities for each class which actually belong to another class.

Due to differences in tree species morphology, even similar stocking density exhibited different responses to the satellite. In the case of dense *Avicennia-Sonneratia* association, they did not register as strongly as the dense *Rhizophora* formation. This phenomenon is attributed to a spreading canopy containing small, whitish and thinner obovate to lanceolate shaped leaves of *Avicennia* compared to the conical crown of dense *Rhizophora* with broader and thicker ovate-shaped leaves. Usually, such *Avicennia-Sonneratia* associations form a belt of only 10 to 50 m width situated near the coastline and sometimes their spectral value is mixed with that of water. This type of mangrove was recognized as a pioneer species that occupied new soil in an estua-

Table 2. Contingency matrix for Signature Evaluation.

Signature /train'g sample (%)	HdR	MdM	LdM	VlM	Np	Rb	Fr	MxL	BrL
HdR	285(87%)	1	0	0	0	0	29	14	0
MdM	0	120(82%)	4	0	0	0	0	0	0
LdM	0	17	97(84%)	1	2	0	0	11	0
VlM	0	0	5	95(99%)	0	0	0	4	22
Np	0	1	0	0	49(96%)	5	0	0	8
Rb	0	6	0	0	0	544(94%)	21	0	8
Fr	38	0	0	0	0	26	102(56%)	43	0
MxL	3	2	10	0	0	0	29	93(56%)	0
BrL	0	0	0	0	0	5	0	0	386(91%)
Total	326	147	116	96	51	575	181	165	424

Where:
HdR = Homogenous Dense Rhizophora
MdM = Medium Dense Mixed Mangrove
LdM = Low Dense Mixed Mangrove
VlM = Very low Mixed Mangrove
Np = Nipa Palm
Fr = Dryland Forest
MxL = Mixed Inland
BrL = Bare land
Rb = Rubber plantation

rine environment. In the digital classified SPOT image, the dense *Avicennia* group was placed into the Medium Density Mixed Mangrove class.

As stated by Hartono (1992) a distinct zonation of mangroves could not be realized from SPOT image analysis because of its spatial resolution of 20 m compared to the size of most mangrove zones (20 to 30 m), but a clear differentiation can be made between the dense and degraded mangroves and shrub formations.

Ground truthing revealed more detailed information where different species compositions are found to be associated with different soil types. It was further found that not only the soil type but also the water availability in terms of tide inundation played an important role in species distribution. However, favorable areas with too much human disturbance due to cutting for charcoal showed a very poor quality stand due to the absence of mother trees.

MOS-1 digital image analysis

MOS-1, an Earth observing satellite of the National Space Agency of Japan, was used for the interpretation of different mangrove land cover categories. Land cover categories determined from analysis of MOS-1 were practically identical with the SPOT results. However, due to its coarser spatial resolution of 50 m, there was an increase in the proportion of heterogenous mangrove forests while groups of small nipa palm did not appear in the final classification. The classified image included *rhizophora* dominant area, Medium Density Mixed Mangrove, Low Density, Very Low, Density, Rubber Plantation, Dryland Forest, Mixed Inland Vegetation and Bare Land. In this regard, MOS-1 has a stronger potential for land use/cover change detection in comparison with SPOT data.

SPOT/JERS1 visual analysis

Visual interpretation in relation to field conditions of combined SPOT and JERS1 imagery showed both **height and density** of the trees. However, mangrove and non-mangrove can not be discriminated using radar imagery alone. Nevertheless, the combination was very useful for identifying even and uneven-aged stands. This can be attributed to the basic sensitivity of radar signals to roughness, which is related to varying tree heights or a single strata of tree canopies.

Variations in tree heights in a two layer fashion with an upper layer occupied by mature trees and a lower strata of young regenerated trees showed a very high radar response. Lower radar signals were observed for the more homogenous canopy texture, compared to those of smaller or taller trees. Debris of dead branches and stumps were brighter than areas without vegetation. However, discrimination of the homogenous nipa species, for instance, was not possible in radar due to its similarity with the uniform height and gregarious nature of young *Rhizophora* saplings.

ERS1/SPOT digital complementary analysis

An ERS-1 (European Remote Sensing Satellite) image taken on October 29, 1992 was used to carry out a complementary analysis of radar with visible/infra-red satellite data. As pointed out by Aschbacher (1991), taking advantage of high spectral information of conventional VIR (visible/infra-red) sensors in combination with radar data is most promising for vegetation monitoring.

The complementary usage of ERS-1 SAR and SPOT increases the ability to classify different features in mangrove ecosystem. In addition, different ages of homogenous *Rhizophora* groups were distinguished (e.g., mature, medium size and young regeneration) from the changing stocking densities of heterogenous stands at varying heights. The uniformity in height of trees occupying a contiguous area has been proven to exhibit lower radar signals than a group of trees of dissimilar height. The tree crown morphology and size of leaves were also part of the above phenomenon.

The bare land category under SPOT classification was further identified as *abandoned* paddy field, areas covered by dead branches and stumps after cutting, soil saturated with water or bare land itself. An error that was occuring in the SPOT analysis for mature *Rhizophora* and *Medium Density Mixed Mangrove* in the mountainous regions of forest or rubber-type vegetation was again noted in their complementary results. Mixed inland vegetation (shrubs and small to medium sized trees) was classified inside the mangrove area, due to the similar roughness of heterogenous low and very low density mangrove stands.

As with other radar scanning systems, the side-looking ERS-1 exhibited a topographic distortion called 'lay-over' where the top of a hill drops in the direction facing the sensor as it moves from pole to pole. As observed in the study area, all the slopes at the western portion have a high return as the satellite scans the earth from the south to the north pole. This caused an increase in the pixel value of the vegetation found on the western slope resulting in an increased density of the mangrove from very low to medium density. In parallel, 'paddy on the top of the mountain' was noted due to the presence of exposed limestone with a mantle of small grasses and shrubs.

Combining ERS-1 and SPOT data yielded 10 mangrove categories (see Table 3). In comparison to SPOT alone which identified 6 mangrove categories. For a more quantitative comparison of each sensor contribution it was suggested that a comparative analysis should be made adopting a common land classification system specifically for mangrove forests or using the maximum number of classes identified in the classified image of the combined ERS-1 and SPOT.

ERS1 multi-temporal overlay

Three dates of filtered ERS1 taken in the same year (i.e. July 16, August 20, and October 29, 1992) were used to carry out multi-temporal analysis. Mangrove as compared with most other types of tropical vegetation did not show any significant seasonal change within a year due to its year-round water availability. The physiological activities, for instance, are neither affected by water stress during the dry season nor by water surplus during the rainy season unlike the inland ecosystems. Mangroves are a typical example of an evergreen forest.

Table 3. Contingency matrix for Signature Evalution of ERS-1 and SPOT Combination

Signature	Training Sample/(%)												
	Dbr	BrL	Pdy	MxL	Rbr	Obk	Mbk	Ybk	Hmm	Mmm	Lmm	VlM	For
Dbr	30 (94)	0	0	0	1	0	0	0	0	0	0	6	0
BrL	0	14 (82)	0	0	0	0	0	0	0	0	0	0	0
Pdy	0	0	73 (95)	0	0	0	0	0	0	0	0	0	0
MxL	0	0	0	104 (76)	88	0	0	0	0	0	2	3	0
Rbr	0	0	0	29	217 (61)	1	1	0	1	0	0	0	17
Obk	0	0	0	0	3	59 (77)	0	0	1	0	0	0	0
Mbk	0	0	0	0	2	5	102 (98)	0	0	0	0	0	0
Ybk	0	0	0	4	0	0	0	102 (99)	0	0	0	0	0
Hmm	0	0	0	0	0	11	0	0	102 (85)	1	20	0	0
Mmm	0	0	0	0	0	0	1	1	14	87 (97)	1	0	0
Lmm	0	0	0	0	0	0	0	0	2	2	190 (84)	27	0
Vlm	2	3	4	0	2	0	0	0	0	0	12	160 (82)	0
For	0	0	0	0	44	1	0	0	0	0	0	0	83 (83)
Total	32	17	77	137	357	77	104	103	120	90	225	196	100

Where:
Dbr = Debris
BrL = Bareland
Pdy = Paddy
Mxl = Mixed Inland
Rbr = Rubber
Obk = Tall Rhizophora
Mbk = Medium Height Rhizophora
Ybk = Low Height Rhizophora
Hmm = High Density Mixed Mangrove
Mmm = Medium Density Mixed Mangrove
Lmm = Low Density Mixed Mangrove
VlM = Very low Density Mixed Mangrove
For = Dryland Forest

In this regard, a multi-temporal overlay of radar images did not yield any significant features to show any pattern or differences occurring in a span of approximately three months. Influences of tide or any human disturbance were not reflected during the specified time frame. The black and white appearance on the color display in reference to the land supported the above statement.

Conclusions

The main objective of this study has been achieved in terms of developing an integrated methodology that will serve as a guideline in the application of remote sensing techniques in assessing the mangrove ecosystem. Based on the results, it can be concluded that:

1. A distinction of mangrove and non-mangrove areas is easily made possible using visual in-

terpretation of SPOT XS data having a spatial resolution of 20 m. Digital image classification provides detailed categories primarily on stocking density of mangroves. In addition, pure and mature Rhizophora and uniform Nipa palm are clearly identified in the produced mangrove map.

2. The use of radar alone can not discriminate mangrove and non-mangrove as exhibited in analyzing both SAR data (i.e., ERS-1 and JERS-1). However, when it is combined with visible and infra-red data of SPOT, it offers an additional dimension in identifying trees of varying heights or groups of mangrove tree species with uniform heights. Areas devoid of vegetation can be further distinguished into mud, abandoned paddy field and those covered by dead branches and stumps in the process of clear-cutting.
   Despite the radar's capabilities for global measurement and imaging that takes place independent of cloud and sunlight conditions, the above results are attributed to the sensitivity of radar signal to the roughness of the target object. This is exemplified by the differences in the canopy architecture in contrast with the sensitivity of visible and infra-red data for vegetation intensity.

It is suggested that advantage should be taken for the Landsat TM multi-spectral capability and its combination with radar data (see Figure 4). The inclusion of a comparative analysis of two set of radar data (ERS-1 and JERS-1) to further enhance their capabilities for this type of ecosystem is also recommended. Related spatial data such as elevation are deemed necessary for the refinement of the above results through the application of GIS analysis, while taking into consideration other spatial data like soil type and level of tidal inundation that affects the existence of this unique type of ecosystem.

Acknowledgement

This study has been carried out under the assistance provided by the Austrian Academy of Sciences, Austrian Association for Development and cooperation, National Research Council of thailand, and the Royal Forestry Department, Thailand. We are grateful to the Remote Sensing Laboratory of AIT for the provision of digital image processing equipment.

References

Aschbacher, J., 1991. Application of Microwave Remote Sensing for Tropical Forest Management. Paper Presented at the 'International Workshop on Conservation and Sustainable Development' 22–26 April, 1991. Khao Yao, Thailand.

Hartono, 1992. Remote Sensing for Mangrove Study in Java, Indonesia. Asian Pacific Remote Sensing Journal. Volume 1, Number 2, January 1992. ESCAP/UNDP Regional Remote Sensing Programme, Bangkok, Thailand.

Annex I. Photographs of various satellite images and from the field.

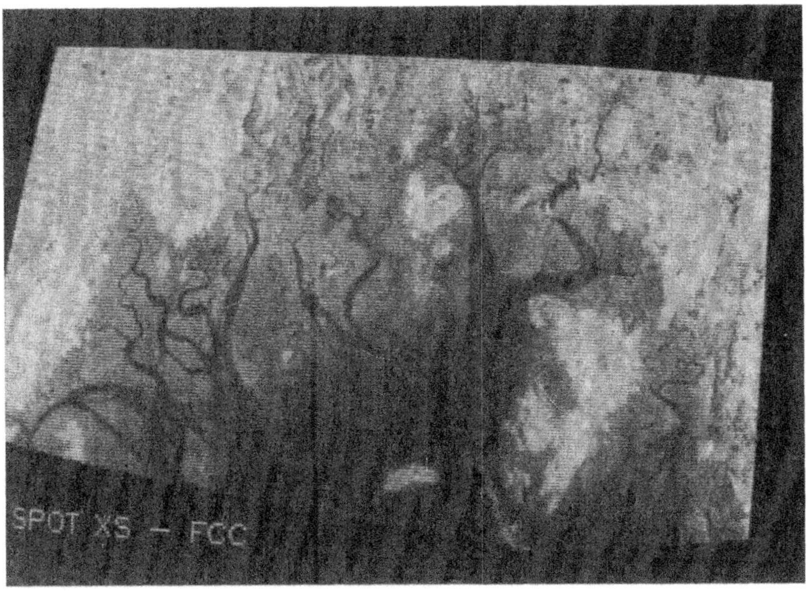

Fig. 1. SPOT XS Standard False color composite of the Study Area (22 March 1992).

Fig. 2. Classified Spot Image (Cyan = Mixed Mangrove, Orange = Low Density, Red = Rubber).

Fig 3 An Overlay of ERS-1 and LANDSAT TM Image

Fig 4 Homogenous Rhizophora stand

Hydrobiologia **295**: 295–302, 1995.
Y. S. Wong & N. F. Y. Tam (eds), Asia-Pacific Symposium on Mangrove Ecosystems.

Diel activity patterns in *Metapenaeus* and *Penaeus* juveniles

J. Honculada Primavera & J. Lebata
Aquaculture Department, Southeast Asian Fisheries Development Center, Tigbauan, Iloilo, Philippines

Key words: Metapenaeus anchistus, Penaeus merguiensis, P. monodon, shrimp, prawn, diurnal periodicity, shelter

Abstract

Small (5–10.9 mm carapace length), medium (11–15.9 mm), and large (16–20.9 mm) juveniles of *Metapenaeus anchistus, Metapenaeus* sp., *Penaeus monodon* and *P. merguiensis* were stocked individually in glass tanks provided with sand substrate, sea water, artificial bamboo shelter, aeration and food. The seven activity types (recorded for each shrimp hourly for 24 h) were classified as below (burrowing) or above substrate (swimming, walking, stationary, in shelter, feeding and cleaning). Shrimp juveniles exhibited a strong diel periodicity – emergence and activity at night and burrowing in the day. The chi-square test showed that type of activity (above/below substrate) was associated with period (light/dark). Diurnal burrowing was greater among *Metapenaeus* than *Penaeus*; inversely, above substrate activities were more frequent for *Penaeus* species compared to *Metapenaeus*. Feeding was the major above substrate and nocturnal activity for *M. anchistus, Metapenaeus* sp. and *P. monodon*. Only *P. monodon* used the shelter consistently. Frequency of the 7 activity types was dependent on juvenile size for *Penaeus*, e.g., the preference for shelters shifted to burrowing with increase in size in *P. monodon*. Results are discussed in relation to the importance of mangrove habitats in providing shelter to penaeids, in particular the mangrove-associated *P. monodon* and *P. merguiensis*.

Introduction

Many penaeid prawn species are associated with specific nursery habitats. In Australia, *Penaeus esculentus* and *P. semisulcatus* juveniles are found in seagrass beds, *P. merguiensis* in mud-mangrove banks, *Metapenaeus endeavouri* in seagrass and other habitats, and *M. ensis* is widespread (Staples *et al.*, 1985). Similarly, in Southeast Africa *P. semisulcatus* is found in macrophyte/seagrass areas, *P. monodon* and *P. merguiensis* along muddy mangrove channels, *P. japonicus* in sandy substrate, *M. stebbingi* in the swash zone with bare substrate, and *M. monoceros* in diverse habitats (de Freitas, 1986).

Mangrove nurseries provide food and/or shelter to penaeid juveniles. As mangrove waterways are devoid of refuges during low tide, Chong *et al.* (1992) argue that food abundance is the likely advantage. On the other hand, Robertson (1988) hypothesized that the importance of mangroves to juvenile *P. merguiensis* is related to protection or shelter from fish predation (provided by prop roots, pneumatophores and other structures) because the main diet of this species can also be satisfied in other habitats.

This study was undertaken to document shelter use and other activities of juveniles belonging to four penaeid species found in mangroves. It also determined the periodicity of such activities, and compared activity patterns among different sizes of juveniles. The results will provide background information for future experiments on the effect of shelters on prawn survival in the presence of predators.

Materials and methods

Experimental animals

Juveniles of *Metapenaeus anchistus, Metapenaeus* sp. novo, and *P. merguiensis* were collected by pocket seine (2–3 mm mesh size) from mangroves in Tandog Island, Nueva Valencia and Sibunag River, Jordan in Guimaras Province, central Philippines. Juveniles of *P. monodon* were produced from a hatchery and grow-out pond in Iloilo Province due to insufficient collections from the riverine mangrove. The animals were held in 300-l fiberglass tanks with sand substrate and conditioned for 1–2 wk.

Experimental tanks

Glass tanks measuring 30.5 × 15.5 × 20.5 cm were used for small and medium juveniles, and 35.5 × 20.0 × 25.5 cm for large juveniles. Tanks were provided with sea water of 20 cm depth, airstones, a black net cover, and a 2–3 cm layer of mangrove sand; some penaeids such as *P. semisulcatus* become arhythmic in the absence of a substrate (Moller & Jones, 1975). To decrease turbidity and facilitate observation, the sand obtained from Tandog Island (10% clay, 32% silt, 58% sand) was sieved to reduce the proportion of fine particles (7% clay, 21% silt, 71% sand). An artificial shelter made of dried bamboo (10–25 cm in length, one end divided into 0.5–1.0 cm wide splits held apart by nylon string) was installed vertically on the substrate.

Chopped fresh squid was fed in excess at 0900–1000 h and 1600–1700 h. It was earlier observed that starved *Metapenaeus* juveniles kept in other tanks (not used in the study) appeared above the sand even during daytime in search of food. Hill & Wassenberg (1987) noted that unfed *P. esculentus* spent more time emerged and moving compared to fed prawns and cautioned against the use of starved animals for behavior studies.

Treatments

Juveniles of four penaeid species of small (5–10.9 mm carapace length), medium (11–15.9 mm CL) and large (16–20.9 mm CL) sizes belonging to eleven species-size combinations at 5–7 replicates per combination were stocked individually in experimental tanks. There were no small *Metapenaeus* sp. The treatments were distributed randomly.

Observations commenced on the third day from stocking; in a dry run, some shrimp burrowed continuously and emerged only 1–2 days after stocking in the glass tanks. Hourly observations of the activities of individual shrimp were recorded for 24 h. A flashlight (two 1.5 volt, size D batteries) with red filter was used for night observations. Red light caused the least disturbance in *P. duorarum* (Bishop & Herrnkind, 1976) and did not inhibit nocturnal emergence in *P. vannamei* (Moctezuma & Blake, 1981). Care was taken not to shine the light directly on the shrimp

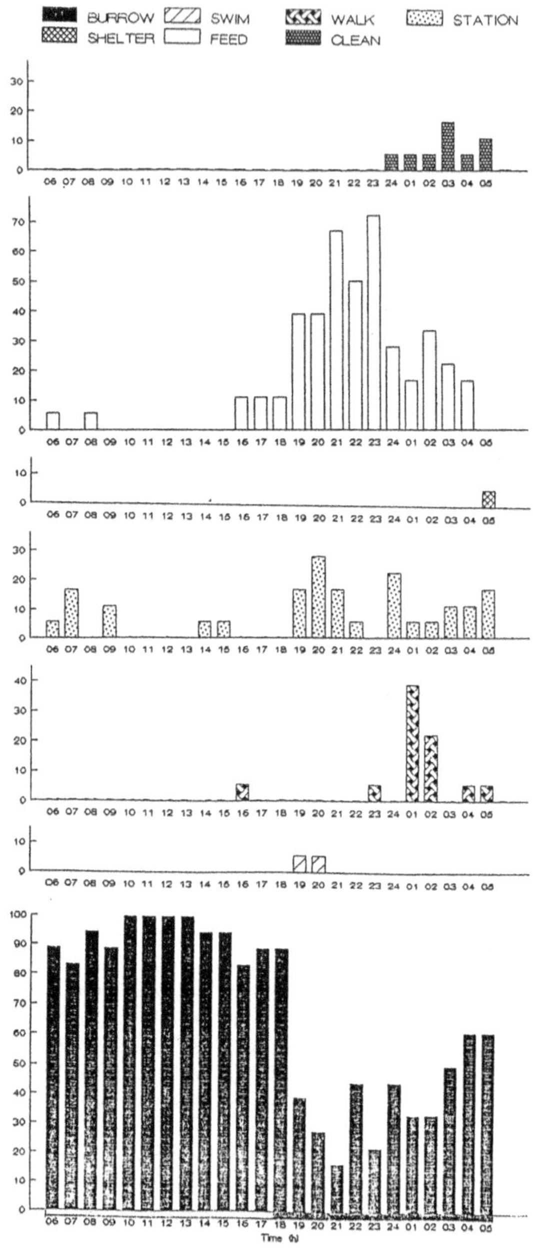

Fig. 1. Diel activity patterns of *Metapenaeus anchistus* juveniles, all sizes ($n = 17$).

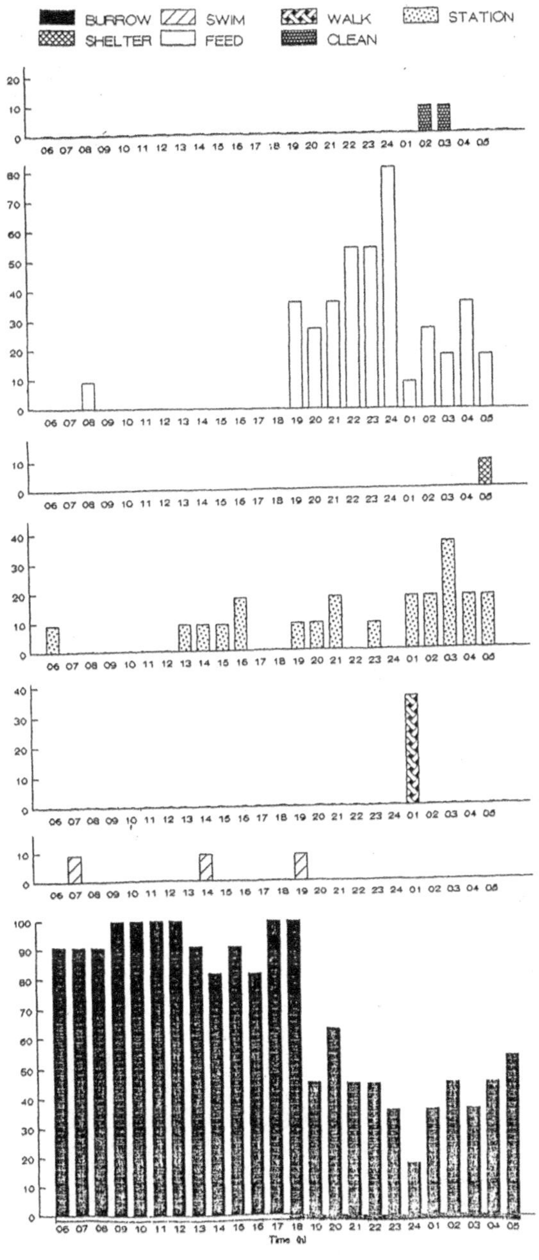

Fig. 2. Diel activity patterns of medium and large *Metapenaeus sp. novo* juveniles ($n = 12$).

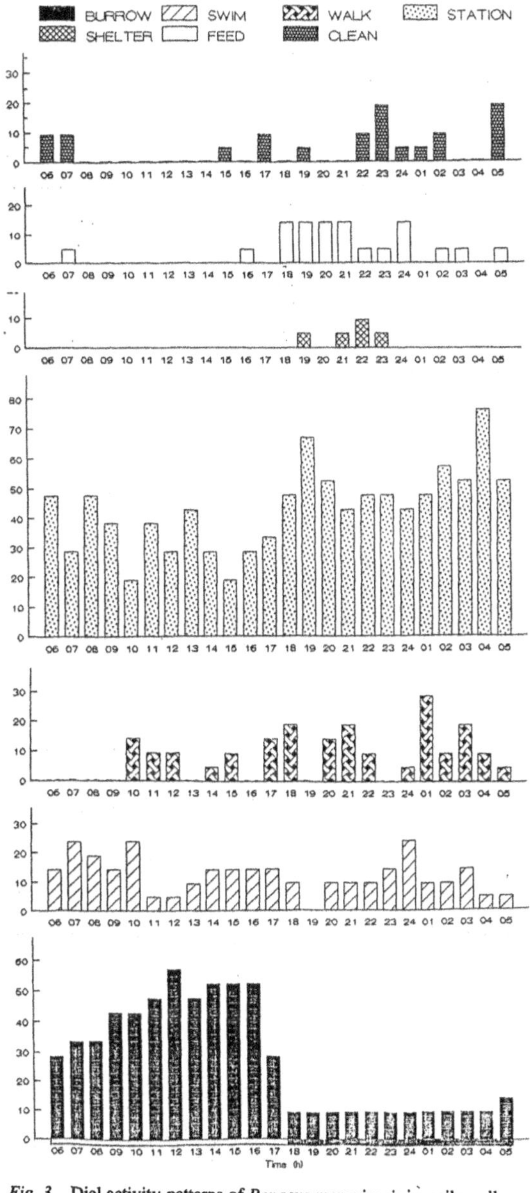

Fig. 3. Diel activity patterns of *Penaeus merguiensis* juveniles, all sizes ($n = 21$).

which could elicit an immediate burrowing response (Hughes, 1966).

The experiment was conducted 2–3 August 1993 (27–29 °C water temperature, 35‰ salinity) for *M. anchistus* and *Metapenaeus* sp. and 9–10 August

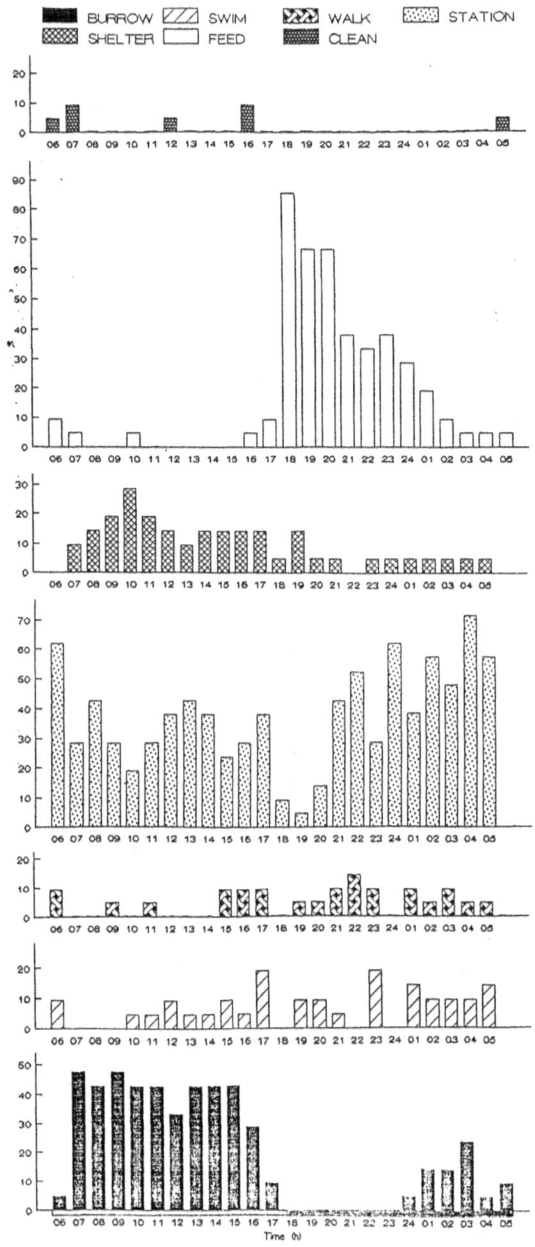

Fig. 4. Diel activity patterns of *Penaeus monodon* juveniles, all sizes ($n = 21$).

1993 (26–27 °C, 35‰) for *P. monodon* and *P. merguiensis*. Sunrise was at 0550 h and sunset at 1820 h.

The chi-square test was used to test the hypothesis of independence between type of activity (below/above substrate) and period (light/dark), and between activity type (burrowing, swimming, etc.) and juvenile size. When necessary, activities were classified as below (burrowing) or above substrate (swimming, walking, stationary, shelter use, feeding and cleaning) to minimize the categories with low counts and which would invalidate the test. Likewise, hourly observations were grouped into light (0600–1700 h) and dark (1800–0500 h) periods.

Results

The observed activity patterns of juveniles showed a strong diel periodicity-shrimp burrow in the substrate at sunrise, remain inactive during the day, emerge at dusk, and stay active throughout the night (Figs. 1, 2, 3, 4). Results of the chi-square test show that type of activity is associated with period of the day for the four species (χ^2: *M. anchistus* = 141, *Metapenaeus* sp. = 81, *P. merguiensis* = 60, *P. monodon* = 53; $p < 0.001$ for all species) such that burrowing is diurnal and above substrate activity nocturnal. However, the two *Metapenaeus* species burrowed more frequently than *Penaeus* juveniles (>90% *vs* 33–41%) (Fig. 5) and emerge mainly during dark (Figs 1, 2). Inversely, *Penaeus* species were more active at night compared to *Metapenaeus* (>90% *vs* 57–61%) (Fig. 5). Swimming, walking and stationary behavior extended even to the daylight hours for *Penaeus* juveniles (Figs. 3, 4).

The process of burrowing follows earlier published descriptions for various penaeid species. The animal creates a depression in the substrate with the help of its pereopods, settles in, and agitates surrounding particles to cover itself (Joshi *et al.*, 1979). The body may be completely buried, marked only by a funnel-shaped depression in the substrate along the sides of the carapace caused by the exhalant respiratory currents (Hughes, 1966) or partially covered with the rostrum, eyes, antennules, antennae and antennal scales, or even parts of the abdomen and telson exposed (Bishop & Herrnkind, 1976).

The seven types of activity were associated with juvenile size for *Penaeus* (χ^2: *P. merguiensis* = 70, *P. monodon* = 61; $p < 0.001$ for both species). (The test was not valid for *Metapenaeus* species because many

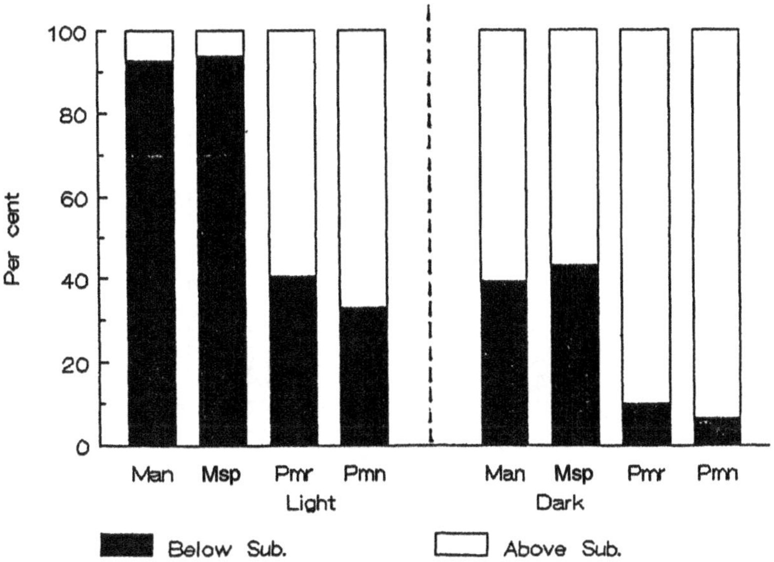

Fig. 5. Frequency of below and above substrate activities of *Metapenaeus anchistus*, *Metapenaeus* sp., *Penaeus merguiensis* and *P. monodon* juveniles (all sizes) during light (0600–1700 h) and dark (1800–0500 h) periods.

of the activity types or categories had zero or low counts.) Large and medium *P. monodon* preferred to burrow while shelter users were mostly small juveniles; burrowers among *P. merguiensis* were mainly large juveniles (Fig. 6).

Among the four penaeid species, only *P. monodon* used the shelter consistently, with a higher frequency in the day than at night (Fig. 4). The other species were rarely found in the shelters and only at night (Figs. 1, 2, 3).

Feeding commenced at or just after sunset (Figs. 1, 2, 3, 4) and was clearly nocturnal – although 20–60% of *P. monodon* and *P. merguiensis* remained above substrate during the day, they hardly fed (Figs. 3, 4). Feeding was also the major above substrate activity for *M. anchistus*, *Metapenaeus* sp. and *P. monodon* with a maximum of 70–90% feeding at night (Figs. 1, 2, 4). Although 90% of *P. merguiensis* had emerged by 1800–2400 h, only a maximum of 15% were observed feeding (Fig. 3).

Nocturnal cleaning activity followed feeding (Figs. 1-4) and mainly involved the mouthparts. In the day, prawns also groomed themselves – they cleaned the head (eyes, antennal scales) using the chelae of maxillipeds and tail fan (telson and uropods), the latter with the posterior abdomen bent forward. (Some of the small juveniles classified as stationary could have been doing cleaning activities.)

Exuviae (indicating ecdysis) were found in the tanks in the early morning – 2 at 0700 h for *M. anchistus*, and 4 at 0300 h and one at 0600 h for *P. merguiensis*.

Discussion

Results of this study should be interpreted within the limitations of the experiment: a) hourly observations which exclude activities between observations, b) a predominantly sand substrate, and c) juvenile sizes >5 mm CL. Nevertheless, a definite pattern of daytime burrowing and nocturnal activity can be seen for the four penaeid species. The same diel periodicity has been observed for *P. esculentus* (Hill & Wassenberg, 1987) and *P. vannamei* (Moctezuma & Blake, 1981). Joshi *et al.* (1979) found that 80% of *M. monoceros* burrowed in mud at 9 a.m. compared to 60% at 3 p.m. In field and laboratory studies, *M. endeavouri*, *P. esculentus* and *P. merguiensis* were observed to be more active at night (Vance, 1992; Vance & Staples, 1992). More than 90% of *P. duorarum* emerged from the substrate within 45 min after the light-dark transition suggesting that light is the major synchronizer

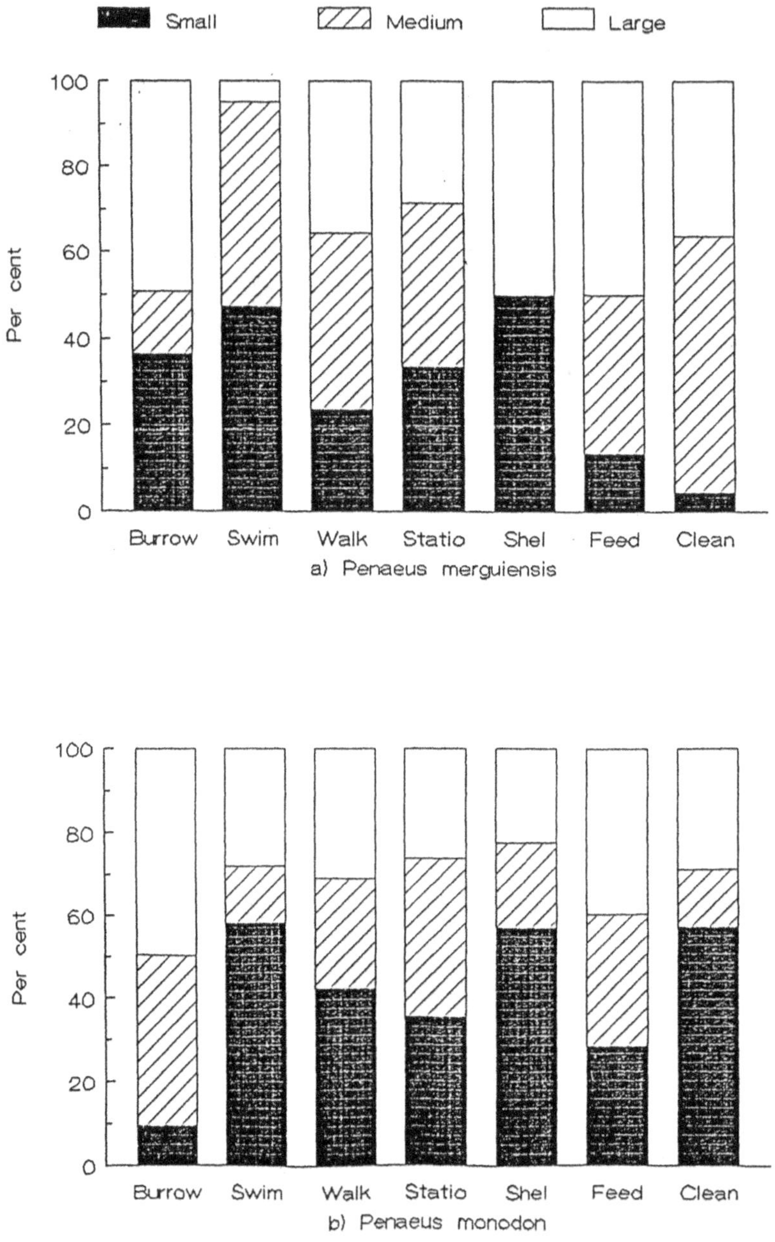

Fig. 6. Frequency of different activity types (burrowing, swimming, walking, stationary, in shelter, feeding and cleaning) in small, medium and large juveniles of a) *Penaeus merguiensis* and b) *P. monodon*.

of the circadian rhythm in penaeids (Moller & Jones, 1975; Bishop & Herrnkind, 1976).

Continuous feeding activity from 1600 to 0600 h in the four species is similar to that observed for *P. esculentus* in the laboratory (Hill & Wassenberg, 1987). The guts of *P. esculentus* and *P. semisulcatus* collected from the wild were partially filled indicating active feeding throughout the night (Wassenberg & Hill, 1987). In this study, a unimodal feeding pattern could be observed with varying peaks for different species – at dusk for *P. monodon* and later towards midnight for *M. anchistus* and *M. ensis*, instead of the bimodal trend found in *P. duorarum* (Reynolds & Casterlin, 1979).

The frequent emergence and relatively low burrowing activity of *P. monodon* could be related to the sandy nature of the experimental substrate as this species prefers the finest medium (Moller & Jones, 1975) and has been observed to burrow in the soft mud of prawn ponds (Delmendo & Rabanal, 1956). In the natural habitat however, *P. monodon* is more frequently seen above the sand (Hughes, 1966). Burrowing in *P. merguiensis* is recorded for the first time here. This species has not been seen to bury under laboratory conditions (Vance & Staples, 1992; Primavera, pers. obs.), neither has its morphospecies *P. indicus* been observed to do so in nature or in aquaria (Hughes, 1966).

There is an ontogenetic shift towards more frequent burrowing in the daytime with increase in size among *Penaeus* juveniles – *P. monodon* and *P. merguiensis* (this study), and *P. vannamei* (Moctezuma & Blake, 1981) – that is not present in *Metapenaeus*. Similarly, most *P. esculentus* and *P. semisulcatus* found buried in the substrate in laboratory tanks were large individuals of 15 mm CL or greater (Hill & Wassenberg, 1993). Perhaps the greater diversity of habitats of some *Metapenaeus* species (Staples *et al.*, 1985; de Freitas, 1986) is related to their efficient burrowing. In contrast, *Penaeus* species, at least the smaller juveniles, depend more on above-ground shelters such as are found in seagrass beds and mangroves, and therefore have more restricted habitats.

Aside from protection against predators, burrowing in penaeids may provide escape from extremes in temperature and, salinity, and lead to energy conservation through decreased oxygen consumption (Moller & Jones, 1975; Lakhsmi *et al.*, 1976; Moctezuma & Blake, 1981).

The relative non-use of shelters by *P. merguiensis*, a mangrove-associated species, may be due to the size (>5 mm CL) of the experimental animals in this study. Vance *et al.* (1990) found that *P. merguiensis* juveniles of 2–4 mm CL were much more abundant in mangrove creeks (where roots and other structures are found) than in the river. Gradual emigration at 5 mm CL from creeks to the river could mean increased tolerance to predation (afforded by morphological and behavioral changes) with greater size of juveniles.

The attachment of small *P. monodon* juveniles to shelters may be due to greater vulnerability to predation than larger shrimp. Artificial shelters in the form of bamboo screens, coconut leaves and plastic sheets significantly increased survival of *P. monodon* juveniles in ponds and nets (Cholik, 1978; Martosudarmo, 1983). Fry collectors in the Philippines use grass bundles, twigs and similar lures to catch wild *P. monodon* fry (Delmendo & Rabanal, 1956; Motoh, 1981). Bamboo or nylon net shelters are installed in commercial nursery ponds and tanks to improve survival rates.

Conclusion

This study establishes the use of shelter under laboratory conditions by small juveniles of *P. monodon*, a species which utilizes mangroves as nursery. The insignificant shelter use by *P. merguiensis*, another mangrove-associated shrimp, could be verified by further experiments testing other kinds of artificial shelter, and juveniles smaller than 5 mm CL known to frequent mangrove creeks. The ontogenetic shift in behavior from shelter use when small to burrowing by larger juveniles of *P. monodon* contrasts with the consistent burrowing pattern in *Metapenaeus* individuals of all sizes. Species of the genus *Metapenaeus* are more widespread in distribution such that efficient burrowing skills would be of greater survival value than shelter use associated with specific vegetation types.

Acknowledgments

We thank Larni Espada for help in data analysis, G. Unlayao for assisting in the experiment, and two anonymous reviewers for their critical comments and suggestions.

References

Bishop, J. M. & W. F. Herrnkind, 1976. Burying and moulting of pink shrimp, *Penaeus duorarum* (Crustacea: Penaeidae), under selected photoperiods of white light and U-V light Biol Bull 156 163–182

Cholik, F, 1978 Study on the effects of different densities of artificial shelters on the survival and growth of sugpo fry (*Penaeus monodon*) in nursery ponds Unpub M S thesis, University of the Philippines in the Visayas, 75 pp

Sasekumar, A, V C Chong, M U C Leh & R D'Cruz, 1990 Mangroves as habitats for fish and prawns Hydrobiologia 247 195–207

de Freitas, A J, 1986 Selection of nursery areas by six Southeast African Penaeidae Estuar coast Shelf Sci 23 901–908

Delmendo, M N & H R Rabanal, 1956 Cultivation of *sugpo* (giant tiger shrimp) *P monodon* Fabricius in the Philippines IPFC Proc 6 424–431

Hill, B J & T J Wassenberg, 1987 Feeding behaviour of adult tiger prawns, *Penaeus esculentus*, under laboratory conditions Aust J mar Freshwat Res 38 183–190

Hill, B J & T J Wassenberg, 1993 Why are some prawns found in seagrass? An experimental study of brown (*Penaeus esculentus*) and grooved (*Penaeus semisulcatus*) tiger prawns Aust J mar Freshwat Res 44 221–227

Hughes, D A, 1966 Investigations of the 'nursery areas' and habitat preferences of juvenile prawns in Mozambique J appl Ecol 3 349–354

Joshi, P K, G K Kulkarni & R Nagabhushanam, 1979 Studies on the behaviour and substratum preference in juveniles of the marine prawn, *Metapenaeus monoceros* (Fabricius) Hydrobiologia 65 195–198

Lakshmi, G, J Venkataramiah & G Gunter, 1976 Effect of salinity and photoperiod on the burying behavior of brown shrimp *Penaeus aztecus* Ives Aquaculture 8 327–336

Martosudarmo, B, 1983 Effects of different substrates and selected feeds on the growth and survival of hatchery produced *Penaeus monodon* postlarvae Unpub M S thesis, University of the Philippines in the Visayas, 83 pp

Moctezuma, M A & B F Blake, 1981 Burrowing activity in *Penaeus vannamei* Boone from the Caimanero-Huizache lagoon system in the Pacific Coast of Mexico Bull mar Sci 31 312–317

Moller, T H & D A Jones, 1975 Locomotory rhythms and burrowing habits of *Penaeus semisulcatus* (de Haan) and *P monodon* (Fabricius) (Crustacea Penaeidae) J exp mar Biol Ecol 18 61–77

Motoh, H, 1981 Studies on the fisheries biology of the giant tiger prawn, *Penaeus monodon* in the Philippines SEAFDEC Aquaculture Dept Tech Rep No 7 128 pp

Reynolds, W W & M E Casterlin, 1979 Diel activity of the pink shrimp *Penaeus duorarum* Hydrobiologia 66 223–226

Robertson, A I, 1988 Abundance, diet and predators of juvenile banana prawns, *Penaeus merguiensis*, in a tropical mangrove estuary Aust J mar Freshwat Res 39 467–478

Robertson, A I & N C Duke, 1987 Mangroves as nursery sites comparisons of the abundance and species composition of fish and crustaceans in mangroves and other nearshore habitats in tropical Australia Mar Biol 96 193–205

Staples, D J, D J Vance & D S Heales, 1985 Habitat requirement of juvenile penaeid prawns, pp 47–54 In Rothlisberg, P C, B J Hill & D S Staples (eds) Second National Prawn Seminar NSP2, Cleveland, Australia

Vance, D J, 1992 Activity patterns of juvenile penaeid prawns in response to artificial tidal and day-night cycles a comparison of three species Mar Ecol Prog Ser 87 215–226

Vance, D J & D J Staples, 1992 Catchability and sampling of three species of juvenile penaeid prawns in the Embley River, Gulf of Carpentaria, Australia Mar Ecol Prog Ser 87 201–213

Vance, D J, M D E Haywood & D J Staples, 1990 Use of a mangrove estuary as a nursery area by postlarval and juvenile banana prawns, *Penaeus merguiensis* de Man, in Northern Australia Estuar coast Shelf Sci 31 689–701

Wassenberg, T J & B J Hill, 1987 Natural diet of the tiger prawns *Penaeus esculentus* and *P semisulcatus* Aust J mar Freshwat Res 38 169–182

Hydrobiologia **295**: 303–309, 1995.
Y. S. Wong & N. F. Y. Tam (eds), Asia-Pacific Symposium on Mangrove Ecosystems.

Mangroves and brackishwater pond culture in the Philippines

J. Honculada Primavera
Aquaculture Department, Southeast Asian Fisheries Development Center, Tigbauan, Iloilo, Philippines

Key words: aquaculture, milkfish, shrimp, mangrove valuation, ecological effects, socioeconomic effects

Abstract

Around 50% of mangrove loss in the Philippines can be traced to brackishwater pond construction. The decrease in mangroves from 450 000 ha in 1920 to 132 500 ha in 1990 has been accompanied by expansion of culture ponds to 223 000 ha in 1990. The history of fishpond development in the country includes a government-sponsored fishpond boom in the 1950-g and 1960s, the proconservation decade of the 1970s followed by a shrimp fever in the 1980s. Production from brackishwater ponds has increased from 15 900 mt worth P7.6 million in 1938 to 267 000 mt valued at P6.5 billion in 1990. On the other hand, the maximum valuation of over \$11 000 ha^{-1} yr^{-1} for unmanaged and managed mangrove forests makes them economically on par with the most profitable pond farming systems. *The loss of mangrove systems and their varied goods and services is the single most important consequence* of brackishwater pond culture in the Philippines. Moreover, intensive shrimp farming is associated with other ecological and socioeconomic effects such as pollution of coastal waters and decline in domestic food crops. New legislation and enforcement of existing laws, conservation of remaining mangroves, massive rehabilitation of denuded mangrove areas, and promotion of sustainable aquaculture and fisheries are recommended.

Introduction

The Philippines is an archipelago of some 7100 islands bordered by 260 million hectares of coastal waters; its fisheries production ranked 11th worldwide in 1989 (BFAR, 1991). Of the country's total 1991 fish production of 26×10^6 mt, 44% came from municipal fisheries, 29% from commercial fisheries and 27% from aquaculture (Table 1). In 1971, aquaculture contributed only 10% to the total production in contrast to 37% from the commercial sector and 53% from the municipal sector.

Nearshore fisheries depends to a greater or lesser degree on mangroves whereas aquaculture is practised predominantly in earthen brackishwater ponds. The history of Philippine mangroves and brackishwater aquaculture is inextricably linked since both systems represent mutually exclusive options.

Mangroves

Species composition

Brown & Fischer (1920) described 17 families, 20 genera and 28 species of mangroves in the country while Arroyo (1979) listed 34 families, 53 genera and 70 species of trees, palms, shrubs and vines. Following the criteria of Tomlinson (1986), 14 families, 16 genera and 26 species of major and minor mangroves are found in the Philippines (Table 2).

The country's premier city of Manila owes its name to the mangrove species *Scyphiphora hydrophyllacea* or *nilad* in Tagalog, a local dialect. When the Spanish came in 1570, the swampy area along the shores of Manila Bay was called *Maynilad* meaning 'There is *nilad*' in reference to the abundance of *S. hydrophyllacea* with its small, white flowers and aniline product popularly used as a cleansing and whitening laundry agent (Juanico, 1983).

Table 1. Comparison of municipal, commercial and aquaculture sectors in Philippine fisheries.

	Municipal[a]	Commercial[b]	Aquaculture[c]
Production (mt)			
1971 (% of total)	542,904	382,276	97,915
	(53.1)	(37.4)	(9.6)
1991 (% of total)	1,146,765	759,815	692,401
	(44.1)	(29.3)	(26.6)
Employment, 1991			
(no. of persons)	675,677	56,715	258,480
Gross monthly income, 1987			
(Phil. Peso)	1,986	19,695	3,721

[a] Vessels of $\leq$3 gross tons, [b] Vessels >3 gross tons, [c] Includes pond, pen, cage and mollusc culture.
Sources: 1971 Fisheries Statistics of the Philippines; 1991 Fisheries Profile of the Philippines; Ibon, 1989.

Area

The total mangrove area in the Philippines estimated at 400 000 to 500 000 ha in 1920 by Brown & Fischer included cultivated plantations in addition to primary and secondary forests based on reports of vast stands of *Rhizophora* and nipa palm grown near Manila for firewood and roof shingles. A SPOT survey revealed that 95% of brackishwater ponds throughout the country were derived from mangroves during the 1952–1987 period (PCAFNRRD, 1991). Approximately half of the mangrove loss of 279 000 ha from 1951 to 1988 (Table 3) can be traced to development of 141 000 ha of culture ponds; the remaining areas were converted to agriculture, salt ponds, industrial and residential areas.

Importance

Among the many amenities and goods from mangroves that have contributed to the well-being of coastal communities are the reduction of coastal and riverbank erosion, dampening of storm surges and high winds associated with typhoons, flood control, pollution abatement and a variety of forestry and fisheries products (Saenger *et al.*, 1983). In the Philippines, the latter include timber and wood products used for fishing, construction, furniture, firewood and charcoal; fish, shrimp, crabs and molluscs for food; and minor items such as tannins, dyes, medicines, and fodder for livestock (Tesoro, 1984). The nipa palm *Nypa fruticans* is one of the most versatile plants in the country with the young leaves used for wrapping food, the midribs bundled into brooms, and the immature kernels eaten fresh and cooked. The processing of thatch and shingles from nipa leaves for roofing and walls of rural houses, and production of alcohol, wine and vinegar from the sap constitute important cottage industries in the countryside.

Valuation

Hamilton & Snedaker (1984) have categorized mangrove goods and services according to their valuation (marketed or not) and location (onsite or offsite). Conventional financial analysis generally covers only marketed goods and ignores nonmarketed items such as medicines and the storm buffer function. Notwithstanding such limitation, a review by Primavera (1993) of available valuation data gives a range of US\$30–2000 ha^{-1} yr^{-1} for both complete systems and managed mangroves, with a maximum figure of \$11 561 ha^{-1} yr^{-1} for sustained harvest of wood products in Malaysia.

A positive correlation between nearshore yields of fish or shrimp and mangrove area has been documented in the Philippines (Camacho & Bagarinao, 1986), Indonesia (Martosubroto & Naamin, 1977), Malaysia (Macnae, 1974; Gedney *et al.*, 1982), and Australia (Staples *et al.*, 1985).

Table 2 Major and minor mangrove species[a] in the Philippines (modified from Brown & Fischer, 1920, Arroyo, 1979)

Family	Species	Common Name
I Major Elements		
A Rhizophoraceae	1 *Rhizophora apiculata*	*Bakauan lalaki*
	2 *R mucronata*	*Bakauan babae*
	3 *Bruguiera cylindrica*	*Pototan lalaki*
	4 *B gymnorrhiza*	*Busain*
	5 *B parviflora*	*Langai*
	6 *B sexangula*	*Pototan*
	7 *Ceriops decandra*	*Malatangal*
	8 *C tagal*	*Tangal*
B Avicenniaceae	9 *Avicennia alba*	*Api api*
	10 *A officinalis*	*Api api*
	11 *A marina*	*Bungalon*
C Sonneratiaceae	12 *Sonneratia alba*	*Pedada*
	13 *S caseolaris*	*Pagatpat*
D Combretaceae	14 *Lumnitzera littorea*	*Tabau*
	15 *L racemosa*	*Kulasi*
E Meliaceae	16 *Xylocarpus granatum*	*Tabigi*
	17 *X moluccensis*	*Piagau*
F Palmae	18 *Nypa fruticans*	*Nipa*
II Minor Elements		
G Aegicerataceae[b]	19 *Aegiceras floridum*	*Tinduk tindukan*
H Euphorbiaceae	20 *Excoecaria agallocha*	*Buta buta*
I Rubiaceae	21 *Scyphiphora hydrophyllacea*	*Nilad*
J Myrtaceae	22 *Osbornia octodonta*	*Taualis*
K Bombacaceae	23 *Camptostemon philippinensis*	*Gapas gapas*
L Lythraceae	24 *Pemphis acidula*	*Bantigi*
M Pteridaceae	25 *Acrostichum aureum*	*Lagolo*
N Sterculiaceae	26 *Heritiera littoralis*	*Dungon late*

[a] Based on criteria of Tomlinson (1986)
[b] Family Myrsinaceae in Tomlinson (1986)

Pond culture

History

In Southeast Asia, the culture of fish in brackishwater ponds started in Java, Indonesia in 1400 (Herre & Mendoza, 1929, Schuster, 1952) The Javanese system of brackishwater pond culture apparently spread to the Philippines (Hora & Pillay, 1962) with fishponds constructed along the shores of Manila Bay (Radcliffe, 1912, Adams *et al*, 1932)

Without active government support, mangrove conversion into ponds proceeded at a slow rate of 760–1200 ha yr^{-1} up to 1940 (Table 4) The Bureau of Fisheries and Aquatic Resources (BFAR) was created in 1947 and US$23 6 million in loans was released by the International Bank for Reconstruction and Development (IBRD) for fishpond construction and operation (Siddall *et al*, 1985) through such conduits as the Rehabilitation Finance Corporation (Villaluz, 1953) Reflecting the widespread view of mangroves as wastelands, Carbine (1948) stated that ' the bangos [milkfish] industry is important because it makes use of

Table 3. Area (in hectares) of mangroves and brackishwater culture ponds in geographical regions of the Philipplnes, 1951-1990.

Region	Mangroves 1951 [a]	Mangroves 1988 [b]	B'water Ponds 1951 [a]	B'water Ponds 1990 [c]
I (Ilocos Region)	771	200	7,599	17,511
II (Cagayan Valley)	7,322	3,400	0	1,635
III (Central Luzon) & Natl. Capital Region	56,799	500	34,172	56,919
IV (Southern Tagalog)	77,997	51,000	2,238	24,429
V (Bicol Region)	42,234	9,900	671	11,607
VI (Western Visayas)	49,035	2,825	27,741	61,863
VII (Central Visayas)	24,213	9,650	2,151	7,642
VIII (Eastern Visayas)	36,501	24,850	1,135	6,400
IX (Western Mindanao)	91,072	19,300	1,399	18,460
X (Northern Mindanao)	18,273 [d]	8,600	623 [d]	4,610
XI (Southern Mindanao)	17,518 [d,e]	7,100	3,223 [d,e]	7,732
XII (Central Mindanao)	6,647 [e]	2,400	1,333 [e]	4,099
Total	418,382	139,725	82,228	222,907
Difference, 1951-1988/90	−278,657		+140,679	
Change (ha yr^{-1})	7,531		3,607	

[a] Fisheries Gazette of the Philippines (data from the Bureau of Fisheries and Bureau of Forestry), 1952
[b] National Mapping and Resource Information Authority, 1988
[c] Fisheries Profile of the Philippines. Bureau of Fisheries and Aquatic Resources, Department of Agriculture, 1990
[d] Of total area for mangroves and fishponds for Surigao in 1951, assume 30% Surigao del Norte (Region X) and 70% Surigao del Sur (Region XI).
[e] Of total area for mangroves and fishponds for Cotabato in 1951, assume 50% South Cotabato (Region XI), 35% Maguindanao (Region XII) and 15% Sultan Kudarat (Region XII).

thousands of acres of otherwise practically valueless land.' Pond development reached a peak of 5000 ha yr^{-1} in the 1950s and 1960s.

With the promulgation of the Fisheries Code which placed mangrove forests under the joint administration of the fisheries and forestry bureaus and disallowed their private ownership, and the Forestry Reform Code which required fishpond developers to retain mangrove buffer zones, pond construction slowed down to 800 ha yr^{-1} in the 1970s (Table 4). Nevertheless, the BFAR continued to list mangroves as 'Swamplands available for development' in its yearly *Fisheries Statistics* up until 1984, revealing its prodevelopment (and anticonservation) stand.

Commodities

As the area devoted to pond culture increased from 61 000 ha in 1940 to 223 000 ha in 1990, production improved by 16-fold from 15 900 mt valued at P7.6 million in 1938 (Philippine Census of 1939) to 267 800 worth P6.5 billion in 1990 (BFAR, 1990) (Table 3). For a long time, the milkfish *Chanos chanos* constituted over 95% of brackishwater pond harvests. As hatchery- produced seed and imported feeds became commercially available in the 1980s, the shrimp industry took off (Primavera, 1992) and the proportion of shrimp, mainly the giant tiger prawn *Penaeus monodon*, increased to 17.8% of total pond production in 1990 (BFAR, 1990) from only 0.8% in 1981 (Auburn University, 1993). Milkfish is a popular food fish for domestic consumption while shrimp is a cash crop grown for urban and export markets.

Another factor that explains the shrimp boom of the 1980s was public and private sector financing such as the 1984 Aquaculture Development Project of the Asian Development Bank (ADB) with a total US$21.8 million including loans for shrimp hatcheries

Table 4. Mangrove area and brackishwater culture pond area, increase and production in the Philippines, 1860-1990 (modified from Primavera, 1993).

Year	Mangrove area (ha)	Brackishwater culture ponds Total area (ha)	Increase ($ha\ yr^{-1}$)	Production (mt)	Remarks
1860	no data (n.d.)	0	762 (18160-1940)	n.d.	First pond recorded in 1863
1920	450,000	n.d.		n.d.	
1940	n.d.	60,998	1,176 (1941-50)	15,936 (1938)	
1950	418,382 (1951)	72,753	5,050 (1951-60)	25,464	*Fishpond boom* Fisheries bureau created; US$23.6 million for pond development
1960	365,324 (1965)	123,252	4,487 (1961-70)	60,120	
1970	288,000	168,118	811 (1971-80)	96,461	*Conservation phase* Natl. Mangrove Committee; Forestry Fisheries Decrees; 79,000 ha mangroves for preservation and conservation
1980	242,000	176,231	4,668 (1981-90)	135,951	*Shrimp fever* Commercial availability of of fry and feeds; $21.8 million ADB shrimp project
1990	132,500	222,907		267,814	

Sources: Brown & Fischer, 1920; Philippine Census, 1939; BFAR, 1970, 1980, 1990; BFD, 1970, 1980; Siddall *et al.*, 1985; NAMRIA, 1988; Auburn University, 1993

and ponds (Primavera, 1993). Pond development again increased to 4700 ha yr^{-1} with much of the hectarage representing converted mangroves (Table 4). Siddall *et al.* (1985) and other observers note that the loss of mangroves in the Philippines and elsewhere has been facilitated by Western economic assistance given the high level of financing from such multilateral development agencies as the IBRD and ADB.

Ecological and socioeconomic effects

The single most important consequence of brackishwater pond culture in the Philippines has been the loss of mangroves; exploitation has been most extensive in Western Visayas and Central Luzon, the regions leading in pond hectarage (Table 3). Multiple-use resources are privatized (Bailey, 1988) and coastal communities especially fisherfolk become marginalized. Municipal fishermen numbered almost three-fold the persons employed in aquaculture but earned only P1986 monthly in 1987 compared to P3721 for the latter (Table 1).

Not only goods but also amenities such as buffer zone protection and flood control disappear along with mangroves. Thousands of deaths and damage to property inflicted by some 20 typhoons that visit the Philippines every year could be reduced by the presence of a coastal greenbelt. Conversion to pond culture (whether of milkfish or shrimp) depreciates the values of mangroves. In addition, intensive shrimp farming has unique consequences related to organic loading and use of chemicals (Primavera *et al.*, 1993).

Recommendations

Measures to mitigate widescale mangrove deforestation and to alleviate the grinding poverty of fishing communities include legislation and law enforcement, mangrove conservation and reforestation, and promotion of ecologically sound aquaculture.

Law enforcement and new legislation

Since the proconservation decade of the 1970s, numerous decrees, proclamations, and orders have been enacted to preserve the country's remaining mangroves (Primavera, 1993) but enforcement has been wanting. In spite of the protected status of mangroves, there has been a yearly loss of around 11 000 ha yr^{-1} from 1981 to 1990 (Table 4). The mangrove buffer zone required by law is absent in the majority of culture ponds throughout the archipelago. While this greenbelt is a forestry requirement, fishponds are under the jurisdiction of the fisheries bureau.

A significant factor is corruption at many levels of the government bureaucracy, particularly the agencies directly responsible for allocation, management or regulation of natural resources. To reduce reliance on rules and administrative discretion and remove corruption, the World Bank (1989) strongly recommended the imposition of fees for use of resources such as ponds to levels commensurate with economic rents, defined as total revenue minus all costs excluding the lease fee (Evangelista, 1992).

The economic rent of ponds converted from mangroves ranges from P515 ha^{-1} to P3296 ha^{-1}, way above the present pond lease fee of P50 ha^{-1} (Evangelista, 1992). Although pond operators have the capacity to pay higher fees, the industry successfully lobbied for a delay in the full implementation of an increase in the pond lease to P1000 ha^{-1} yr^{-1} (Primavera, 1993).

Conservation and reforestation of mangroves

The conservation of the country's remaining mangroves appears to be a losing battle because of ineffective law enforcement and the entry of powerful political and business interests in the pond industry.

In the absence of a major economic policy change that would increase resource use fees (World Bank, 1989), efforts should focus on the replanting of denuded areas. The earliest mangrove reforestation projects in the provinces of Bohol, Cebu and Negros Oriental were initiated by schoolchildren and the community (Yao, 1986). Reforestation is a major feature of the 1984 Central Visayas Regional Project which covers 650 ha in five nearshore fisheries sites. Another government project is located in Kalibo, Aklan in northern Panay with 50 ha of foreshore area planted into *Rhizophora* and nipa palm. Under contract reforestation, the government provides seedlings, technical assistance, cash and a Stewardship Contract which requires permanent forest cover in the area while allowing selective harvest of forest and fisheries products.

As of 1990, 8705 ha of mangroves have been replanted throughout the country with funding from the World Bank, Asian Development Bank, Overseas Economic Cooperation Fund of Japan, and the national government (PCAFNRRD, 1991). Still this area is insignificant compared to a total 300000 ha of mangroves destroyed since the 1920s. If rent or profits now accruing to pond owners can be collected by the government (through increased resource use fees), such collections can fund mangrove rehabilitation programs (World Bank, 1939; Evangelista, 1992). Reforestation should include underutilized or abandoned ponds and the perimeter dikes of operational ponds.

Environment-friendly aquaculture and fisheries

Artisanal fishing communities harvest a variety of food products from mangroves. With a little capital and labor input, income can be increased by the introduction of rock mounds known as *amatong* for fish aggregation, fish cage culture, crab fattening in cages, seaweed culture, and mollusc culture on stakes, rafts and ropes (Primavera, 1993). These technologies are amenable to small-scale, family level operations in mangrove-lined estuaries, bays and coves.

Moreover, culture ponds may not necessarily preclude the presence of mangroves. Dikes and tidal flats fronting early Indonesian *tambak* were planted with *Avicennia*, *Rhizophora* and other species for firewood, fertilizers (from decaying leaves), and protection from wave action (Schuster, 1952). At the turn of the century, similar minded Filipino pond operators planted rows of *Rhizophora*, *Sonneratia* and nipa palm especially along river banks for protection against wind, waves and soil erosion (Adams *et al.*, 1932). Ironically, their present-day counterparts do not retain or plant a mangrove greenbelt although it is now required by law

References

Adams, W, H R Montalban & C Martin, 1932 Cultivation of bangos in the Philippines Phil J Sci 47 1–37

Arroyo, C A, 1979 Flora of the Philippines mangrove Biotrop Spec Pub 10 33–44

Auburn University, 1993 Philippine Prawn Industry Policy Study Prepared for the Coordinating Council of the Philippines Assistance Program and U S Agency for International Development International Center for Aquaculture and Aquatic Environments, Auburn University

Bailey, C, 1988 The social consequences of tropical shrimp mariculture development Ocean Shore Mgmt 11 31–44

BFAR, 1970 Fisheries Statistics of the Philippines Bureau of Fisheries and Aquatic Resources, Department of Agriculture, Quezon City, Philippines

BFAR, 1971 Fisheries Statistics of the Philippines Bureau of Fisheries and Aquatic Resources, Department of Agriculture, Quezon City, Philippines

BFAR, 1980 Fisheries Statistics of the Philippines Bureau of Fisheries and Aquatic Resources, Department of Agriculture, Quezon City, Philippines

BFAR, 1990 Fisheries Profile of the Philippines Bureau of Fisheries and Aquatic Resources, Department of Agriculture, Quezon City, Philippines

BFAR, 1991 Fisheries Profile of the Philippines Bureau of Fisheries and Aquatic Resources, Department of Agriculture, Quezon City, Philippines

BFD, 1970 Forestry Statistics of the Philippines Bureau of Forest Development, Department of Environment and Natural Resources, Quezon City, Philippines

BFD, 1980 Forestry Statistics of the Philippines Bureau of Forest Development, Department of Environment and Natural Resources, Quezon City, Philippines

Brown, W H & A F Fischer, 1920 Philippine mangrove swamps In W H Brown (ed), Minor Products of Philippine Forests, 1 Bureau of Printing, Manila 9–126

Camacho, A S & T Bagarinao, 1986 Impact of fishpond development on the mangrove ecosystem in the Philippines In Mangroves of Asia and the Pacific Status and Management Tech Rep UNDP/UNESCO Research and Training Pilot Programme on Mangrove Ecosystems in Asia and the Pacific (RAS/79/002) Natural Resources Management Center and National Mangrove Committee, Ministry of Natural Resources, Philippines 383–405

Carbine, W F, 1948 Bangos culture in the Philippines Prog Fish Cult 10 187–197

Evangelista, D L, 1992 Management of mangrove areas in Calauag Bay, Quezon Province, Philippines Naga ICLARM Q 15 47–49

Gedney, R H, J M Kapetsky & W W Kuhnhold, 1982 Training on assessment of coastal aquaculture potential, Malaysia (regional) SCS/GEN/83/35 South China Sea Fisheries Development and Coordinating Programme, Manila, Philippines

Hamilton, L S & S C Snedaker (eds), 1984 Handbook for Mangrove Area Management United Nations Environment Programme, Kenya and Environment and Policy Institute, East-West Center, Hawaii, 123 pp

Herre, A C W T & J Mendoza, 1929 Bangos culture in the Philippine Islands Phil J Sci 38 451–509

Hora, S L & T V R Pillay, 1962 Handbook on fish culture in the Indo-Pacific Region FAO Fish Bio Tech Pap 14 1–204

Ibon, 1989 A fishy predicament Ibon Facts and Figures 12 1–3

Juanico, M B, 1983 Metro Manila a travel and historical guide M J Editorial Consultants, Quezon City

Macnae, W, 1974 Mangrove forests and fisheries IOFC/DEV/74/34 Food and Agriculture Organization, Rome, 35 pp

Martosubroto, P & M Naamin, 1977 Relationship between tidal forests (mangroves) and commercial shrimp production in Indonesia. Mar Res Indonesia 18 81–86

National Mapping Resource and Information Authority (NAMRIA) 1988

PCAFNRRD, 1991 The Philippines recommends for mangrove production and harvesting Philippine Council for Agriculture, Forestry and Natural Resources Research and Development, Department of Science and Technology, Los Banos, Laguna Forestry Research Series No 74, 96 pp

Primavera, J H, 1992 Prawn/shrimp culture industry in the Philippines In A Fast & L J Lester (eds), Marine Shrimp Culture Principles and Practices Elsevier, Amsterdam 701–728

Primavera, J H, 1993 A critical review of shrimp pond culture in the Philippines Rev Fish Sci 1 151–201

Primavera, J H, C L Pitogo, M R dela Pena & J M Ladja, 1993 A survey of chemical and biological products used in intensive prawn farms in the Philippines Mar Poll Bull 26 35–40

Radcliffe, L, 1912 Notes on pond-culture in the Philippines Trans am Fish Soc 1911 289–293

Saenger, P, E J Hegerl & J D S Davie, 1983 Global Status of Mangrove Ecosystems IUCN Commission on Ecology Papers No 3, Gland, Switzerland

Schuster, W H, 1952 Fish culture in the brackish water ponds of Java IPFC Spec Pub 1 1–143

Siddall, S E, J A Atchue III & P L Murray, Jr, 1985 Mariculture development in mangroves a case study of the Philippines, Panama and Ecuador In J R Clark (ed), Coastal Resources Management Development Case Studies Renewable Resources Information Series, Coastal Management Pub No 3 Prepared for the National Park Service, U S Dept of the Interior, and the U S Agency for International Development Research Planning Institute, Inc, Columbia, South Carolina

Staples, D J, D J Vance & D S Heales, 1985 Habitat requirement of juvenile penaeid prawns In P C Rothlisberg, B J Hill & D S Staples (eds), Second National Prawn Seminar NSP2, Cleveland, Australia 47–54

Tesoro, F O, 1984 Traditional utilization of mangrove forests and management implications in the Philippines Bakawan 3 6–10

Tomlinson, P B, 1986 The botany of mangroves Cambridge University Press, 413 pp

Villaluz, D K, 1953 Fish Farming in the Philippines Bookman, Manila, 336 pp

World Bank, 1989 World Bank Country Report Philippines Environmental and natural resources management study The World Bank, Wash, D C, 173 pp

Yao, C E, 1986 Mangrove reforestation in Central Visayas Canopy Intl 12 6–9

Hydrobiologia **295**: 311–321, 1995.
Y. S. Wong & N. F. Y. Tam (eds), Asia-Pacific Symposium on Mangrove Ecosystems.

Mangroves as filters of shrimp pond effluent: predictions and biogeochemical research needs

A. I. Robertson[1] & M. J. Phillips[2]
[1] *Australian Institute of Marine Science, PMB No 3, Townsville MC Qld 4810, Australia*
[2] *Division of Food and Agricultural Engineering, Asian Institute of Technology, GPO Box 2754, Bangkok 10501, Thailand*

Key words: mangrove forest, nitrogen, nutrient budgets, phosphorus, shrimp pond effluent

Abstract

Preliminary estimates of the ratio of mangrove forest: shrimp pond area necessary to remove nutrients from shrimp pond effluent are made using budgets of nitrogen and phosphorus output for semi-intensive and intensive shrimp ponds combined with estimates of total net primary production in *Rhizophora*-dominated mangrove forests in tropical coastal areas. If effluent is delivered directly to mangrove forest plots, it is estimated that, depending on shrimp pond management, between 2 and 22 hectares of forest are required to filter the nitrogen and phosphorus loads from effluent produced by a 1 hectare pond. While such ratios may apply to small scale, integrated shrimp aquaculture - mangrove forestry farming systems, the variability in mangrove hydrodynamics makes it difficult to apply such ratios at a regional scale. Before mangroves can be used to strip shrimp pond effluent more research is required on the effects that high ammonia and particulate organic matter loads in pond effluent have on nutrient transformations in mangrove sediments and on forest growth.

Introduction

Mangrove systems trap, transform and export materials. Whether mangroves are sources or sinks of materials depends on several factors, induding the element or substance under consideration, soil and vegetation types, geomorphic setting, hydrodynamics and the time frame of observation (Woodroffe, 1992; Wolanski *et al.*, 1992: Alongi *et al.*, 1992; Boto, 1992). Mangroves form an important part of the 'aquatic continuum' (Billen *et al.*, 1991) from river catchments to the sea, acting as filters for land-derived materials before they enter the ocean (e.g. Nedwell, 1974, 1975; Clough *et al.*, 1983), but it is difficult to generalize about the assimilative capacity of particular mangroves with respect to added nutrients.

Mangrove forests and their waterways have long been used as convenient sites for the disposal of sewage and wastewater in tropical countries. There have been several past attempts to evaluate the ability of mangroves to assimilate nutrients derived from sewage (Nedwell, 1975: Clough *et al.*, 1983; Boto, 1992), but recent losses of mangroves in many developing countries necessitates an assessment of an emerging serious threat to tropical mangroves - shrimp aquaculture. There has been a rapid expansion of marine and brackish water shrimp aquaculture in many tropical countries, particularly in South East Asia (Chua & Tech, 1990). Shrimp aquaculture presents an interesting challenge to coastal managers and scientists because most shrimp ponds are constructed in or adjacent to mangrove systems; large areas of mangroves have been lost to this form of development (Hatcher *et al.*, 1989; Barg, 1992). The more intensive shrimp aquaculture practices produce large quantities of effluent with high particle loads and ammonia concentrations (Lin, 1993), which in many cases pollutes the water used for future shrimp production. There are many problems associated with using mangrove areas for shrimp ponds, including direct loss of mangrove forests with subsequent loss of dependent capture fisheries, and development of acid-sulphate soils (e.g. Ong, 1982; Hong & San, 1993). In this paper, we ask whether retention of existing mangrove forests around sites of shrimp aquaculture production, or planting of new mangroves in mixed shrimp-forest farming models, can ameliorate the impacts of effluent derived from intensively managed shrimp ponds.

First, we provide a brief review of the range of shrimp farming practices currently used in coastal region of southeast Asia, particularly regarding potential effluent loads. Secondly, we review briefly some recent advances in mangrove biogeochemistry that are relevant to the issue of mangroves as effluent filters. Finally some predictions are made about the area of mangroves required to remove nutrients from the effluent produced by different shrimp pond farming practices. Acknowledging that many factors influence the nutrient-stripping capabilities of mangrove systems, we suggest research that is required to help aid management decisions about the balance between mangroves and shrimp ponds in tropical coastal regions.

Table 1. Nutrient and particle concentrations and bacterial cell densities and production in effluent water from intensively-managed shrimp ponds. For comparison, the range of values in pristine tidal and estuarine mangrove waters are also provided. Data from Kanit & Putha (1992), Lin (1993), Alongi *et al.* (1992), Robertson *et al.* (1993) and unpubl. data.

Variate	Intensive shrimp-pond effluent water	Pristine mangrove waterways
Salinity (‰)	10.0–35.0	0–37
NH_3 (μM)	1.97–73.15	0.10–1.42
NO_3 (μM)	0.05–1.54	0–11.75
PO_4 (μM)	0.53–4.21	0–5.26
TSS (mg l^{-1})	119–225	67–3312
Chl*a* (μg l^{-1})	20–250	0.2–5.07
Bacterial numbers (cells ml^{-1})	8.8–25.7 $\times 10^{6}$*	0.85–4.70 $\times 10^{6}$
Bacterial production (μg C l^{-1} d^{-1})	39–87*	0–18

* Based on semi-intensive ponds (Moriarty, 1986).

Shrimp aquaculture systems and effluent production

Shrimp aquaculture practices vary from traditional extensive culture systems practiced for centuries (Kungvankij, 1985) to the super-intensive culturing systems currently being used in Thailand. In traditional extensive systems, relatively large ponds (up to 50 ha, but usually less than 5 ha) are constructed in intertidal (usually mangrove) regions. Wild shrimp fry enter the ponds either during tidal water exchange, or are intentionally gathered from the wild and stocked directly in to ponds. Growth of individual shrimp in these systems relies on natural pond fertility, since fertilizers and feeds are not generally used. Retention time of water in the ponds varies, but may be up to two months. Yields of shrimp from such systems are low, in the range 100-300 kg ha^{-1} y^{-1} (Kungvankij, 1985). Since there is no supplementary feeding in these systems, their contribution to total nutrient loading in coastal waters is low. For instance, maximum concentrations of ammonia, nitrate and phosphate of 6.2, 1.0 and 1.8 μM recorded in extensive shrimp ponds in Dat Mui District of southern Vietnam (Hong & An, 1992) were not greatly different to those in adjacent mangrove creeks, or other pristine mangrove systems (see Table 1).

During the past two decades, improvements in extensive shrimp culture systems have led to increased shrimp production. Among improvements are liming of ponds prior to stocking to condition the soil, application of organic or inorganic fertilizers to enhance natural food production, increased stocking densities due to pumping, and removal of pests and predators through chemical techniques.

Semi-intensive pond systems have achieved higher yields of shrimp (14 tonnes ha^{-1} y^{-1}) through better control of pond size and construction to facilitate water exchange, higher stocking rates (often using hatchery reared post-larvae, particularly *Penaeus monodon*), application of fertilizers and regular supplementary feeding (often trash fish) and water exchange in ponds. Stocking density of postlarval and juvenile shrimp in semi-intensive systems range from 2–5 per m^2.

In the last decade there has been a rapid shift to intensive culture systems in many parts of southeast Asia (e.g. Chua & Tech, 1990). These systems combine high initial stocking densities (>30 per m^2) and feeding rates with pond management techniques that aim to maintain a water-column algal bloom beneficial to the behaviour and growth of *Penaeus monodon*, the preferred culture species. Depending on local conditions, feeding with pelletized food takes place at least once per day and there is regular water exchange (up to 30% of pond volume per day) near the end of the grow-out period. Yields from such systems are regularly 10 tonnes ha^{-1} y^{-1} (Lin, 1993).

Recently, 'super'-intensive culture systems have been used in Thailand. Mainly in response to very poor water quality in intake waters for ponds, and subsequent disease risks, these systems do not rely on high rates of water turnover. Rather, by treatments of the ponds with lime before and during the grow-out period, the pH of the pond water is maintained at a constant level. Together with fertilization, this procedure helps to maintain a constant water-column algal bloom in the ponds. Post-larval shrimp are stocked at densities >70 per m^2 and pelletized food is fed to prawns several times per day. Careful placement and constant use of mechanical stirrers ensures that waste products accumulate in the centres of ponds, and that aerobic conditions are maintained in a region bordering the pond walls where shrimp feed. Claims of yields of 20 t ha^{-1} y^{-1} have been made for such systems.

As the level of intensity in shrimp farming systems increases, so do the concentrations of nutrients in effluent water and the total load of nutrients entering coastal waterways (Table 1). The largest increases are in the concentrations of ammonia (by 1 to 2 orders of magnitude) and in the particulate loads in the forrn of chlorophyll *a* and bacterial cell densities (Table 1). This is not surprising, considering that algal blooms are maintained within the ponds. With the high levels of nutrients available from the breakdown of added food pellets and fertilizer, these blooms are not nutrient limited, and the decomposition of organic matter and the excretion by shrimp and other organisms in the pond maintain the high ammonia concentrations. Much of the food added to intensive ponds supports the growth of water-column bacteria; bacterial numbers and production are an order of magnitude greater than in mangrove waterways (Table 1).

Not all excess nutrients in semi-intensive and intensive shrimp ponds are in the dissolved or suspended particulate phases. Most accumulate in pond sediments (see Table 4). The high organic content of pond sediments supports high bacterial cell densities and production (Moriarty, 1986). There are no published records of the nutrient concentrations in pond sediments at harvest-time in shrimp ponds, but our budgeting exercises (see later) indicate that sediments must be a major site of nutrient accumulation.

The usual management practice following harvest in semi-intensive and intensive shrimp ponds is to remove the organic-rich sediments prior to initiation of the next production cycle. Pond sediments are either stored in settlement area, spread on supra-tidal waste dumps or placed in adjacent waterways.

Biogeochemistry of mangrove systems: potential for nutrient filtering

There have been two recent reviews of nutrient cycles in mangrove-dominated systems (Alongi *et al.*, 1992; Boto, 1992) and we wish to highlight some of the most recent research findings that are relevant to the preliminary models of the nutrient stripping capabilities of mangrove systems which we construct later in the paper.

Nutrient requirements for growth of Rhizophora *forests*

There have been previous estimates of the nutrient requirements to support primary production in *Rhizophora*-dominated forests (e.g. Clough *et al.*, 1983; Boto & Wellington, 1988). As Clough (1992) has pointed out, no study has yet measured total net primary production in a natural or managed mangrove forest. While estimates of total net above-ground production (litter fall + wood accumulation) are available for some mangrove forests, there are no published estimates of root production, or estimates of release of dissolved organic material from living mangrove roots.

Estimates of wood accumulation in *Rhizophora* forests vary from 6–45 t ha^{-1} y^{-1}, and average 20 t ha^{-1} y^{-1} (see review by Clough, 1992). Litter fall in the same forests averages 8 t ha^{-1} y^{-1} (Table 2).

Root biomass is very high in mangrove forests (Komiyama *et al.*, 1987; Gong & Ong, 1990) when compared to terrestrial forests (Vogt *et al.*, 1986). A recent study of fine root turnover rates in *Rhizophora* spp. forests of northeastern Australia (Robertson and Dixon, 1993 and manuscript in preparation) has shown that fine root production is equivalent to the total above-ground net production. There may therefore be 20+8 = 28 t ha^{-1} y^{-1} of fine root production in typical *Rhizophora* forests (Table 2). By converting dry mass production to nitrogen (N) and phosphorus (P) equivalents (from Clough *et al.*, 1983; Spain & Holt, 1980; Gong & Ong, 1990), it is possible to estimate N and P requirements to support total net forest production of 219 kg N ha^{-1} y^{-1} and 20 kg P ha^{-1} y^{-1}, respectively (Table 2).

These estimates are likely to be very conservative if used to calculate the ratio of forest:shrimp pond necessary to strip nutrients from aquaculture effluent because they are based on an average of natural and managed forests. Plantations of *Rhizophora* can be managed to

Table 2. Average values for annual production of different components of humid, tropical *Rhizophora*-dominated forests in terms of dry mass, nitrogen and phosphorus (for sources, see text).

Component of production	Dry mass Production ($t\ ha^{-1}\ y^{-1}$)	Tissue Nutrients N (%)	P (%)	Nutrients required for primary production ($kg\ N\ ha^{-1}\ y^{-1}$)	($kg\ P\ ha^{-1}\ y^{-1}$)
Litterfall	8	1.00	0.10	80	8
Wood accumulation	20	0.12	0.03	24	6
Root production	28	0.41	0.02	115	6
TOTALS	56			219	20

increase tree growth rates (Haron, 1981), and mixed shrimp pond plantation farming system (e.g. Hong & San, 1993) may have greater production potentials than presently realised. In addition, in response to fertilization, *Rhizophora* trees not only increase their growth rates, but also exhibit increased nutrient concentrations in their tissues, thus enhancing nutrient removal from the environment (Clough *et al.*, 1983).

On the other hand, canopy production is lost from trees as litter fall, and litter may or may not be retained within forests, depending on the degree of tidal flushing and leaf burial by crabs (Robertson *et al.*, 1992). Inclusion of litter fall in total net production estimates may overestimate the nutrient stripping capabilities of *Rhizophora* forests.

Finally, it is worth noting that mangrove forests can only be used as nutrient filters if mass accumulated during growth is harvested (for timber, firewood or charcoal), or at least retained or recycled within mangrove sediments (as is the case with fine root production).

Sediment bacteria

Bacteria in mangrove sediments are more productive than in temperate marine sediments, and they play an essential role as sinks for dissolved and particulate nutrients (Alongi, 1988a; Alongi, 1993a). Bacterial productivities ranging from 0.2 to 5.1, averaging 1.5 g $C\ m^{-2}\ d^{-1}$, have been recorded in *Rhizophora* forests in tropical Australia (Alongi, 1988a) and on mudflats adjacent to mangrove forests (Moriarty, 1986; Alongi, 1988b, 1991). These highly productive populations are able to sequester much of the dissolved nutrients available in sediment porewaters and in overlying waters. Flux chamber experiments conducted by Stanley *et al.* (1987) and Boto *et al.* (1989) in northern Australia, revealed negligible DON (as amino acids) and DOC flux in or out of the sediment, despite very large concentration differences between porewaters (high levels) and overlying waters (low levels). However, when sediments were poisoned to kill bacteria, 27–69 mg $N\ m^{-2}\ d^{-1}$ and 0.4–2.4 g $C\ m^{-2}\ d^{-1}$ was liberated from sediments to the water. It was estimated that such fluxes support between 9 and 38% of the nitrogen and carbon requirements for sediment bacterial production. Alongi (1991) performed similar flux studies on bare mud banks adjacent to mangrove forests in the Fly Delta of Papua New Guinea. He found that in all cases there were fluxes of nutrients (DOC, dissolved inorganic nitrogen and phosphorus, dissolved organic nitrogen and phosphorus) into the sediments from overlying waters, indicating that sediments and their microbiota are sinks for dissolved material.

Work in mangrove forests and adjacent mudflats shows a significant positive correlation between sediment bacterial growth rates and dissolved organic phosphorus in porewaters and total phosphorus (Alongi, 1991; Alongi, 1993b). However, the difficulty of conducting nutrient enrichment experiments with intertidal sedimentary bacteria (Alongi, 1993a) precludes definitive tests of nutrient limitation of bacterial growth, although this is likely.

Sewage or aquaculture effluent added to mangrove areas may be sequestered and processed by sediment bacteria (Nedwell, 1974; Moriarty, 1986; Alongi, 1993a). However, processing efficiency tends to decrease with increasing input. High organic loadings usually shift balanced aerobic-anaerobic systems to complete anaerobiosis. Anaerobic systems are less efficient in nutrient recycling. Little work has been done on anaerobic bacterial processes in mangrove sediments. Rates of sulphate reduction in tropical mangroves are similar to those in temperate intertidal sediments, but lower than in saltmarsh soils (e.g. Kristensen *et al.*, 1991), owing to moderate redox levels and high rates of aeration of soil by burrowing crabs. However, rapid turnover of the NH_4 pool in some mangrove sediments indicates subsurface activity of other (non sulphate-reducing) anaerobes (Blackburn *et al.*, 1987).

Few estimates of denitrification are available for mangrove sediments. Nitrate discharged from a sewage treatment plant in Fiji mangroves, and not taken up by phytoplankton, decreased by 30% by the time it passed through the estuary, with estimated reduction rates of 26.2 - 87.6 mg N m^{-2} d^{-1} (Nedwell, 1975). By comparison, in an unpolluted mangrove system in tropical Australia, the mean rate of denitrification was 0.18 mg N m^{-2} d^{-1} (Alongi *et al.*, 1992).

Mangrove sediment chemistry

When soluble reactive phosphorus is added to soils, it is usually rapidly immobilized by adsorption reactions depending to a large extent on the soil clay mineralogy, iron content and redox status (Holford & Patrick, 1979). Additions of 400 kg P ha^{-1} (in the form of superphosphate) over 1 year to mangrove sediments in northern Australia increased the concentration of plant-available phosphorus by a factor of 10 (Boto & Wellington, 1988). However, highly reduced mangrove soils have far less capacity to absorb phosphorus, owing to the greater solubility of iron sesquioxides (Boto, 1992). Adsorption maxima for mangrove soils may range from 250–700 μg P g^{-1} dry weight (Clough *et al.*, 1983; Boto, 1992) and are lower in soils which have naturally high P concentrations, or which have received high nutrient effluent for years.

Total nitrogen levels in pristine mangrove sediments range from 600–2000 μg N g^{-1} dry weight, mainly in the form of organic nitrogen. Ammonium is the main forrn of inorganic nitrogen in the mostly anaerobic mangrove soils. The availability of cation exchange sites in mangrove soils is often low because of the generally high levels of sodium in sediments. The result is that most ammonium occurs in the pore-water dissolved phase in concentrations ranging from 0–760 μM (Clough *et al.*, 1983; Alongi *et al.*, 1992). The loss of porewater ammonium from mangrove soils depends on the degree of soil permeability (low loss with heavy clays), the degree of tidal flushing (higher losses with higher flushing), the degree of denitrification occurring in the sediments and the rate of uptake by plants (Clough *et al.*, 1983; Alongi *et al.*, 1992). Boto (1992) queried whether mangrove sediments could be viewed as long term sinks for nitrogen added from nutrient pollution, if most effluent was delivered in the form of ammonia rather than nitrate. In the case of high nitrate concentrations, Nedwell (1975) has argued that denitrification could significantly reduce the nitrate levels in treated sewage effluent applied to mangrove soils.

Effluent loads from coastal shrimp ponds

Enough data are available to prepare some preliminary budgets for nutrient loads in the effluent derived from different types of shrimp ponds (Tables 3 and 4). Because there are no data available for DOC in shrimp pond effluent we have restricted our budgeting exercise to nitrogen (N) and phosphorus (P).

Based on recent interviews with farmers and managers in Tra Vinh Province of southern Vietnam, we were able to construct approximate N and P budgets for a semi-intensive shrimp pond managed in the way detailed in Table 3. Unfortunately, no data were available on the nutrient concentration in pond inflow or effluent water. For the purpose of this exercise we assumed that N and P concentrations in intake water were similar to those for dissolved inorganic materials measured in creeks and improved-extensive culture systems of nearby Minh Hai Province. Hong & An (1992) reported concentrations of 0.174 mg N l^{-1} and 0.054 mg P l^{-1} for creek water, with pond concentrations being approximately twice those in creek water. These are likely to be underestimates for semi-intensive ponds, but are the best figures available at present. The budget (Table 3) shows that the major input of N and P to the ponds is via trash fish used as shrimp feed. There appears to be a food conversion ratio (food input/shrimp yield) of 10. The harvest of shrimp is a minor part of the N and P budgets (Table 3). The total annual effluent load of N from a 1 ha semi-intensive pond is estimated to be 565 kgN, of which dissolved inorganic N in pond discharge water is estimated to be 23% of the total (Table 3). The remaining N would be in the particulate form, and presumably in the pond sediments that would be removed at the end of each culture period (i.e. 0.5 y). Total annual P load in effluent is estimated to be 70 kgP, of which 57% is estimated to be in the dissolved inorganic form in discharge waters.

Our budgets for intensive shrimp ponds in north-eastern Thailand are likely to be more reliable, as they are based on a 5 month detailed study of pond inputs

Table 3. Nitrogen and phosphorus budgets for semi-intensive shrimp (*Penaeus merguiensis* and *P. monodon*) ponds. Based on information collected from 3 sources in Tra Vinh Province, Vietnam, August 1993 (A. Robertson and M. Phillips, unpubl. data). Remainder component of outputs is the difference between total inputs and (shrimp harvest + discharge water) outputs.

Pond management		
Stocking density	4 juv m^{-2} (2 g FW m^{-2})	
Culture period	0.5 y	
Size of pond	1.0 ha	
Depth of pond	1.0 m	
Feed input (trash fish)	10,000 kg FW ha^{-1} y^{-1}	
Yield of shrimp	1,000 kg FW ha^{-1} y^{-1}	
Water turnover	10% per day	
Fertilizer	No	
Nutrient concentrations		
N content of shrimp	2.94% FW	
P content of shrimp	0.19% FW	
N content of feed (trash fish)	5% FW	
P content of feed (trash fish)	0.5% FW	
Budgets (for 1 ha pond)		
	N (kg ha^{-1} y^{-1})	P (kg ha^{-1} y^{-1})
Inputs		
Feed	500	50
Pond inflow	64	20
Shrimp juveniles	1	trace
Total input	565	70
Outputs		
Shrimp harvest	29	2
Pond discharge water	128	40
Remainder (sediment)	408	28
Total output	565	70

and outputs (Table 4). The inputs of N and P to these ponds via feed pellets are 3 and 8 times higher than inputs of N and P to semi-intensive ponds from shrimp feed. Total annual effluent output from 1 ha of intensive ponds is 1570 kgN and 433 kgP, most of which (87% for N, 91% for P) is in the form of pond sediments at the end of the harvest periods (Table 4). Total effluent loads produced by intensive shrimp ponds are 3 and 8 times greater than semi-intensive ponds for N and P, respectively (Tables 3 and 4).

Area of mangrove needed to remove N and P from effluent

Using estimates of the N and P required for primary production in an average, humid tropical *Rhizophora* forest (Table 2) as a guide, we have attempted to provide a rough 'rule-of-thumb' for the area of mangrove forest that would be necessary to remove N and P from the effluent of shrimp ponds (Table 5). For semi-intensive ponds 2 to 3 ha of forest would be required for each 1 ha pond while, for intensive ponds, up to 22 ha of *Rhizophora* forest might be needed.

Table 4. Nitrogen and phosphorus budgets for intensive shrimp (*Penaeus mondon*) ponds. Based on a study of five ponds in Chantaburi Province, Thailand, January May 1993 (M. Phillips, unpubl. data). Remainder component of outputs is the difference between total inputs and (shrimp harvest + discharge water) outputs.

Pond management		
Stocking density	52 post larvae m^{-2}	
Culture period	122 d	
Size of pond	0.4 ha	
Yield of shrimp	13.79 t FW ha^{-1} y^{-1}	
Feed input (pellets)	26.42 t FW ha^{-1} y^{-1}	
Fertilizer	Yes (see below)	
Nutrient concentrations		
N content of shrimp	2.94% FW	
P content of shrimp	0.19% FW	
N content feed	7.07% FW	
P content feed	1.64% FW	
Pond Budgets (normalized to 1.0 ha pond size)		
	N (kg ha^{-1} y^{-1})	P (kg ha^{-1} y^{-1})
Inputs		
Feed	1868	433
Inflow water	43	8
Fertilizer	62	17
'Tea seed cake'	1	1
Shrimp stock	1	trace
Total input	1975	459
Outputs		
Shrimp harvest	405	26
Pond discharge water	199	39
Remainder (sediment)	1371	394
Total output	1975	459

Assumptions and implications for management and research

To make these first-order estimates we assumed that 1) effluent is delivered directly and evenly dispersed in the mangrove forest; 2) all N and P is available for plant uptake; 3) plant uptake is the only sink for effluent N and P, and; 4) the average values for *Rhizophora* forest production are appropriate for a general model. Of these assumptions, the last is the least critical. High variation in mangrove primary production is caused mainly by the degree of aridity (which controls soil wetness and salinity) and air temperature (Clough, 1992). Shrimp aquaculture is centred on the humid tropics (Chua & Tech, 1990) and our estimates of average *Rhizophora* production come from such regions.

Not all N and P in pond effluent will be available for primary production. Depending on the nature of the receiving sediments, a proportion of effluent P would be immobilized as iron, calcium or aluminium phosphates (see earlier). Likely, some P would already be adsorbed on sediments on pond bottoms, before removal. If some effluent P becomes bound in the sediments, our estimates of forest area required for nutrient filtering are overestimates. The most serious (perhaps simplistic) assumption concerns delivery of effluent to mangrove forests. As Clough *et al.* (1983) and Boto (1992) have pointed out, delivery of effluent to mangrove forests, rather than mangrove creeks appears to be the most preferable option owing to the

Table 5. First-order estimates of the ratio of *Rhizophora* mangrove forest area to shrimp pond area that would be required to remove nitrogen and phosphorus loads produced during the operation of semi-intensive and intensive shrimp pond operation. Based on figures in Tables 2-4. See text for assumptions and drawbacks with these estimates.

Element in effluent	Pond management type	
	Semi-intensive	Intensive
Nitrogen	2.4:1	7.2:1
Phosphorus	2.8:1	21.7:1

influence of high nutrient loads on aquatic food chains. In the integrated shrimp-forest farming models currently used in some parts of South East Asia (e.g. Hong & San, 1993), with their mixture of *Rhizophora* forest and semi-intensive shrimp culture, it may be possible to engineer delivery of effluent water and sediment from ponds to adjacent forests. Finally, the potential for high rates of denitrification in mangrove sediments (see earlier) imply that our estimates of removal of nitrogen from effluent by mangrove forests is likely an underestimate.

Can our rule-of-thumb on forest area:pond area (Table 5) be scaled up to the regional level (i.e. areas of 10^2 km^2) to provide guidelines for coastal managers who seek the correct balance of shrimp pond and mangrove forest areas that will allow sustainable shrimp farming? A major source of variability in estimating appropriate ratios of forest to pond area on larger scales will be the residence time of water in mangrove waterways, particularly if effluent is delivered directly to mangrove creeks, as is the case in many semi-intensive and intensive shrimp farming operations.

The residence time of water in mangrove systems is controlled by a number of synergistic factors (Fig. 1). Water will have long residence times in the upper reaches of intricately branched creek systems in very flat, wide mangrove forests, made longer by the dense root systems of *Rhizophora* forests (Fig. 1). In the dry season in northern Australia, water may be trapped for days to weeks in such tidal mangrove creeks (Wolanski & Ridd, 1986). Even in the wet season, when estuaries are flushed, the residence time in secondary creeks may be measured in days (Wolanski & Ridd, 1986). The area of forest required to strip nutrients from shrimp pond effluent will therefore increase with increasing complexity of mangrove waterways, since large loads of effluent may accumulate in these regions. Conversely, in short, unbranched creeks with a relatively narrow band of mangroves, residence time of water will be short, and effluent may be delivered rapidly to the coastal oceans if this is desirable. However, depending on coastal topography, material delivered from mangrove waterways may be trapped in embayments by coastal boundary layers (Wolanski & Ridd, 1990) and thus pollute inshore regions.

In our calculations, we have also assumed that there are no other sources of N and P loading other than shrimp pond effluent. Clearly, this is not usually the case in the coastal regions of south east Asia, where a cocktail of agricultural, industrial and urban waste is often delivered to nearshore habitats (e.g. Chua *et al.*, 1989).

We have based our arguments on budgets of N and P, but have not considered the organic carbon load in shrimp pond effluent. As far as we know there are no data on DOC concentrations in shrimp pond effluent. POC concentrations are likely to be high, given the high concentrations of chlorophyll a and bacteria in the effluent from intensive ponds (Table 1). This organic matter will have a low C:N ratio, and would therefore be likely to support rapid rates of bacterial turnover in mangrove sediments and waterways. This is turn may have major implications for oxygen levels in mangrove sediments and water, which are characterized by naturally low oxygen concentrations (Boto, 1992). In Table 6, we have summarized some of the questions that require further research if mangroves are to be used in stripping nutrients from effluent. As with an earlier review of mangroves and sewage, (Clough *et al.*, 1983), we stress the need for monitoring of existing shrimp pond effluent systems and the short- and long-term impacts they have had on sediments, fauna and forest production.

Conclusions

If the effluent from shrimp ponds could be delivered directly to mangrove forests, such as in small-scale, integrated shrimp-forest farming systems, we estimate that somewhere between 2 to 22 hectares of forest would be required to filter the nitrogen and phosphorus from pond waste, depending on the pond management regime. However, the probable impact of high concentrations of ammonia and particulate organic matter on mangrove sediments and plant life requires further

Table 6. Research requirements for sustainable management of mangroves to strip nutrients from shrimp pond effluent. We have separated mangrove plantations managed purely for timber or firewood production in small scale farming systems (= small scale plantations) and forests, natural or otherwise, which support a variety of system functions.

Type of mangrove system	Research questions
Small scale plantations	
	• what loads of ammonia and particulate and dissolved organic matter (DOM and POM) cause complete anaerobiosis of mangrove sediments, and tree mortality?
	• what is the long term 'assimilative capacity' of sediments for phosphorus?
	• how do different sediment types influence the retention and transformation of effluent nitrogen?
	• what silviculture practices will maximize forest yields, and nutrient uptake rates?
Forests (on a regional scale)	
	• what loadings of ammonia, DOM and POM have significant negative effects on benthic micro-, meio- and macrofauna?
	• what are the indirect effects of the loss of burrowing fauna on system functions?
	• what combinations of effluent loads and hydrodynamic regimes cause anoxic conditions in mangrove waterways?

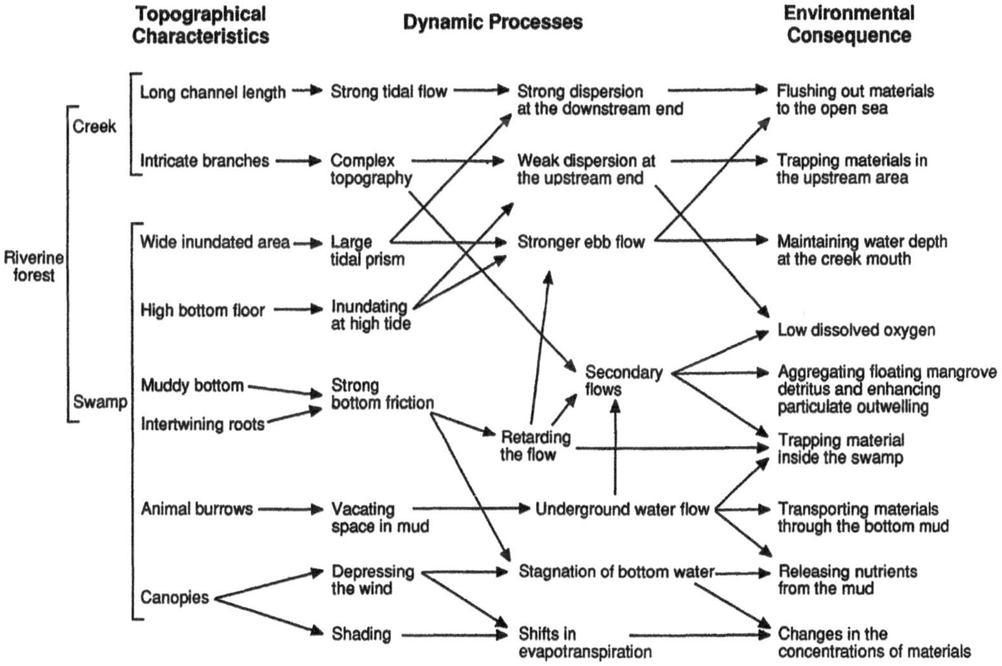

Fig. 1. How topographical features control physical processes and the trapping of materials in mangrove ecosystems (waterways plus forested swamp areas). Based on Wolanski *et al.* (1992).

research before it is known whether such practices are sustainable (Table 6)

Whether the ratios of forest area shrimp pond area can be used on a more regional scale, to indicate the correct balance between areas devoted to shrimp aquaculture in mangrove ecosystems depends, to a large degree, on two factors the hydrodynamics in mangrove waterways and the nutrient loading from other pollution sources As with previous reviews of the role of mangroves in the tertiary treatment of sewage (Clough *et al*, 1983, Boto, 1992), we believe that if mangroves are to be used to strip off aquaculture-derived nutrients, where possible, effluent should be delivered to higher intertidal mangrove areas, rather than directly to mangrove waterways This would minimize the effects on food chains and fisheries in tidal waters

Acknowledgements

We would like to thank the Chaiyong Limthongkul Foundation for funding a trip by Alistar Robertson to Thailand in February 1993 and the Australian Centre for International Agricultural Research for funding for Alistar Robertson and Michael Phillips to visit Vietnam in August 1993 Alistar Robertson would like to thank the organizers of the Asia-Pacific Symposium on Mangrove Ecosystems for the invitation to present this paper Dan Alongi and John Benzie read and provided useful comments on an earlier draft of the paper This is contribution no 696 from the Australian Institute of Marine Science

References

Alongi, D M, 1988a Bacterial productivity and microbial biomass in tropical mangrove sediments Microb Ecol 15 59–79

Alongi, D M, 1988b Microbial-meiofaunal interrelationships in a tropical mangrove and sandflat ecosystem J mar Res 46 349–365

Alongi, D M, 1991 The role of intertidal mudbanks in the diagenesis and export of dissolved and particulate materials from the Fly Delta, Papua New Guinea J exp mar Biol Ecol 149 81–107

Alongi, D M, 1993a The role of bacteria in nutrient recycling in tropical mangrove and other coastal benthic ecosystems Hydrobiologia in press

Alongi, D M, 1993b The relationship of forest type on microbial nutrient relationships in tropical mangrove sediments J exp mar Biol Ecol in press

Alongi, D M, K G Boto & A I Robertson, 1992 Nitrogen and phosphorus cycles In A I Robertson & D M Alongi (eds), Tropical Mangrove Ecosystems American Geophysical Union Press, Washington 251–292

Barg, U C, 1992 Guidelines for the promotion of environmental management of coastal aquaculture development FAO Fisheries Technical Paper No 328, Rome, FAO, 122 pp

Blackburn, T H, D Christensen, A M Fanger, K Henriksen, H Iizumi, N Iversen & P Limpsaichol, 1987 Mineralization processes in mangrove and seagrass sediments In J Hylleberg (ed), Ao Yon a mangrove in the Andaman Sea Institute of Ecology and Genetics, University of Aarhus, Denmark 22–32

Billen, G, C Lancelot & M Meybeck, 1991 N, P and Si retention along the aquatic continuum from land to ocean In R F C Mantoura, J M Martin & R Wollast (eds), Ocean Margin Processes in Global Change John Wiley and Son Ltd 1944

Boto, K G, 1992 Nutrients and mangroves In D W Connell & D W Hawker (eds), Pollution in tropical aquatic systems CRC Press, Boca Raton 130–145

Boto, K G, D M Alongi & A L J Nott, 1989 Dissolved organic carbon-bacteria interactions at the sediment water interface in a tropical mangrove system Mar Ecol Prog Ser 51 243–251

Boto, K G & J T Wellington, 1988 Seasonal variations in concentrations and fluxes of dissolved organic and inorganic materials in a tropical, tidally dominated mangrove waterway Mar Ecol Prog Ser 50 151–160

Chua, T E & E Tech, 1990 Aquaculture in Asia Quo Vadis? In M M Joseph (ed), Aquaculture in Asia Asian Fisheries Society, Indian Branch, Mangalore 13–30

Chua, T E, J N Paw & F Y Guarin, 1989 The environmental impact of aquaculture and the effects of pollution on coastal aquaculture development in Southeast Asia Mar Pollut Bull 20 335–343

Clough, B F, 1992 Primary productivity and growth of mangrove forests In A I Robertson & D M Alongi (eds), Tropical Mangrove Ecosystems American Geophysical Union Press, Washington 225–250

Clough, B F, K G Boto & P M Attiwill, 1983 Mangroves and sewage a reevaluation In H J Teas (ed), Biology and ecology of mangroves Dr W Junk Publishers, The Hague 151–161

Gong, W K & J E Ong, 1990 Plant biomass and nutrient flux in a managed mangrove forest in Malaysia Estuar coast shelf Sci 31 519–530

Hatcher, B G, R E Johannes & A I Robertson, 1989 Review of research relevant to conservation of shallow tropical marine ecosystems Oceanogr mar Biol ann Rev 27 337–414

Haron, H A H, 1981 A working plan for the second 30 year rotation of the Matang Mangrove Forest Reserve, Perak Perak State Forestry Department, Ipoh, Malaysia, 115 pp

Holford, I C R & W H Patrick, Jr, 1979 Effects of redox potential and pH on phosphate removal from wastewater during land application Prog Water Techn 11 215–225

Hong, P N & L D An, 1992 Preliminary study on the environmental characteristics and the utilization of the coastal accretions of Ngoc Hien District - Minh Hai Province Report to State Committee for Sciences, Hanoi, 77 pp

Hong, P N & H T San, 1993 Mangroves of Vietnam IUCN, Bangkok, 173 pp

Kanit & Putha, 1992 Qualitative and quantitative properties of wastewater discharged from intensive shrimp ponds at Amphoe Ranod, Songkhla Province National Institute of Coastal Aquaculture, Department of Fisheries, Thailand, Technical Paper No 5, 26 pp

Komiyama, A, K Ogino, S Aksornkoae & S Sabhasri, 1987 Root biomass of a mangrove forest in southern Thailand 1 Estimation by the trench method and the zonal structure of root biomass J trop Ecol 3 97–108

Kristensen, E, M Holmer & N Bussarawit, 1991 Benthic metabolism and sulphate reduction in a south-east Asian mangrove swamp Mar Ecol Prog Ser 73 93–103

Kungvankij, P, 1985 Overview of penaeid shrimp culture in Asia In Y Taki, J H Primavera & J A Llobrera (eds), Proceedings of the first international conference on the culture of penaeid shrimp/prawns South East Asia Fisheries Development Centre, Iloilo 11–22

Lin, C K, 1993 Resource recovery from wastewater of intensive shrimp farming In Positive impacts of shrimp farming on the environment, coastal quality and socioeconomic conditions in Thailand Department of Fisheries Thailand Seminar, Songkhla Province, Thailand 126–136

Moriarty, D J W, 1986 Bacterial productivity in ponds used for culture of penaeid prawns Microb Ecol 12 259–269

Nedwell, D B, 1974 Sewage treatment and discharge into tropical coastal waters Search 5 187–190

Nedwell, D B, 1975 Inorganic nitrogen metabolism in a eutrophicated tropical mangrove estuary Wat Res 9 221–231

Ong, J E, 1982 Mangroves and aquaculture in Malaysia Ambio 11 252–257

Robertson, A I, D M Alongi & K G Boto, 1992 Food chains and carbon fluxes In A I Robertson & D M Alongi (eds), Tropical Mangrove Ecosystems American Geophysical Union Press, Washington 293–326

Robertson, A I, P A Daniel, P Dixon & D Alongi, 1993 Pelagic biological processes along a salinity gradient in the Fly Delta and adjacent river plume (Papua New Guinea) Cont Shelf Res 13 205–224

Robertson, A I & P Dixon, 1993 Separating live and dead fine roots using colloidal silica an example from mangrove forests Plant and Soil 157 151–154

Robertson, A I & P Dixon, in preparation Fine root mass, decomposition and production in three tropical mangrove forests

Spain, A V & J A Holt, 1980 The elemental status of the foliage and branchwood of seven mangrove species from northern Queensland Divison of Soils Divisional Report No 49 CSIRO, Australia 8 pp

Stanley, S O, K G Boto, D M Alongi & F T Gillan, 1987 Composition and bacterial utilization of free amino acids in tropical mangrove sediments Mar Chem 22 13–30

Vogt, K A, C C Grier & D J Vogt, 1986 Production, turnover and nutrient dynamics of above- and belowground detritus of world forests Adv ecol Res 15 303–377

Wolanski, E, Y Mazda & P Ridd, 1992 Mangrove hydrodynamics In A I Robertson & D M Alongi (eds), Tropical Mangrove Ecosystems American Geophysical Union Press, Washington 43–62

Wolanski, E & P Ridd, 1986 Tidal mixing and trapping in mangrove swamps Estuar coast shelf Sci 23 759–771

Wolanski, E & P Ridd, 1990 Mixing and trapping in Australian tropical coastal waters In R T Cheng (ed), Residual currents and long-term transport Coastal and Estuarine Studies, Vol 38, Springer-Verlag, N Y 165–183

Woodroffe, C, 1992 Mangrove sediments and geomorphology In A I Robertson & D M Alongi (eds), Tropical Mangrove Ecosystems American Geophysical Union Press, Washington 7–42

Hydrobiologia **295**: 323–333, 1995.
Y. S. Wong & N. F. Y. Tam (eds), Asia-Pacific Symposium on Mangrove Ecosystems.

Conservation evaluation of nine Hong Kong mangals

M. W. Yipp[1], C. H. Hau[2] & George Walthew[1]
[1] *Department of Applied Biology & Chemical Technology, Hong Kong Polytechnic*
[2] *World Wide Fund for Nature, Hong Kong*

Key words: conservation evaluation, mangal, Hong Kong

Abstract

A comparative analysis of the conservation value of nine Hong Kong mangals spread across Hong Kong has been undertaken. The analysis is based on both biological and socio-economic attributes. Biological attributes used in the analysis include plant species richness, benthic and arboreal gastropod species richness, occurrence of rare species within each site, and site representativeness. A review of the past and present development and development plans around and within each site has been undertaken to assess the comparative degree of hazard to each mangal. The drastic decline of mangal habitats throughout Hong Kong highlights the need for conservation of what remains today. Priorities and strategies for conservation of these nine mangals are discussed with particular emphasis on the preservation of biodiversity and the management of these habitats as education and research sites.

Introduction

Mangals are essentially tropical coastal plant formations which typically occur on the estuaries of rivers that run over a shallow continental shelf (Tomlinson, 1986). Species diversity in equatorial regions is highest, and declines with increasing latitude. At 21°30′N, Hong Kong is part of the China coast and well within the cline of mangrove distribution of the Western Indo-Pacific region. The highest mangrove plant diversity of this region is found in Malaysia near the equator (Chapman, 1984) where 27 species occur. Species richness decreases northwards to just 1 species in southern Japan at 31°30′N (Walter & Breckle, 1986). Along the coast of China and closer to Hong Kong, 11 species have been recorded in Taiwan at 22–25°N (Walter & Breckle, 1986) and 6 species in Fujian at 23 °46′N (Lin & Wei, 1983). Morton & Morton (1983) listed 7 species of mangrove in addition to *Acanthus ilicifolius* which was considered as a mangrove-associate species. Thus, Hong Kong cannot be considered as depauperate in terms of mangrove species representation in relation to its geographical situation.

Historically, the largest areas of Hong Kong mangals have been found at Deep Bay and inside Tolo Harbour, but both places have been greatly altered over the past seventy years. The Deep Bay mangroves have been first reclaimed and used as *gei wais* (traditional shrimp farming ponds), most of which were later changed to fish ponds (Irving & Morton, 1988). Today, only the seaward fringe remains, representing approximately 15% of the original mangal. Part of this is now a conservation area managed by the World Wide Fund for Nature Hong Kong (WWF HK). The largest Tolo Harbour mangals have been destroyed and reclaimed into large town areas and a race course, and only two mangals of reasonable size are left. It was estimated in 1988 (Holmes, 1988) that there has been a 42% reduction of natural mangrove coastlines in Tolo Harbour. However, about 50 ha more of mangrove habitats will be destroyed by the building of a new airport and associated developments in the north shore of Lantau Island (Greiner Maunsell, 1991a, b; Consultants in Environmental Sciences (Asia), 1991; Mott McDonald, 1991a, b). In the meantime, the remaining mangrove habitats in Hong Kong are threatened by other urban developments.

The above events highlight the importance of conservation measures which must be taken to protect and sustain the remaining mangals in Hong Kong. Although Morton & Morton (1983) have written a general account of Hong Kong's mangals, there has been no comparative and systematic studies covering a diversity of sites, so that there is no baseline information on which to compare sites for the purpose of conservation evaluation and prioritization.

This work reports on data collected as part of a comparative study of nine Hong Kong mangals, and attempts to use these as a basis for assessing the comparative conservation potential of each site. Current conservation status, approved and proposed plans for development at each mangal and its hinterland are also reviewed, and a conservation strategy proposed for the nine sites.

Description of sites studied

Nine mangals from around the coast of Hong Kong were selected for study after a preliminary tour of sites containing mangrove vegetation. A number of very small stands (<0.5 ha) containing sparse trees were not included in this survey. These nine sites include three from the west coast and six on the east, separated by the island of Hong Kong and the Kowloon peninsula (Fig. 1). Of the three west coast sites, only Tsim Bei Tsui (TBT) was situated in the mainland; the other two (Tung Chung (TC) and Tai Ho Wan (THW)) being on the north shore of Lantau Island. The east coast sites were all on the mainland, and situated within enclosed bays and inlets, with the exception of Inner Port Shelter (IPS) in the south.

Materials and methods

The distribution of flora and gastropod fauna within each site was investigated by transect studies. A total of 5 transect lines were laid out in each site, covering the extent of mangrove vegetation from land to sea in a direction perpendicular to the coastline. For vegetation analysis, the transect was divided into 6 equally spaced zones, and a 3 m × 3 m quadrat placed in each zone. Within each quadrat, the incidence of mangrove trees and shrubs were recorded, together with a visual estimate of percentage canopy cover by each canopy species. Mean percentage cover occupied by each of the canopy species were calculated for each site. The incidence of vines and ferns in each site were recorded, though not quantitatively assessed. In each of the 6 zones within the transect, benthic gastropod species and numbers were recorded from 2 quadrats (0.5 m × 0.5 m), while arboreal gastropods on all surfaces (stems, leaves and exposed roots) of two trees of each species within the zone were also counted.

Assessment of species richness

A species richness score was calculated for each site by expressing species number present as a proportion of the total species encountered in this survey. This was performed separately for the flora and gastropod fauna, and then combined to give a site species richness rank.

Assessment of rarity

A number of rare plants and gastropods of Hong Kong mangals were used to compute the rarity rank of each of the nine sites for evaluation of the need for conservation. Species were considered to be rare if they occurred within the transects in 2 sites or less. Since these nine sites represented most of the total area of remaining mangrove stands in Hong Kong at the time of survey, species found in 2 sites or less would represent a rare occurrence for the geographical area of Hong Kong. However, several species fulfilling this criterion were not included as they have been recorded as common in habitat types other than mangrove stands around Hong Kong. These included the vine *Canavalia maritima* recorded as a common beach plant by Thrower & Tang (1976), and the gastropods *Nerita yoldii*, *Thais clavigera*, and *Melanoides tuberculata*. *N. yoldii* has been recorded from three boulder shores by Hill (1980), *T. clavigera* from 4 sites within Tolo Harbour alone (Taylor, 1980) and *M. tuberculata* from more than 10 sites (Yipp, 1991). Plant rarity in each site was ranked according to the number of rare species encountered, and gastropod rarity (R) for each site was computed using a modification of Jefferson's method (1984):

$$R = \sum 1/q_n$$

where q = number of quadrats containing a rare species; and n = number of rare species in the site.

Assessment of representativeness

In assessing the representativeness of each mangal, plants species were considered representative if they occurred in 5 sites or more. A 'representative plants score' was calculated for each site by expressing the number of representative plants present as a proportion of the total number of these. Likewise, a similar score was obtained for representative gastropods,

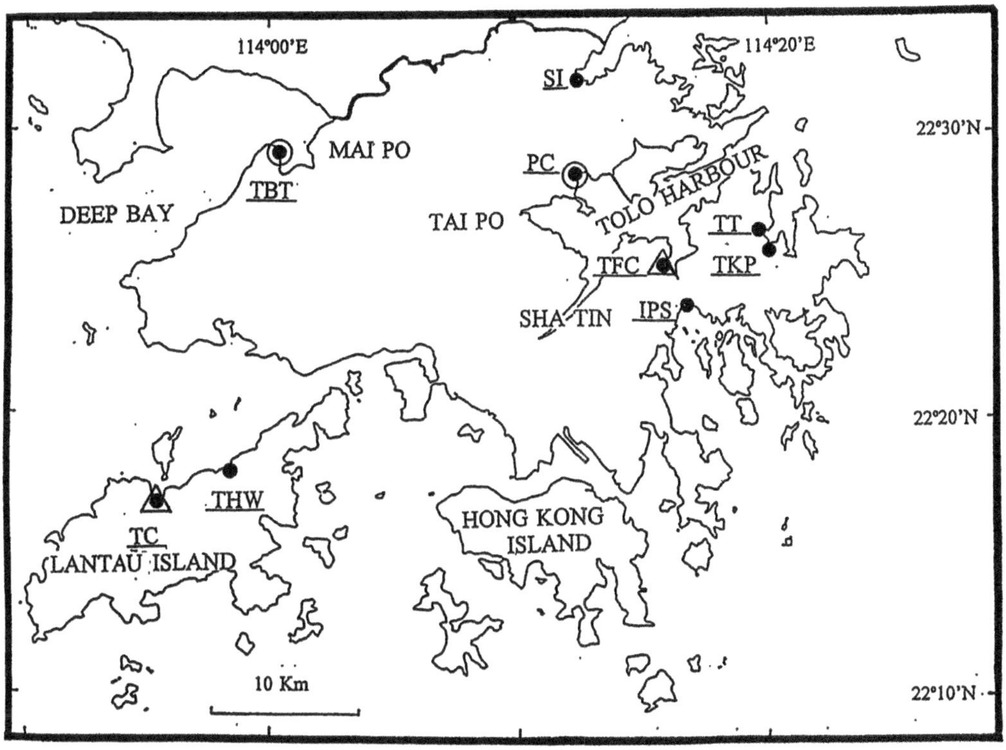

Fig. 1. Map of Hong Kong, showing the nine mangals in this study. An open circle around a closed circle indicates a SSSI, while an open triangle surrounding a closed circle a proposed SSSI.

which occurred in 5 sites or more. From the estimate of percent cover in the thirty 3 m × 3 m quadrats surveyed at each site, the mean % canopy occupied by each canopy species was calculated. The number of representative mangrove trees which had an overall mean canopy cover of >5% in the site was also expressed as a fraction of the total number of representative species, to give the 'tree canopy score'. The lower limit of 5% was used to take into account the normally low canopy fraction occupied by species characteristic of the landward and seaward fringes, and also to prevent exclusion of mangals which had a more open type of canopy (see below). Similarly, the proportion of representative gastropods which occurred at a density $>0.5\ m^{-2}$ were calculated to give the 'gastropod density score'.

To assess the typicalness of plant distribution patterns in each site, reference was made to Walthew and Yipp (in prep.) where analysis of mangrove plant distribution by multivariate methods showed that *Excoecaria agallocha* was a landward zone species, *Kandelia candel* a mesomangal species, and *K. candel*, *Aegiceras corniculatum* and *Avicennia marina* foreshore species. One point each was awarded to each transect for the following:

1. the presence of *E. agallocha* in the landward two quadrats followed by its absence elsewhere;
2. the dominance of the canopy by *K. candel* in the middle two quadrats;
3. the presence of either one of *K. candel*, *A. corniculatum* or *A. marina* in the seaward two quadrats.

The total points for each site was expressed as a fraction of the maximum possible 15 points, to give the 'transect typicalness score' for each site.

The 'mangal structure score' was also calculated with reference to Walthew and Yipp (in prep.) in which Hong Kong mangals were classified into two types: tall trees and closed canopy growing in silt/clay substrates, and short trees and open canopy growing on sandy substrates. A decreasing scale of 3 to 1 points were awarded to sites according to their degree of fit into this classification scheme, and the score calculated as a proportion of the maximum value.

Degree of hazard

A review of current town planning, development and conservation proposals in and around the nine mangals was conducted. Based on the likelihood of development, the direct and indirect effects of such developments and the current conservation status, the nine sites were each assigned a degree of hazard rating, which ranks the degree of threat to the survival of each site.

Results

Floral and gastropod species richness

Table 1 shows the distribution of mangrove vegetation across the nine sites surveyed. A total of 9 species of trees and shrubs, 4 vines and 1 fern are recorded. However, not one site was found to contain all 14 plants. The highest plant species richness was found at To Kwa Peng (13 species), and the lowest at Starling Inlet (5 species). The majority of sites contained 8 to 9 species. Of the mangrove trees and shrubs, *Kandelia candel*, *Aegiceras corniculatum* and *Avicennia marina* were found in all sites. However, *A. corniculatum* occurred in very low density in Starling Inlet, and failed to be included within the transects. Likewise, *A. marina* was not recorded in the transects at Starling Inlet, Inner Port Shelter and Tai Ho Wan. *K. candel* was the most locally abundant species within 5 sites, and was co-dominant with *A. corniculatum* in 3 sites. *Bruguiera gymnorrhiza* was the second most widespread species and occurred in 8 sites, while *Excoecaria racemosa* was found in 7, and *Lumnitzera racemosa* was restricted to only 3 sites. These last three species were rarely abundant within any site, the only exception being a high incidence of *B. gymnorrhiza* at Tai Ho Wan.

By reason of its low-growing habit, *Acanthus ilicifolius* did not take up any significant proportion of the canopy within the quadrats, although it was quite widespread throughout Hong Kong (7 sites). By contrast, *Heritiera littoralis* was only found at To Kwa Peng, while several young trees of a species of *Rhizophora* was found at Tsim Bei Tsui. As no reproductive organs had been developed in these young trees, their specific identity could not be confirmed. This genus has been documented as not occurring in Hong Kong by Morton & Morton (1983) and Catt (1986) although *R. mucronata* has been recorded in the Checklist of Hong Kong Plants (Hong Kong Herbarium, 1974).

Three species of vines were quite widespread over 5 to 7 sites, while *Canavalia maritima* and the fern *Acrostichum aureum* were restricted in their distribution to 1 and 2 sites respectively.

A total of 51 gastropod species were recorded from the nine mangals, belonging to 17 families. The numbers of species from each family are given in Table 2, together with the total species counts from each site. The majority of sites contained over 20 species, with Inner Port Shelter and Tai Ho Wan being exceptional in having only 8 species.

To Kwa Peng had the highest species richness rank (Table 3) contributed by a high count of both plants and gastropods, while the second and third rank of Tai Tan and Tsim Bei Tsui respectively was mainly contributed by high gastropod species counts. In contrast, the low ranking of Tai Ho Wan and Port Shelter were mainly due to low gastropod species counts, while the remaining sites had comparable floral and gastropod scores.

Floral and gastropod rarity

Table 4 shows the distribution of rare plants and gastropods among the nine sites. Among the mangrove flora, only *Rhizophora* sp. and *Heritiera littoralis* and the fern *Acrostichum aureum* fulfilled the rarity criteria, while 23 species of gastropods were rare. Of the latter, 6 species belonged to the family Assimineidae, and most of these were found at Tsim Bei Tsui, and were thus regionally rare in Hong Kong. 9 species were ellobiids and these were locally rare in a few sites, occurring in very low numbers frequently outside the transects. Since most of the mangrove flora were well represented across the nine sites, the number of rare species was small, and floral rarity contributed less to the overall site rarity ranking than rare gastropods. As a result, Tsim Bei Tsui ranked highest in rarity assessments, followed by To Kwa Peng and Tai Tan.

Mangal representativeness

Table 1 shows the frequency of occurrence of the flora in each of the nine sites. According to the criteria set out previously, 9 species were representative of Hong Kong mangals. These include *Excoecaria agallocha*, *Aegiceras corniculatum*, *Bruguiera gymnorrhiza*, *Kandelia candel*, *Avicennia marina*, *Acanthus ilicifolius*, *Caesalpinia crista*, *Derris trifolia-*

Table 1 Distribution of mangrove plants in nine Hong Kong mangals, showing mean percent cover of canopy by each species (+/- indicates presence or absence of species without estimate of canopy cover)

	Starling Inlet	Tai Tan	To Kwa Peng	Plover Cove	Three Fathoms Cove	Inner Port Shelter	Tung Chung	Tai Ho Wan	Tsim Bei Tsui
Trees and Shrubs									
Acanthus ilicifolius L	–	+	+	+	+	–	+	+	+
Lumnitzera racemosa Willd	–	–	+	3 6	11 6	–	–	–	–
Excoecaria agallocha L	–	7 8	3 4	10 6	4 9	4 7	9 8	–	17 0
Aegiceras corniculatum (L) Blanco	+	35 7	43 5	10 2	22 9	0 8	43 1	39 1	18 6
Bruguiera gymnorrhiza (L) Lamk	0 4	13 0	9 8	–	0 4	2 1	8 5	29 6	5 0
Kandelia candel (L) Druce	99 6	32 3	40 7	68 2	59 4	92 5	30 8	31 3	51 1
Rhizophora sp	–	–	–	–	–	–	–	–	+
Heritiera littoralis Dryand in Aiton	–	–	+	–	–	–	–	–	
Avicennia marina (Forsk) Vierh	+	11 2	2 6	7 4	0 8	+	7 8	+	8 3
Vines									
Caesalpinia crista L	–	+	+	–	–	+	+	+ –	
Canavalia maritima (Aubl)	–	–	+	–	–	–	–	–	–
Derris trifoliata Lour	–	+	+	+	–	–	–	+	+
Clerodendrum inerme (L) Gaertn	–	+	+	+	+	–	+	+	+
Ferns									
Acrostichum aureum L	+	–	+	–	–	–	–	–	–

Table 2 Distribution of gastropods in nine Hong Kong mangals, showing number of species recorded within each family in transect studies Numbers in brackets are additional species observed outside the transects

	Starling Inlet	Tai Tan	To Kwa Peng	Plover Cove	Three Fathoms Cove	Inner Port Shelter	Tung Chung	Tai Ho Wan	Tsim Bei Tsui	Total number of species
Trochidae	0	1	1	1	1	0	1	0	0	1
Turbinidae	0	1	1	1	1	1	0	0	0	1
Neritidae	3	4	4 (1)	4 (1)	3 (1)	0	5	2	2 (1)	6
Littorinidae	3	4	4	3	3	3	3	3	3	4
Iravidiidae	0	0	0	0 (1)	1	0	0	0	1 (1)	2
Assimineidae	1	0 (1)	0 (1)	0	1	0	3	0	4 (3)	7
Thiaridae	0	0	0	0	1	0	0	0	1	2
Stenothyridae	1	0	0	0	0	0	1	0	0 (1)	2
Planaxidae	0	1	0	0 (1)	1	0	0	0	0	1
Potamididae	3 (1)	3 (1)	3 (1)	5	4 (1)	2 (1)	5	2	1	5
Cerithidae	0	1	1	2	2	1	0	0	0	2
Muricidae	0	0	0	0	0	0	1	0	0	1
Nassariidae	0	1	1	0	0	0	0	0	0	1
Haminocidae	1	0	0	0	0	0	0	0	0	1
Ellobiidae	2	2 (5)	0 (6)	2 (3)	6	0	1	0	4	11
Amphibolidae	0	0	0	0	0	0	0	0	1	1
Onchididae	1	0 (3)	0 (3)	0 (1)	0	0	0	1	2 (1)	3

Table 3. The species richness of vegetation and gastropods in nine Hong Kong mangals and the associated conservation priority ranking of each site.

	Starling Inlet	Tai Tan	To Kwa Peng	Plover Cove	Three Fathoms Cove	Inner Port Shelter	Tung Chung	Tai Ho Wan	Tsim Bei Tsui
Total plant species	5	9	13	8	8	6	8	8	9
Total gastropod species	16	28	27	25	26	8	20	8	26
Plant species richness score	0.36	0.64	0.93	0.57	0.57	0.43	0.57	0.57	0.64
Gastropod species richness score	0.31	0.55	0.53	0.49	0.51	0.16	0.39	0.16	0.51
Site species richness score	0.67	1.19	1.46	1.06	1.08	0.59	0.96	0.73	1.15
Site species richness rank	8	2	1	5	4	9	6	7	3

Table 4. Distribution of rare plant species and gastropod families in nine Hong Kong mangals, showing the number of rare gastropod species.

	Starling Inlet	Tai Tan	To Kwa Peng	Plover Cove	Three Fathoms Cove	Inner Port Shelter	Tung Chung	Tai Ho Wan	Tsim Bei Tsui
Rhizophora sp.									+
Heritiera littoralis			+						
Acrostichum aureum	+		+						
Plant rarity rank	2.5	5.5	1.0	5.5	5.5	5.5	5.5	5.5	2.5
Littorinidae		1	1						
Iravidiidae				1	1				2
Assimineidae	1				1		2		6
Stenothyridae	1						1		1
Cerithidae				1	1				
Ellobiidae		6	5	3	4		1		3
Onchididae	0	2	2	1					2
Amphibolidae									1
Gastropod rarity index	1.50	9.00	7.50	5.50	5.58	0	4.33	0	10.92
Gastropod rarity rank	7	2	3	5	4	8.5	6	8.5	1
Site rarity rank	4.5	3	2	6	4.5	8.5	7	8.5	1

ta, and *Clerodendrum inerme*. In addition, 12 representative gastropods were also identified: *Monodonta labio, Lunella coronata, Nerita chamaeleon, Clithon oualaniensis, C. faba, Littoraria articulata, L. ardouiniana, L. melanostoma, Cerithideopsilla djajariensis, Cerithidea rhizophorarum, Terebralia sulcata*, and *Clypeomorus moniliferum*. The various representative attributes of each site were calculated and given in Table 5. Starling Inlet and Inner Port Shelter were notable in their low scores for most attributes of representativeness, resulting in overall lowest ranks. Tai Tan, Plover Cove and Tung Chung ranked highest in representativeness.

Table 5. Representativeness attributes and respective scores for each of nine Hong Kong mangals, and the associated conservation priority ranking of each site.

	Starling Inlet	Tai Tan	To Kwa Peng	Plover Cove	Three Fathoms Cove	Inner Port Shelter	Tung Chung	Tai Ho Wan	Tsim Bei Tsui
Representative plants score	0.22	1.00	1.00	0.78	0.78	0.67	0.89	0.89	0.89
Representative gastropods score	0.58	1.00	1.00	1.00	1.00	0.67	0.75	0.58	0.33
Tree canopy score	0.17	0.83	0.50	0.67	0.50	0.17	0.83	0.50	0.83
Gastropod density score	0.17	0.75	0.67	0.58	0.67	0.17	0.67	0.25	0.17
Transect typicalness score	0.33	0.40	0.40	0.80	0.67	0.47	0.53	0.80	0.80
Mangal structure score	1.00	1.00	1.00	1.00	1.00	0.67	1.00	0.33	0.67
Site Representativeness score	2.47	4.98	4.57	4.83	4.62	2.82	4.67	3.35	3.69
Site Representativeness Rank	9	1	5	2	4	8	3	7	6

Table 6. Current conservation status of nine Hong Kong mangals, and the estimated degree of hazard in each site.

	Starling Inlet	Tai Tan	To Kwa Peng	Plover Cove	Three Fathoms Cove	Inner Port Shelter	Tung Chung*	Tai Ho Wan*	Tsim Bei Tsui**
Approved project(s)							X	X	
Application(s) in progress				X	X				
Susceptible to development	X					X			
Unlikely to be developed		X	X						
Direct Destruction	X			X	X	X	X	X	X
Indirect Effects	X	X	X	X	X	X			
SSSI				X					X
Proposed SSSI					X		X		
Overall degree of hazard	3	4	4	2	2	3	1	1	4

* North Lantau Development
** Susceptible to illegal destruction

Degree of hazard

The existing conservation status of the nine mangals are presented in Table 6, together with development plans associated with each. Only two sites enjoy the Site of Special Scientific Interest (SSSI) status: Tsim Bei Tsui and Plover Cove.

Tsim Bei Tsui is part of the Inner Deep Bay SSSI (Agriculture and Fisheries Department, 1986) and is also included in a government Development Permission Area Plan (Planning Department, 1991a) where land use is strictly controlled. It is also part of the Inner Deep Bay Buffer Zone 1 and the planning intention is primarily to protect the natural surroundings, including intertidal biological communities. New developments within this zone should not occur unless they are required to support the conservation of the area's natural features and scenic qualities (Town Planning Board, 1992). Despite these efforts, illegal dredging of mangroves has occurred recently in a tidal pond close to the site.

The Plover Cove mangal is a designated conservation area referred to as the Ting Kok SSSI (Agriculture and Fisheries Department, 1984). It is an important field study site for Hong Kong students and also a feeding ground for egrets and herons nesting in the nearby egretries. However, industrial encroachment and destruction caused by highway construction has occurred despite its SSSI status. Furthermore, the land around the eastern end of the mangal is an 'Unspecified Use Area' (U-zone) under the Development Permission Area Plan (Planning Department, 1991b). Although an application for developing part of the U-zone into a low-rise residential area has been rejected, the applicant is allowed to apply for a review, and other, new, applications are also likely.

Upon completion of the Chek Lap Kok Airport development in North Lantau Island, about 50 ha of mangrove will be destroyed. The Tai Ho Wan mangal will be totally destroyed by coastal reclamation for the North Lantau Expressway (Mott McDonald, 1991a) and the Tai Ho New Town (Mott McDonald, 1991b). The Tung Chung mangal is comprised of three distinct patches at the eastern, southern and western side of Tung Chung Bay, and the former two patches will be reclaimed for the Tung Chung New Town development (Mott McDonald, 1991b). While the western patch will be adversely affected by such works, it will not be destroyed directly (Greiner Maunsell, 1991). A proposal for designation the of this patch into a SSSI by World Wide Fund for Nature Hong Kong (WWF HK) in 1991 is currently still being considered by the government.

A similar application has been made by WWF HK to designate Three Fathoms Cove as a SSSI in 1992. This is an important field study site for students and for scientific research. However, the presence of a U-zone surrounding this mangal (Planning Department, 1991c) has placed this site under serious threat. Proposed developments include an 18-hole golf course and club house facility north of the mangal, and a medium-rise residential development in a valley on the west. Though no direct mangrove destruction is planned, increased water pollution, in particular, fertilizers and pesticides from the golf course, will affect this mangal.

There are no known development plans at the remaining four sites. Tai Tan and To Kwa Peng are unlikely to be developed since they are situated in small areas adjacent to the Sai Kung Country Park, which is a conservation area. Furthermore, To Kwa Peng has no direct road access, while Tai Tan has a low development potential due to its very small area. By contrast, Starling Inlet and Inner Port Shelter are in close proximity to major roads and susceptible to development. Although Starling Inlet is immediately adjacent to the Yim Tso Ha Egretry SSSI (Agriculture & Fisheries Department, 1974) and about 300 m away from the A Chau SSSI (Agriculture & Fisheries Department, 1985), the present land use proposal is a U-zone (Planning Department, 1991d), and only intended to restrict development to agricultural and recreational uses which are generally compatible with the rural environment. Finally, Inner Port Shelter is susceptible to sewage pollution because there is a traditional village behind the mangal which only has minimal sewerage facilities.

Discussion

While vegetation distribution pattern is frequently used as the only biotic component in conservation evaluation programmes (Kirby, 1986) the inclusion of many more taxa undoubtedly improves the resolution power of any subsequent site value rankings. However, the lack of background data on the faunal distribution across Hong Kong's mangals, and the imminent destruction of two of the present study sites call for an urgent survey of all the remaining mangals throughout Hong Kong and to formulate a conservation strategy.

In reviewing the biological value and wildlife potential of the nine study sites, the provision of feeding, nursery and breeding sites for coastal fish and shellfish by mangrove habitats should also be taken into account. Numerous studies have indicated the importance of this type of coastal formation in providing food and shelter for fish fry and shellfish larvae and their adults (Odum, 1979; Griffin, 1985). Thus although sites like Starling Inlet and Inner Port Shelter have low rankings in terms of plant and gastropod species richness, the large extent at Starling Inlet (study area estimated at 4.3 ha), and the presence of well-developed, tall trees in both sites (Walthew & Yipp, in prep.) are good indicators of their potential in nutrient generation, especially for juvenile fish and shellfish.

Habitat use by coastal birds is another factor contributing to the biological value of these mangals, e.g. the trees providing roosting and/or nesting sites, and the foreshores providing valuable foraging grounds. Using information gathered from Chalmer's account of Hong Kong birds (1986), we have estimated that as many as 29 species could be using the area in and around Starling Inlet in the above manner. In close proximity to the Mai Po marshes, Tsim Bei Tsui is another site of high potential value to coastal birds. Finally, a number of resident and winter visiting species have been reported in sites around Tolo Harbour, in which Plover Cove and Three Fathoms Cove are situated. Bird sightings were not included in the evaluation for species richness as these were not conducted quantitatively across the nine study sites. In view of the large number of shore birds recorded from Starling Inlet, it is likely that the paucity of sightings in mangals such as Tai Tan and To Kwa Peng, which are otherwise biologically diverse, could be due to their geographical remoteness and poor accessibility to bird watchers.

While criteria were set up to determine both regional (across Hong Kong) and local (within site) rarity of species based on data collected, nevertheless such criteria are somewhat arbitrary. A border-line species in this assessment was *Lumnitzera racemosa* which was present in three sites with a maximum overall canopy cover of 11.6% in Three Fathoms Cove. This is a back-mangal species like *Excoecaria agallocha* and both are highly sensitive to encroachment by extending villages and roadworks. In addition, *Bruguiera gymnorrhiza*, though widespread (8 out of 9 sites), frequently occurs in low density and failed to attain dominance in any one site. Walthew & Yipp (in prep.) found this species to have a narrow habitat requirement for particular soil conditions. Both *L. racemosa* and *B. gymnorrhiza* are thus susceptible to environmental perturbations, and would demand special care to ensure their survival in Hong Kong.

In addition, many of the rare gastropods show both regional and local rarity, being confined to one or two sites and occurring at low densities. Conservation of these species would require a proper understanding of their habitat requirements and provision thereof. Especially for those species which do not produce planktonic veliger larvae, local extinction from a site would render subsequent recolonization through dispersal highly difficult.

The present study did not attempt to survey the foreshore seaward of mangrove plant distribution, yet this part of the seashore is an integral part of the habitat, and not only contribute to the biological value of the site, but are also affected by changes to the mangal. The seagrasses *Zostera nana* and *Halophila ovata* occur in the foreshore of the western section of Tung Chung (Mott McDonald, 1991b) and is a very rare incidence in Hong Kong as seagrasses have only been documented from four sites (Morton & Morton, 1983). Recent visits to these sites confirm that *H. ovata*, which had been present in all four sites, is now locally extinct from three. Another yet unidentified species has also been found at Tsim Bei Tsui. Such information, though incomplete for all sites, is nevertheless useful in helping to determine conservation priorities until such a time when more baseline data becomes available.

The fast pace of urban development in Hong Kong over the years has significantly reduced its mangrove coastline, the most significant losses occurring in Tolo Harbour and Deep Bay. Today, only one mangal, the Mai Po marshes in Deep Bay is being actively managed for conservation. While two of the nine sites in this study are designated SSSIs and two more applications for designation are being considered, no active management programmes are in place to ensure the sustained existence of these sites. Our study has highlighted some of the attributes of each site which are commonly included in conservation evaluation procedures elsewhere (Usher, 1986). By combining plant and gastropod species richess, occurrence of rare species, and representativeness of each site, the overall conservation value can be ranked, in decreasing order, as follows: Tai Tan, To Kwa Peng, Tsim Bei Tsui, Three Fathoms Cove, Plover Cove, Tung Chung, Starling Inlet, Tai Ho Wan and Inner Port Shelter.

By the time the Chek Lap Kok Airport and associated urban development in North Lantau Island have been completed, another 18% of Hong Kong's remaining mangal heritage would have disappeared. The case for conservation of all the remaining mangals in Hong Kong is extremely strong, especially in view of the present and future rapid development of coastal economic zones throughout Southern China, whose mangrove habitats will come under sustained threats of destruction. Thus, the priority ranking that we put forward in this study should not be used as a guideline to destroy certain mangals because of their low ranking, but rather, it should be used as a guideline to determine the priority of conservation work if the resources for doing so are in short supply.

In the following paragraphs, we set out a proposed conservation strategy for the nine mangals in this study, which is based on biological attributes as obtained through the transect studies, together with other incidental documents on flora and faunal occurrence. Due consideration is also given to the degree of hazard facing each site.

Every effort should be made to expedite the designation of west Tung Chung into a SSSI, to monitor the changes in environmental quality caused by reclamation works, and to provide a large facility to house a reserve population of the *Zostera* and *Halophila* seagrasses to prevent their local extinction, and as a stock for future environmental restoration work.

Tai Tan, To Kwa Peng and Tsim Bei Tsui, though not under any development threats at present, are nevertheless susceptible to localized encroachments. These three sites have high biological values, particularly in species richness and rarity scores, and since all of them have close physical proximity to existing conservation areas, should be considered for annexation with these areas. Tai Tan and To Kwa Peng are both coastlines of a country park, and could be considered for re-zoning into the country park to enjoy the same conservation status. Designation into SSSI is a weak alternative. For Tsim Bei Tsui, its current conservation status is considered to be satisfactory, and improved on-site monitoring to prevent further illegal destruction should be carried out.

Plover Cove and Three Fathoms Cove are widely used as educational field sites for Hong Kong's student population and should continue to remain so. Both had intermediate rankings in terms of overall biological value, but Plover Cove in particular is a highly representative mangal and thus of high educational potential. The designation of Three Fathoms Cove into SSSI is both timely and appropriate. In the event that both development proposals for the hinterland around this mangal be approved, a buffer zone should be set up and adequate mitigation measures adopted to reduce the impact of the project on the environment, especially on the water quality of the bay.

Although Starling Inlet and Inner Port Shelter scored low ranks in biological attributes, the former mangal is close to two other SSSIs, and is of substantial size to support a diversity of birds and fish and shellfish. Owing to the large decline in Hong Kong's mangal areas, these should also be designated as SSSIs with further investigations into their biological potential being carried out at the same time.

In view of the destruction that has occurred at Tsim Bei Tsui and Plover Cove even after designation as SSSIs, strengthening the enforcement of land use ordinances is needed, and requires increased local government administration vigilance.

The success of the Mai Po marshes as a conservation area for Hong Kong could be partly attributed to its large extent and its value to migrant birds. None of the mangals in this study are comparable to Mai Po in terms of their total area, bird diversity and *in situ* habitat diversity which includes managed ponds, mangroves and mudflats. As such, the potential for management of the nine sites along the Mai Po model is very low. However, the mangal habitat is an important teaching resource for high schools throughout Hong Kong, and it should be possible to set up a monitoring network among local schools, and to construct a long-term database to which schools can report upon certain key features and representative species which they note during their field visits. Changes can thus be monitored to enable early detection of indirect effects of development in hinterlands around these mangals to assist in management decisions. A field studies centre should be established to co-ordinate this effort, preferably at Plover Cove or Three Fathoms Cove, where most student activities are currently centred. The additional, obvious benefits to conservation education and practice for Hong Kong are apparent.

Acknowledgments

This work received funding support from research grants 341/068 and 340/912, Hong Kong Polytechnic.

References

Agriculture & Fisheries Department, 1974 Site of Special Scientific Interests S2/75 – Yim Tso Ha Egretry Hong Kong Government

Agriculture & Fisheries Department, 1984 Site of Special Scientific Interests S3/84 – Ting Kok Hong Kong Government

Agriculture & Fisheries Department, 1985 Site of Special Scientific Interests S1/85 – A Chau Hong Kong Government

Agriculture & Fisheries Department, 1986 Site of Special Scientific Interests S1/86 – Inner Deep Bay Hong Kong Government

Catt, P, 1986 Vegetation In A geography of Hong Kong (T N Chiu & C L So, eds) Oxford University Press, Hong Kong

Consultants in Environmental Sciences (Asia) Ltd , 1991 New airport master plan study recommendations on ecological mitigation measures Provisional Airport Authority, Hong Kong

Chalmers, M L , 1986 Annotated checklist of the birds of Hong Kong, 4th ed Hong Kong Bird Watching Society

Chapman, V J , 1984 Mangrove biogeography In Hydrobiology of the Mangal Dr W Junk Publishers, The Hague

Greiner Maunsell, 1991a New airport master plan study environmental assessment of construction impacts – summary and recommendations Provisional Airport Authority, Hong Kong

Greiner Maunsell, 1991b New airport master plan environmental impact assessment – final report Provisional Airport Authority, Hong Kong

Griffin, R K, 1985 The importance of mangrove/coastal wetland to three commercial fisheries in the N T, particularly for Barramundi (*Lates calcifer*) In Coasts and tidal wetlands of the Australian monsoon region (K N Bardsley, J S Davie & C D Woodroffe, eds) Australian National University Mangrove Monograph No 1, pp 277–284

Hill, D S, 1980 The Neritidae (Mollusca Prosobranchia) of Hong Kong (B Morton, ed) The Malacofauna of Hong Kong and Southern China Hong Kong University Press, Hong Kong pp 85–100

Holmes, P R , 1988 Tolo Harbour – the case for integrated water quality management in a coastal environment J Inst Water & Environmental Management, Volume 2, No 2

Hong Kong Herbarium, 1974 Checklist of Hong Kong Plants Government Printer, Hong Kong

Irving, R & B Morton, 1988 A geography of the Mai Po Marshes Hong Kong University Press, Hong Kong

Jefferson, R G , 1984 Quarries and wildlife conservation in the Yorkshire Wolds, England Biol Cons 20 363–380

Kirby, K , 1986 Forest and woodland evaluation In Wildlife conservation evaluation (M B Usher, ed) Chapman & Hall

Lewis, R R III , 1982 Mangrove forests In Creation and restoration of coastal plant communities (R R Lewis, III ed) CRC Press, Florida 219 pp

Lin, P & X M Wei, 1983 Ecological notes on the mangroves of Fujian, China In Tasks for vegetation science, Vol 8 (H J Teas, ed) Dr W Junk Publishers, The Hague

Morton, B & J Morton, 1983 The seashore ecology of Hong Kong Hong Kong University Press, Hong Kong

Mott MacDonald Hong Kong Ltd, 1991a North Lantau Expressway environmental assessment report, Yam O and Tai Ho sections final report Highways Department, Hong Kong Government

Mott MacDonald Hong Kong Ltd, 1991b North Lantau development environmental assessment (revised) Territory Development Department, Hong Kong Government

Odum, W E , 1979 Pathways of energy flow in a south Florida estuary, Sea Grant Tech Bull No 7 University of Miami, Coral Gables, Fla

Planning Department, 1991a Draft Lau Fau Shan and Tsim Bei Tsui development permission area plan No DPA/YL LFS/1 Hong Kong Government

Planning Department, 1991b Draft Ting Kok development permission area plan No DPA/NE–YK/1 Hong Kong Government

Planning Department, 1991c Draft Shap Sze Heung development permission area plan No DPA/NE-SSH/1 Hong Kong Government

Planning Department, 1991d Draft Luk Keng and Wo Hang development permission area plan No DPA/NE LK-1 Hong Kong Government

Thrower, S L & H Tang, 1976 Hong Kong Herbs and Vines The Hong Kong Government Printer, Hong Kong

Tomlinson, P B , 1986 The Botany of mangroves Cambridge University Press, Cambridge

Town Planning Board, 1992 Consideration of Section 16 application for development in areas around Mai Po nature reserve Hong Kong Government

Usher, M B (ed), 1986 Wildlife Conservation Evaluation Chapman & Hall 394 pp

Walter, H & S W Breckle, 1986 Ecological systems of the Geobiosphere Vol 2 Tropical and subtropical zonobiomes Springer-Verlag, Berlin

Walthew, G & M W Yipp (in prep) The mangal vegetation of Hong Kong

Yipp, M W , 1991 The relationship between hydrological factors and distribution of freshwater gastropods in Hong Kong Verh int Ver Limnol 24 2954–2959

Hydrobiologia **295**: 335–342, 1995.
Y. S. Wong & N. F. Y. Tam (eds), Asia-Pacific Symposium on Mangrove Ecosystems.

The distribution of mangrove-associated gastropod snails in Hong Kong

George Walthew
Department of Applied Biology & Chemical Technology, Hong Kong Polytechnic, Hong Kong; Present address: Department of Zoology, University of Hong Kong, Hong Kong

Key words: mangrove, gastropod, conservation, Hong Kong

Abstract

The distributions of mangal gastropod snails in Hong Kong were investigated by transect studies in nine mangals from around the coast. Fifty-one species of mangal associated gastropods were found, of which 25 were considered common. Gastropod faunas varied considerably between sites, both in terms of number of species and species composition. No one gastropod assemblage was found to be typical of the Hong Kong mangal. Local habitat differences were considered to have a major influence over species distributions, both within and between mangals. It is concluded that if species diversity is to be preserved then a representative sample of mangals needs protection.

Introduction

Hong Kong lies at the northern edge of the tropics between latitudes 22 ° 09′ and 22 ° O′ N. It is situated at the mouth of the Pearl River and the western coastline is therefore estuarine in character (Fig. 1). The Pearl River discharges 308.26 billion m^3 of water annually, 80% of it in summer (Morton, 1990), lowering salinities and increasing turbidity of the sea west of Hong Rong (Ho *et al.*, 1991). The east coast is not affected, as the westward flowing Philippine current keeps the river water to the west in the summer. However, the salinity of coastal waters does fall in summer when monsoonal rains and typhoons cause considerable local run-off from the steep hillsides.

Although mangals occur along the south China coast and reach their northernmost limit in southern Japan the number of mangrove plant species decreases northwards.

Gastropod molluscs form an important component of mangal faunas (MacNae, 1968) and so it is possible to use them as an indicator to compare faunas at different locations.

This paper considers these questions:

1. In terms of fauna, are the Hong Kong mangals representative of other tropical mangals in the region?
2. Do the mangals in Hong Kong differ in their faunas and if so to what extent?
3. Hong Kong is a small place so is it sufficient to protect a single (large) mangal, such as Mai Po, to preserve a representative mangal fauna?

Methods

Nine mangals from around the coast of Hong Kong were selected for study, three on the west coast and six on the east (Fig. 1). These were sampled during the winter periods, November to March, 1990/91 and 1991/92. Sites were chosen to include variations in hydrology arising both from local runoff and the general west to east transition from estuarine to more oceanic conditions across the coastal waters of Hong Kong. In each mangal five transects were run from the seaward edge, landward to where the back-of-beach vegetation begins (usually denoted by the presence of *Hibiscus tiliaceus* L. and/or *Pandanus tectorius* Soland ex Rark.). At equal intervals along each transect six sampling stations were laid out with the first at the seaward edge and the last just before the back-of-beach vegetation.

At each transect station a visual estimate was made of the percentage of each quadrat covered with stones. Samples were taken of all ground dwelling gastropods by collecting from two quadrats, of 0.25 m^2 area, situated on opposite sides of the transect line. From the

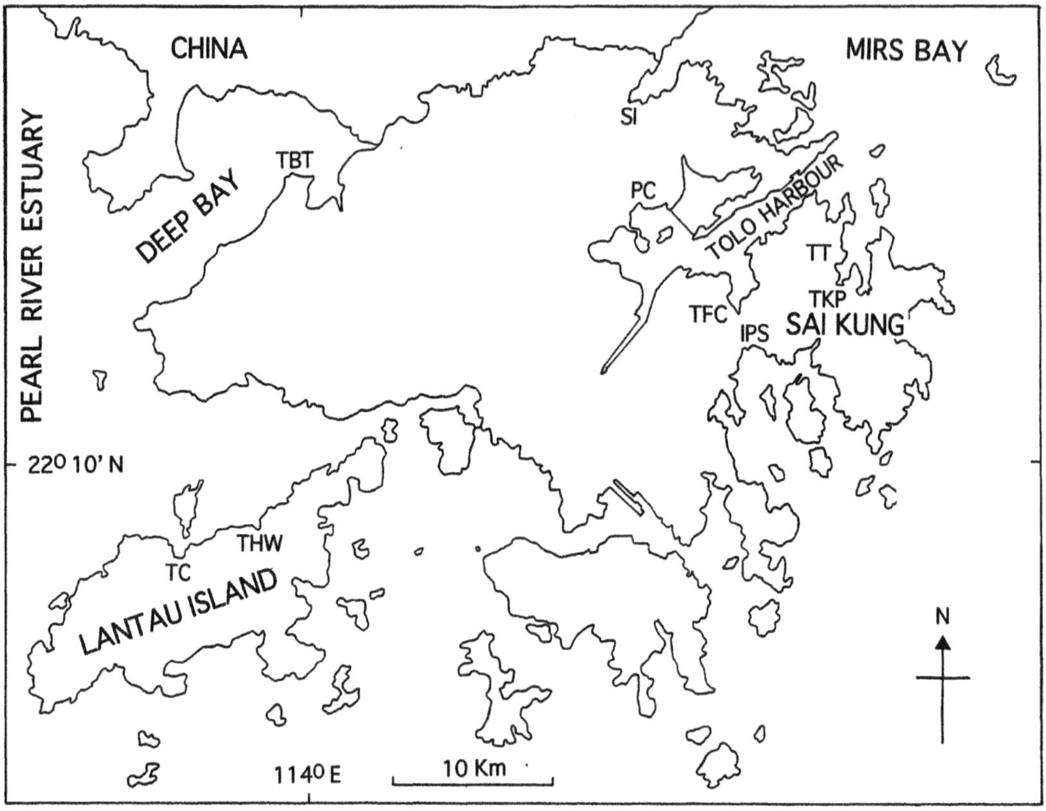

Fig. 1. Map of Hong Kong showing the location of the nine study sites. SI = Starling Inlet, PC = Plover Cove, TFC = Three Fathoms Cove, TT = Tai Tan, TKP = To Kwa Peng, IPS = Inner Port Shelter, TC = Tung Chung, THW = Tai Ho Wan, TBT = Tsim Bei Tsui.

two gastropod samples a mean density for the station was estimated.

The arboreal gastropods at each transect station were also sampled. Two trees of each species present in a 3 m × 3 m quadrat were chosen and all the gastropods present on the leaves, branches, trunk and roots were collected. An estimate of density was made by multiplying the mean of the number of snails on the sample trees by the number of trees in the quadrat and then converting to numbers m^{-2}.

The height above datum of each quadrat was estimated by measuring the water depth at high tide and by levelling with an Abney level. The measurements were then compared against datum values in tide tables for Hong Kong.

Statistical analysis

The dissimilarity of the gastropod snail assemblages at the nine mangal sites was tested using cluster analysis of both the presence-absence and frequency data of the species at each site. The analysis was carried out using the combinatorial linear model of Lance & Williams (1967).

Results

Fifty-one species of gastropods belonging to 17 families were found in the nine mangals investigated (Table 1). The most important families, in terms of the numbers of species and distribution, were; the Neritidae (6 spp.), Littorinidae (4 spp.), Potamididae (5 spp.) and Ellobiidae (11 spp.). However, far fewer than 51 occurred at any individual site. The number of species ranged from a minimum of eight at both Inner Port Shelter and Tai Ho Wan, to a maximum of 28 at Tai Tan, with the median for all nine sites being 25 species (Table 2).

Species densities were found to vary widely within the mangals and the maxima tended to differ considerably from the mean values. In Hong Kong, two species were found to exceed a density of 500 m^{-2}. *Cerithidea djadjariensis* reached the highest density with a maximum of 901 m^{-2} recorded at Plover Cove while *Clypeomorus moniliferum* was recorded at a maximum of 658 m^{-2} at Tai Tan. At least one species exceeded a density of 100 m^{-2} at every site except Tai Ho Wan where the total ground dwelling gastropod fauna has a density of less than one snail per square metre (Table 3). Mean species densities, however, were much lower because the density distribution of most species was extremely variable.

No one assemblage of gastropod molluscs was found to be typical of the nine Hong Kong mangals. Only 14 species occurred in more than 50 percent of sites and just three of the *Littoraria* species were common to all nine. Thirteen species were found at only a single site each. Apart from four species these were all from either the seaward or landward fringes.

Similarities amongst the gastropod assemblages of the nine mangals was investigated using cluster analysis of the presence-absence data of species. Both group average and centroid methods placed Tai Tan, To Kwa Peng, Plover Cove and Three Fathoms Cove together in a single group (Fig. 2). As may have been anticipated from their geographical proximity and similar locations (Fig. 1), the two Sai Kung sites (Tai Tan and To Kwa Peng) showed the greatest similarity, followed by the two Tolo Harbour sites (Plover Cove and Three Fathoms Cove). The Tsim Bei Tsui mangal proved to be the least similar to all the others, whatever clustering method was used, and was always the last to join the cluster. The remaining four sites (Starling Inlet, Inner Port Shelter, Tung Chung and Tai Ho Wan) clustered differently depending on the clustering method. However, the levels at which they joined the cluster were close together in all cases and these sites would appear to be equally dissimilar from each other and also from the main Sai Kung/Tolo Harbour group.

Cluster analysis of the frequency of occurrence of the gastropod species within the 30 sampling units at each site was also made. The result of this differed somewhat from that of the presence-absence analysis. It placed the two Tolo Harbour sites (Plover Cove and Three Fathoms Cove) further apart in the Sai Kung/Tolo Harbour group and this was further joined by Tung Chung which merged with the cluster before Plover cove (Fig. 3). This proved to be true in both group average and centroid clustering methods.

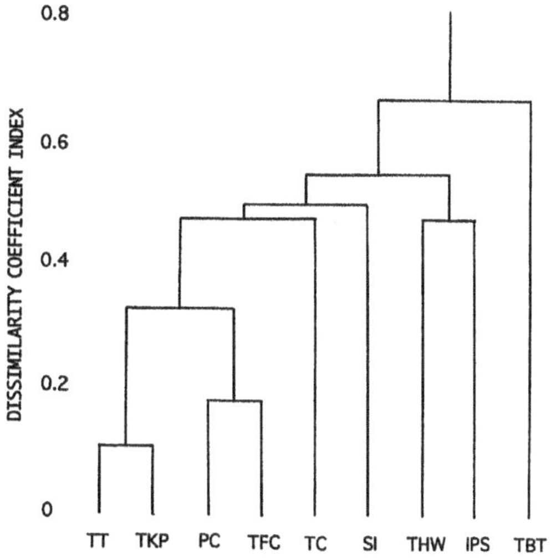

Fig. 2. Cluster analysis of gastropod species presence-absence in nine Hong Kong mangals using the group average method. For site abbreviations see Fig. 1.

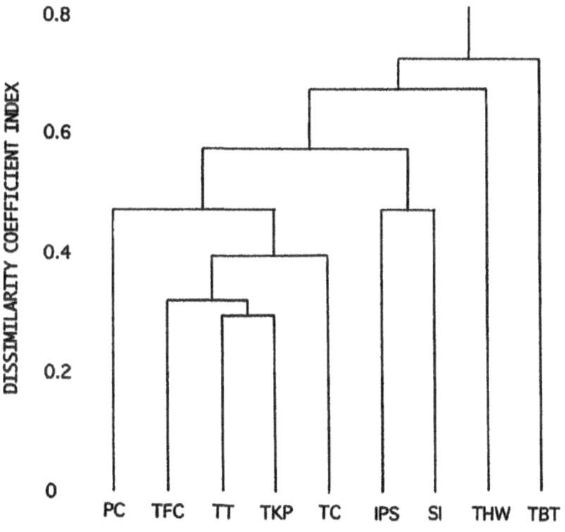

Fig. 3. Cluster analysis of gastropod species frequency in nine Hong Kong mangals using the group average method. For site abbreviations see Fig. 1.

Table 1 A list of the gastropod molluscs found at nine Hong Kong mangals

CLASS GASTROPODA SUBCLASS PROSOBRANCHIA	SUBCLASS OPISTHOBRANCHIA
ORDER ARCHAEOGASTROPODA	ORDER CEPHALASPIDEA
FAMILY Trochidae	FAMILY Haminocidae
Monodonta labio (Linnaeus)	*Haloa* sp
FAMILY Turbinidae	
Lunella coronata (Gmelin)	SUBCLASS PULMONATA
FAMILY Neritidae	
Nerita chamaeleon Linnaeus	ORDER BASOMMATOPHORA
Nerita lineata Gmelin	
Nerita yoldii Récluz	FAMILY Ellobiidae
Clithon oualaniensis (Lesson)	*Auriculastra* sp
Clithon faba (Sowerby)	*Cassidula aurisfelis* (Bruguièra)
Dostia vioacea (Gmelin)	*Cassidula plecotrematoides* Möllendorff
	Cassidula schmackeriana Möllendorff
ORDER MESOGASTROPODA	*Ellobium chinensis* (Pfeiffer)
	Ellobium polita (Metcalfe)
FAMILY Littorinidae	*Laemodonta exaratum* (H & A Adams)
Littoraria articulata (Philippi)	*Laemodonta punctatostriata* (H & A Adams)
Littoraria ardouiniana (Heude)	*Laemodonta punctigera* (H & A Adams)
Littoraria melanostoma (Gray)	*Melampus triticeus* (Küster)
Littoraria pallescens (Philippi)	*Pythia fimbriosa* Möllendorff
FAMILY Iravadiidae	FAMILY Amphibolidae
Iravadia bombayana Stoliczka	*Salinator fragilis* (Lamarck)
Iravadia ornata Blanford	
FAMILY Assimineidae	ORDER SYSTELLOMMATOPHORA
Assiminea brevicula (Pfeiffer)	
Assiminea dohrneana (Nevill)	FAMILY Onchidiidea
Assiminea nitida (Pease)	*Onchidium hongkongensis* Britton
Assiminea woodmasoniana Nevill	*Paraoncidium reevesii* (Gray)
Assiminea sp 5	*Platevindex mortoni* Britton
Assiminea sp 6 (lutea?)	
Assiminea sp 7	
FAMILY Thiaridae	
Melanoides tuberculatus (Müller)	
Sermyla tornatella (Lea)	
FAMILY Stenothyridae	
Stenothyra sp 1	
Stenothyra sp 2	
FAMILY Planaxidae	
Planaxis sulcatus (Born)	
FAMILY Potamididae	
Cerithidea djadjariensis (Martin)	
Cerithidea ornata A Adams	
Cerithidea rhizophorarum A Adams	
Terebralia sulcata (Born)	
Batillaria multiformis (Lischke)	
FAMILY Cerithidae	
Cerithium coralium Kiener	
Clypeomorus moniliferum (Kiener)	
ORDER NEOGASTROPODA	
FAMILY Muricidae	
Thais clavigera Küster	
FAMILY Nassariidae	
Nassarius festivus (Powys)	

Table 2 Frequency of occurrence of gastropod species at each of nine Hong Kong mangals (percentage of 30 quadrats) + = species present but not found in the quadrats

	SI	TT	TWP	PC	TFC	IPS	TC	THW	TBT
M labio		7	13	17	27		10		
L coronata		37	17	10	20	3			
N chamaeleon		47	13	17	7		23	3	3
N lineata	7	13	3	+			7		+
N yoldii							3		
C oualaniensis	23	17	7	13	13		17	3	
C faba		10	7	10	27				
D violacea	7		+	3	+		13		50
L articulata	3	20	3	7	20	0	80	10	43
L ardouiniana	3	67	53	20	40	10	93	17	30
L melanostoma	13	87	53	23	40	23	100	30	67
L pallescens		3	7						
I bombayana				+					10
I ornata					3				+
A brevicula		+	+				13		27
A dohrueana	8								8
A nitida					3				20
A woodmasoniana									+
Assiminea sp 5							3		+
Assiminea sp 6									3
Assiminea sp 7							3		+
M tuberculatus					3				
S tornatella									3
Stenothyra sp 1	3								+
Stenothyra sp 2							3		
P sulcatus		3		+	10				
C djadjariensis	67	43	70	73	47	77	60	17	
C ornata	7	+	+	47	+		20		7
C rhizophorarum	+	43	10	27	40	+	27	3	
T sulcata	10	63	87	93	77	57	57		
B multiformis				27	10		10		
C coralium				7	10				
C moniliferum		20	47	33	3	7			
T clavigera							3		
N festivus		10	3						
Haloa sp	7								
Auriculastra sp					3				
C aurisfelis	13		+	13	10				
C plecotrematoides		+	+	+	3				
C schmackeriana		+	+	+	3				
E chinensis	3	+		+	3				3
E polita									20
L exaratum		+	+						
L punctatostriata							10		13
L punctigera		+	+						3
M triticeus		3			13				
P fimbriosa		3	+	3					
S fragilis									3
O hongkongensis		+	+						+
P reevesii	7	+	+					3	3
P mortoni		+	+	+					7
No of species	16	28	27	25	26	8	20	8	26

Table 3 Mean densities of the more numerous gastropod species in nine Hong Kong mangals (individuals m^{-2})

	SI	TT	TKP	PC	TFC	IPS	TC	THW	TBT
M labio		0 1	1 1	0 7	2 3		0 5		
L coronata		2 0	2 5	2 3	2 5				
N chamaeleon		3 0	0 7	0 4	0 3		2 2	0 1	
N lineata	0 0	0 1	0 0				0 1		
C oualaniensis	6 0	13 6	0 2	16 5	11 8		0 4	0 1	
C faba		0 6	0 2	0 1	15 7				
D violacea	0 1			0 1			0 1		2 0
L articulata	0 0	0 2	0 0	0 0	0 1		18 2	2 6	2 3
L ardouiniana	0 0	1 7	1 1	0 1	0 4	0 1	4 0	0 1	0 2
L melanostoma	0 1	4 4	1 1	0 2	0 5	0 3	20 0	5 0	1 2
I bombayana									2 6
A brevicula							0 4		9 2
A dohrneana	0 2								0 2
A nitida					0 1				3 8
P sulcatus		0 0			0 5				
C djadjariensis	21 9	41 4	31 4	54 4	14 2	29 7	29 8	0 5	
C ornata	0 1			1 0			0 3		0 8
C rhizophorarum	19 4	0 3	8 1	8 8		5 4	0 1		
T sulcata	0 3	14 5	18 5	21 1	8 8	2 6	5 0		
B multiformis				24 8	3 1		2 3		
C coralium				0 6	1 1				
C moniliferum		30 5	16 1	4 9	0 4	0 2			
N festivus		0 4	0 0						
Haloa sp	0 3								
C aurisfelis	0 3			0 1	0 2				
E polita									0 5
L punctatostriata							0 6		1 3
all ground-dwelling snails	32 5	29 4	126	61 0	135	74 8	46 6	0 7	19 6

Discussion

The gastropod families typical of tropical mangals; the Littorinidae, Neritidae, Potamididae, Cerithidae and Ellobiidae (MacNae, 1968); were all well represented in Hong Kong. Berry (1972) listed 32 species of gastropods from the mangals of the Selangor coast of West Malaysia and although his list is not exhaustive, the 51 Hong Kong species compare well with those of West Malaysia. The median of 25 species from the nine Hong Kong sites is the same as found on transects through the mangal at Port Swettenham in West Malaysia by Sasekumar (1974). Macnae (1968) observed that the north-eastern extension of the Indo-west Pacific region undergoes a progressive subtraction of mangrove associated molluscs. This is not the case at the latitude of Hong Kong and any subtractions from the fauna must occur further north.

The maximum densities of gastropod species in the Hong Kong mangals were found to be comparable to data published for densities in mangals elsewhere. At a West African mangal in Ivory Coast, *Pachymelania aurita* was recorded at a density of 712 m^{-2} (Binder, 1968) and *P. fusca* at 1024 m^{-2} in Cameroon (Plaziat, 1974), while *Cerithidea cingulata* reached a density of 579 m^{-2} on the Red Sea coast of Saudi Arabia (Price *et al.*, 1987).

Between site differences

Between the nine mangals, the gastropod faunas differed both in the number of species and the species composition. The sites with the most similar gastropod faunas in terms of the species present were also those which were closest together. Tai Tan and To Kwa Peng on the Sai Kung peninsula and in adjacent inlets, separated by little more than one kilometre of rocky coastline (Fig. 1), were found to be only 10% dissimilar (Fig. 2). Similarly Plover Cove and Three Fathoms Cove, on opposite sides of the 8 km wide Tolo Harbour, were 18% dissimilar. When species abundances are considered and not just the presence or absence, the assemblages appeared far less similar (Fig. 3) although the main species grouping of Tai Tan, To Kwa Peng, Plover Cove, Three Fathoms Cove and Tung Chung remain less than 50% dissimilar from each other. This grouping coincides with the stony nature of these sites all of which were found to have stones in more than a third of the 30 sample plots at each site. The remaining four sites all had less than 15% of plots with stones and were all notable for their soft, ill consolidated substrates.

Lunella coronata which was otherwise widely distributed and common at the seaward edge of the mangal, was not found on the west coast even at Tung Chung which was otherwise similar in environment to the stony mangals of the east coast. This distribution was attributed to the estuarine influence on the west coast. *Clypeomorus moniliferum* had a similar distribution to *L. coronata* and may also be excluded by this. *Littoraria pallescens* was only found on the east coast. Reid (1986) classified this last species as marine and avoiding estuarine waters.

The restriction of the two ellobiids, *Cassidula aurisfelis* to the east coast and *Ellobium polita* to the west, is not explained by such physical parameters. Both are large species of the back mangal having the free-swimming veliger stage suppressed (Macnae, 1968) and, unlike their congeners (*C. plecotrematoides*, *C. schmackeriana* and *E. chinensis*), both species are very much restricted to the mangrove habitat. It is probable that localised destruction of habitat eliminated these species from many of the sites. On recent regrowth of the mangal such species would find it difficult to recolonise quickly due to the lack of a marine veliger stage. Interestingly, the prosobranch *Cerithidea ornata*, which was also restricted to the mangal, was not so limited in distribution and would suggest that it was able to recolonise quickly due to its veliger stage. Brandt (1980) in fact found *E. polita* in mangals in Tolo Harbour, but these sites have since been lost to development.

Few of the gastropod species of the Hong Kong mangals were found exclusively in this habitat. Of the 51 species found in the mangal, only *Littoraria pallescens*, *Cerithidea ornata*, *Cassidula aurisfelis* and *Ellobium polita* were found exclusively in the mangal. Many of the prosobranch species could be found on sheltered shores in the absence of mangroves. This suggests that the mangrove associated gastropod molluscs are not dependent on the plants but rather that they both have a similar requirement for sheltered soft shores. *Monodonta labio*, *Lunella coronata*, *Nerita chamaeleon* and *Planaxis sulcatus* are species of the seaward zone of the mangal but are more ubiquitous, occurring commonly on sheltered rocky shores attached to rocks and boulders, rather than to mangrove trees (Chambers, 1980; Morton & Morton, 1983; Takenouchi, 1983; McMahon & Cleland, 1990). Much of the back-mangal gastropod fauna in Hong Kong is shared with saltgrass meadows (Morton & Morton, 1983) and many species (particularly the pulmonates) seemed to use it as an alternative habitat. Large populations of *Cerithidea rhizophorarum* occurred among the grasses and the pulmonates *Ellobium chinensis*, two *Auriculastra* species and *Melampus triticeus* were all more common here than in the mangal.

Conclusion

Hong Kong mangals are as diverse as those of others in the tropics, while the individual mangals differ considerably amongst themselves in their gastropod faunas.

Despite the influence of the Pearl River on the west coast, no clear east-west division was found between the ground-dwelling gastropod faunas although a few species were affected and in these cases distribution was the result of estuarine influence. However, some species did exhibit distributions which required further explanation, particularly that of the mangal dependant pulmonates *C. aurisfelis* and *E. polita*. It is thought that the distribution of these species may be due to local extinctions caused by habitat destruction and difficulty in recolonisation.

Mangals are not just a collection of mangrove plants, no two have the same fauna. In order to preserve a representative fauna it is necessary to protect a selection of sites which must be chosen with care.

Acknowledgments

I thank Dr May Yipp for advice during the course of this research. This work was supported by research grant 341/068 from the Hong Kong Polytechnic. Attendance at the Symposium on Mangrove Ecosystems was made possible by sponsorship from the Caltex Green Fund

References

Berry, A J, 1972 The natural history of West Malaysian Mangrove Faunas Malay Nat J 25 135–162

Binder, E, 1968 Répartition des mollusques dans la lagune Ebrié Cahiers de l'Office de la Recherche Scientifigue et Technique Outre-Mer, Ser Hydrobiologie 2 3–34

Brandt, A J, 1980 An annotated checklist of the non-marine molluscs of Hong Kong In B Morton (ed), Proceedings, First International Workshop on the Malacofauna of Hong Kong and Southern China, 23 March–8 April 1977 Hong Kong University Press, Hong Kong

Cantera, J, P M Arnaud & B A Thomassin, 1983 Biogeographic and ecological remarks on molluscan distribution in mangrove biotopes J Moll Stud, Suppt 12A 10–26

Chambers, M R, 1980 Zonation, abundance and biomass of gastropods from two Hong Kong rocky shores In B Morton (ed), Proceedings, First International Workshop on the Malacofauna of Hong Kong and Southern China, 23 March–8 April 1977, Hong Kong Hong Kong University Press, Hong Kong

Ho, K C, M Smith-Evans & M F Kong, 1991 Marine Water Quality in Hong Kong Annual Report, Marine Water Quality Monitoring Programme for 1990 Government Printer, Hong Kong

Lance, G N & W T Williams, 1967 A general theory for classificatory sorting strategies 1 Hierarchical systems Computer Journal 9 373–380

Li, F X & S H Gao, 1985 The ground-dwelling molluscan faunas of mangrove swamps in Hong Kong and Xiamen In B Morton & D Dudgeon (eds), Proceedings of the Second International Workshop on the Malacofauna of Hong Kong and Southern China, Hong Kong, 1983 Hong Kong University Press, Hong Kong

MacNae, W, 1968 A general account of the fauna and flora of mangrove swamps and forests in the Indo-West Pacific region Adv mar Biol 6 73–270

McMahon, R F & J D Cleland, 1990 Thermal tolerance, evaporative water loss and behaviour during prolonged emergence in the high zoned mangrove gastropod *Cerithidea ornata*, evidence for atmospheric water uptake In B Morton & D Dudgeon (eds), Proceedings of the Second International Marine Biological Workshop The Marine Flora and Fauna of Hong Kong and Southern China, Hong Kong, 1986 Hong Kong University Press, Hong Kong

Morton, B, 1990 Pollution and the sub-tropical inshore hydrographic environment of Hong Kong In B Morton & D Dudgeon, (eds), Proceedings of the Second International Marine Biological Workshop The Marine Flora and Fauna of Hong Kong and Southern China, Hong Kong, 1986 Hong Kong University Press, Hong Kong

Morton, B & J Morton, 1983 The Seashore Ecology of Hong Hong Hong Kong University Press

Plaziat, J C, 1974 Répartition des mollusques amphibies de quelques littoraux et estuaires à mangrove (Nouvelle Calédonia et Cameroun) Rôle de la salinite dans les modifications locales des peuplements de mangrove Haliotis, Paris 4 167–177

Price, A R G, P A H Medley, R J McDowall, A R Dawson-Shepherd, P J Hogarth & R F G Ormond, 1987 Aspects of mangal ecology along the Red Sea coast of Saudi Arabia J Nat Hist 21 449–464

Reid, D G, 1986 The littorinid molluscs of mangrove forests in the Indo-Pacific Region British Museum, London

Sasekumar, A, 1974 Distribution of macrofauna on a Malayan mangrove shore J anim Ecol 43 51–69

Takenouchi, K, 1985 A boulder shore gastropod fauna in Hong Kong In B Morton & D Dudgeon (eds), Proceedings of the Second International Workshop on the Malacofauna of Hong Kong and Southern China, Hong Kong, 1983 Hong Kong University Press, Hong Kong

Wells, F E, 1985 The Potamididae (Mollusca Gastropoda) of Hong Kong, with an examination of habitat segregation in a small mangrove system In B Morton & D Dudgeon (eds), Proceedings of the Second International Workshop on the Malacofauna of Hong Kong and Southern China, Hong Kong, 1983 Hong Kong University Press, Hong Kong

Wells, F E, 1990 Distribution of marine invertebrates in a Hong Kong mangrove, with emphasis on molluscs In B Morton & D Dudgeon (eds), Proceedings of the Second International Marine Biological Workshop The Marine Flora and Fauna of Hong Kong and Southern China, Hong Kong, 1986 Hong Kong University Press, Hong Kong

Yipp, M W, 1982 The distribution of ground-dwelling gastropods in a small mangrove stand in Hong Kong In B Morton & C K Tseng (eds), Proceedings of the First International Marine Biological Workshop The Marine Flora and Fauna of Hong Kong and Southern China, Hong Kong, 1980 Hong Kong University Press, Hong Kong

Hydrobiologia **295**: 343–351, 1995.
Y. S. Wong & N. F. Y. Tam (eds), Asia-Pacific Symposium on Mangrove Ecosystems.

The ecology of mangrove conservation & management

Ong Jin-Eong
Centre for Marine and Coastal Studies, Universiti Sains Malaysia, 11800 Penang, Malaysia

Key words: mangrove, ecology, conservation, management, climate change

Abstract

Despite the recent better understanding and awareness of the role of mangroves, these coastal forest communities continue to be destroyed or degraded (or euphemistically reclaimed) at an alarming rate. The figure of 1% per year given by Ong (1982) for Malaysia can be taken as a conservative estimate of destruction of mangroves in the Asia-Pacific region. Whilst the Japanese-based mangrove wood-chips industry continues in its destructive path through the larger mangrove ecosystems of the region, the focus of mangrove destruction has shifted to the conversion of mangrove areas into aquaculture ponds and the consequences of the unprecedented massive addition of carbon dioxide to the atmosphere by post industrial man.

Mangroves are non-homogeneous; characterised by distinct vegetative zones that occupy the interface between land and sea and dynamically interacting with the atmosphere above as well as with the influences of the adjacent land and sea. The conservation of mangroves should thus include not only the various vegetation and tidal inundation zones but also the adjacent marine and terrestrial areas (including the water catchment area).

On the current concern with global climate change, it is pointed out that relative sea level change is very much site dependent. For effective planning and management, it is vital to know if a particular site is stable, rising or sinking so efforts should be directed to find suitable methods for determining this. However, should rapid relative sea level rise take place, there is very little likelihood of saving mangroves whose landward margins have been developed by man, a fact to bear in mind when selecting sites for conservation. The Matang mangroves of Malaysia is a rare case of successful sustainable management of a tropical rain forest. Although the tools of management are available they are not widely applied. We particularly urge the Japanese mangrove wood-chips industry to look to long term sustainable use rather than short term gains. A suggestion is made to appeal to the new Government of Japan to take the lead in environmental friendliness especially to the rain forests of the Asia-Pacific region.

Introduction

A SCOR/UNESCO biosphere inventory of mangrove lands in the late seventies came up with a total figure of some 162 210 km^2 of mangroves in the world (Saenger *et al.*, 1983). Although the distribution of mangroves stretches into the higher latitudes (to about 40 ° north and south where their spread is prevented essentially by ground frost) their main distribution is in the tropical areas. This coincides with present area of rapid human development activities and so mangroves are presently highly threatened. Even in these areas mangrove forests occupy no more than 1 or 2 percent of the total land area and so are not given much attention by most nations.

About a third of the world's mangroves are located in Asia and over the last twenty years or so these are being increasingly 'reclaimed' or degraded. Much of the present pressure is from conversion to aquaculture ponds and non-sustainable harvesting of timber for the wood-chip industry. It is not the intention here to expound on the importance and usefulness of mangroves (for this see e.g. Saenger *et al.*, 1983; Hamilton & Snedaker, 1984).

Despite a recent better understanding and awareness of the role of mangroves, these coastal forest communities continue to be destroyed or degraded (or euphemistically 'reclaimed') at an alarming rate. The figure of 1% per year given by Ong (1982) for Malaysia can be taken as a conservative estimate of destruction of mangroves in the Asia-Pacific region. Whilst the Japanese-based mangrove wood-chips industry continues in its destructive path through the larger mangrove ecosystems of region, the focus of mangrove destruction has shifted to the conversion of mangrove areas into aquaculture (mainly prawn) ponds.

The aim of this paper is mainly to present an ecological basis for the conservation and management of mangroves. The mangrove ecosystem has to be considered as a special case, different from most other ecosystems. This is mainly because mangroves represent interface ecosystems that straddle the land and the sea, from freshwater to seawater, often with distinct zones of plant and animal species and are closely linked not only with the atmosphere above (like most other ecosystems) but also with the adjacent landward catchment and the seaward marine system. Thus it is unrealistic to just conserve a patch of mangrove as we would a patch of terrestrial rain forest. Many of the examples here will be from the mangroves of Asia but the principles are applicable worldwide.

Conservation

We need to explain what we mean by the term conservation and the following is essentially taken from Ong & Gong (1991). The term conservation connotes different concepts to different people. Apparently the term, used in its ecological sense, was originally derived from the phrase 'conservation by wise use' (see Darling, 1964). The Oxford Concise Dictionary (6th Edition, 1976) defines conservation as 'preservation, especially of natural environment', and conserve as 'keep from harm, decay or loss, especially with view to later use'.

It becomes apparent that depending on the situation, there may be differences in degree of preservation: from a total 'hands-off' situation to one of 'later use', 'wise use', 'sustainable use' and a host of other more liberal uses.

In many situations it is necessary to adopt a pragmatic rather than idealistic approach. For our present discussion, we would like to confine discussion to two broad categories of conservation: the first is one of conservation for posterity and the second for long term prosperity. The two have to go hand in hand. In other words, we should have some areas of representative mangroves legislated as National Parks or World Natural Heritage Areas (*i.e.* conserved for posterity) and most of the rest of our mangroves conserved through sustained yield management (*i.e.* for prosperity). A certain amount of mangroves would inevitably be lost as a result of strategic uses (e.g. airports and ports).

In Malaysia, Ong & Gong (1991) have suggested that about 5% (or some 30 000 hectares) of the total area of mangroves should be conserved as National Parks (or its legal equivalent) on a nationwide basis. This is certainly not too much to ask as a long term investment for posterity. At the same time this will ensure that the areas set aside are large enough and adequately buffered to survive in their natural undisturbed state.

Since mangroves represent a very dynamic ecosystem and a representative one is not usually homogeneous but showing a gradation from the freshwater landward aspect to the seawater seaward aspect, a whole strip of land from the freshwater to the seawater aspect will be required. The boundary of the mangrove National Park should (where possible) extend into the sea as well as into its freshwater catchment so as to allow space for natural growth. To ensure maximum viability it is also best that these mangrove National Parks be located within mangroves identified for sustained yield management of mangrove trees.

Major threats to mangroves

Strategically, to effectively protect or conserve our mangrove resources, it would be wise to identify and understand the main threats to these resources. The following are what we perceive as the major threats:

Population pressures

Burgeoning populations are possibly the biggest cause of mangrove destruction and degradation. This can be clearly seen, all over Asia: where populations in coastal areas are high, little good mangroves are left. Like all systems, mangroves and coastal areas have a certain carrying capacity. Once this capacity is exceeded, the system will be stressed and will start to break down. In areas where population pressures are great there appears to be little that can be done to retain the mangroves. All we can say here is that it is heartening to note that many of the countries in the Asia Pacific region are beginning to control their population growth. All we can do is hope that the rate of population growth can be brought down fast enough before irreversible damage to the environment results.

This population problem is the basis of all the other threats that we discuss below.

Aquaculture development

Much of the mangroves in the Philippines was destroyed or degraded in the sixties as a result of extensive conversion to fish ponds (largely milkfish culture). Most of the fish ponds were abandoned after a few years due to the drop in fertility. It seems that a lesson was not learnt as seen from neighbouring countries converting their mangroves to prawn ponds. Thailand has now lost or degraded most of its mangroves as a result of extensive conversion to prawn ponds. One major problem with the use of mangroves for aquaculture ponds is the acid sulphate conditions associated with most mangroves (e.g. Ong, 1982), yet such development continues (most likely the result of a band of unscrupulous aquaculture consultants). The tiger prawn (*Penaeus monodon*) culture industry is now plagued with disease and nutrition problems. Yet there is still tremendous pressure to use mangroves for prawn ponds.

Wood-chips operation

The mangrove wood-chips industry is monopolised by Japan. Vast areas of mangroves in the Asia Pacific region have been aggressively cleared of timber and many areas are still being exploited by this industry. Yet the economic returns to the countries whose mangroves have been exploited are meagre (Nair, 1977; Ong, 1982). From their *modus operandi* in Sabah and Sarawak the operators appear only interested in a quick harvest and then moving on to the next site rather than the sustained use of a particular site. The operation in Sabah (involving a mean annual clear felling of some 4000 ha) has been stopped and the one in Sarawak (mean annual clear felling of some 1600 ha) has virtually ground to a halt due to the exhaustion of timber from the extensive mangroves (of the Rajang River system) within a short period of about 25 years. Yet these could have been easily managed on a sustained yield basis with a little bit of care and less of greed. It is past time that so-called advanced nations like Japan should have both respect and responsibility for the world's environment. The mangrove wood-chips industry continues happily with its operations in other countries in the Asian Pacific region. Perhaps the new and apparently more open Government of Japan will be receptive to using their influence to bear on those in their mangrove wood-chips industry to be more responsible for their actions. They may even be persuaded to make more positive contributions towards maintaining a more healthy environment, especially in the Asia-Pacific region.

Land reclamation

This is especially related to population pressure as well as less than well informed (but more likely, less than responsible) or unenlightened authorities 'reclaiming' mangrove lands for industrial or housing estates. It is a quick way to make a fast buck where demand for land (as a result of population pressure) is high.

Minimum size of conservation area

The question of what constitutes the minimum size necessary for conservation is a difficult one to answer. If only the flora is considered, a smaller area may be needed but if fauna is also considered then the area needed will have to be larger.

For the tropical rain forest, Whitmore (1984) suggested that in the short term some 50 breeding adult animals may be adequate but to prevent continuous loss of genetic material, some 500 breeding adults may be necessary. Whitmore, taking figures from Medway & Wells (1971) for various species of hornbill (birds) and from Chivers & Davies (1979) for species of apes and monkeys, showed that 500 individuals would require areas ranging from 20 km^2 (for the Bushy-crested Hornbill) to 186 km^2 (for the siamang or the great gibbon, *Hylobates syndactylus*). The areas given are those occupied by 500 individuals so the area needed for 500 breeding adults will be considerably larger.

Mangroves may have to be considered differently. We know that discrete communities of mangroves that occupy bays and inlets can often be a hectare or less in area. On the other extreme, a single mangrove ecosystem may extend thousands of square kilometres. So, in part at least, the size would very much depend on the mangrove system we are looking at. If the total area of the system is small then it is reasonable to conserve the whole ecosystem but if the ecosystem covers thousands or tens of thousands of square kilometres, then in many cases it may not be reasonable to have the entire ecosystem set aside for conservation. In such a situation we are back to determining a minimum self sustaining area.

Let me give you an example from a recent observation in the Kuala Selangor Nature Park, which consists of just over 300 hectares of partly degraded mangroves. The seaward side of the mangroves are 'pristine' but there is a bund separating the seaward section from the landward section. The area next to the landward section is a mix of land used for housing and farmland so there is basically no buffer area on the landward side of the mangroves. The mangrove area landward of the bund is relatively dry (tidal inundation class 5 of Watson, 1928, before the bund was built) and with the bund there is now no tidal inundation and the only water received by these mangroves is from rain and freshwater runoff. What we noticed was that many of the *Rhizophora* and some of the *Bruguiera* trees were dead. This could be due to the area drying up but the other species, including mangroves like *Avicennia* and *Excoecaria*, looked healthy enough and since the bund has been there a long time, it is unlikely that the deaths were due to disruption of the water regime of the area. We eventually noticed a few highly defoliated *Rhizophora* trees but there were no caterpillars or insects on the few remaining leaves. There was however a large troop of the silvered leaf monkey *Presbytes cristatus*. These leaf-eating monkeys had been responsible for the defoliation. The monkeys are apparently selective in their choice of leaves, preferring the larger leafed *Rhizophora* to the smaller leafed *Bruguiera*. They also seem to select a particular tree at a time to feed on and to almost defoliate that tree before moving on to the next tree. Gill & Tomlinson (1971) had shown that, in *Rhizophora mangle* L., no new foliage will appear if branches greater than 2.5 cm in diameter are pruned (see Snedaker *et al.*, 1992 for details of susceptibility of *Rhizophora mangle* to pruning and defoliation). It is apparent from our observations that the removal of most of the leaves (the monkeys appear to prefer the meristematic tips of the shoots to the older leaves) of *Rhizophora apiculata* Bl. and *R. mucronata* Lamk. had killed them.

The leaf monkeys have been in that area for a long time (Harrison,1974, reported the troop living in raintrees just outside the park) but why has this mass kill of *Rhizophora* only occurred recently (most of the trees would not have been dead for more that about a year)? It appears likely that development in the surrounding areas has destroyed or reduced their usual leaf source and this has forced them to turn to *Rhizophora* and *Bruguiera* in the park. The lesson to be learnt is that surrounding areas may be as important as the mangrove areas being conserved and if surrounding buffer areas are not given equal protection then the whole conservation exercise will not be effective.

On a similar note, a population of proboscis monkeys, *Nasalis larvatus*, is associated with a small patch (certainly no more than a couple of 100 hectares) of mangroves in the Bako National Park in Sarawak. These rare monkeys, endemic to Borneo, supposedly feed on *Sonneratia* mangrove leaves and retreat into the adjacent terrestrial rain forest to roost. Here again we see an area adjacent to the mangroves playing a vital role.

Response to sea level change

It is not possible, much as I would like, for me to avoid this current but highly controversial topic. Mangroves straddle the land and the sea so it is vital to understand what happens should sea level rise or fall. This understanding is important for both the management and the conservation of mangroves. I will also be presenting a blinkered view if I do not at least dwell briefly on the relation between the increase in carbon dioxide and other greenhouse gasses that has taken place in the last 200 years and sea level change.

Carbon dioxide, methane and oxygen isotopic ratio

'Post-industrial' man has added a very significant amount of carbon dioxide and methane to the atmosphere. There is unequivocal evidence of this. Carbon dioxide, for example, has increased from about 280 ppmv to the present level of about 350 ppmv. The trend in the increase in carbon dioxide is very clearly and convincingly shown in the famous Mauna Loa (Hawaii) data series. There is no argument that carbon dioxide and methane have increased at a tremendous rate over the last 200 years. There is also little or no argument that increase in carbon dioxide (the source of methane complicated by release from rice fields and farm animals) concentration is due mainly to the burning of fossil fuels. From here we move into uncertainty (speculation and prediction). Since the greenhouse gasses retain heat, increase in carbon dioxide and methane could lead to a rise in the surface temperature of the earth, followed by such things as expansion of the waters of the oceans, melting of polar ice and the consequent rise in sea level (e.g. see Schneider, 1992 & Lindzen, 1990 for two different views).

The most dramatic correlation between carbon dioxide and methane and rise in temperature is seen in ice core studies and the now famous Vostok ice core which goes to just over 2000 metres or equivalent to about 160 000 years of continuous record (e.g. Raynaud *et al.*, 1993). By looking at the oxygen isotopic ratio in the core it is possible to estimate the temperature at which the snow (which becomes incorporated into the ice core) forms. The Vostok ice core has been analysed for carbon dioxide, methane and stable oxygen isotope ratio (from which the surface temperature in the Antarctic, over where the Vostok ice core was obtained, can be estimated). The data give a very convincing correlation between increase in concentrations of carbon dioxide and methane and increase in surface temperature. There is agreement that the increase in temperature follows the rise in carbon dioxide but there is uncertainty if the increase in methane concentration is the cause or consequence of temperature rise. On top of this, the rise in carbon dioxide, methane and temperature correlates very well to changes in sea level. Still, not everybody is convinced.

Sea level changes

This is a very complex phenomenon and leaves many scientists not directly in the field confused or perplexed. The perception of sea level change is very much dependent on where the measurement is made. If the measurement is made on sites where the land is rising faster (as in sites where glaciers have recently melted or are melting, thus reducing load on the land and letting it bounce back up) than the actual rising water level (as caused by increased volume of the oceans), then what is seen is a relative fall in sea level. Glaciers are not the only thing that can cause the land to rise or sink. The earth's plates are in constant motion so that on one side a plate can tilt downwards causing the opposite side to tilt upwards. So unless the rise or fall of sea level (as caused by volume increase) is extremely rapid, relative sea level change is very much site dependent. So even though the oceans' volume may be increasing (as a result of thermal expansion of water or increase in mass from melting polar ice caps), sea level rise will not be seen universally. In many areas land masses may be rising at a faster rate so a relative fall in sea level is seen.

Let me take you to a specific site: the Straits of Malacca. Let us go back 20 000 years (20 ka) to the period between the end of the Quaternary and the beginning of the Holocene period (into the last glacial period). By 10 ka, the sea level had risen some 70–100 metres to some 53 metres below the present sea level (a rising rate of about 1 cm per year over a 7–8000-year period). The sea level rose to reach the present level around 7 ka and continued until a peak of about 5 metres above the present by 5–4 ka and in the last 4000 years has fallen about 5 metres to the present level (Geyh *et al.*, 1979). So, for the Straits of Malacca (located on the relatively tectonically stable, according to Tjia, pers. comm., Sunda Shelf), relative sea level has been falling at a rate of slightly under 1 mm per year over the past 4000 years. On the other hand, Peltier & Tushingam (1989) have estimated (from isostatically filtered tide gauge data) the global sea level change (over the past 50 years) to be $+2.4 \pm 0.90$ mm per annum. What I like to emphasise is that the rate for the geophysical study is a mean over four thousand years and is located in the Straits of Malacca whereas the study from 40 tide gauges, spread over the world, is only over fifty years. If indeed the Straits of Malacca is on a tectonically stable plate, then the discrepancy in rates would be difficult to reconcile. Such is the uncertainty we have to contend with although this caveat is often not pointed out.

We need to take one more step into the realm of the uncertain. What is the expected rise in sea level as a result of the increase in greenhouse gasses? The predicted (how good the various predictions are is again a matter of conjecture) rate centres around 5 mm per annum over the next 100 years or so (UNESCO, 1990; IPCC, 1990; IGBP, 1992).

The Straits of Malacca scenarios

We will look at two scenarios based on the Straits of Malacca and in particular the mangroves on the Malaysian side of the Straits (mainly because we have more readily available data).

As discussed earlier the mangroves (at the same relative sea level as we know them today) have been very recently (4.5 ka) subjected to a 2–3 mm per year relative sea level rise and an even greater rising rate of 10 mm per annum earlier on (when relative sea level was much below the present level) so we would expect that our present day mangroves will be able to handle a 5 mm per annum predicted sea level rise. There is however one very major difference: post-industrial *Homo sapiens*. When the sea level rose 4500 years or so ago, nature was able to operate unimpeded. Mangroves spread into the plains now occupied mainly by rice fields (it certainly was not rice then). Now, most of the land behind our mangroves are mainly rice fields dotted with towns and cities 'belonging' to man: man, who has the technology to set up defences against this aqueous intrusion. The landward part of much of the Malaysian mangroves bordering the Straits of Malacca is already bunded to prevent saltwater intrusion and to convert the land to agriculture and other uses. It is a simple matter of raising the level of these earth bunds by a mere half a metre and we will be safe for the next 100 years. The mangroves on the seaward side will slowly drown and we would expect a change in species composition in those areas that allow mangrove to still thrive.

Let us proceed to a less extreme scenario. Taken as a whole the mangroves on the Straits of Malacca are growing on accreting coasts. Macnae (1968) gives the vivid example of Palembang, a town on the Sumatran side of the Straits of Malacca as still a river-mouth port 400 years ago but is now 50 km inland (accretion have moved at a rate of around 125 m per annum!). On the Malaysian side, the palynological works of Kamaludin (1989, in press) have shown that mangrove deposits extend to about 10 metres in depth. There would have been two periods of deposition: the first was when the sea level was on the rise (a period between 7000 and 4.5 ka) and the second was when the sea level was falling between about 4.5 ka and present). The total period of deposition will not have been more than 5000 years. This would give us a sedimentation rate of about 2 mm per annum (also see Ong, 1993). This was during a period when there was no significant human perturbation of the environment. With the present activity of land clearing, sedimentation would probably be higher so it is reasonable to double the sedimentation rate to 4 mm per annum which is not much different from the predicted 5 mm sea level rise. In such a situation we will not feel the effects of the predicted sea level rise.

There are a few main points to be made from the above scenarios. The first is that no matter whether sea level rises or falls (at the rates we think we are seeing or generally predicting) the relative sea level at any particular site could be rising, falling or be stable depending on the nature of tectonic movement. If we understand this we are in a position to select the relevant sites that are stable, rising or sinking with respect to relative sea level. If we are interested in monitoring changes in the mangrove ecosystem, as some international organisations are, this is a necessary first step.

The second point is that should rapid relative sea level rise take place, there is very little likelihood of saving mangroves whose landward margins have been developed by man. It is vital that those responsible for the selection (or those who formulate the guidelines) of conservation sites for posterity fully understand this.

The third point I like to make is that for effective planning and management, it is vital to know if a particular site is stable, rising or sinking so efforts should be directed to finding suitable methods for determining this.

So if we have to select only one site in the Straits of Malacca for conservation for posterity it would certainly be more rational to move over to the Sumatran (Indonesia) side of the Straits where human activity is minimal and where there are still extensive areas of pristine mangroves as well as freshwater swamp forests and other natural ecosystem behind the mangroves. Potentially the mangroves here may migrate inland and survive as they did some 4500 years ago. The question is then why the Indonesian people should be asked to contribute their mangroves for the world to keep in posterity. That is an important political and socioeconomic question that must be addressed, but at another forum.

Management

From a purely forestry point of view, many of the mangroves in Asia (especially those formerly under British administration) are relatively well managed. Many of these mangrove forests represent rare examples of sustained yield management of a tropical forest ecosystem. One main reason for this success is perhaps the occurrence of mangroves in almost mono-specific stands so that the simpler plantation management system can be applied. Another reason is perhaps the excellence of some of the colonial foresters. Anybody who has read Watson's (1928) work on the mangrove forest of Malaya will realise how thoroughly he understood (and how much time he must have spent in) these mangrove forests.

The Matang mangroves that Watson worked on is today probably the best managed (based essentially on the rule of thumb management plans that he, together with other British foresters in this part of the world, developed) mangroves in the world. This simple management plan (e.g. see Haron, 1981) is based on clear-felling of small patches (each no more than a few hectares). A number of good trees are left to act as seed trees. The slash takes about 2 years to decompose and the area is then assessed for natural regeneration by seedlings. If natural regeneration is poor, the area is manually planted with seedlings collected from elsewhere. The forest is than left to grow until about 15 years when trees have reached a size suitable for the harvest of poles. A 4-foot stick is used for thinning which on average removes about half the total number of stems. This thinning is carried out for commercial rather than silvicultural reasons (see Gong & Ong, this volume, for more details). A second thinning is carried out at 20 years, this time using a 6-foot stick and again about half the number of stems is removed. In the past, a third thinning (using an 8-foot pole) was carried out at 25 years (to encourage growth of seedlings) but this third thinning is no longer carried out. Only the third thinning was designed to be silvicultural. Clear-felling takes place at 30 years so completing the rotation. The Matang mangroves are now in its third rotation. The thinnings described here are unique to the Matang mangrove management system.

In Thailand and in Indonesia a similar system is used, except that the clear-felling is done in narrow strips of about 50 metres or so (perpendicular or at 45 ° to the water's edge). In Thailand the rotation period is 15 years. In Sabah and Sarawak, a different system is used based on minimum girth (e.g. Chai & Lai, 1984). There is no clear-felling and trees below a certain girth are left behind. Large areas (rather than small patches or narrow strips) are felled in Sabah and Sarawak so it is not possible to categorically state if the problem is with the method or with the large areas cleared at any one time. However, from the fact that most mangroves occur as almost monospecific stands of almost even age we are of the opinion that the small patch/narrow strip clear-fell system is the more successful and thus to be recommended.

There is an apparent decline in the production in Matang (Tang *et al.*, 1984; also see Gong & Ong, this volume). If this decline is real, one cause may be the fact that areas (particularly those under tidal inundation classes 2, 4 & 5 of Watson, 1928) previously occupied by other species [e.g. main ones like *Avicennia alba* Blume, *A. marina* (Forsk) Vierh., *A. officinalis* L., *Bruguiera parviflora* Wight & Arn., *B. cylindrica* (L.) Bl., *B. gymnorhiza* (L.) Lamk., *B. sexangula* (Lour.) Poir., *Excoecaria agallocha* L., *Heritiera littoralis* Buch.-Ham., *Sonneratia alba* J. Smith, *S. caseolaris* (L.) Engler, *S. ovata* Backer, *Xylocarpus granatum* König and *X. mollucensis* (Lam.) M. J. Roemer] are being replaced (through silvicultural treatment) by *Rhizophora apiculata* (the preferred species for the production of charcoal). There is thus an ecological price to pay! It would thus be ecologically more efficient not to plant areas under tidal inundation classes 2, 4 and 5 with *Rhizophora apiculata* as presently practised. This would mean that from the forestry point of view, the area under tidal inundation class 2 will have to be considered as non-productive since *Avicennia alba* and *Sonneratia alba* that often dominate this zone are presently not considered commercial timbers. This zone would still have its role to play, mainly as the area where plant colonisation occurs as well as an erosion protective zone. Areas under tidal inundation class 4 are often dominated by the *Bruguiera* species and these could be very efficiently managed for the production of poles. Areas under tidal inundation class 5 usually has the highest diversity of species, as would be expected since this is where the mangrove/true terrestrial plant ecotone is located. There are still many species here that are commercially very valuable (like *Heritiera littoralis, Xylocarpus mollucensis* and *X. granatum*) but these very rarely occur in large numbers in pure or almost pure stands so that commercial exploitation of their timber has its problems. Nonetheless these are or can be exploited commercially for furniture grade timber. In terms of sensitivity to sea level rise this zone will be the most ecologically sensitive. This is especially so if this zone is a narrow one. If mangroves are to be preserved managers must very seriously consider leaving a buffer strip of the terrestrial zone intact. In areas where the area under tidal inundation class 5 is extensive [usually dominated by the palm *Nypa fruticans* (Thumb.) Wurmb.] the ecological sensitivity may not be as great.

So essentially it would make good ecological sense if the uppermost (together with a strip from the terrestrial zone) and the lowermost zones (including the mudflats) are left intact or put to minimal exploitation. As to which zone is more important would depend on whether the relative sea level at the site is rising or falling. For sites with relative rising sea level, the landward zone merits more attention whilst at sites where relative sea level is falling, the seaward end merits more attention. Keeping the appropriate zone open will ensure that the dynamics of the system operates unheeded.

Conclusions

Mangroves are non-homogeneous open ecosystems that are extremely dynamic, interacting with the atmosphere above as well as with the processes of the adjacent land and sea. The conservation of mangroves should thus include not only the various vegetation and tidal inundation zones but also the adjacent marine and terrestrial areas (including the water catchment areas). Determining the minimal size necessary for effective conservation is site specific and requires a thorough understanding of the ecology of its flora and fauna.

The burgeoning population with its increasing pollution (particularly the unprecedented massive addition of green house gases in the last 200 years), extensive exploitation by the Japanese mangrove wood-chips industry and the conversion to prawn aquaculture ponds are the main threats to mangroves.

The possible threat posed by the unprecedented huge increase in greenhouse gases (by post-industrial man's energy guzzling activities) is of great concern to most ecologists. One possible consequence of the rapid increase in greenhouse gases is the rise in sea level. There are many uncertainties but it was pointed out that relative sea level change is very much site dependent. For effective planning and management, it is vital to know if a particular site is stable, rising or sinking so efforts should be directed to finding suitable methods for determining this. However, should rapid relative sea level rise take place, there is very little likelihood of saving mangroves whose landward margins have been developed by man: an important conservation strategy to bear in mind.

We see in the Matang mangroves of Malaysia a rare case of successful sustainable management of a tropical rain forest. The tools of management are available and we suggest that these be more widely applied. We particularly urge the Japanese mangrove wood-chips industry to look to long term sustainable use rather than short term gains. An appeal is made to the new Government of Japan to be much more environmental friendly to the rain forests of the Asia-Pacific region.

Acknowledgments

I thank Dr Gong Wooi Khoon for discussions and critical reading of the manuscript. The infrastructure and time provided by the Universiti Sains Malaysia is also gratefully acknowledged.

References

Chivers, D. J. & G. Davies, 1979. Abundance of primates in the Krau Game Reserve, Peninsular Malaysia. In A. G. Marshall (ed.), The Abundance of Animals in Malesian Rain Forests. Department of Geography, University of Hull.

Chai, Paul P. K. & K. K. Lai, 1984. Management and utilisation of mangrove forests in Sarawak. In E. Soepadmo, A. N. Rao & D. J. Macintosh (eds), Proceedings of the UNESCO Asian Symposium on 'Mangrove Environment: Research and Management', Universiti Malaya, Malaysia: 785–795.

Darling, F., 1964. Conservation and ecological theory. J. Ecol. 52: 39–46.

Geyh, M. S., H. R. Kudrass & H. Streif, 1979. Sea-level changes during the late Pleistocene and Holocene in the Straits of Malacca. Nature 278: 441–443.

Gill, A. M. & P. B. Tomlinson, 1971. Studies on the growth of red mangrove (*Rhizophora mangle* L.). 3. Phenology of the shoot. Biotropica 3: 109–124.

Hamilton, L. S. & S. C. Snedaker (eds), 1984. Handbook for mangrove area management. UNEP/East-West Centre, Environment and Policy Institute, 123 pp.

Harrison, J., 1974. An introduction to mammals of Singapore and Malaya. Malayan Nature Society, Singapore, 340 pp.

Haron, Hj. A. Hassan, 1981. A working plan for the second 30-year rotation of the Matang mangrove forest reserve Perak. State Forestry Department, Perak. Rajan & Co. (Printers) Sdn. Bhd., Ipoh, Perak, 109 pp.

IGBP, 1992. Global change: reducing uncertainties. The International Geosphere-Biosphere Programme, The Royal Swedish Academy of Science, Stockholm.

IPCC (Intergovernmental Panel on Climate Change), 1990. Climate Change. The IPCC Scientific Assessment 9. Sea level rise. WMO/UNEP, Cambridge University Press, Cambridge: 257–281.

Kamaludin bin Hassan, 1989. Palynology of the lowland Seberang Prai and Kuala Kurau areas North-West Peninsular Malaysia. Geol. Soc. Malaysia, Bull. 23: 199–215.

Kamaludin bin Hassan, in press. The changing mangrove shorelines in Kuala Kurau, Peninsular Malaysia. Paper submitted to Sedimentary Geology.

Lindzen, R. S., 1990. Some coolness concerning global warming. Bull. Amer Meteor. Soc. 71: 288–299.

Macnae, W., 1968. A general account of the fauna and flora of the mangrove swamps and forests in the Indo-West-Pacific Region. Adv. mar. Biol. 6: 73–270.

Medway, Lord & Wells, D. R., 1971. Diversity and density of birds and mammals at Kuala Lompat, Pahang. Malay. Nat. J. 24: 238–247.

Nair, M. Y., 1977. An appraisal of the economic potential of mangrove swamps. M.S. thesis, Universiti Pertanian Malaysia, Serdang, 134 pp.

Ong, J. E., 1982. Mangroves and aquaculture in Malaysia. Ambio 11: 252–257.

Ong, J. E., 1993. Mangroves – a carbon source and sink. Chemosphere 27: 1097–1107.

Ong, J. E. & W. K. Gong, 1991. Mangroves. In R. Kiew (ed.), The State of Nature Conservation in Malaysia. Malayan Nature Society, Kuala Lumpur: 22–28.

Peltier, W. R. & A. M. Tushingam, 1989. Global sea level rise and the greenhouse effect: might they be connected. Science 244: 80–810.

Raynaud, D., J. Jouzel, J. M. Barnola, J. Chappellaz, R. J. Delmas & C. Lorius, 1993. The ice record of greenhouse gases. Science 259: 92–933.

Saenger, P. S., E. J. Hegerl & J. D. S. Davie (eds), 1983. Global Status of Mangrove Ecosystems. IUCN Commission of Ecology, Netherlands, 88 pp.

Schneider, S. H., 1992. The climatic response to greenhouse gases. Adv. ecol. Res. 22: 1–32.

Snedaker, S. C., M. S. Brown, E. J. Lahmann & R. J. Araujo, 1992. Recovery of a mixed-species mangrove forest in South Florida following canopy removal. J. coast. Res. 8: 919–923.

Tang, H. T., H. A. H. Haron & L. C. Cheah, 1984. Mangrove forests of Peninsular Malaysia – a review of management and research objectives and priorities. In E. Soepadmo, A. N. Rao & D. J. Macintosh (eds), Proceedings of the UNESCO Asian Symposium on 'Mangrove Environment: Research and Management' Universiti Malaya, Malaysia: 796–808.

UNESCO, 1990. Relative Sea-level Change: A Critical Evaluation. UNESCO Rep. Mar. Sci. No 54.

Watson, J. G., 1928. Mangrove forests of the Malay Peninsula. Malay. For. Rec. 6: 1–275.

Whitmore, T. C., 1984. Tropical Rain Forests of the Far East. 2nd edn. Clarendon Press, Oxford.

Hydrobiologia **295**: 353–362, 1995.
Y. S. Wong & N. F. Y. Tam (eds), Asia-Pacific Symposium on Mangrove Ecosystems.

An initial assessment of mangrove resources and human activities at Mahout Island, Arabian Sea, Oman

M. M. Fouda & M. Al-Muharrami
Sultan Qaboos University, College of Agriculture, P.O. Box 34, Al-Khod 123, Muscat, Sultanate of Oman

Key words: mangroves, *Avicennia marina*, community structure, resources, human impacts

Abstract

This study was initiated to provide information on mangrove structure, communities, wildlife, socio-economics and human impacts on one of the most important areas of mangroves in Oman at Mahout Island, located on the Arabian Sea coast. The island is fringed with a luxuriant vegetation of mangroves formed by *Avicennia marina* (Forsk.) Vierh. Tree height, diameter at breast-height (dbh) and basal area (m^2) show progressive increases from the upper to the lower level of the shore. Density (number of stems/0.1 ha) is variable, showing three peaks in transects across western and southern mangroves. The northern mangroves are the most developed, characterized by muddy substrate, followed by the western on a sandy bottom, and the southern surrounded by rock-flat. The mangroves support a large number of fishes, shrimps, crabs and molluscs. Wildlife are represented by shorebirds, waterfowls, turtles and dolphins.

The Mahout population consists of bedouins, mostly supported by fishing. It is semi-settled with seasonal migration in summer to large towns. Fishing activities are usually between September and May. Shrimps and fish are the main natural resources. Human impacts were identified and qualitatively assessed: overutilization of mangroves by woodcutting or livestock grazing; litter and pollution; killing of turtle and other wildlife; removal of benthic marine organisms; the proposed construction of a fishery harbor; and the potential of mariculture and tourism. Management, based on sustainable development, should be undertaken to protect Mahout resources.

Introduction

Mangroves of the Arabian region are significant not just biologically, but also in a historical context (Sheppard *et al.*, 1992). They were the first mangroves to be reported in the world literature by Nearchus and Theophrastus over 2000 years ago (Baker & Dicks, 1982). Of the four mangrove species known from the region, *Avicennia marina* (Forsk.) Vierh. is by far the commonest. It is unusually tolerant of harsh environmental conditions, in particular low temperature and high salinity (Clough, 1993). The mangroves of Arabia represent mosaic habitat containing both hard and soft substrates (Por *et al.*, 1977; Price *et al.*, 1987). They are inhabited by other species typical of muddy, sandy or rocky shores devoid of mangrove vegetation. Compared with Indian Ocean mangals, the number of mangrove and associated species in Arabia is low, although most of the characteristic fauna are still present (Price *et al.*, 1987). Low species diversity is attributed to the severe climatic and environmental conditions in conjunction with the limited range of suitable habitats and niches (Sheppard *et al.*, 1992).

The Sultanate of Oman has a coastline of 1700 km which contains a wide diversity of coastal habitats, including mangroves. However, very little is known of the functional roles of these communities. The Island of Mahout lies on the Arabian coast of Oman about 400 km south of Muscat, within a sheltered bay, Ghubbat al-Hashish, about 3.5 km offshore from the small village of Film (Fig. 1). The island is fringed with a luxuriant development of mangroves and the Sultanate's shrimp fishery center is also in this area. Barr al Hikman wetlands, including Film, Mahout and Ghubbat al-Hashish, hold internationally important concentrations of shorebirds and waterfowls (Green *et al.*, 1992). The Sultanate of Oman has embarked on a major plan to develop the fishing industry. This plan includes the construction of a number of modern fishery harbors and establishment of mariculture projects at strategic locations including Mahout. In addition, the importance of this region for wildlife and its potential for tourism have drawn the attention of development planners. However, the environmental impacts of these developmental projects on marine habitats and wildlife have been little studied.

Therefore, the objectives of this study are: to provide information on Mahout mangrove structure in terms of tree diameter, height, density, and basal area; brief description of zonation and faunal assemblages; review the present knowledge of wildlife of the Mahout region; provide a preliminary study on the socio-economics of Mahout; and finally identify and qualitatively assess human impacts on the marine environment of the Mahout area.

Description of Mahout Island

The Island of Mahout is pear-shaped, located at the Gulf of Masirah close to the mainland at Film (Fig. 1). A sandy beach is developed on the east side whilst the remainder of the island is fringed with a luxuriant development of mangroves, formed by only one single species, *Avicennia marina.* Immediately behind the beach crest there is a tidal depression behind which land rises to above high water mark. The people of the island live in huts and cabins of timber and palm fronds on the beach crest and on the area inshore of the tidal depression.

The northern side of Mahout Island is characterized by mudflats covered with water from the numerous mangrove creeks. This mudflat area is used by a large number of birds, mostly waders. Beyond the mudflat ia a channel which is very difficult to cross with a boat except at high tide. The mangrove of this northern side is well developed.

On the western side of the island, the bottom is sandy and is partly exposed at low tide. However, it is very difficult to walk through the mangrove from the sea side due to the dangerous quicksands and the numerous mangrove creeks with muddy bottoms. In addition, the dense mangroves of this western side make it impossible to walk through, and blood sucking insects and mites which keep biting all the time cause discomfort to any one. However, this area of mangrove is well developed and is extending seaward.

The southern side of the island is characterized by rockflat and small colonies of coral reefs, limiting the continuous seaward growth of mangroves. However, the mangroves of this southern side are well developed and can be traversed from either sea side or from the upper level of the shore which is mostly sand mixed with mud. Large mangrove creeks exist among these mangroves.

The mangroves cover a large area in Mahout Island. Only a small part is occupied by the fishing village, followed by salt marsh or sabkha, then the mangroves. Based on the available limited number of aerial photographs as well as a survey by boat around the island and by walking along the island, it seems that the vegetation cover is approximately 60% of the island (about 300 ha).

Materials and methods

The mangrove structure of Mahout Island was studied by the Point-Center Quarter Method (Cottam & Curtis, 1956). Three transect lines, each 200 m long, were established to represent the north, west and south sides of the mangrove. Transects were selected in sites to allow easy and safe crossing along the mangrove from the upper level of the shore to the low level. These sites were usually at places where mangrove creeks can be traced from the shore.

At each sampling point, four quarters were established by crossing the compass direction of the transect line with a perpendicular line. The distance from the sampling point to the mid point of the nearest tree in each quadrant was measured. The four distances were averaged. The mean of the four distances measured has been found empirically to be the square root of the mean area per tree (Cottam & Curtis, 1956). A total of 20 sampling points were taken, each at intervals of 10 m along the transect line. The total density (stems m^2) was obtained by dividing the mean area per individual into the unit area on which density is to be expressed. It was computed as the reciprocal of the square of the mean distance (1 d^2). Number of stems per 0.1 ha was obtained by multiplying stem density/m^2 by 1000. Tree diameter at breast-height (dbh) was measured at 1.3 m above the ground level. Basal area (m^2) was calculated by converting the diameter data (dbh) to basal area using the following formula (Cintron & Novelli, 1984):

$$\text{Basal area } (m^2) = 0.00007854 \times (\text{dbh})^2$$

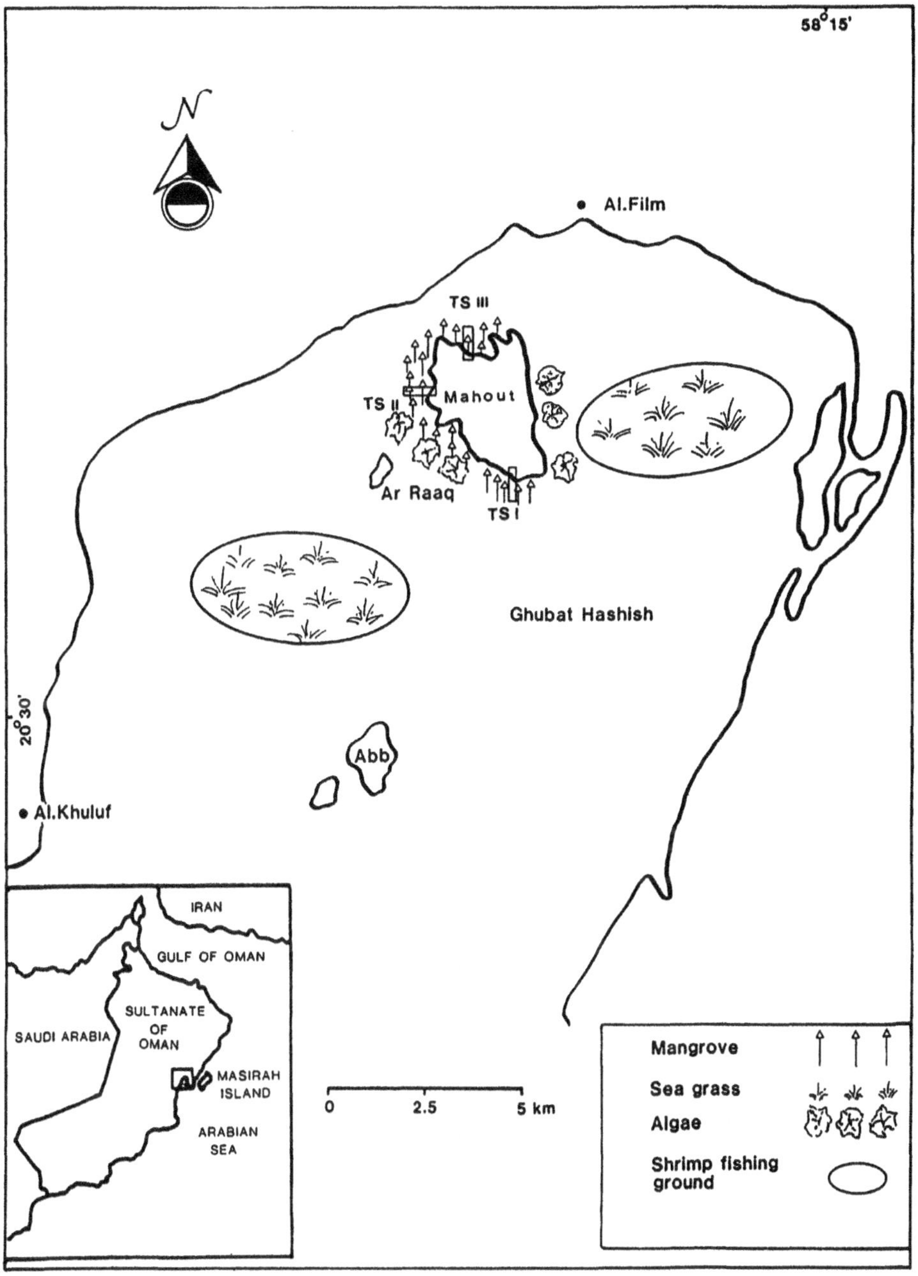

Fig. 1. Maps of the Sultanate of Oman (inset) and Mahout Island showing its main habitats and study sites (TS I, II, III).

Fishes and shrimps were collected by a small seine net, 4 m × 1.5 m, from mangrove creeks and the surrounding shallow water. The mesh size was small, 1 mm × 1 mm, and effectively retained small fish and shrimps. On capture, fishes and shrimps were immediately killed and preserved in 10% formalin. Invertebrates, mostly crabs and molluscs, were collected at low tide by hand from different places across a transect from the upper level of shore to the lower level about 30 cm in water depth. They were killed and preserved in 10% formalin for further analyses. Population densities of molluscs, mostly Cerithidae, were estimated (number of individuals/m^2) and for crabs number of burrows/m^2 were counted (Warren, 1989). Zonation of different habitats along the northern and western sides of the mangroves were also studied where the dominant species were recorded. In the laboratory, samples were sorted and identified to species level (when possible) using available literature. During the regular field visits to Mahout between December 1990 and July 1992, local residents were interviewed to understand the importance of mangroves to their livelihood. At the same time observations were made regarding any sign of disturbance in the mangroves and associated fauna. Furthermore, discussions were held at Muscat with officials at the Ministries of Agriculture and Fisheries and Regional Municipalities and Environment regarding the current and future developmental plans at Mahout and the surrounding areas. Meanwhile reports concerning wildlife and fisheries of Mahout were consulted.

Results and discussion

Mangrove structure

Data on mangrove structure are shown in Figs 2 to 5. Height varied from a minimum average of 0.9 m at the upper level of the shore to a maximum of 8.0 m near seaward edge. At all three transects, there were progressive increases in tree height from the shore to seaward. The greatest heights occurred at a distance about 140–200 m from the shore. The mean height of *Avicennia marina* was 4.9 m for the northern transect (3), and this decreased to 3.7 m for the western transect (2) and further decreased to 2.7 m for the southern transect (1). This corresponds to the nature of the substratum surrounding the mangroves, being muddy at the north where mangroves were best developed, sandy at the west and a rock flat at the south.

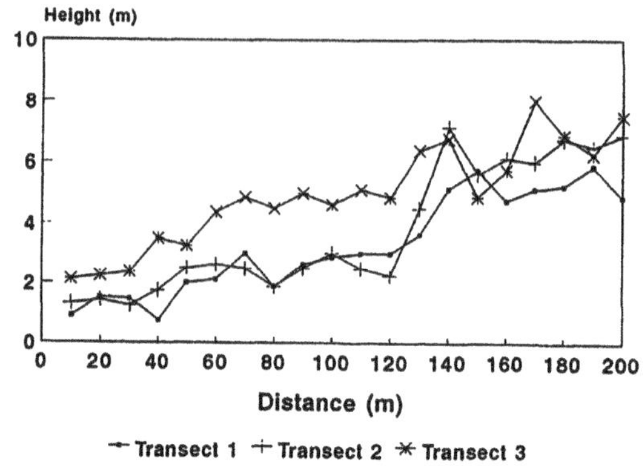

Fig. 2. Changes in mangrove height (m) across three transect lines (200 m) from the upper to the lower shore level. Transect 1 represents southern side, 2 western side and 3 northern side of Mahout mangrove.

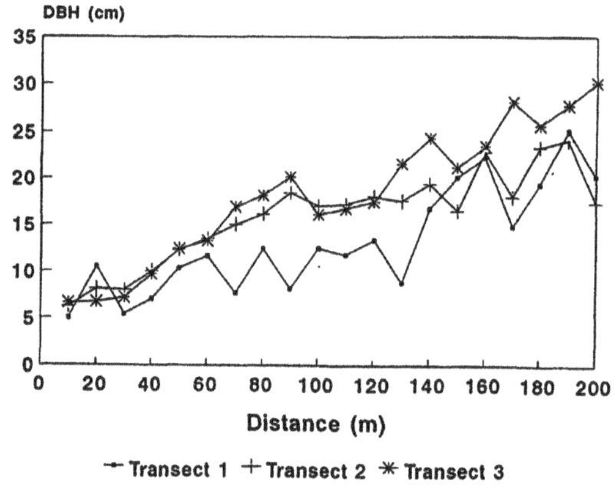

Fig. 3. Changes in mangrove tree diameter at breast-height (dbh) with the changes in distance (m) from the upper to the lower level of Mahout mangrove.

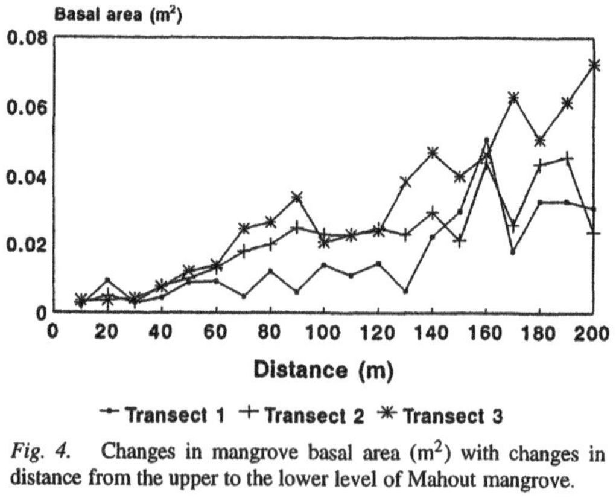

Fig. 4. Changes in mangrove basal area (m^2) with changes in distance from the upper to the lower level of Mahout mangrove.

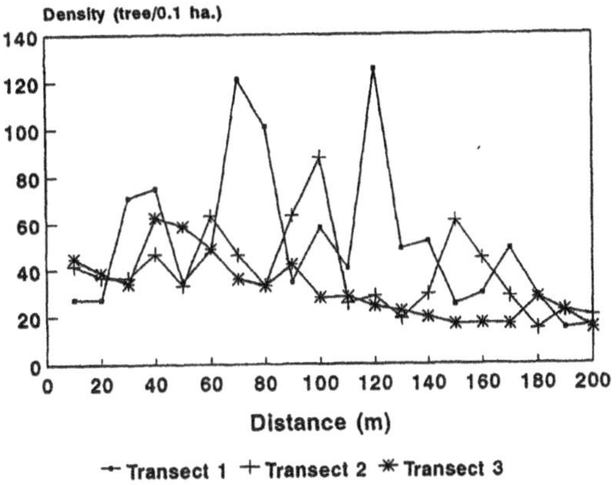

Fig. 5. Mangrove density (stem/0.1 ha) across the mangrove transect lines.

These results confirm that the mangroves of Mahout are the most developed in the region. In the Arabian Gulf stunted individuals (1–2 m) of the same species (Basson *et al.*, 1987) occur where winter temperature falls to nearly 0 °C (Price *et al.*, 1987). Salinity has also been suggested as a limiting factor affecting mangrove height as Price *et al.*, (1987) found a significant inverse correlation between the two along the Red Sea coast of Saudi Arabia. Similarly, Por *et al.*, (1977) suggested that in Sinai (Gulf of Aqaba, Red Sea) salinity is probably the main factor restricting the number of mangrove species, as well as some plant and animal species normally associated with mangal. On the other hand, Chapman (1984) suggested that low temperature, rather than salinity is the main limiting factor.

Observations in other parts of the world also suggest that climate may affect mangrove size (Lear & Turmer, 1977) in *Avicennia marina.* In Mahout, neither salinity nor temperature seem to be the limiting factors restricting the mangrove development as salinity varies from 34 to 42‰ and mean temperature from 19.1 °C in January to 29.5 °C in June. Aridity is possibly the main factor influencing the growth of mangroves here. It is influenced by many factors but chiefly by rainfall, cloud cover, humidity, wind and solar radiation (Clough, 1993). Rainfall at Mahout is less than 100 mm y^{-1}, humidity 80–90%, and solar radiation up to 600 mwh cm^{-2} (Department of Meteorology, 1992). Other limiting factors seem to be the water level and the nature of bottom surrounding these mangroves.

Data for diameter at breast-height (dbh) and basal area (m^2) followed the same pattern observed for height (*i.e.* there are progressive increases in dbh and basal area with distance from the upper level of the shore to the lower level). Positive significant correlations ($r = 0.8$–0.9) exist between these variables and the distance from the shore. On the other hand, the average density (number of stems/0.1 ha) varied from one transect to another, being quite high (36.98 stems/0.1 ha) for the southern transect and moderate for the northern (27.8 stems/0.1 ha) and western (25.75 stems/0.1 ha) transects. This may be due again, to the nature of bottom surrounding these mangroves. In the southern side, where the rock-flat exists, mangroves are restricted in development due to the lack of space, but have higher densities in the available space. On the other hand, in the northern side large areas are available for development where the bottom is muddy and water is shallower. Both the southern and western sides exhibit at least three peaks of density along the transects: the southern transect showed peaks at 30–40 m, 70–80 m, and 120 m, and the western transect at 60, 100, and 150 m.

The present data in general represent the lowest density of Mahout mangrove, because measurements were taken in areas of mangrove creeks and dense areas were avoided due to the difficulties in crossing them. Therefore, further sampling is clearly required for more detailed characterization of the mangrove structure of Mahout Island.

Associated biota

The mangrove community of Mahout Island includes a faunal assemblage of many species. The northern side of the island is of fine muddy substratum. At low tide, subtidal shallow areas are dominated by sea-grass beds mostly *Halophila ovalis* (R. Brown) Hook.f. These grass beds shelter many species of bivalves including pinnids and venerids. The intertidal zone is characterized by extensive mud flats which are dominated by a dense population of snails mostly Potamididae (e.g. *Pirinella conica* Blain) and Cerithidea (*Cerithidea cingulata* (Gmelin) and *C. scabridum* Philippi). Mud crabs, *Scylla serrata* (Forsk.), are also common in this mud flat area together with many birds, mostly waders, which feed on snails and other invertebrates. Crabs were the most dominant group among the mangrove trees and the surrounding mud and sabkha. They belong to the family Ocypodidae (mainly the fiddler crabs *Uca sp.*, *Macropthalmus sp.* and *Dotilla sp.*), Grapsidae (*Grapsus spp.*), Xanthidae and Calappidae *Matuta lunaris* (Forsk.).

Mangrove creeks contained many juvenile shrimps of *Penaeus indicus* H. Milne Edwards, *P. semisulcatus* de Haan and *Metapenaeus monoceros* (Fabricius), as well as small fish including the mudskipper *Periophthalmus spp.*, the Cyprinodont, *Aphanius dispar* (Ruppell) and the glassfish *Ambassis natalensis* Gilchnist and Thompson. These fishes together with gobies, blennies, tigerfish *Terapon jarbua* (Forsk.), flatfish, *Bothus spp.*, halfbeak, *Hemiramphus sp.*, silver biddy *Gerres oyena* (Forsk.) are typical residents of mangroves (Por, 1984). Closely associated with mangroves are mullet, *Mugil cephalus* L., milkfish, *Chanos chanos* (Forsk.) and croaker *Otolithes ruber* (Schneider) whereas the loosely associated fish include mangrove snapper, *Lutjanus argentimaculatus* (Forsk.), gizzard shad *Nematolosa nasus* (Bloch), anchovy *Thryssa mystax* (Schnieder), bluefish *Pomafomus saltatrix* (L) carangid *Caranx ignobilis* (Forsk.) and seabream, *Acanthopagrus sp.*

At the southern side of the island, a different habitat was observed. The shore was mostly sandy except at the mangroves. At the upper level of the shore, ghost crabs, *Ocypode saratan* (Forsk.) were very common, characterized by the numerous pyramids at the entrance of their burrows. Crabs belonging to the Grapsidae were also common. Gastropods, *Murex spp.* was frequent and caused painful injury due to their sharp spines. Green turtles, *Chelonia mydas* (L.) were also seen at this sandy shore. Mangrove trees were inhabited by a large number of crabs which were numerous in density and species. The sesamid crab, *Chiromanthus sp.* was quite common in burrows especially around the roots of mangroves. Where the substrate was soft, *Helice leachi* Hess and other graspids were abundant together with *Microphthalmus spp.*, especially where the mud was sticky and most fluid. The swimming crabs (Portunidae) were represented in mangrove swamps by their largest and heaviest species, *Scylla serrata*, a valuable edible crab which can reach up to more than one kg in weight.

Two transect lines each of 100 m were established in intertidal zone areas to count the abundance of crab burrows (no m^2 as described by Warren (1989). Density was counted at 10 m intervals along the transect. In the northern area of the mangroves (transect 3), density varied between 45 and 87 with an average of 66.5 burrows m^2. In the southern side (transect 1), it varied between 28 and 80 with an average of 46.5 burrows m^2. The difference in the averages of the two transects was highly significant ($P<0.001$). However distribution of density across the transect was not consistent and no correlation was found between abundance and distance from the upper to the lower level of the shore.

Large snails belonging to the family Potamididae particularly *Terebralia palustris* L. were also found in considerable numbers. However, the most abundant snails were *Pirinella conica* and *Cerithidea spp.* which formed dense populations (varying from 330 to 678 ind. m^{-2}) in areas abundant with filamentous green algae as well as blue-green algae usually at low tide. At high tide, most of these small snails migrate elsewhere.

Large wildlife

A recent study by Green *et al.* (1992) indicated the importance of Mahout mangroves and the surrounding coastal zone of Barr al Hikman, for wintering shorebirds and other water fowls. This study reported a total of 220 676 shorebirds of 53 species, among which waders totalled 133 679 of 26 species, or 60% of all birds. Cormorants, herons, egrets, spoonbills, flamingos, and many species of waders, gulls, and terns were very common. Barr al Hikman and surrounding areas, including Mahout and Ghubbat al Hashish hold internationally important concentrations of shorebirds, notably crabplovers, sandplovers, dunlins and redshank.

Green turtles were observed at several places at Mahout Island. A total of 37 turtles were counted along the sandy beach (the eastern side of the island) and the muddy bay facing the mangroves at the northern side, and a few dead specimens were seen trapped among mangrove trees at the western side of the island. Most of these turtles have been killed by fishermen who caught them for meat. The killing of sea turtles for food is an old tradition practised by fishermen of Mahoūt Island. Fishermen confirmed that there are many sea turtles at Ghubbat al-Hashish and the surrounding seagrass beds which are feeding ground for sea turtles. Few turtles nest at the island and fishermen collect their eggs, usually in early summer.

During the field study, skeletons of a few dolphins were also seen trapped at the muddy bottom of the mangroves of the northern side of Mahout Island. Many dolphins were also observed from the shore swimming near the southern side of Mahout Island. During the survey by boat around the island many dolphins were seen chasing fish. Fishermen confirmed that dolphins are present in the area all year round. They occur in groups from 2 to 12 individuals often with one or 2 juveniles. Although no attempt was made to determine dolphin species, fishermen claimed that there are several species of dolphins and even whales in the surrounding waters. It is possible that the Indo-Pacific Humpack Dolphin, *Sousa chimensis* (Osbeck) is the most common dolphin in the area. This species is known to live in shallow waters and has been reported several times in the Mahout area. Terrestrial mammals observed at Mahout mangroves included the bushy-tailed jird, *Sekeetamys calurus* (Thomas), and the sand fox, *Vulpes ruppelli* Pocock.

Socio economic survey

The Mahout population displays typical characteristics of Islamic and Arabic culture. Everyday life is strongly influenced by the bedouin life style. The rapid socio economic development in the Sultanate during the last two decades has not fully trickled down to this area due to its relative isolation from the rest of the Sultanate. No precise information on population size was available but the Ministry of Agriculture and Fisheries estimated that about 150 fishing fiberglass boats are in use, which would indicate about the same number of families with perhaps three to four times this number of people in total.

The population in Mahout Island consists of bedouins, mostly fishing-based, some pastoralists. It is semi-settled with seasonal migration, when a large part of the population shifts to large towns, mostly Sinaw, during the summer months when fishing virtually stops. The main fishing grounds for shrimps are in Ghubbat al-Hashish whereas the landing sites are at Film and Khaluf. Fishing is done very early in the morning where almost all boats leave the island to the fishing grounds. By 8:00 or 9:00 a.m. fishing stops and the boats go back to the island. Shrimp catch fluctuated between 150 and 590 MT annually (Ministry of Agriculture and Fisheries, 1991). The catch during the early part of the season (September–October) is mostly sub-adults (around 100 shrimp kg^{-1}). As the season progresses through January to May, the average size increases to medium and large (20–40 shrimps kg^{-1}). When the shrimp catch is low, fishermen shift their activities to shark and other pelagic fishes (e.g. kingfish) and the catch is sold, either fresh or sun-dried, to the fishing companies at Film, and then sold to the capital area and other large towns. Fishermen's income varies from 2000 to 5000 Omani Riyals/year (1 O.R. = 2.58 US$). Fishermen's average expenses for a fishing season can add up to 1000 R.O. (this however may not include depreciation on boat/motor, etc.).

Fishing is a seasonal activity. During June, July and August fishing virtually stops because it gets too hot to go to the sea with open boats, and because, this coincides with the breeding seasons and knowledgeable elders for centuries have stopped fishing during the summer. Almost the entire population of Mahout with all their savings from fishing move to market/oasis towns (e.g. Sinaw). It is believed that they work in date gardens during the summer, but a few individuals confirmed that summer is for rest and leisure.

Human impact on the environment

The negative influence of people in Mahout has increased recently. Up to the end of the seventies camels, donkeys and sometimes boats were used to bring the catch from the coast to the villages. At this time a large part of Mahout must have been almost untouched, balanced systems, stable for a long time. By national development programs, the local fishermen and also the nomads got cars and motor boats. Thus, travel times were reduced and working on sea and on land became more efficient. The following human impacts have been identified and qualitatively assessed:

Over-utilization of mangroves

Fishermen claim that mangroves were abundant in the past. Elders mentioned that the whole island was completely covered with mangroves. There is no other vidence to support this view and further geological analyses are needed. Mangroves at Mahout are affected greatly by fishermen. First of all wood cutting for building houses has been a traditional practice for a long time. With the current rate of population growth at Mahout (3.5%), it is likely that more trees will be used in the near future. Livestock grazing and browsing on mangrove leaves and seeds are problems. In 1991, about 700 camels were seen by the authors grazing at the mangroves. Fishermen bring camels for grazing in winter. This practice was stopped in 1992 after recommendations by the Ministry of Regional Municipalities and Environment, but fishing families have a lot of livestock including sheep and goats which still graze on mangroves. Furthermore, people from the mainland, mostly Hayy, often remove mangrove leaves from Mahout to feed their livestock. The negative impact by grazing and browsing of domestic livestock puts severe pressure on the plant cover, and alternatives should be provided (e.g. Ministry of Agriculture and Fisheries should provide food for the livestock).

Litter and pollution
Litter and garbage are everywhere along the sandy beach of Mahout. They are brought in either by the waves or are simply accumulations of fishermen's garbage. In addition, fishermen discard along the beach a large number of unwanted fish (some of them are quite large such as bluefish) and remains of others, mostly sharks. This is responsible for environmental pollution along the beach and may endanger the resources they are so heavily depend upon. Therefore, it is recommended that the municipality should install refuse bins at various places and beach refuse should be properly disposed or incinerated. Meanwhile, public awareness programs on environmental pollution should be initiated at Mahout.

Turtles
Oman is considered one of the few countries which have dense populations of several species of turtles (Green turtle, *Chelonia mydas* (L.); Olive Ridley turtle, *Lepidochelys olivacea* (Eschsholtz); Hawksbill turtle, *Eretmochelys imbricata* (L.) and loggerhead turtle, *Caretta caretta*, (L.). However, there are several kinds of human impact which threaten the sea turtle populations directly or indirectly.

Hunting of adult turtles for meat, shell and oil is an old tradition in Mahout. Individual turtles spotted by fishermen are immediately caught and killed. Accidental capture of turtles in fishing nets was observed several times during field visits. Fishermen never hesitate to catch a sea turtle in their cruises.

Collection of turtle eggs for food and medicinal purposes is still a habit of the local people. Although turtle nesting was not observed during the field survey, few turtles were seen dead in some sandy places among mangroves. It is known that only 25% of the clustres will develop into hatchlings and of these only 8–10% will reach the juvenile stage (Salm & Salm, 1991).

The increasing motorization of the fishing boats enables fishermen not only to drive into areas they could not reach in former times but also to be independent of wind. Thus, locations frequented by sea turtles are nowadays more often visited by boats with outboard engines. The animals are heavily disturbed in their mating and feeding grounds. The outboard engines which are mostly of the short sword type generate turbulence above the sediment surface, rips out the fragile seagrass shoots and damage the seagrass beds which are feeding grounds.

Removal of shells and sea-cucumbers
Women and children often go at low tide to the shallow water reefs, close to the southern side of the island where they collect sea-cucumbers and shells. They are exported to the far east, after cooking at the island. The removal of shells and sea-cucumbers has indirect effects on mangroves: continuous woodcutting for fuel to cook these organisms is likely to lead to degradation of mangroves in the near future.

Fishery harbour at Mahout Island

The Sultanate of Oman has embarked on a major plan to develop its fishing industry. The plan includes the construction of modern fishing harbors at strategic locations on the coastline together with the development of ancillary facilities, infrastructure, social facilities and training. The scope of the overall development is based on a ten year sector development plan (1990–2000) which aims to establish a modern fishing industry by the year 2000 (Ministry of Agriculture and Fisheries, 1990). An inception report (Ministry of Agriculture and Fisheries, 1992) summarized the initial project appraisal for the fishery harbor at Mahout. Two locations were considered for constructing the harbour: at Film on the mainland north of Mahout Island; or on the east shore of Mahout Island. Seven alternative schemes were also considered, including mooring and berthing facilities for the boats and several alternative rockfill causeways linking facilities on the mainland at Film with Mahout Island. This report lacks critical environmental data and the following detailed studies must be considered prior to the approval of this project: (a) the impact on tidal flats; (b) turbidity of the water column, especially in the shallows; (c) impact on sea-grass beds; (d) impact on migrant bird feeding habitats, especially of winter migrants; and (e) Impact on the unique Ghubbat al-Hashish shrimp fishery.

Any plan involving construction of a causeway connecting the mainland and construction on the island is not preferable because: (a) it will cause serious environmental impact to the island by permitting access to vehicles including trucks and animals (e.g. camels) which will destroy the important mangrove of Mahout; (b) the causeway will affect natural circulation patterns of water around the island, and this is likely to upset the sea-grass beds in the area. It will also lead to the accumulation of sediments and mud around the causeway

Mariculture
Within the framework of developing the fishing industry in the Sultanate, the Ministry of Agriculture and Fisheries plans to develop a shrimp culture industry Consequently, a study of the prospect for the development of a shrimp culture industry in the Sultanate was initiated at Mahout Island It is likely that very soon mariculture will commence at Mahout and this will require infrastructure which may be beyond the carrying capacity of Mahout Island The environmental conditions for the survival of mangroves in Mahout are very harsh, and the potential negative impact of human activity is very large It is recommended that before establishing the shrimp culture in Mahout, environmental impact assessment should be made

Tourism
The master plan for the coastal areas of Barr al-Hikman and Masirah Island (Ministry of Regional Municipalities and Environment, 1992) recommended tourism development including tourist reception/dispatch center at al Hayy (20 km from Film), Marine/Scientific research center at Film/Khaluf, turtle and bird watching sites, and camping areas at selected places including tents/adopted lodges These recommendations have not yet been considered for implementation However, the Sultanate is encouraging tourism which is taking place mainly at Muscat and Salalah It is likely that the tourist industry will be expanded in the future to include other coastal areas such as Barr al-Hikman A delicate balance between conservation and full exploitation of Barr al-Hikman including Mahout, Film and Khaluf has to be found Previous experience elsewhere has shown that tourism has led to serious environmental impacts on coastal resources including wildlife

Conclusion

Although the mangroves of Mahout are still the largest in Oman and support a large number of fish, shrimp, other invertebrates and wildlife, they are threatened by human impacts Monitoring research programs should be carried out and also public awareness programs should be encouraged Conservation management should be undertaken to protect the Mahout Island mangroves Finally, sustainable development, a concept to which Oman subscribes, means utilization of resources at a level which ensures their continued productive existence The management of all these aspects and the role of Mahout's mangroves in this, must keep to this central concept if Oman's natural resources are not to be wasted

Acknowledgments

The authors would like to thank the following Dr Rodney Salm who made his time available for a lengthy discussion on Coastal Zone Management, Dr Adel N Gindy of the Marine Science & Fisheries Center, and Mr Gregorio Hermosa, Jr (SQU), Mr Luther Schillak who presented and discussed the masterplan for Barr al-Hikman, Mr Hilal M Al-Mukhaini, Director of Fishery Statistics, Ministry of Agriculture and Fisheries for providing fisheries annual reports, Mr Ali Amer Al-Kiyumi, Director of National Conservation Strategy, and Mr Nagieb Ahmed Al-Rawas, Director of Regional Planning, Ministry of Regional Municipalities and Environment for all the help extended and for providing access to reports available on Conservation and Development

References

Baker, J M & B Dicks, 1982 The Environmental Effects of Pol lution from the Gulf of Oil Industry IUCN/MEPA report for the Expert Meeting of the Gulf Co ordinating Council to review environmental issues

Basson, P W, J E Burchard, J T Hardy & A Price, 1987 Biotopes of the Western Arabian Gulf Marine Life and Environments of Saudi Arabia, Dhahran, Saudi Arabia, Aramco

Chapman, V J, 1984 Mangrove Biogeography In Hydrobiology of the Mangal (F D Por and I Dor, eds) Dr W Junk Publishers, The Hague pp 15–24

Cintron, G & Y S Novelli, 1984 Methods for studying mangrove structure The Mangrove Ecosystem, Research Methods UNESCO

Clough, B F, 1993 Constraints on the growth, propagation and utilization of mangroves in arid region In Towards the rational use of high salinity tolerant plants (H Lieth & A Al-Mossom, eds) Kluwer Academic Publishers, Netherlands

Cottam, G & J J Curtis, 1956 The use of distance measures in phytosociological sampling Ecology 37 451–460

Department of Meteorology, 1992 Annual Report Directorate General of Civil Aviation and Meteorology, Ministry of Communication, Oman

Green, M G, M McGrady, S Newton & J D Uttley, 1992 The shorebirds of Bal al Hikman and Ghubbat al Hashish, Oman Report of a visit by the West Asian Shorebird Survey 23 December– 12 January 1990

Lear, R. & T. Turner, 1977. Mangroves of Australia. Queensland: University of Queensland press.
Ministry of Agriculture and Fisheries, 1990. Ten Years Fisheries Sector Development Plan (1991–2000). Main Report, Volume 1. Ministry of Agriculture and Fisheries. Muscat, Sultanate of Oman.
Ministry of Agriculture and Fisheries, 1991. Annual Report. Department of Fisheries Statistics, Muscat, Sultanate of Oman.
Ministry of Agriculture and Fisheries, 1992. Fishery Harbour at Mahout. Inception Report No. 202761 (01/01). Muscat, Sultanate of Oman.
Ministry of Regional Municipalities and Environment, 1992. Study for Wildlife and Conservation Areas Master Plan for the Coastal Areas of Barr al Hikman and Masirah Island. Final Report.
Por, F. D., 1984. Editor's notes on mangal fishes of the World. In Hydrobiology of the Mangal. (F. D. Por & I. Dor, eds), Dr W. Junk publishers. The Hague, pp. 207–209.
Por, F. D., I. Dor & Amir, 1977. The mangal of Sinai. Limits of an ecosystem Helgolanderissenschaftliche Meeresuntersuchungen, 30: 295–314.
Price, A. R. G., P. A. H. Medley, R. S. McDowall, D. Awson Shepherd, P. J. Hogarth & R. F. G. Ormond, 1987. Aspects of mangal ecology along the Red Sea Coast of Saudi Arabia. J. Nat. Fish. 21: 449–464.
Salm, R. V. & S. Salm, 1991. Marine turtles in Oman. Historical Association of Oman, Muscat.
Sheppard, C., A. Price & C. Roberts, 1992. Marine Ecology of the Arabian Region, Patterns and processes in extreme environments. Academic Press, UK.
Warren, J. H., 1989. The use of burrows to estimate abundance of intertidal crabs. Aust. J. Ecol. 15: 277–286.

GPSR Compliance

The European Union's (EU) General Product Safety Regulation (GPSR) is a set of rules that requires consumer products to be safe and our obligations to ensure this.

If you have any concerns about our products, you can contact us on ProductSafety@springernature.com

In case Publisher is established outside the EU, the EU authorized representative is:

Springer Nature Customer Service Center GmbH
Europaplatz 3
69115 Heidelberg, Germany

Batch number: 10816681

Printed by Printforce, the Netherlands